Climatic Changes on a Yearly to Millennial Basis

Climatic Changes on a Yearly to Millennial Basis

Geological, Historical and Instrumental Records

edited by

N. -A. MÖRNER
Geological Institute, Stockholm University, Sweden

and

W. KARLÉN
Department of Physical Geography, Stockholm University, Sweden

D. Reidel Publishing Company

A MEMBER OF THE KLUWER ACADEMIC PUBLISHERS GROUP

Dordrecht / Boston / Lancaster

Library of Congress Cataloging in Publication Data
Main entry under title:

CIP

Climatic changes on a yearly to millennial basis.

Proceedings of the Second Nordic Symposium on Climatic Changes and Related Problems, held in Stockholm, Sweden, May 16–20, 1983.
Bibliography: p.
Includes index.
1. Climatic changes–Congresses. 2. Climatic changes–Scandinavia–Congresses. 3. Geology, Stratigraphic–Recent–Congresses. I. Mörner, Nils-Axel. II. Karlén W. III. Nordic Symposium on Climatic Changes and Related Problems (2nd : 1983 : Stockholm, Sweden)
QC884.2.C5C575 1984 551.6 84-8214
ISBN 90-277-1779-6

Published by D. Reidel Publishing Company
P.O. Box 17, 3300 AA Dordrecht, Holland

Sold and distributed in the U.S.A. and Canada
by Kluwer Academic Publishers,
190 Old Derby Street, Hingham, MA 02043, U.S.A.

In all other countries, sold and distributed
by Kluwer Academic Publishers Group,
P.O. Box 322, 3300 AH Dordrecht, Holland

Printed in The Netherlands

TABLE OF CONTENTS

III. THE LAST 1,500 YEARS

PREFACE

Nils-Axel Mörner & Wibjörn Karlén

Organizers of the Stockholm Symposium in 1983
Stockholm University, Sweden

This book is the Proceedings of the SECOND NORDIC SYMPOSIUM ON CLIMATIC CHANGES AND RELATED PROBLEMS held in Stockholm, Sweden, May 16-20, 1983 (Frydendahl et al., 1983; Mörner, 1983). This was "an international interdisciplinary symposium with special reference to Nordic records and their relation to global climatic changes".

The first NORDIC SYMPOSIUM ON CLIMATIC CHANGES AND RELATED PROBLEMS was held in Copenhagen in 1978 (Frydendahl, 1978) and had a very broad scientific program. The SECOND NORDIC SYMPOSIUM was specifically directed to certain problems, time ranges and sources of information, as discussed below.

Scandinavia is a classical area for the study of climatic fluctuations during the last 20,000 years. A major part of the basic data for the international theories and synthesis originate from Scandinavian data. This applies, for example, to:
(1) The drastic climatic fluctuations in connection with the recession of the land-ice giving rise to the alternation of interstadials (warm periods with rapid ice recession) and stadials (cold periods with readvances or retardations in the ice recession).
(2) The climatic changes during the Holocene with its well-known sub-division according to the Blytt-Sernander system (nowadays being used far outside the boreal region for which it was originally defined).
(3) The Holocene climatic optimum and the subsequent climatic deterioration at around 2500 BP (well-established already at the beginning of this century).

The Nordic countries, therefore, have a great responsibility for critically examining available data, improving the data, synthesizing them and placing them in their correct global context. Of course, the Nordic countries, too, have a lot that must be revised and up-dated. It is of fundamental importance to compare and distinguish between local, regional, hemispheral and global climatic changes; i.e. between the climatic changes in the Nordic countries and the corresponding changes elsewhere over the globe (within Europe, within the northern hemisphere and between the two hemispheres).

We, therefore, decided that the SECOND NORDIC SYMPOSIUM ON CLIMATIC CHANGES AND RELATED PROBLEMS should be focused on the processes and data that are typical for the Nordic countries and that this material should be placed in its correct global perspective, so that the driving mechanisms could be investigated and, maybe even, deciphered. A large group of international experts were therefore invited to the symposium.

During the last decade, paleoclimatology has mainly been focused on the so-called Milankovitch variables (several symposia, many papers, much discussion, etc.), whilst climatology has naturally been focused on the present "weather machine". The climatic and paleoclimatic changes of intermediate length or cyclicity have, because of this, largely been forgotten. We, therefore, put the emphasis on the yearly to millennial type of climatic changes.

Our program - and hence this proceeding - is, therefore, highly interdisciplinary and addressed to people working with geological (and other paleo-records), archaeological, historical and instrumental data, as well as geophysisists (including astronomers) and modellists trying to find the causation mechanisms.

Out of the original 64 contributions (Frydendahl et al., 1983), we are here happy to be able to print 52 of them, plus 3 additional ones (from invited speakers who could not attend). These papers can be divided into four main groups; viz. one on paleoclimatologic records, one on historical records, one on instrumental records and one on theories, models and geophysical explanations.

REFERENCES

Frydendahl, K. (Ed.), 1978. Proceedings of the Nordic Symposium on Climatic Changes and Related Problems. Dansk Meteorol. Inst., Climatol. Pap. 4, 260 pp.

Frydendahl, K., Karlén, W., Mörner, N.-A., & Rasmussen, E. (Eds.), 1983. Abstracts, Second Nordic Symposium on Climatic Changes and Related Problems. Dansk Meteorol. Inst., 163 pp.

Mörner, N.-A., 1983. Stockholm Excursion Guide. Geol. Inst., Stockholm Univ., 16 pp.

ACKNOWLEDGEMENTS

The SECOND NORDIC SYMPOSIUM ON CLIMATIC CHANGES AND RELATED PROBLEMS was organized by the two Editors in co-operation with Knud Frydendahl and Erik Rasmussen from Denmark, who organized the first Nordic Symposium in 1978, and under the scientific auspicies of the International Commission on Climate (ICCL) of IAMAP, the Paleoclimatic and Holocene Commissions of INGUA and Project 158 of IGCP.

The main symposium was held at the Wennergren Center. Each day ended with a concluding summary and general discussion. One day was devoted to a field excursion in the Stockholm area (and evidence of deglaciation, shorelevel displacement, neotectonics, etc.) and a boat trip to the island of Björkö with a visit to the old city or village of Birka from the Viking Age (often termed "capital" of Sweden at that time) guided by Mörner (geology) and Ambrosiani (archeology). "Geologklubben", the organization of students in geosciences at Stockholm University, hosted an evening in the Old Observatory of Stockholm.

The symposium was financed by a major contribution from "Nordisk Kulturfond", a grant from the Swedish Department of Education and a small contribution from UNESCO via ICCL.

We are indepted to Ulla Mörner for preparing everything for our out-door reception-meeting, Jonas Mörner for taking care of the registration problems, Jessie Karlén for linguistic correction of several manuscripts, Ylva Brolin for re-typing several manuscripts, and, above all, Gunnar Johansson who - with neverfailing skill and enthusiasm - took care of most of the practical problems.

LIST OF CONTRIBUTORS

ALENIUS, P. Institute of Marine Research, P.O. Box 33, SF-00931 Helsinki 93, Finland

ALM, G. Department of Physical Geography, University of Stockholm S-106 91 Stockholm, Sweden

AMBROSIANI, B. Swedish Historical Museum, P.O. Box 5405, S-114 84 Stockholm, Sweden

AMMANN B. Systematisch-Geobotanisches Institut Uni Bern, Altenbergrain 21, CH-3013 Bern, Schweiz

ANIOL, R.W. Universität Hamburg, Ordinariat für Holzbiologie, Leuschnerstrasse 91, D-2050 Hamburg 80, F.R. Germany

BARRY, R.G. CIRES, University of Colorado, P.O. Box 449, Boulder, Co. 80309, U.S.A.

BARTHOLIN, T.S. Laboratory of Quaternary Biology, University of Lund, Tornavägen 13, S-223 63 Lund, Sweden

de BEAULIEU, J.L. Faculté des Sciences et Techniques de Saint-Jérôme, rue Henri Poincaré, F-133 97 Marseille, Cedex 13, France

BERGLUND, B.E. Laboratory of Quaternary Biology, University of Lund, Tornavägen 13, S-223 63 Lund, Sweden

BIRNIE, J. University of Aberdeen, Department of Geography, St. Mary's High Street, Old Aberdeen AB9 2UF, Scotland, U.K.

BJÖRCK, S. Department of Quarternary Geology, University of Lund, Tornavägen 13, S-223 63 Lund, Sweden

BUCHA, V. Geophysical Institute of Czechoslovakia, Academy of Science, CS-141 31 Praha 4 - Sporilov, Bočni 11, CP. 1401, Czechoslovakia

CARTER, T.R. The International Institute for Applied Systems Analysis, Schlossplatz 1, A-2361 Laxenburg, Austria

CHAIX, L. Museum d'Histoire Naturelle, Genève, Schweiz

CLERÇ, J. Faculté des Sciences et Techniques de Saint-Jérôme, rue Henri Poincaré, F-133 97 Marseille, Cedex 13, France

COÛTEAUX, M. Faculté des Sciences et Techniques de Saint-Jérôme, Laboratoire de Botanique Historique et Palynologie, rue Henri-Poincaré, F-133 97 Marseille, Cedex 4, France

DIGERFELDT, G. Department of Quarternary Geology, University of Lund, Tornavägen 13, S-223 63 Lund, Sweden

ECKSTEIN, D. Universität Hamburg, Ordinariat für Holzbiologie, Leuschnerstrasse 91, D-2050 Hamburg 80, F.R. Germany
EICHER. U. Systematisch-Geobotanisches Institut, Uni Bern, Altenbergrain 21, CH-3013 Bern, Schweiz
ELIAS A.S. INSTAAR, University of Colorado, Boulder, Co. 80309, U.S.A.
FAIRBRIDGE, R.W. Columbia University in the City of New York, Department of Geological Sciences, N.Y. 10027, U.S.A.
FAURE, H. Laboratoire de Géologie du Quarternaire, CNRS-Luminy, Case 907, F-132 88 Marseille, Cedex 9, France
FLOHN, H. Meteorologisches Institut der Universität Bonn, Auf dem Hügel 20, D-5300 Bonn 1, F.R. Germany
FREDSKILD, B. Greenland Botanical Survey, Botanical Museum, University of Copenhagen, Gothersgade 130, DK-1123 Copenhagen K, Denmark
GAC, J.Y. Office de la Recherche Scientifique et Technique Outre-Mer O.R.S.T.O.M., Centre de Hann, B.P. 1386, Dakar - Sénégal
GAILLARD, M.-J. Laboratory of Quarternary Biology, University of Lund, Tornavägen 13, S-223 63 Lund, Sweden
HEIKKINEN, O. Department of Geography, University of Helsinki, Hallituskatu 11-13, SF-00100 Helsinki 10, Finland
HEINE, K. Geographische Institut der Universität Regensburg, Universitätsstrasse 31, D-8400 Regensburg, F.R. Germany
HEINO, R. Finnish Meteorological Institut, Climatological Division, Vuorikatu 24, SF-00100 Helsinki 10, Finland
HILLAIRE-MARCEL, C. I.D.R.C., Université du Quebec à Montréal, BP 8888, Succ. A, Montréal H3C 3P8, Canada
HOFMANN, W. Max Planck Institut, Limnologie, Plön, F.R. Germany
HOLMGREN, B. Abisko Research Station, S-980 24 Abisko, Sweden
HSIUNG, J. Department of Earth, Atmospheric, and Planetary Sciences, Massachusett's Institute of Technology, 54-1522 Cambridge, Mass. 02139, U.S.A.
IREGREN, E. Museum of National Antiquities, P.O. Box 5405, S-114 84 Stockholm, Sweden
JÄKEL, P. Free Universität of Berlin, Institut für Physische Geographie, Grunewaldstrasse 35, D-1000 Berlin 41, F.R. Germany
KAHMA, K. Institute of Marine Research, P.O. Box 166, SF-00141 Helsinki 14, Finland
KARLÉN, W. Department of Physical Geography, University of Stockholm, S-106 91 Stockholm, Sweden
KELLOGG, T.B. SPECMAP, Quarternary Institute, University of Maine at Orono, Orono, Maine 04473, U.S.A.
KERÄNEN, R. Department of Geography, University of Oulu, SF-905 70 Oulu 57, Finland
LAMB, H.H. School of Environmental Sciences, University of East Anglia, Block F, University Village, Norwich NR4 7T3, England
LANDSCHEIDT, T. Schroeter Institut, Research Cycles Solar Activity Im Dorfe 14, D-2804 Lilienthal-Klostermoor, F.R. Germany

LARSEN, S.E. Physics Department, Risø National Laboratory, DK-4000 Roskilde, Denmark
LAUNIAINEN, J. Institute of Marine Research, P.O. Box 166, SF-00141 Helsinki 14, Finland
LEAN, J.L. CIRES, University of Colorado, NOAA/ERL/ARL, R/E /ARX2, 325 Broadway, Boulder, Co. 80302, U.S.A.
LEMDAHL, G. Laboratory of Quarternary Biology, University of Lund, Tornavägen 13, S-223 63 Lund, Sweden
LEZINE, A.M. Laboratoire de Géologie du Quarternaire, C.N.R.S., Faculté des Sciences de Luminy, Case 907, F-132 88 Marseille, Cedex 9, France
LIEDBERG-JÖNSSON, B. Laboratory of Quarternary Biology, University of Lund, Tornavägen 13, S-223 63 Lund, Sweden
LINDGREN S. Department of Atmospheric Sciences, Hebrew University, Jerusalem, Israel
MAKKONEN, L. Institute of Marine Research, P.O. Box 166, SF-00141 Helsinki 14, Finland
MATTHEWS, J.A. Sub-Department of Geography, Department of Geology, University College, Cardiff, P.O. Box 78, Cardiff CF1 1XL, England
MONTEILLET, J. I.F.A.N, Université de Dakar, B.P. 206, Dakar - Sénégal
MÖRNER, N.-A. Department of Geology, University of Stockholm, S-106 91 Stockholm, Sweden
MUNZAR J. Geographical Institute Čsav, Czechoslovenská Academia of Science, Mendlovo Nàmi, CZ-662 82 Brno, Czechoslovakia
NEWELL, R.E. Department of Earth, Atmospheric, and Planetary Sciences, 54-1522 Massachusetts Institute of Technology, Cambridge, Mass. 02139, U.S.A.
NEUMANN, J. Department of Meteorology, University of Helsinki, Hallituskatu 11-13, SF-00100 Helsinki 10, Finland
NGOM, P.M. Direction des Mines et de la Géologie, B.P. 1238, Dakar, Sénégal
OGILVIE, A.E.J. Climatic Research Unit, School of Environmental Sciences, University of East Anglia, Norwich NR4 7T3, U.K.
OLSSON, I.U. Department of Physics, University of Uppsala, P.O. Box 530, S-751 21 Uppsala, Sweden
PARRY M.L. The International Institute for Applied Systems Analysis, Schlossplatz 1, A-2361 Laxenburg, Austria
PAYETTE, S. Centre d'Etudes Nordiques, Université Laval, Cité Universitaire, Quebec G1K 7P4, Canada
PERSSON, T. Laboratory of Quarternary Biology, University of Lund, Tornavägen 13, S-223 63 Lund, Sweden
PETERSEN, G.H. Zoological Museum, University of Copenhagen, Universitetsparken 15, DK-2100 Copenhagen Ø, Denmark
PETERSEN, K.S. Geological Survey of Denmark, Thoravei 31, DK-2400 Copenhagen NV, Denmark
PETERSON, E.W. Oregon State University, Corvallis, Or. 97331, U.S.A.
PEZERIL, G. Direction des Mines et de la Géologie, B.P. 1238, Dakar, Sénégal

PFISTER, C. Historical and Geographical Institute, University of Bern, Engehaldenstrasse 4, CH-3012 Bern, Schweiz
PHILIP, A. Department of Physical Geography, University of Uppsala, P.O. Box 554, S-751 22 Uppsala, Sweden
van de PLASSCHE, O. Institute of Earth Sciences, Free University de Boelelaan 1085, NL-1007 MC Amsterdam, The Netherlands
REILLE, M. Faculté des Sciences et Techniques de Saint-Jérôme, rue Henri Poincaré, F-133 97 Marseille, Cedex 13, France
RENBERG, I. Department of Ecological Botany, University of Umeå, S-901 87 Umeå, Sweden
RINGBERG, B. Geological Survey of Sweden, University of Lund, Tunavägen 35, S-223 63 Lund, Sweden
SAOS, J.L. Office de la Recherche Scientifique et Technique Outre-Mer O.R.S.T.O.M., Centre de Hann, B.P. 1386, Dakar, Sénégal
SAVIJÄRVI, H. Department of Meteorology, University of Helsinki, Hallituskatu 11-13, SF-00100 Helsinki 10, Finland
SCHOVE, D.J. St. David's College, 29, South Eden Park Road, Beckenham, Kent BR3 3BQ, U.K.
SEGERSTRÖM, U. Department of Ecological Botany, University of Umeå, S-901 87 Umeå, Sweden
SERVANT, M. Services Scientifiques Centraux de L'O.R.S.T.O.M., 70,74 Route d'Aulnay, F-931 40 Bondy, France
SIEGENTHALER, U. Systematisch-Geobotanisches Institut, Uni Bern, Altenbergrain 21, CH-3013 Bern, Schweiz
SKYE, E. Institute of Floral Biology, University of Uppsala, P.O. Box 559, S-751 22 Uppsala, Sweden
STARKEL, L. Polish Academy of Sciences, Institute of Geography and Spatial Organisation, Physical Geography Department, ul. sw. Jana 22, PL-31-018 Krakow, Poland
TOBOLSKI, K. University of Poznan, Poznan, Poland
WALLIN, J.-E. Department of Ecological Botany, University of Umeå, S-901 87 Umeå, Sweden
WILKINSON, B. University of Birmingham, P.O.Box 363, Birmingham, B15 2TT, U.K.

CLIMATIC CHANGES ON A YEARLY TO MILLENNIAL BASIS
AN INTRODUCTION

Nils-Axel Mörner

Stockholm University, Sweden

1. INTENTIONS, GOALS AND MEANS

The climate changes and shifts in a broad spectrum of cycles and time ranges (Fig. 1). The effects of such changes may be local, regional, hemispherical or global, as illustrated in Figs. 1 and 2.

At the Stockholm Symposium, we did not want to discuss the Milankovitch variables or the present weather machine, but rather to concentrate on the climatic changes and shifts that cover decades, centuries and millennia. We wanted to study the past, present and future by means of geological, historical and instrumental records, and finally to be able to separate and distinguish changes that are induced locally, regionally, hemispherically and globally, as illustrated in Fig. 2.

The instrumental records give evidence of local as well as regional and hemispherical changes. The last glaciation maximum at about 20,000 BP seems to represent a globally (directly or indirectly) induced climatic deterioration. During the symposium, we wanted to investigate the character of the main Late Glacial and Holocene climatic changes and shifts by analysing and comparing our Nordic records with respect to available global information. As illustrated in Figs. 1 and 2, it was established that the Late Glacial and Holocene climatic changes and shifts, besides local effects, seemed to give evidence of regional as well as hemispherical effects, but probably no strictly global effects.

Everyone who works with paleoclimate must be aware of the fact that we are all, so to say, "short-term visitors in the past" and bring back impressions that, however true they are, may not necessarily be significant over a longer time or space range (as was the case with the sudden thunderstorm we were all subjected to when we walked among the graves from the Viking Age at Birka on the island of Björkö).

N.-A. Mörner and W. Karlén (eds.), Climatic Changes on a Yearly to Millennial Basis, 1–13.

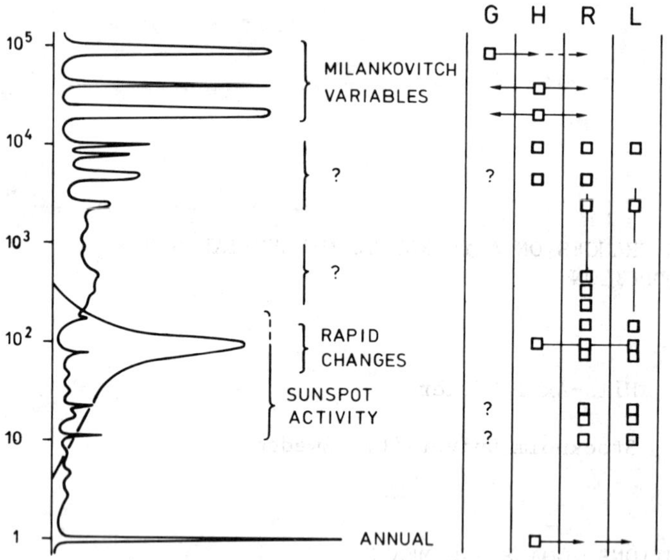

Fig. 1. Semiquantitative spectrum of climatic cycles and changes and their tentative geographic scale of equal effects; G (global), H (hemispheric), R (regional) and L (local). Rapid climatic changes in the order of around 100 years are particularly investigated.

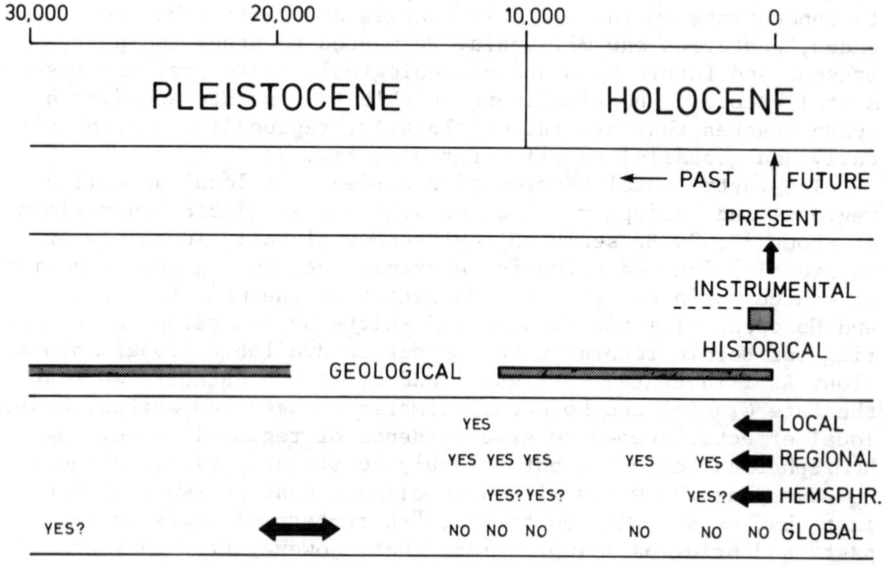

Fig. 2. Time scale and range of analysis, type and range of methods applied, and geographical range of the paleoclimatic changes and shifts (Fig. 1) studied.

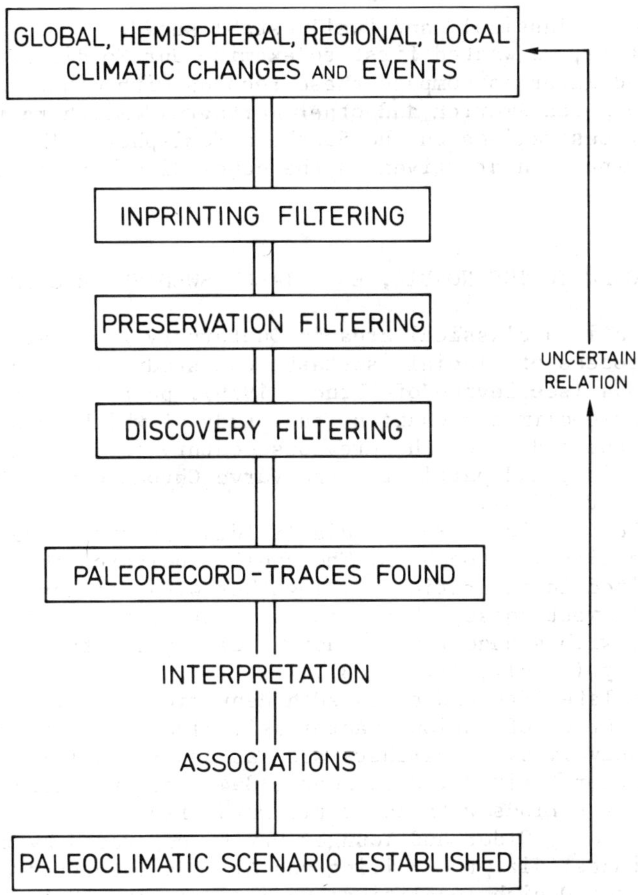

Fig. 3. Relation between actual paleoclimatic changes and the recorded traces and interpreted scenarios.

As illustrated in Fig. 3, a paleoclimatic change or event cannot be directly recorded. The records we happen to find and are able to interprete are the remains after several filtering processes; the inprinting of the signal on a recordable parameter, the preservation of this parameter and the discovery or analysis of this parameter. The established paleorecord-signal must be interpreted so that a paleoclimatic scenario of the actual change or event is established. This scenario has, of course, an uncertain relation to the change or event that actually occurred; it may be true, partly true or wrong. This is a dilemma that can only be overcome by a never-failing search for new records and documents and an open attitude to revisions and reinterpretations, even (and, maybe, especially) when the interpretations have become "classical" and "well-established".

Therefore, we wanted first to examine our Nordic records critically and later to compare these records with other records in Europe, in North America and other Northern Hemisphere regions, and in various regions on the Southern Hemisphere. The results of this comparison are given in the concluding chapter of this book.

2. BACKGROUND TO THE NORDIC, ESPECIALLY SWEDISH, RECORDS

Scandinavia is a classical area in Quaternary geology; partly for the process of glacial isostasy, the study of which goes back to 1694 (see review of Mörner, 1979), partly for the evidence of paleoclimatic changes, the study of which goes back, at least, to the middle of the previous century (e.g. review of Mörner, 1976a), and partly for the Varve Chronology of De Geer (e.g. De Geer, 1940).

In 1858, shells of Portlandia (Yoldia) arctica were found in the varved clay in Stockholm. The shells indicated that the clay was deposited in an arctic sea. Together with similar finds on the Swedish west coast, this lead Otto Torell to his theory of an ice age with a huge Fennoscandian ice cap all the way down into Germany (Torell, 1872-1873).

In the late 19th and early 20th centuries, the immigration and distribution of various macrofossil plant elements were studied intensively by researchers such as Andersson (1896), Blytt (1876), Nathorst (1872), Sernander (1894) and Steenstrup (1841). Three tundra periods with an arctic Dryas flora were established; viz, the Oldest, Older and Younger Dryas, separated by periods of milder climate; viz. the Bölling and Alleröd Interstadials (named after two Danish localities; Harz & Milters, 1901; Iversen, 1942). These five periods were included under the main period labelled the Late Glacial (with a poorly defined or undefined beginning). The subsequent period, the Postglacial or Holocene was subdivided into the Subarctic, Boreal, Atlantic, Subboreal and Subatlantic periods by Blytt and Sernander (e.g. Sernander, 1894), the last four of which were extended to a regional usage and, nowadays, often a global usage. These periods were originally defined from the plant macrofossil variations in sediments, bogs and travertine-tufa beds. During the Atlantic and Subboreal, the climate was found to have been significantly warmer than today. Andersson (1902) showed that Corylus (hazel) once had a much more northernly distribution than today and calculated the corresponding temperature difference at 2.5°C. Nordmann (1904) identified the same climatic optimum in the marine fauna in northern Jylland. At the International Geological Congress in Stockholm in 1910, Sernander (1910), Andersson (1910) and Nordmann (1910) presented detailed paleoclimatic records. At the same time, De Geer (1910) presented a map of the main deglaciation features in southern Sweden.

In 1884, De Geer recognized the varved clay in the Stockholm region as annual deposits formed in front of the receding ice margin. The geochemical evidence of the annual cyclicity was given by Högbom (1889). In 1904, De Geer started an intensive investigation of the varved clays in Scandinavia with the aim of building up an absolute time scale and reconstructing the recession of the ice margin in great detail (De Geer, 1912, 1940).

a) The Late Glacial

After the 20,000 BP glaciation maximum when the Fennoscandian ice cap reached down into Denmark and northern Germany, the ice margin receded with numerous oscillations (e.g. Mörner, 1969, Fig. 42) up to about 13,300 BP when there was a distinct readvance and/or stagnation in the ice melting and the deposition of the Luga-Low Baltic-Lista moraines.

At around 13,000 BP, there was a drastic change in the mode of uplift and subcrustal mass transfer (Mörner, 1979, 1980a), which must reflect a change from peripheral melting to a considerable thinning of the entire ice dome. At about the same time, there was a significant change in climate.

The onset of the Bölling Interstadial is not well dated within the Nordic region. Furthermore, it is not necessarily fully synchronous (Mörner, 1971a) with the 3-parted "Bölling" zone identified and dated in more continental areas like southern Germany and the East Baltic region.

At around 12,700 BP (or 12,600-13,000 BP), there was a marked change in the marine fauna, however. This was first observed and dated by Mörner (1969, 1971b) in the Kattegatt region, and later by Mangerud (1977) in southwestern Norway. It is well-established in Denmark (e.g. S. Petersen, this volume) and in the North Sea (Carlsen, 1983). This sudden immigration of a new marine fauna and flora corresponds to a general change of the circulation in the North Atlantic (Ruddiman & McIntyre, 1981).

The Older Dryas Stadial represents a short cooling in the order of 150 years (Mörner, 1969). Both its age (Mörner, 1969; Mangerud, 1977; Berglund et al., this volume) and correlation with certian moraine ridges (Berglund & Mörner, this volume) remain problematic, however.

The Alleröd Interstadial stands out as a very distinct and warm period. The climatic characteristics seem well-established by Kolstrup & Buchardt (1982) and Berglund et al. (this volume). Its duration seems to have been in the order of 800 years.

The Younger Dryas Stadial represents a sudden and drastic cooling that sat in very sharply at about 11,000 C14-years BP. Its subdivision and proper ending remain problematic. The so--called Middle Swedish Endmoraines (in southern Sweden) are not traceable over long areas. Björck & Digerfeldt (this volume) have given an important revision of the relation between the Younger

Dryas Stadial and the moraine zones and the so-called "drainage of the Baltic Ice Lake". The varve chronology does not yet provide a reliable solution to the age and correlation problems. It seems clear, however, that the ingression of saltwater at varve -1073 represents a synchronous event but that the "drainage" of the Baltic Ice Lake occurred about 300 varves (= years) earlier when the ice receded from Mt. Billingen (Berglund & Mörner, this volume). It is of fundamental importance for the discussion of the causation mechanism that the Younger Dryas Stadial stands out as a distinct paleoceanographic event in the North Atlantic (Ruddiman & McIntyre, 1981).

At around 10,000 BP, the mode and rate of recession changed and became so rapid (about 200 m/year) that the entire ice cap was gone in less than 1500 years. The rate of glacial isostatic uplift reached its maximum at around 10,000 BP (Mörner, 1979). As a matter of fact, it is not clear whether it is the melting that triggered the uplift or the rapid uplift that triggered the rapid ice recession or a combination of both. At around 10,000 or 10,200 BP, the climate experienced a drastic amelioration. This was recorded over a centruy ago. It is seen in the faunal and floral elements as well as in the lithology (clay/gyttja boundary). It is not until the investigation by the Norwegian OSKAP group (1983), however, that we have real evidence of a change in the marine fauna and flora that concur with the drastic change in oceanic circulation in the North Atlantic as recorded by Ruddiman & McIntyre (1981).

Southern Scandinavia has been chosen as the type area for the Pleistocene/Holocene boundary (Mörner, Ed., 1976; Olausson, Ed., 1982).

b) The Holocene

The subdivision of the Holocene according to the Blytt-Sernander system is classic. The early recording of a significant postglacial climatic optimum - in the terrestrial (Andersson, 1902) as well as the marine (Nordmann, 1904) records - is outstanding. The varve-chronological dating of certain palynological levels and events of species immigration (Fromm, 1938) provided palynology with fairly precise dating basis that remained unique up to the time of radiocarbon.

All since the development of the method by von Post in 1916, palynology has been the main method of studying and reconstructing the past climatic and environmental changes in the Nordic countries (e.g. Nilsson, 1935; Iversen, 1944; Berglund, 1966). It has also become an important method of tracing the settlement history (e.g. Vuorela, 1975).

The variations in peat-growth and the occurrence of so-called recurrence surfaces (e.g. Granlund, 1932) provide a datable register of past changes in temperature and precipitation (e.g. Nilsson, 1964; Aaby, 1976).

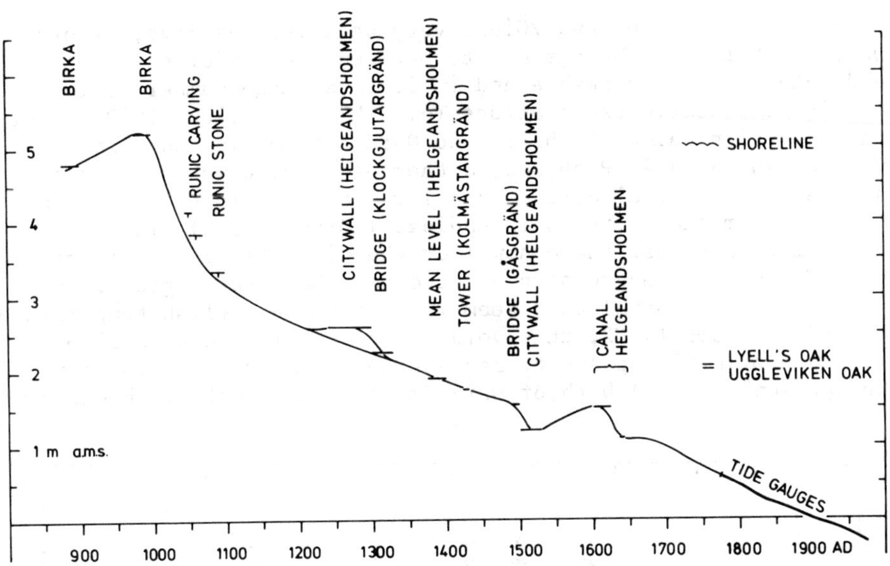

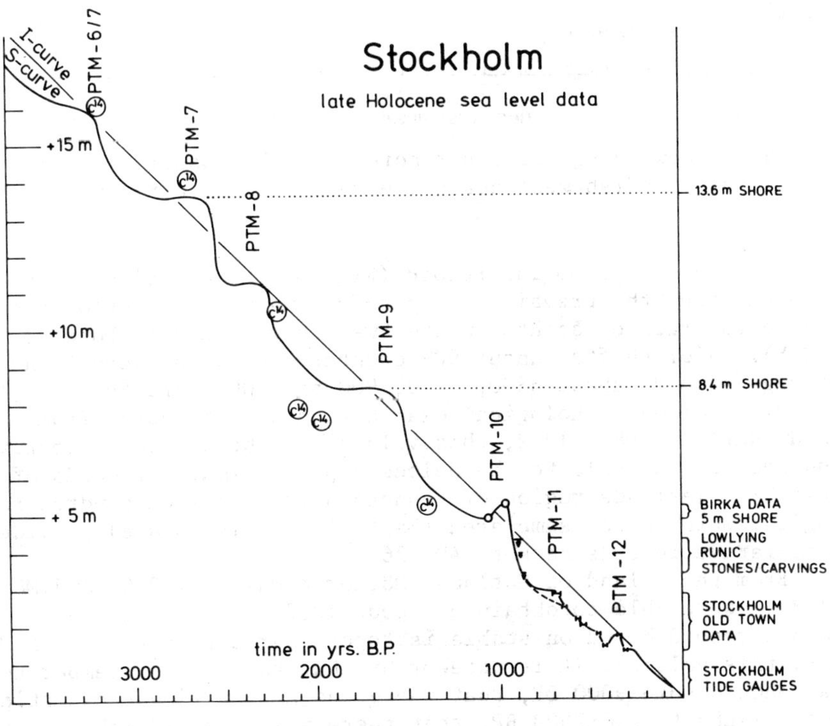

Fig. 4. Shorelevel displacement curves for the Stockholm region (from Mörner, 1983). Above: the last 1200 years´ record. Below: the last 3500 years´ record (I = absolute uplift).

During the 60ies and 70ies, very detailed sea level records were established. The agreements between the different records and authors are remarkable and indicate a large number of low-amplitude eustatic oscillations (e.g. Mörner, 1976b, 1980b). For the excursion guide of the Second Nordic Symposium on Climatic Changes and Related Problems, a thorough review was made of all available historical data on sea level changes in the Old Town of Stockholm and in the surroundings (Mörner, 1983). It is of great interest that there now (for the first time) is evidence of a Medieval transgression and two post-Medieval regressions (Fig. 4, upper curve) that seem to match the English temperature records of Lamb (1982, this volume) very well. There is also evidence of a Viking Age transgression followed by a distinct regression (Fig. 4) both of which must have a eustatic background.

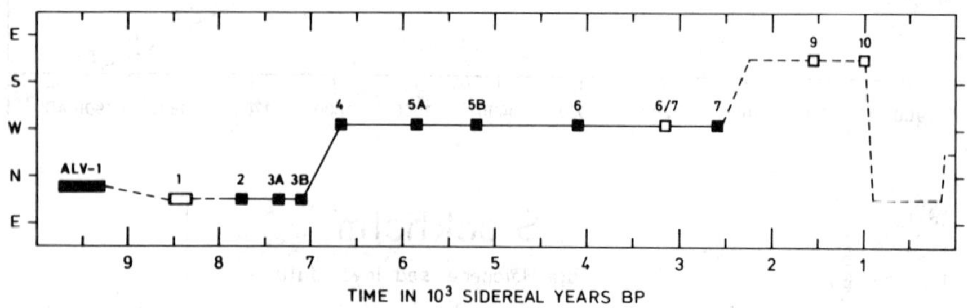

Fig. 5. Paleowind directions recorded in the fossil beaches on the Swedish west coast (redrawn from Mörner, 1980b).

A tentative paleowind record (Fig. 5) was established by investigating the prevailing beach-forming wind-direction recorded in the raised beaches on the Swedish west coast (Mörner, 1980b). Major shifts (about 90° counter-clockwise) were found or suggested at about 6800, 2400, 900 and 100 years BP.

By detailed lithological examination of lake cores from Lapland, Karlén (1981, 1982, this volume) has been able to improve and add many details to the Holocene paleoclimatic records of this high latitude region of Fennoscandia. Detailed dendrochronological work in the same area (Bartholin, this volume) provides comparative records back to AD 436.

From the island of Gotland, Mörner & Wallin (1976) and Mörner (1980c) were able to obtain an about 10,700 years long paleotemperature record based on stable isotopes (curve B in Fig. 6 of Mörner, this volume). It is noteworthy that the present temperature is reached some 9000 BP, that there was a significant climatic optimum about 8000-2500 BP, that there was a considerable climatic deterioration of about 2.5^{o} C at about 2500 BP (the Subboreal/Subatlantic transition) and that the temperature during the Subatlantic has remained significantly lower than during the climatic

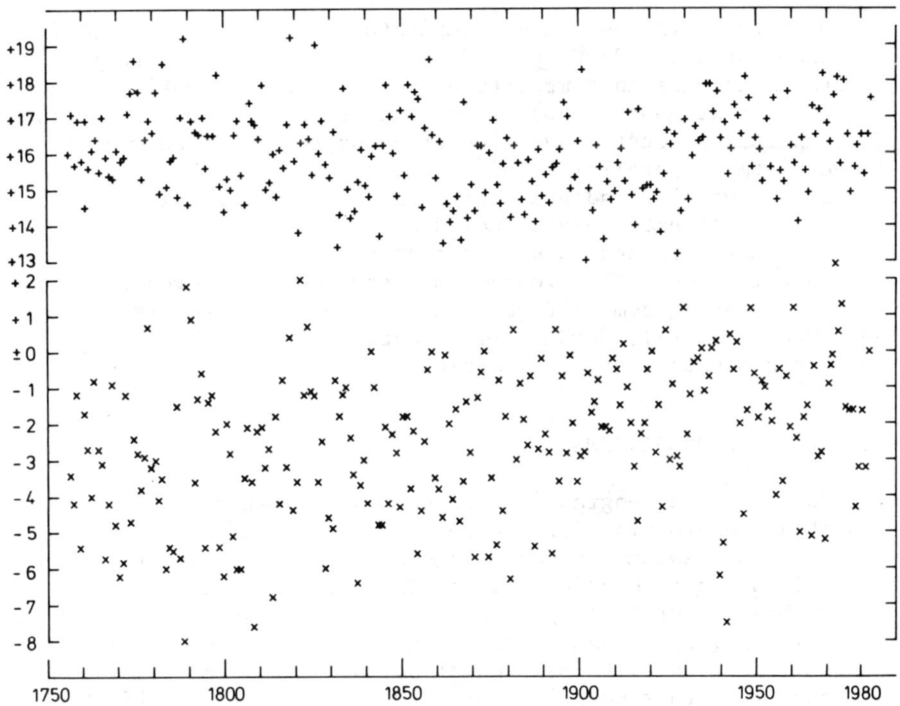

Fig. 6. Stockholm temperature records 1756-1983. Upper graph (+) gives mean summer (June-August) temperature and lower graph (x) mean winter (December-March) temperature.

optimum.

The long, annually varved, lake cores studied by Renberg and Segerström (e.g. this volume) open new possibilities of establishing long continual paleoclimatic records of the Holocene with a sharp dating control.

c) Instrumental Records

Stockholm - the place of our symposium - has several long records of instrumental measurments. Whilst sporadic temperature measurements began in the 1720ies in Uppsala, continual temperature measurements started in 1756 at the Old Observatory in Stockholm (Liljequist, 1943, 1949). The Lovön observatory for geomagnetic field variations began in 1929 (Wollin et al., 1973). A permanent tide gauge station for continual recording of the relative uplift or sea level changes began to operate in 1774 (Mörner, 1973, 1979).

Fig. 6 gives individual temperature plots of mean winter and summer temperature in Stockholm form 1756 to 1983. The main trend in the winter temperature seems to be a rise in the order of 2° C.

The main trend in the summer temperature seems to be a fall in the order of about 2° C up to about 1920 from when there is a general mean rise in temperature (with a smooth transition, this rise starts at about 1890). This means that the summer and winter data exhibit a clear converging trend up to 1890-1920 when the trend becomes "parallel". This could perhaps be taken as a break of the "natural" trend and a sign of environmental disturbances. The scatter of data points includes both different cycles of secular climatic changes and random noice.

In this book, other long-term instrumental records from Scandinavia covering some 100 years are discussed by Peterson & Larsen (weather data), Keränen (lake data; see also: Koutaniemi, 1983) and Makkonen et al. (Baltic data).

3. CAUSATION MECHANISMS

The duration and geographic extension of climatic changes are essential information for the deciphering of the origin and causation mechanisms of the recorded changes as illustrated in Table 1 in the concluding chapter (Mörner, this volume).

We have, therefore, paid a lot of consideration to the character - in age, space and correlations - of the recorded climatic changes on a yearly to millennial basis. Concluding remarks, especially on this question, are given at the end of this book.

REFERENCES

Aaby, B., 1976. Cyclic climatic variations in climate over the past 5500 yr reflected in raised bogs. Nature 263, 281-284.

Andersson, G., 1896. Svenska växtvärldens historia. Stockholm, 136 pp.

Andersson, G., 1902. Hasseln i Sverige fordom och nu. Sveriges Geol. Undersökn. Ca-3, 1-168.

Andersson, G., 1910. Swedish climate in the Late-Quaternary period. In: Die Veränderungen des Klimas seit dem Maximum der letzten Eiszeit (XI Intern. Geol. Congr.). Stockholm, 247-294.

Bartholin, T.S., this volume. Dendrochronology in Sweden.

Berglund, V.E., 1966. Late-Quaternary vegetation in eastern Blekinge, south-eastern Sweden. A pollen-analytical study. I. Late-Glacial time. II. Post-Glacial time. Opera Botanica 12:1, 1-180, 12:2, 1-190.

Berglund, B.E. & Mörner, N.-A., this volume. Late Weichselian deglaciation and chronostratigraphy of Southern Scandinavia: problems and present "state of the art".

Berglund, B.E., Lemdal, G., Liedberg-Jönsson, B. & Persson, T., this volume. Biotic responses to climatic changes during the time span 13,000-10,000 BP - A case study from SW Sweden.

Björck, S. & Digerfeldt, G., this volume. Climatic changes at

Pleistocene/Holocene boundary in the Middle Swedish endmoraine zone, mainly inferred from stratigraphic indications.
Blytt, A., 1876. Immigration of the Norwegian Flora. Alb. Cammermeyer. Christiania (Oslo), 89 pp.
Carlsen, R., 1983. Sedimentologiske og stratigrafiske undersökelser av kvartaere sedimenter fra Gullfaks, Blokk 34/10, Nordsjöen, Hovudupgave, Inst. Geol. Univ. Oslo.
De Geer, G., 1910. Södra Sverige i senglacial tid. Öfversitkskarta med åsar, ändmoräner och räfflor. Sveriges Geol. Undersökn. Ba-8.
De Geer, G., 1912. A geochronology of the last 12000 years. Compte Rendu XI Congr. Geol. Intern., Stockholm 1910, 241-253.
De Geer, G., 1940. Geochronologia Suecia Principles. Kungl. Sv. Vet. Akad. Hand., Ser. 3, Bd. 18:6, 1-367.
Granlund, E., 1932. De svenska högmossarnas geologi; deras bildningsbetingelser, utvecklingshistoria och utbredning jämte sambandet mellan högmossebildning och försumpning. Sveriges Geol. Undersökn. C-373, 1-193.
Hartz, N. & Milters, V., 1901. Det senglaciale ler i Alleröd tegelverksgrav. Medd. Dansk Geol. Foren., II:8, 31-60.
Högbom, A.G., 1889. Om relationen mellan kalcium- och magnesiumkarbonat i qvartära aflagringar. Geol. Fören. Stockh. Förh. 11, 263-273.
Iversen, J., 1942. En pollenanalytisk tidsfaestelse av ferskvandslagene ved Nörre Lyngby. Medd. Dansk Geol. Fören. 10, 130-151.
Iversen, J., 1944. Viscum, Hedera and Ilex as climatic indicators. Geol. Fören. Stockholm Förh. 66, 463-483.
Karlén, W., 1981. Lacustrine sediment studies. Geogr. Ann. 63A, 273-281.
Karlén, W., 1982. Holocene glacial fluctuations in Scandinavia. Striae 18, 26-34.
Karlén, W., this volume. Dendrochronology, mass balance and glacier front fluctuations in northern Sweden.
Keränen, R., this volume. Certain relationships between lake level variations and some climatic factors in Finland.
Kolstrup, E. & Buchardt, B., 1982. A pollen analytical investigation supported by ^{18}O-record of a Late Glacial lake deposit at Graenge (Denmark). Rew. Paleobot. Palynol. 36, 205-230.
Koutaniemi, L., 1983. Complexity of solar variability, hydrology and climatic conditions as evidenced in the case of the Oulujoki and Kemijoki river basin, northern Finland. Fennia 161, 289-301.
Lamb, H.H., 1982. The development of climate and its relevance to human affairs. Proc. Royal Inst. 54, 65-95.
Lamb,, H.H., this volume. Climate and history in northern Europe and elsewhere.
Liljequist, G.H., 1943. The severity of the winters at Stockholm 1757-1942. Geogr. Ann. 1943, 1-24.
Liljequist, G.H., 1949. On fluctuations of the summer mean temperature in Sweden. Geogr. Ann. 1949, 159-178.

Makkonen, L., Launiainen, K.K. & Alenius, P., this volume. Long--term variations in some physical parameters of the Baltic Sea.
Mangerud, J., 1977. Late Weichselian marine sediments containing shells, foraminifera, and pollen, at Ågotnes, western Norway. Norsk Geol. Tidsskr. 57, 23-54.
Mörner, N.-A., 1969. The Late Quaternary history of the Kattegatt Sea and the Swedish West Coast; deglaciation, shorelevel displacement, chronology, isostasy and eustasy. Sveriges Geol. Undersökn. C-640, 1-487.
Mörner, N.-A., 1971a. A Late Weichselian climatic zone system for southern Scandinavia and related areas. Geol. Fören. Stockh. Förh. 93, 236-238.
Mörner, N.-A., 1971b. The age of the Ågård Interstadial and the Fjärås Stadial. Geol. Fören. Stockh. Förh. 93, 768-773.
Mörner, N.-A., 1973. Eustatic changes during the last 300 years. Palaeogeogr. Palaeoclim. Palaeoecol. 13, 1-14.
Mörner, N.-A., 1976a. Paleoclimatic records from South Scandinavia global correlations, origin and cyclocity. Paleolimn. Lake Biwa and the Japanese Pleistocene 4, 499-528.
Mörner, N.-A., 1976b. Eustatic changes during the last 8,000 years in view of radiocarbon calibration and new information from the Kattegatt region and other northwestern European coastal areas. Palaeogeogr. Palaeoclim. Palaeoecol. 19, 63-85.
Mörner, N.-A., 1979. The Fennoscandian uplift and Late Cenozoic geodynamics: Geological evidence. GeoJournal 3, 278-318.
Mörner, N.-A., 1980a. The Fennoscandian uplift: geological data and their geodynamical implication. In: Earth Rheology, Isostasy and Eustasy (N.-A. Mörner, Ed.). Wiley, 251-284.
Mörner, N.-A., 1980b. The northwest European "sea-level laboratory" and regional Holocene eustasy. Palaeogeogr. Palaeoclim. Palaeoecol. 29, 281-300.
Mörner, N.-A., 1980c. A 10,700 years' paleotemperature record from Gotland and the Pleistocene/Holocene boundary events. Boreas 9, 283-287.
Mörner, N.-A., 1983. Stockholm Excursion Guide. 2nd Nordic Symp. Climatic Changes Related Problems, Stockholm 1983, 16 pp.
Mörner, N.-A., this volume. Planetary, solar, atmospheric, hydrospheric and endogene processes as origin of climatic changes on the Earth.
Mörner, N.-A., this volume. Climatic changes on a yearly to millennian basis. Concluding remarks.
Mörner (Ed.), N.-A., 1976. The Pleistocene/Holocene boundary: A proposed boundary-stratotype in Gothenburg, Sweden. Boreas 5, 193-275.
Mörner, N.-A., & Wallin, B., 1976. A 10,000 year temperature record from Gotland, Sweden. Palaeogeogr. Palaeoclim. Palaeoecol. 21, 113-138.
Nathorst, A.G., 1872. Om arktiska växtlämningar i Skånes sötvattensbildningar. Ofversigt Kongl. Vet. Akad. Förhandl. 29, 123-142.

Nilsson, T., 1935. Die pollenanalytische Zonengliderung der spät- und postglazialen Bildungen Schonens. Geol. Fören. Stockh. Förh. 57, 385-562.
Nilsson, T., 1964. Enwicklungsgeschichte Studien im Ageröd Mosse, Schonen. Lunds Univ. Årsskr. NF2, 59:8, 1-34.
Nordmann, V., 1904. Dosinalagene ver Kattegatt. Medd. Dansk Geol. Foren. II:10, 23-40
Nordmann, V., 1910. Post-Glacial climatic changes in Denmark. In: Die Veränderungen des Klimas seit dem Maximum der letzten Eiszeit (XI Intern. Geol. Congr.). Stockholm, 313-328.
Olausson (Ed.), E., 1982. The Pleistocene/Holocene boundary in south-western Sweden. Sveriges Geol. Undersökn. C-794, 1-288.
OSKAP group member, 1983. Upper Quaternary marine Skagerrak (NE North Sea) deposits: stratigraphy and depositional environment. Inst. Geol. Univ. Oslo, Intern Skr. 43, 164 pp. (Also: Norsk Geol. Tidsskr., 1984, in press).
Petersen, K.S., this volume. Late Weichselian sea-levels and fauna communities in northern Vendsyssel, Jutland, Denmark.
Peterson, E.W., & Larsen, S.E., this volume. Climate variation in northern Europe during the past century. Evidence form a Danish record.
Ruddiman, W.F., & McIntyre, A., 1981. The North Atlantic Ocean during the last deglaciation. Palaeogeogr. Palaeoclim. Palaeoecol. 3.5 145-214.
Sernander, R., 1894. Studier övfer den Gotländska vegetationens utvecklingshistoria. Thesis, Uppsala, 112 pp.
Sernander, R., 1910. Die schwedische Torfmoore als Zeunen postglazialer Klimaschwankungen. In: Die Veränderungen des Klimas seit dem Maximum der letzten Eiszeit (XI Intern. Geol. Congr.). Stockholm, 197-246.
Steenstrup, J.J., 1841. Geognostisk-geologisk undersögelse af Skovmoserne Vildnesdam og Lillemost i det nordlige Sjaelland, ledsaget af sammenlignende Bemaerkninger hendet fra Danmarks Skov-, Khaer- och Lyngmoser i almindlighed. Copenhagen, 104 pp.
Torell, O., 1872-1873. Undersökningar öfver istiden. I & II. Öfvers. Kongl. Sv. Vet. Akad. Förhandl. (I) 1872:10, 25-66 (1972) & 1973:1, 47-64 (1973). (Also: Deutsch. geol. Gesell. 40, 250-257, 1888).
Wollin, G., Kukla, G.J., Ericson, D.B. & Wollin, J., 1973. Magnetic intensity changes 1925-1970. Nature 242, 34-37.
Vuorela, I., 1975. Pollen analysis as a mean of tracing settlement history in Finland. Acta Bot. Fennica 104, 1-48.

I. The Deglaciation Period

LATE WEICHSELIAN DEGLACIATION AND CHRONOSTRATIGRAPHY OF SOUTHERN SCANDINAVIA: PROBLEMS AND PRESENT "STATE OF THE ART"

Björn Berglund and Nils-Axel Mörner

Lund University and Stockholm University, Sweden

Southern Sweden has been a classical area for detailed registration of paleoclimatic fluctuations during the Late Weichselian. We still believe that this region is unusually favourable for deciphering the paleoclimatic history as recorded in the ice recession and biozonation, as well as for establishing a chronology of this history via radiocarbon dating and varve counting. However, the hitherto proposed reconstructions have to be revised (e.g. Lundqvist, 1961; Nilson, 1968; Mörner, 1969; Berglund, 1979). We realize that there are still serious problems connected with the identification and correlation of deglaciation features and their proper dating and relation to biozonation and paleoclimatic cold/warm periods.

We, therefore, found it necessary to attempt a critical analysis of "the present state of the art" (Berglund & Mörner, 1983).

This leads to a revised chronology for the traditionally used paleoclimatic time units and we, therefore, do not strictly use the chronozone system proposed by Mangerud et al. (1974).

After the biostratigraphical model of Berglund (1966) and the paleoclimatic deglaciation model of Mörner (1969), much new data have accumulated, especially thanks to detailed local mapping in selected key areas as a part of the Swedish contribution to the IGCP-24 project (Lagerlund et al., 1983). This has produced a new base of starting to reorganize the interpretation of ice marginal zones, deglaciation chronology and biostratigraphical correlations.

DISCUSSION

Our discussion is illustrated by two figures; Fig. 1 giving a map

N.-A. Mörner and W. Karlén (eds.), Climatic Changes on a Yearly to Millennial Basis, 17–24.

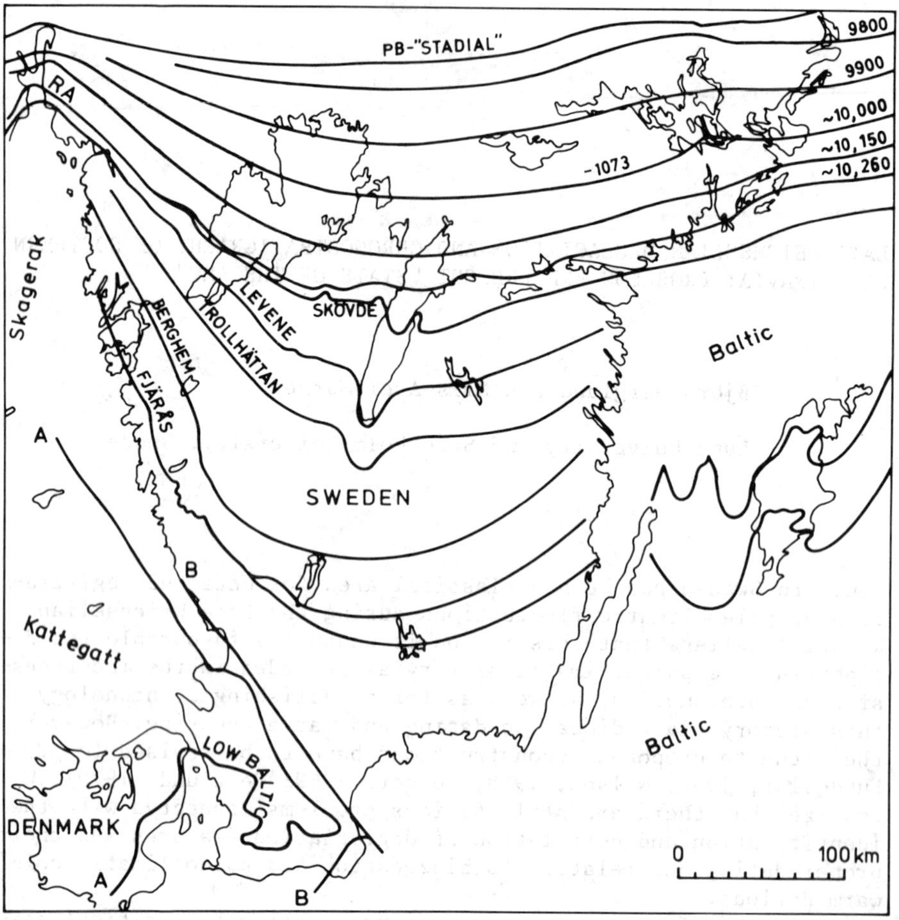

Fig. 1. Suggested ice marginal lines across southern Sweden. Names of major moraines (on the west coast) are given. Varve chronological dates from the Stockholm region are given in the upper right area. Ice marginal positions A (about 13,200 BP) and B (about 13,500 BP) refer to the two alternatives discussed in the text.

of the suggested ice marginal lines (only slightly altered from that of Berglund & Mörner, 1983) and Fig. 2 giving a chronological scheme summarizing the data discussed.

The first key point is the "Older Dryas" line. When new radiocarbon dates were obtained from the area inside the Berghem moraine zone (by Mörner correlated with the Older Dryas period), it seemed as if the position of the moraine-line equivalent to the Older Dryas cold phase was to be found in the areas south of Lake

Vättern and around the southwestern bay of Lake Vänern (Berglund, 1979). This seems now to have been confirmed by the work of Björck and Digerfeldt (1982) on Mt. Hunneberg.

As we neither can see reasons to advocate a separate ice left over the Highland of Småland or a more or less instantaneous collapse of the ice cap, we are left with the alternative to move the position of the ice margin during the Older Dryas cold phase from the Berghem moraine-line to the Levene or Trollhättan moraine-lines. Doing so, we have to ignore some data from the Swedish West Coast (e.g. Klinehögkärret of Mörner, 1969, 1979).

We realize that it is very hard, sometimes impossible, to date ice marginal lines directly and acurately by radiocarbon dating. Some guidance (e.g. Lagerlund et al., 1983) is obtained, however, from the floating varve chronology established along the Swedish East Coast (Kristiansson, 1982).

The age of the Older Dryas biozone/chronozone has been given at about 11,800-12,000 BP (Mörner, 1969; Tauber, 1970; Mangerud et al., 1974; Berglund, 1979). New radiocarbon dates do, however, open the possibility that it, in fact, may be about 12,200 years old (Björck, 1984; Berglund et al., this volume), which would better agree with the dating of a glacial readvance in western Norway (Mangerud, 1977, 1980; Mangerud et al., 1979). Further studies are necessary.

With and uncertainty in the proper dating of the Older Dryas cold phase ranging from about 12,200 to 11,800 BP, it is not clear which moraine zone that really correlates with this cold phase: the Levene, Trollhättan and Berghem lines are all potential candidates.

Having so revised the ice marginal position at about 12,000-11,800 BP, i.e. during the Older Dryas cold phase according to the original chronology, we review the younger positions as follows:

The initial phase of the Younger Dryas Stadial we place along the Skövde Moraine and its continuation to the Ra moraines in Norway. We note that the change from Alleröd to Younger Dryas is very drastic and that the period of optimal Alleröd-temperature (with well developed birch forests and high-organic limnic production in Skåne) seems to fall at the end of the Alleröd zone. The age of Alleröd/Younger Dryas (AL/YD) boundary is well established at about 11,000 radiocarbon years BP. This agrees with the results of Björck and Digerfeldt (this volume).

A period of severe climate and slow ice recession prevailed for some 500-600 years. (We note that there are 660 varves between the onset of the Salpauselkä I and the end of the Salpauselkä II zones in Finland). During this period, the ice receded up to, or rather north of, the northern slope of Mt. Billingen.

At about 10,500 radiocarbon years BP, there is a noticeable improvement in the vegetational records in the southernmost Sweden (Björck, 1981; Berglund et al., this volume). This may correlate with the rapidly increased rate of ice recession on Södertörn at around varve -1370, i.e. 10,262 varves BP (Mörner, 1977) and in the area north of Mt. Billingen (Mörner, 1970). According to the new results of Björck and Digerfeldt (this volume), there was a final "drainage" of the Baltic Ice Lake at about 10,500 BP.

Björck and Digerfeldt (1982) have shown that there was a very rapid, not to say "instantaneous", drop in relative sea level in the Hunneberg area at about 10,200 radiocarbon years BP. (Note the fact, that this occurred west of Mt. Billingen and must, hence, not be confused with any "drainage" of the Baltic Ice Lake). At about 10,200 radiocarbon years BP, the ice margin must have been somewhere north of Mt. Billingen. Exactly where, we do not know.

The sudden ingression of saltwater to Stockholm at varve -1073 (connected with a rapidly increased rate of ice recession and various tectonic effects that Mörner (1980) interpreted as evidence of a very large earthquake) most likely correlates with the arrival of saltwater in Finland at their varve +292. Apparently, however, the ice margin was considerably north of Mt. Billingen at this occasion. The ice margin may have been somewhere along northern Vänern and south of Hjälmaren and Mälaren. This extension drawn by us in 1983 (Berglund & Mörner, 1983) agrees well with the recently presented varve chronological work northeast of Lake Vättern (Perhans, 1984).

The two periods of retarded rates of ice recession on Södertörn, south of Stockholm (Mörner, 1976a, 1977) at around varves -1255 and -1120 may perhaps correlate with the Ås and Ski moraines in Norway (Sörensen, 1979), respectively. We do not, however, have any morphological criteria for such a correlation, so it should only be considered a possibility.

Varve -1073 corresponds to 9965 varves BP. The conversion to radiocarbon years is not known.

At about varve -840 (i.e. 9730 varves BP), there was a retarded ice recession. This may be traced westwards via the Riddarhyttan and Brattforsheden deltas (Mörner, 1976a) to the Aker moraine in southern Norway. It represents a short Preboreal cooling (Mörner, 1976a).

Having so discussed the period after the Older Dryas period, we now turn to the period prior to and outside the Older Dryas ice marginal position.

Lagerlund et al. (1983) have attempted a correlation of the ice marginal lines and zones across southern Sweden, combining the ice marginal features on the West Coast with the varve chronology (still floating, however) along the East Coast.

This suggests that there is about 200 varves between the Levene and Trollhättan lines, somewhat less than 200 varves between the Trollhättan and Berghem lines and some 200-300 (-400) varves between the Berghem and Fjärås (= Göteborg) lines.

If the age of the Levene moraine line is locked at about 11,800 BP, this would date the other moraine lines as follows: Trollhättan at 12,000 BP, Berghem at 12,200 and Fjärås at about 12,400-12,500 (12,600) BP. If the age of the Trollhättan line is locked at 12,200 BP (as done by Lagerlund et al., 1983), the other moraine lines would be dated as follows: Levene at 12,000 BP, Berghem at 12,400 BP, and Fjärås at 12,600-12,700 BP.

Back to the ice marginal position along the Fjärås (Göteborg) line, we have, so far, been able to agree and propose a mutual revised deglaciation system. For the period prior to the position along the Fjärås line, we have, however, not yet been able to agree. This holds for the interpretation of available radiocarbon dates, the identification of cold and warm phases, and the configuration of the ice margins.

At around 12,700-13,000 BP, there was a distinct change in the molluscan fauna from high arctic to boreo-arctic (Mörner, 1969, 1971). This marks a drastic change in the oceanic conditions; a change from a meltwater-dominated bay with pack-ice during most of the year, to a more or less open bay with a new circulation pattern and having more direct influence of Atlantic water (both more saline and warmer). We both agree on this.

Mörner (1969, 1979) believes that the paleoceanographic changes connected with the sudden shift from high-arctic to boreo-arctic faunas (the onset of the Ågård interstadial) are likely to have caused so great changes in the isotopic composition of the water that the ages of the two faunas cannot be directly compared. He finds the ages of the high-arctic fauna to be too old in relation to those of the boreo-arctic fauna, and claims that the stratal sequences (often varved) indicate a short time difference. He believes that there was a real, paleoclimatically caused, readvance at around 13,200 BP both of the Öresund lobe (the classic "Low Baltic Readvance") reaching up to Glumslöv and of the main ice cap readvancing up against Mt. Hallandsåsen (the G4-line of Mörner, 1969). He accepts the date of the Lockarp mammuth of about 13,200 BP (Lagerlund et al., 1983) as a proper age of this event. Finally, he believes that the general cold phase persisted for some 300 years, during which period the ice receded from Mt. Hallandsåsen to the coast of Halland.

Berglund (1979; Berglund & Lagerlund, 1981; Lagerlund, 1983) considers the ages of about 13,500 BP of some high-arctic faunal elements as proper ages comparable with the younger ages of the boreo-arctic fauna. Lagerlund identifies a younger, about 600 years long, cold phase which had its initial ice marginal position along the so-called Halland Coastal Moraines (Berglund, 1979). Lagerlund interpretes a Baltic ice lobe readvance in the southern Öresund area as a surge of short duration some time after 13,500 BP.

Further studies are needed to evaluate the significance of these two models. We, therefore, leave this question open here. Intense stratigraphical studies will continue in Skåne, which is a key area for the understanding of the Weichselian glaciation dynamics.

Correlations between ice marginal positions on the West Coast and the East Coast must also explain the paleomagnetic differences recorded. In the West Coast region the glacial clays generally

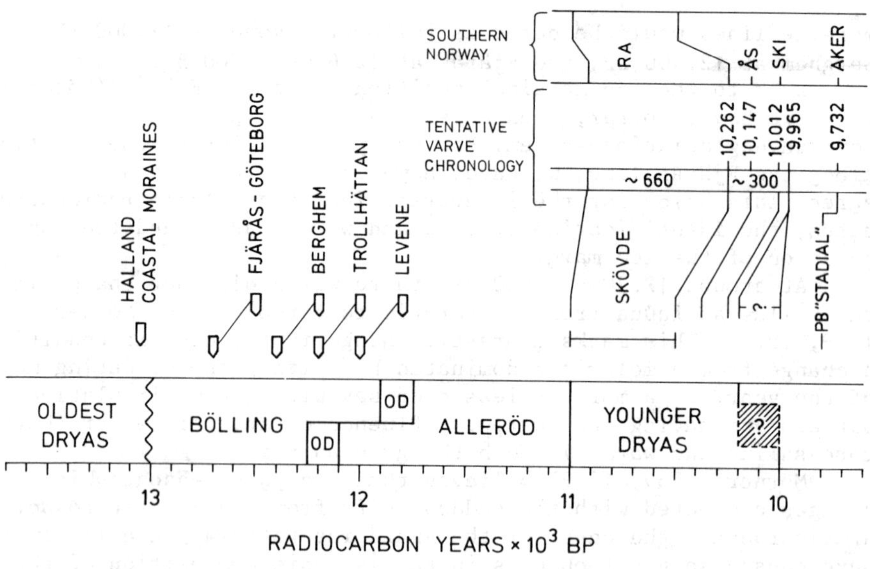

Fig. 2. Chronological summary of the data and discussion given in the text (cf. Fig. 1).

have high inclination values of some 60-70° (except for in the Ågård clay pit where it is 30-40°), whilst they in Blekinge and on the Kristianstad Plain have low values of about 30-40° (Ising, 1942; Mörner, 1975, 1976b; Noel 1975).

We agree that the first area of Sweden to be deglaciated was the Ängelholm area (Mörner, 1969) and that this deglaciation had the form of an opening interlobe area between the main (NE) ice and the Öresund lobe.

We do not want to discuss all the possible and hypothetic ice marginal positions prior to the "Halland Coastal Moraines" or the "Low Baltic Readvance" (whether a surge or a real readvance, we leave open).

We want to stress, however, that the classical D-line in Denmark is completely undated. It has been correlated with the Pommeranian line in Germany and Poland. We have no reasons to doubt this. An age of about 15,000 BP has often been suggested. This may be true, but it is based on assumptions and not on datings.

The C-line across Jutland, i.e. the classical maximum of the Late Weichselian glaciation, is not well dated. Some Polish dates of the corresponding ice marginal position give an age of younger than 22,000 (Kozarski, 1980). There seems to be good reasons to believe that the age is about 20,000-18,000 BP, but we cannot really substantiate this by proper dates within southern Scandinavia. We find no reasons to doubt that the Late Weichselian maximum really reached the C-line.

With more work, an ever searching attitude, and mutual collaboration and attempts of understanding each other, we think the Late Weichselian history as registered in southern Sweden can keep and consolidate its position of classical area. We deeply believe it deserves it.

REFERENCES

Berglund, B.E., 1966. Late-Quaternary vegetation in eastern Blekinge, southeastern Sweden. A pollen-analytical study. I. Late Glacial time. Opera Botanica 12:1, 1-180.

Berglund, B.E., 1979. The deglaciation of southern Sweden 13,500-10,000 BP. Boreas 8, 98-118

Berglund, B.E. & Lagerlund, E., 1981. Eemian and Weichselian stratigraphy in South Sweden. Boreas 10, xxx-xxx.

Berglund, B.E., Lemdahl, G., Liedberg-Jönsson, B. & Persson, T., this volume. Biotic responses to climatic changes during the time span 13,000-10,000 BP - A case study from SW Sweden.

Berglund, B.E. & Mörner, N.-A., 1983. The Late Weichselian deglaciation and chronostratigraphy of Southern Scandinavia: problems and present "state of the art". Abstracts, 2nd Nordic Symp. Clim. Changes Related Problems (Stockholm Univ.), 9-13.

Björck, S., 1984. Bio- and chronostratigraphic significance of the Older Dryas Chronozone on the basis of new radiocarbon dates. Geol. Fören. Stockh. Förh. 105, in press.

Björck, S. & Digerfeldt, B., 1982. The Late Weichselian shore displacement at Hunneberg, southern Sweden, indicating complex uplift. Geol. Fören. Stockh. Förh. 104, 132-155.

Björck, S. & Digerfeldt, G., this volume. Climatic changes at Pleistocene/Holocene boundary in the Middle Swedish endmoraine zone, mainly inferred from stratigraphic indications.

Ising, G., 1942. On the magnetic properties of varved clay. Ark. Mat. Astr. Fys. K. Sven. Vetenskapsakad. 29A:5, 1-37.

Kozarski, S., 1980. An outline of the Vistulian stratigraphy and chronology of the Great Poland Lowland. Quat. Studies Poland 2, 21-35.

Kristiansson, J., 1982. Varved sediments and the Swedish time scale. Geol. Fören. Stockh. Förh. 104, 273-275.

Lagerlund, E., 1983. Preliminary synthesis of the deglaciation of South Sweden. IGCP-24, Report 8.

Lagerlund, E., Knutsson, G., Åmark, M., Hebrand, M., Jönsson, L.-O., Karlgren, B., Kristiansson, J., Möller, P., Robison, J.M., Sandgren, P., Terne, T. & Waldermarsson, D., 1983. The deglaciation pattern and dynamics in South Sweden, a preliminary report. Lund Univ., Dept. Quat. Geol., LUNDQUA Rep. 24, 1-7.

Lundqvist, G., 1961. Karta över landisens avsmältning och högsta kustlinjen i Sverige (med beskrivning). Sveriges Geol. Undersökn. Ba-18.

Mangerud, J., 1977. Late Weichselian marine sediments containing shells, foraminifera, and pollen, at Ågotnes, western Norway. Norsk Geol. Tidsskr. 57, 23-54.
Mangerud, J., 1980. Ice-front variations of different parts of the Scandinavian ice sheet, 13,000-10,000 years BP. In: Studies in the Lateglacial of North-West Europe (J.J. Lowe, J.M. Gray & J.E. Robinsson, Eds.), Pergamon Press, 23-30.
Mangerud, J., Andersen, S.T., Berglund, B.E. & Donner, J.J., 1974. Quaternary stratigraphy of Norden, a proposal for terminology and classification. Boreas 3, 109-128.
Mangerud, J., Larsen, E., Longva, O. & Sönstegaard, E., 1979. Glacial history of western Norway 15,000-10,000 BP. Boreas 8, 179-187.
Mörner, N.-A., 1969. The late Quaternary history of the Kattegatt Sea and the Swedish West Coast; deglaciation, shorelevel displacement, chronology, isostasy and eustasy. Sveriges Geol. Undersökn. C-640, 1-487.
Mörner, N.-A., 1970. The Younger Dryas Stadial. Geol. Fören. Stockh. Förh. 92, 5-20.
Mörner, N.-A., 1971. The age of the Ågård Interstadial and the Fjärås Stadial. Geol. Fören. Stockh. Förh. 93, 768-773.
Mörner, N.-A., 1975. Palaeomagnetism and the relation between the Bredåkra delta and the Fjärås Stadial and the Gothenburg Magnetic Excursion. Geol. Fören. Stockh. Förh. 97, 298-301.
Mörner, N.-A., 1976a. The Pleistocene/Holocene boundary in southern Sweden. Boreas 5, 197-204.
Mörner, N.-A., 1976b. The Gothenburg Magnetic Excursion. Quatern. Res. 7, 413-427.
Mörner, N.-A., 1977. Varve chronology on Södertörn: recording and dating of the "drainage" of the Baltic Ice Lake and correlation with the Finnish varve chronology. Geol. Fören. Stockh. Förh. 99, 423-425.
Mörner, N.-A., 1979. The deglaciation of southern Sweden: a multi--parameter consideration. Boreas 8, 189-198.
Mörner, N.-A., 1980. A 10,700 years´ paleotemperature record from Gotland and Pleistocene/Holocene boundary events in Sweden. Boreas 9, 283-287.
Nilsson, E., 1968. Södra Sveriges senglaciala historia.; geokronologi, issjöar och landhöjning. Kungl. Vet. Akad. Handl., Ser. 4, Bd 12:1, 1-117.
Noel, M., 1975. The palaeomagnetism of varved clays from Blekinge, southern Sweden. Geol. Fören. Stockh. Förh. 97, 357-367.
Perhans, P.-E., 1984. Lervarvskronologiska studier i den norra delen av mellansvenska israndzonen i Östergötland och Södermanland. Medd. Stockh. Univ. Geol. Inst. 255, 167.
Sörensen, R., 1979. Late Weichselian deglaciation of the Oslofjord area, south Norway. Boreas 8, 241-246.
Tauber, H., 1970. The Scandinavian varve chronology and C-14 dating. In: Radiocarbon variations and absolute chronology (I. Olsson, Ed.), Wiley, 173-196.

BIOTIC RESPONSE TO CLIMATIC CHANGES DURING THE TIME SPAN 13,000-10,000 B.P. - A CASE STUDY FROM SW SWEDEN

Björn E Berglund, Geoffrey Lemdahl, Bodil Liedberg-Jönsson and Thomas Persson
Lund University, Sweden

Palaeoecological studies based on analysis of pollen, plant macrofossils and insect remains at a stratigraphic reference site, covering the time span 13,000-10,000 yrs B.P., in SW Sweden (Håkulls Mosse, province of Skåne) are interpreted against the two dominant palaeoclimatic models: the palaeobotanical model (Iversen) and the palaeoentomological model (Coope). An important time lag is found in the vegetation response to climatic improvement after deglaciation, which means that the biotic changes are best explained against the palaeoentomological model. This implies optimal summer temperatures 13,000-12,500 followed by a gradual deterioration until 11,000 when a sudden drop of temperature leads to a minimum between 11,000 and 10,500. A distinct rise of summer temperature around 10,500 is confirmed although the time lag in the vegetation response is ab. 300 years.

1. INTRODUCTION

Biostratigraphic studies of deposits from the Weichselian lateglacial in South Sweden have been descriptive - the main aim being chronological correlation and documentation of vegetational changes in relation to the deglaciation pattern, either local studies for selected lake basins or regional studies for selected provinces. The climatic background has been interpreted in rather simple terms, and dependent on the assumption that plants can be used directly as palaeoclimatic indicators. Integrated studies of a few palaeoecological reference sites within South Sweden are supposed to give us a better understanding of the relation biota and climate. We are demonstrating this by a synthetic description of stratigraphic studies at Håkulls Mosse, an ancient lake in SW

N.-A. Mörner and W. Karlén (eds.), Climatic Changes on a Yearly to Millennial Basis, 25–36.

Sweden. The study is coordinated by B E Berglund. The special investigations include sedimentology (Dearing), sediment chemistry (Malmer), radiocarbon dating (S Håkansson), pollen (Persson, Berglund), diatoms (H Håkansson), plant macrofossils (Liedberg-Jönsson) and insect remains (Lemdahl). The preliminary biostratigraphic results are interpreted climatically against the two dominant climatic models for this period.

2. PALAEOCLIMATIC MODELS FOR THE PERIOD 13,000 TO 10,000 YEARS B.P.

The palaeoclimate during the Weichselian lateglacial of continental NW Europe have been reconstructed on the basis of mainly pollen analysis and beetle studies of lake sediments. In addition to these two palaeobiological methods oxygen isotope analysis may be added as an independent palaeoclimatic tool, but very little is known about the oxygen isotope stratigraphy in South Scandinavia (cf. Kolstrup & Buchardt, 1982). It seems relevant to speak about two dominant palaeoclimatic models:

A) The palaeobotanical model, or the Iversen model, which goes back to the classical works on Danish lateglacial sequences, fully described by Iversen (1954, cf. 1973) but applied by many other palaeobotanists in NW Europe (e.g. Berglund, 1966). Approximate temperatures have been reconstructed by using the indicator species approach (cf. Birks, 1981), which means that e.g. past July temperatures were transferred from the knowledge of the modern relation between summer temperature and the distribution of a specific taxon. The traditional temperature curve obtained in this way is shown as curve A in Fig. 1. It shows the well-known climatic optima in late Bølling time, 12,300-12,000 B.P., and in Allerød time, 11,800-11,000 B.P., with the maximum temperature just before 11,000 B.P., followed by a sudden decrease correlated with the Younger Dryas stadial climate 11,000-10,200 B.P.

B) The palaeoentomological model, or the Coope model, which is founded on numerous well-documented beetle faunas mainly in British lateglacial sequences described by Coope and his collaborators (e.g. Coope, 1975, 1977). Fossil finds of beetles are compared with the modern distribution of these taxa in a way similar to that used for plants, which also makes it possible to obtain climatic parameters for the past. It is assumed that insects are generally dispersed more rapidly than plants, but especially during a deglaciation period. One would therefore expect that a climatic curve reconstructed on the basis of insect finds would present a more accurate curve. Curve B of Fig. 1 shows such a curve based on the studies by Coope. A thermal maximum is indicated slightly after 13,000 B.P. with temperatures around the present ones. It is followed by a climatic deterioration in two steps, first ab. 12.400 to 12,200 B.P., then ab. 11,000 B.P. with a thermal minimum ab. 10,500 B.P. Possibly the time resolution is

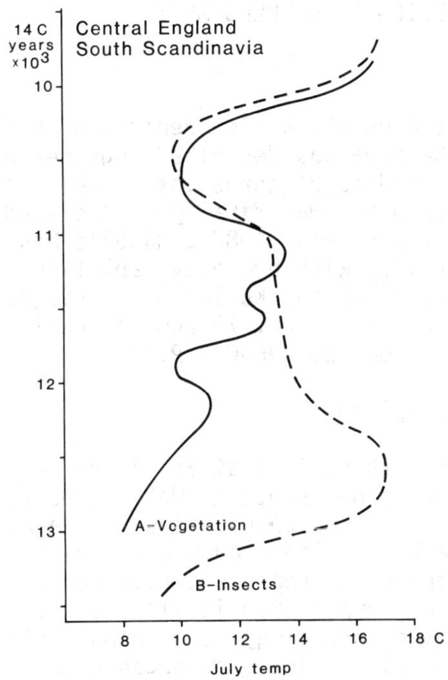

Figure 1. Average July temperatures for the time span 13-10,000 B.P. reconstructed from A) Palaeobotanical evidences (mainly acc. to Iversen 1954, 1973) and B) Palaeoentomological evidences (Coope, 1977). Both areas are supposed to be representative for areas with similar climatic regimes (Central England and South Scandinavia).

less accurate for British sequences with beetle stratigraphy in the time span 11-10,000 B.P.

It is obvious that an important time lag is involved in the vegetational response to a climatic improvement (cf. Davis, 1978; Birks, 1981; Delcourt et al., 1983; Prentice, 1983). This was also found in the British material and further discussed - the controversy between the interpretations on palaeobotanical and palaeoentomological material (cf. Coope & Brophy, 1972; Pennington, 1975, 1977; Coope, 1977). However, this discussion has not yet been applied on the Scandinavian environment and its palaeoecological material, although the revised deglaciation chronology seemed to support the palaeoclimatic model based on insect studies in Britain (Berglund, 1979, p. 113).

3. PALAEOECOLOGICAL STUDIES OF HÅKULLS MOSSE

3.1. Location

Håkulls Mosse is situated on the Kullen Peninsula in SW Sweden (cf. map of Fig. 2). The area was described together with the first biostratigraphic studies of three late-glacial sites by Berglund (1971). The peninsula was finally deglaciated between 14,000 and 13,500 B.P. (Lagerlund, 1980). Håkulls Mosse is situated in a fissure valley with hummocky, ablation till morphology ab. 125 m a.s.l. on the Kullen horst ridge. The basin was occupied by an ancient lake (ab. 70x200 m) until its overgrowing in Early Boreal time (ab. 9000 B.P.).

3.2. Stratigraphy and sampling

A regular sampling grid with corings at ab. 70 points forms the basis for the stratigraphic description. Main cores for detailed sedimentological and biostratigraphic analysis have been taken in the central part of the basin. The sequence at point A5, situated 40 m from the nearest shore, is representative for the main part of the lake. The sediment composition is illustrated by Fig. 3. When using the radiocarbon chronology (see below) this sequence shows the following general pattern. Low organic content (but increasing values upward) occurs before 11,400, high organic content 11,400 to 11,050, low organic content 11,050-10,200 with the very lowest values in the lake and its surroundings 11,050-10,700 B.P. The organic content is correlated with the organic production. High silt content occurs in the bottom, partly varved clay in layers 1 and 2 deposited before 12,300, in the thin layers 3 and 4 deposited 12,300-12,000 B.P. and in layers 10 and 11 deposited 11,050-10,700 B.P. The silt content is correlated mainly with erosion of exposed soils in the catchment. A percentage and influx pollen diagram is based on this sequence. Plant macrofossils and insect remains have been studied in a nearby core (X4), but until now only covering layers 2-14. The pollen diagram in Berglund (1971) was derived from an earlier core close to A5. A 12 m long transect close to the shore of the basin has also been cored in detail in order to trace sediment changes which may be due to shore displacements (water-level lowerings). In a main core of this transect (ab. 10 m from the ancient shore) pollen, plant macrofossils and insect remains

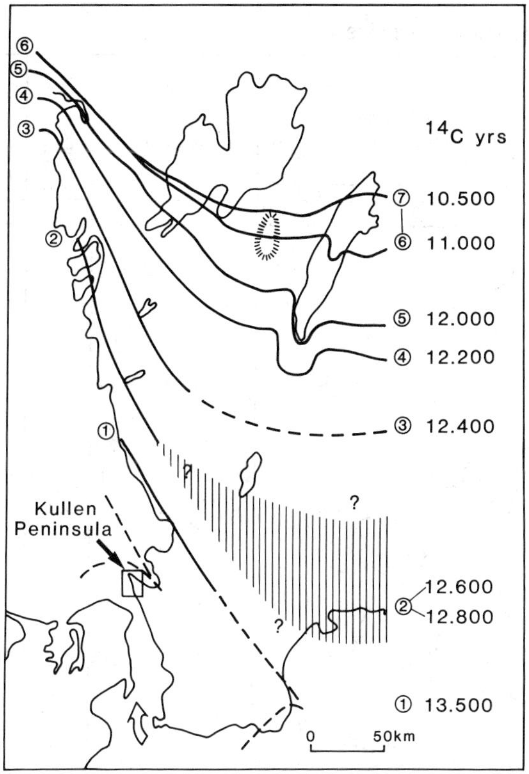

Figure 2. The location of the investigation area, the Kullen Peninsula in SW Sweden, in relation to radiocarbon dated ice-marginal zones (Berglund, 1979, revised after Johansson, 1982; Lagerlund et al., 1983; Björck & Digerfeldt, 1984).

have been analyzed (core B8:6).

3.3. Chronology

The chronology is based on 20 radiocarbon datings of the master core A5. The sediment accumulation rate curve seems to be reliable when using the dates from layers 7-16. Layers older than 11,700 B.P. seem to have been influenced by the hard-water error. When assuming a constant accumulation rate before 11,200 B.P. a tentative chronology has been obtained for the whole sequence, indicating that deposition started ab. 12,500 B.P. and that the more silty layers 3-5, just below the "classical" Allerød, are of the age 12,100-12,300 B.P. If these layers

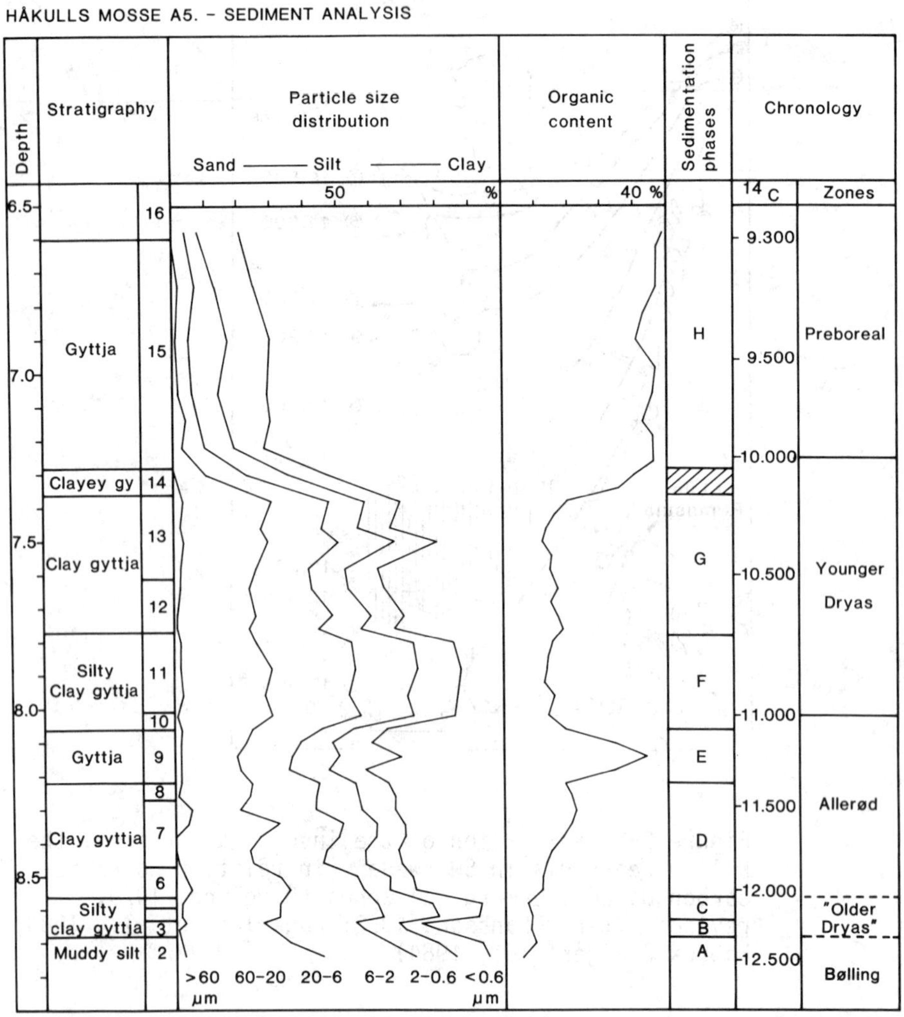

Figure 3. Sediment analysis of the master core A5 from Håkulls Mosse. Particle size of the minerogenic fraction is measured by X-ray technique in a sedigraph analyzer. Organic content corresponds to loss on ignition at 550°C. Sedimentation phases are distinguished on the basis of the mentioned parameters. The chronological scheme follows Mangerud et al. (1974) except that Older Dryas is given the proposed revised age 12,300-12,100 B.P. (Björck, 1984 and this study).

correspond to the so-called Older Dryas phase the dating is in good agreement with the redating of Older Dryas in SE Sweden by Björck (1984). Absolute ages and chronozone boundaries are indicated in Fig. 3, however with the tentative age of the Older Dryas interval 12,100-12,300 B.P. (cf. Mangerud et al., 1974; Berglund, 1979, p. 110). Possibly the deposition starts slightly earlier in the margin of the basin than in the central part (correlation A5/B8:6).

3.4. Palaeoecological results

The main ecological changes are described on the basis of two complete pollen diagrams (A5, B8:6), one complete plant macrofossil diagram (X4) and two almost complete, but preliminary, beetle diagrams (X4, B8:6), besides the stratigraphical information. Fig. 4 is a schematic illustration to the changes documented.

Bølling time (before 12,500 B.P.). Melting of dead-ice, formation of the lake basin, unstable soil conditions due to solifluction. No real vegetation cover, organic production of the lake extremely low. Edaphic conditions were influenced by local conditions: the decaying of ice-bodies and the inland-ice being maximum 100 km away. The period 13,500-12,500 B.P. is possibly the time-span during which periglacial nivation occurred in SW Sweden (Rapp, 1982).

Bølling time (12,500-12,300 B.P.). More stable soil conditions than before, organic production increasing, lake vegetation characterized by Chara and Nitella, open steppe-tundra vegetation with dwarf-willows, Betula nana and Hippophaë may indicate subarctic climate with dry summers. Single finds of Betula alba nutlets may indicate tree birch in the area. The beetle-fauna indicates temperate conditions through the occurrence of subalpine-boreal species. There are no finds of cold-demanding beetles.

Older Dryas time (12,300-12,100 B.P.). Increased soil erosion with silt transport to the lake basin, lake conditions as before. Open steppe-tundra vegetation with higher frequencies of nitrogen-fixating taxa such as Dryas and leguminous plants than before and after. High grass frequencies may indicate dry conditions. The beetles indicate temperate conditions.

Early and middle Allerød time (12,100-11,400 B.P.). Decreasing soil erosion, low lake macrophyte production, low lake level 11,700-11,400. Open tundra vegetation with dwarf-shrub heaths of Betula nana and after 11,700 also Empetrum. Tree-birch of some slight importance. The beetle remains show a mixed composition of temperate as well as cold-demanding species. Open landscape character is indicated by Amara alpina.

Late Allerød time (11,400-11,050 B.P.). Very stable soil conditions probably caused by a more continuous humus cover, increased lake macrophyte production (Chara, Nitella), low lake-

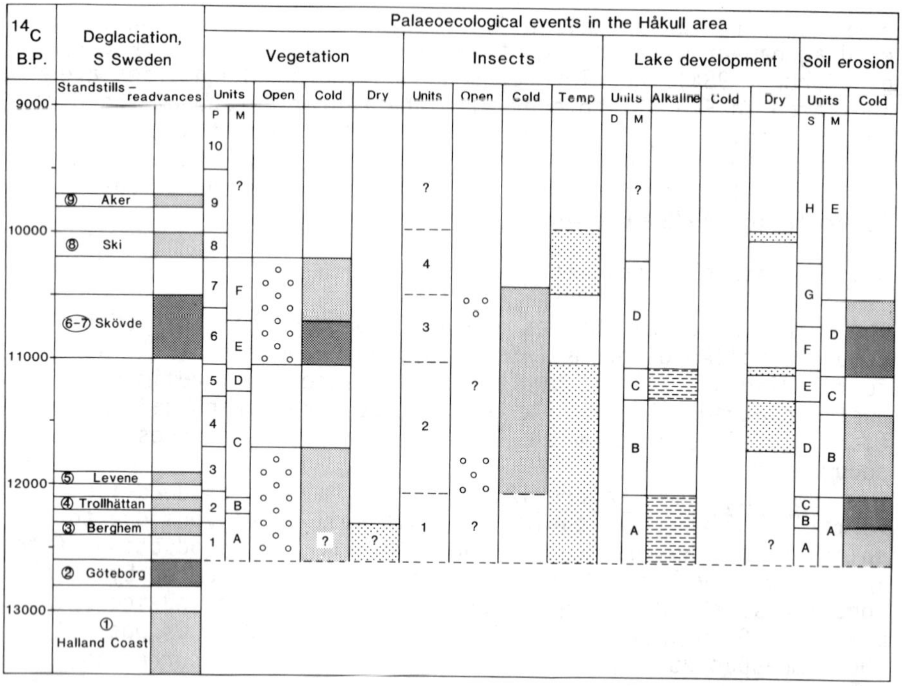

Figure 4. Simplified scheme presenting the preliminary interpretation of palaeoecological events in the Håkull area related to radiocarbon time scale and deglaciation pattern. In the columns for "Cold" the shading intensity illustrates cold and very cold conditions.

Deglaciation. The ice-marginal zones 3-5 are not directly dated (cf. Berglund & Mörner, this volume). Zones 6-7 are mainly one complex ice-marginal zone, the Skövde zone (Björck & Digerfeldt, this volume). Readvances indicating very cold conditions occur mainly during formation of zone 2 and 6.

Vegetation. Biostratigraphic units are given for pollen (P) and plant macrofossils (M).

Insects. The investigation and the interpretation of the Coleoptera fauna is preliminary so far. Cold-demanding species include Amara alpina, A. torrida, Arpedium brachypterum, Boreaphilus henningianus, Acidota quadrata and Simplocaria metallica. Temperate conditions are indicated by the occurrence of subalpine-boreal species such as Gyrinus areatus, Donacia sp., Cetonia sp., Simplocaria semistriata, Selatosomus melancholicus. All these species have a more southern distribution. Open land conditions are indicated by the occurrence of Amara alpina.

Lake development. Diatom assemblages (D) are not yet identified. Plant macrofossils (M) are the basis for a subdivision of the aquatic flora development. Mass occurrence of Chara and Nitella

indicate alkaline conditions. Sandy-silty layers at the shore indicate low water-levels.
Soil erosion. Sedimentation phases (S) are based mainly on the results from core A5 (Fig. 3). The macrofossil analyses (M) reveal erosion of terrestrial soils through rich occurrence of the fungi sclerotia *Cenococcum*.

level at the very end of the period. Woodland tundra vegetation characterized by *Betula alba* trees, juniper shrubs and *Empetrum* heaths. The beetle composition is about the same as before.

Early Younger Dryas time (11,050-10,500 B.P.). Intense soil erosion, especially before 10,700, rather high lake macrophyte production (*Batrachium*), high lake-level. Subarctic-oceanic tundra vegetation with less steppe elements than in the pre-Allerød tundra but with high frequencies of ferns (*Lastrea dryopteris*) and club-mosses (*Lycopodium* sp.). The beetle composition is characterized by the total dominance of cold-demanding species. The vegetation and the sediment changes around 11,050 indicate a sudden climatic deterioration.

Middle Younger Dryas time (10,500-10,200 B.P.). Rather stable soil conditions caused by the gradual revegetation and formation of a continuous humus cover, lake vegetation and lake conditions as before. Subarctic tundra vegetation with gradual expansion of heath ecosystems (with sedges, grasses, ferns, club-mosses, *Betula nana* and *Empetrum*). Possibly scattered birch trees occurred in the area. The beetle compositions show a sudden change around 10,500, being dominated by temperate species again.

Late Younger Dryas time (10,200-10,000 B.P.). Very stable soil conditions, high organic production in the lake as well as in the total catchment causing deposition of high-organic gyttja as in late Allerød time. Low water-level is indicated just around 10,000 B.P. Woodland tundra-prealpine woodland characterizes the landscape, which means dominance of *Betula alba* trees and juniper shrubs together with some heath elements (*B. nana*, *Empetrum*). Tall herb meadows with *Filipendula* and *Angelica* are expanding. The fossil beetle record clearly indicates temperate conditions.

4. CONCLUSIONS

The following palaeoclimatic conclusions may be drawn from this study.
1) There is a contradiction in the biotic response to the climatic development when vegetation and beetle fauna are compared. This may be explained by assuming a palaeoclimatic development similar to curve B in Fig. 1 for the period 13,000 to 10,000 B.P., as has been proposed for Coope for Britain (Coope, 1977) and in general terms discussed for NW Europe (van Geel & Kolstrup, 1978). However, our final results may differ at least in details (cf. points

2 and 4 below).
2) Due to our geographic situation in South Sweden it is hard to trace the regional climatic trend for the period before 12,300 B.P. There are some indications of a cooling, or at least increased soil erosion, after 12,300 although the general vegetation development shows a progressive successional development until the woodland expansion in late Allerød time ab. 11,400 B.P. Not until then is there a balance between summer temperature and vegetation. This means a time lag of about 1000 years for the biotic response to the climate.
3) The transition Allerød/Younger Dryas at about 11,000 B.P. indicates a sudden and distinct climatic deterioration with response in soils and vegetation (cf. Usinger, 1981 and earlier papers). This is also reflected in the distinct readvance of the Scandinavian inland ice (Berglund, 1979; Mangerud, 1980; Björck & Digerfeldt, this volume).
4) In the middle of Younger Dryas time ab. 10,500 the beetle remains indicate a distinct warming, which means that the subarctic-arctic phase of Younger Dryas has a length of only ab. 500 years. This is in agreement with several other studies in southern Sweden (Björck, 1981, pp. 90-91; Björck & Digerfeldt, this volume). However, there is also here a distinct time lag for the response of the terrestrial vegetation which can be estimated to ab. 300 years.

ACKNOWLEDGEMENTS

We are indebted to Christin Andréasson for drawing the illustrations, Sören Håkansson for radiocarbon analyses and Karin Price for typing the manuscript. We are also grateful to several colleagues at the Department of Quaternary Geology for discussions on Late Weichselian correlations in southern Scandinavia. - The work has been financially supported by the Swedish Natural Science Research Council (NFR).

REFERENCES

Berglund, B.E., 1966. Late-Quaternary vegetation in eastern Blekinge, southeastern Sweden. A pollen-analytical study. I. Late-Glacial time. Opera Botanica 12, 1. 180 pp. Lund.
Berglund, B.E., 1971. Late-Glacial stratigraphy and chronology in South Sweden in the light of biostratigraphic studies on Mt. Kullen, Scania. Geologiska Föreningens i Stockholm Förh. 93, 11-45.
Berglund, B.E., 1979. The deglaciation of southern Sweden 13,500-10,000 B.P. Boreas 8, 89-118.

Birks, H.J.B., 1981. The use of pollen analysis in the reconstruction of past climates: a review. In: Wigley, T.M.L., Ingram, M.J. & Farmer, G. (eds.): Climate and History. Cambridge. 111-138.

Björck, S., 1981. A stratigraphic study of Late Weichselian deglaciation, shore displacement and vegetation history in south-eastern Sweden. Fossils and Strata 14. 93 pp.

Björck, S., 1984. Bio- and chronostratigraphic significance of the Older Dryas Chronozone - on the basis of new radiocarbon dates. Geologiska Föreningens i Stockholm Förh. 105, in print.

Björck, S. & Digerfeldt, G., 1984. Climatic changes at Pleistocene/Holocene boundary in the Middle Swedish endmoraine zone, mainly inferred from stratigraphic indications. This volume.

Coope, G.R., 1975. Climatic fluctuations in northwest Europe since the Last Interglacial, indicated by fossil assemblages of Coleoptera. In: Wright, A.E. & Moseley, F. (eds.): Geol. Journ. Special Issue 6, 153-168.

Coope, G.R., 1977. Fossil coleopteran assemblages as sensitive indicators of climatic changes during the Devensian (Last) cold stage. Phil. Trans. R. Soc. Lond. B. 280, 313-340.

Coope, G.R. & Brophy, J.A., 1972. Late Glacial environmental changes indicated by a coleopteran succession from North Wales. Boreas 1, 97-142.

Davis, M.B., 1978. Climatic interpretation of pollen in Quaternary sediments. In: Walker, D. & Guppy, J.C. (eds.): Biology and Quaternary Environments. Austral. Acad. of Scienc., Canberra. 35-51.

Delcourt, H.R., Delcourt, P.A. & Webb, T., III, 1983. Dynamic plant ecology: The spectrum of vegetational change in space and time. Quaternary Science Reviews 1, 153-175.

van Geel, B. & Kolstrup, E., 1978. Tentative explanation of the Late Glacial and Early Holocene climatic changes in North-Western Europe. Geologie En Mijnbouw 57, 87-89.

Iversen, J., 1954. The Late-Glacial flora of Denmark and its relation to climate and soil. Danm. Geol. Unders. II, 80, 87-119.

Iversen, J., 1973. The development of Denmark's nature since the Last Glacial. Danm. Geol. Unders. V, 7 C. 126 pp.

Johansson, B., 1982. Deglaciationen av norra Bohuslän och södra Dalsland. Geol. inst., Chalmers tekn. högskola, Publ. A 38, 180 pp.

Kolstrup, E. & Buchardt, B., 1982. A pollen analytical investigation supported by an ^{18}O-record of a Late Glacial lake deposit at Graenge (Denmark). Rev. Palaeobot. and Palynol. 36, 205-230.

Lagerlund, E., 1980. Litostratigrafisk indelning av Västskånes Pleistocen och en ny glaciationsmodell för Weichsel. Univ. of Lund, Department of Quaternary Geology. Report 21, 120 pp.
Lagerlund, E., Knutsson, G., Åmark, M., Hebrand, M., Jönsson, L.-O., Karlgren, B., Kristiansson, J., Möller, P., Robison, J.M., Sandgren, P., Terne, T. & Waldemarsson, D., 1983. The deglaciation pattern and dynamics in South Sweden, a preliminary report. Lund University, Department of Quaternary Geology, LUNDQUA Report 24, 7 pp.
Mangerud, J., 1980. Ice-front variations of different parts of the Scandinavian ice sheet, 13,000-10,000 years BP. In: Lowe, J.J., Gray, J.M. & Robinson, J.E. (eds.): Studies in the lateglacial of North-West Europe. Pergamon Press. 23-30.
Mangerud, J., Andersen, S.T., Berglund, B.E. & Donner, J.J., 1974. Quaternary stratigraphy of Norden, a proposal for terminology and classification. Boreas 3, 109-128.
Pennington, W., 1975. A chronostratigraphic comparison of Late-Weichselian and Late-Devensian subdivisions, illustrated by two radiocarbon-dated profiles from western Britain. Boreas 4, 157-171.
Pennington, W., 1977. The Late Devensian flora and vegetation of Britain. Phil. Trans. R. Soc. Lond. B. 280, 247-271.
Prentice, C., 1983. Postglacial climatic change: vegetation dynamics and the pollen record. Progress in Physical Geography 7, 273-286.
Rapp, A., 1982. Periglacial nivation cirques and local glaciations in the rock canyons of Söderåsen, Scania, Sweden. A discussion and new interpretation. Geografisk Tidskrift 82, 95-98.
Usinger, H., 1981. Pollen- und Grossrestanalysen an limnischem Spätglazial aus dem Scharnhagener Moor, Schleswig-Holstein. Schr. Naturw. Ver. Schlesw.-Holst. 51, 85-105.

CLIMATIC CHANGES AT PLEISTOCENE/HOLOCENE BOUNDARY IN THE MIDDLE SWEDISH ENDMORAINE ZONE, MAINLY INFERRED FROM STRATIGRAPHIC INDICATIONS

Svante Björck and Gunnar Digerfeldt

University of Lund, Sweden

The Billingen area in the Middle Swedish endmoraine zone has been studied with respect to climatic changes and their effects between c. 11,300 and 9,500 yr B.P. With litho-, bio- and chronostratigraphic investigations combined with various physical and chemical analyses a picture of the Late Weichselian/Early Holocene glacial, vegetational and, climatic history has been obtained. In the later part of the rather warm Allerød Chronozone the Baltic Ice Lake was drained and the glacier retreated north of Mt. Billingen. A significant glacier readvance occurred in the early, cool part of the Younger Dryas Chronozone once again damming the huge glacial lake in the east. Around 10,500 yr B.P. a significant climatic amelioration created prerequisites for a normal vegetational succession and the final retreat of the glacier from the area. The last drainage of the Baltic Ice Lake occurred and after another c. 600 years the area for a short time was influenced by marine water some hundred years before the whole area was lifted above sea level. Since the climatic amelioration in the middle of the Younger Dryas Chronozone no significant climatic change seems to have occurred during the time-period studied. At least not according to the records from this very ice-proximal area.

INTRODUCTION

The chronology of the Middle Swedish endmoraines and the drainage of the Baltic Ice Lake are two classical topics in Swedish Quarternary studies. As the interpretations of these problems greatly affect our picture of the environmental changes at the

N.-A. Mörner and W. Karlén (eds.), Climatic Changes on a Yearly to Millennial Basis, 37–56.

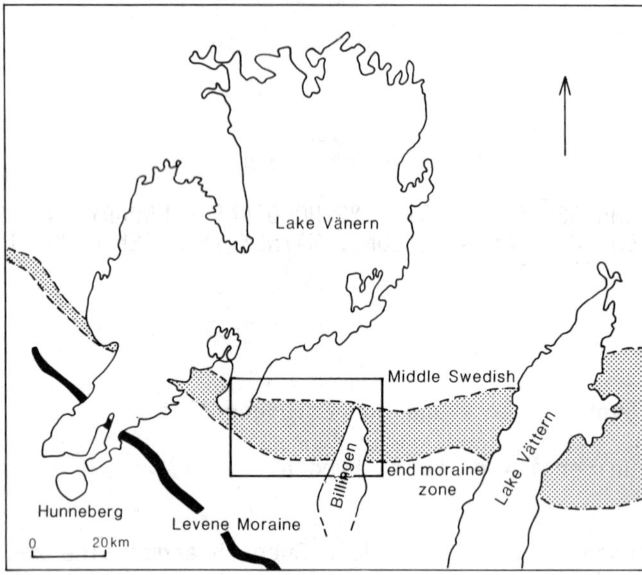

Fig. 1. The Lake Vänern-Lake Vättern region showing the positions of the study area.

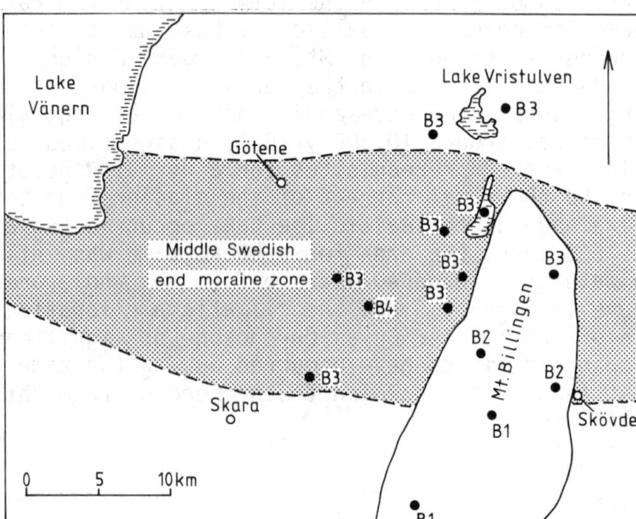

Fig. 2. The study area with the Middle Swedish end-moraine zone. Black dots show investigated lakes with the pollen or C14-dated age (B1-B4) of the oldest sediments.

Pleistocene/Holocene boundary, the area investigated (Fig. 1) must be regarded as a key area for paleoclimatic studies.

All authors (Munthe, 1940; Lundqvist, 1961; Nilsson, 1953, 1968; Mörner, 1969, 1970; Berglund, 1976, 1979; Fredén, 1982; Berglund and Mörner, this volume) engaged in the timing of the Skövde and Billingen lines (Fig. 1) of the Middle Swedish end-moraines place these moraines in the Younger Dryas Chronozone (Mangerud et al., 1974) and thus having been formed during the Younger Dryas stadial, recognized in the whole of NW Europe as a period of significant cooling.

As the Baltic Ice Lake, according to many authors, is supposed to have been drained north of Mt. Billingen (Munthe, 1910, 1928a, 1928b, 1940; Lundqvist, 1921, 1931, 1961; Ramsay, 1924; Johansson, 1926, 1937, 1943; Westergård, 1926; Sauramo, 1934, 1937, 1958; Caldenius, 1944; Nilsson, 1968; Donner, 1969, 1978, 1982; Strömberg, 1974, 1977; Björck, 1979, 1981; Fredén, 1982) there is a close relationship between the ice-marginal positions, the timing of the Middle Swedish endmoraines and any drainage of the Baltic Ice Lake (Fig. 1). Most of the previous works have not, however, been concentrated on the actual deposits around Mt. Billingen. A more complete stratigraphic study has now been carried out, combining new litho-, bio- and chronostratigraphic data with older stratigraphic descriptions. New and old shoreline and morphology data have then been added to the stratigraphic material to present a model which, at least for the moment being, seems to be the most probable one. It will be shown that any possible model is mainly ruled by rather dramatic climatic changes and in their turn resulting in quite dramatic geological events.

BIOSTRATIGRAPHY AND CHRONOLOGY

Altogether lake sediments from 15 lakes (Fig. 2) have been analysed in one way or another in order to date the oldest sediments of these lakes. In spite of some difficulties with low organic content 25 C14-dates from 7 different lakes have been carried out. In addition to that the oldest sediments from 13 of the lakes have been pollen analysed and in some of those lakes also diatom, oxygen isotope, and geochemical analyses have been performed. Finally these analyses have been related to different lithological parameters in order to relate bio-, and chronostratigraphy to lithostratigraphy.

Based on pollen and radiocarbon analyses a dated pollen stratigraphy has been set up for the Billingen area (Fig. 3). The main purpose for this stratigraphy is to use it as a dating tool for sediments with low organic content.

The oldest lake sediments found in the area are of Allerød age (Fig. 3), which means that they are older than 11,000 yr B.P. This is based on pollen evidences and C14-dates.

Like in Blekinge (Björck, 1981) and at Hunneberg (Björck

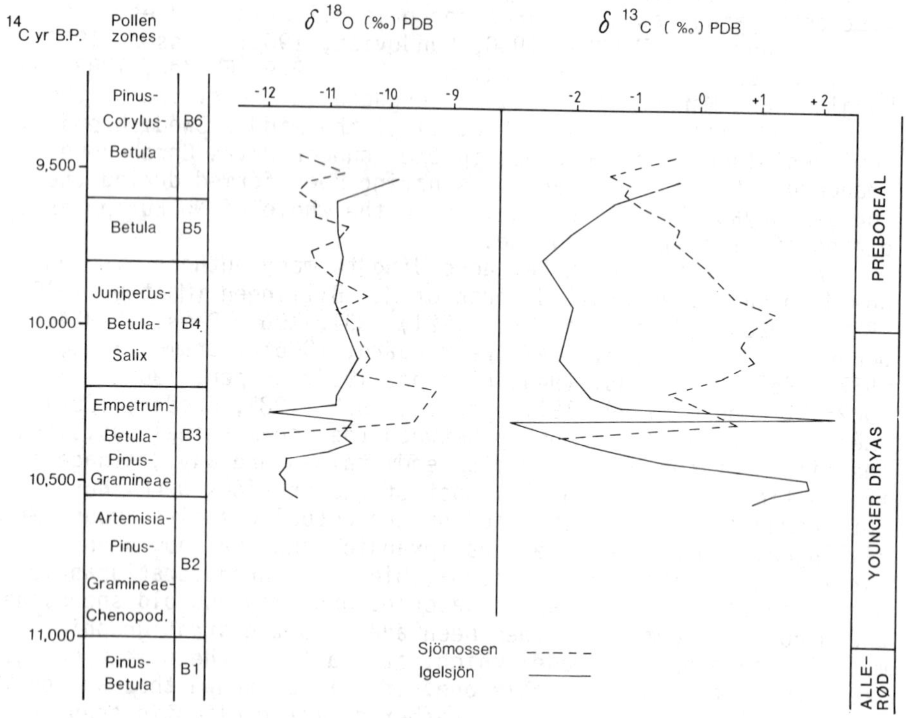

Fig. 3. The six (B1-B6) discerned pollen zones between 11,300 and 9,000 yr B.P. at Billingen related to radiocarbon chronology and isotope records of two lakes in the area.

and Digerfeldt, 1982) 70 km WSW of Mt. Billingen, the Younger Dryas Chronozone at Mt. Billingen is divided into three (B2-B4), rather distinct, pollen assemblage zones. The oldest of these zones is characterized by the abundance of Artemisia pollen. Pinus, Gramineae and Chenopodiaceae pollen do also reach significant values. Parts of, or the whole of this zone is found at six sites. This zone is supposed to represent the "classical" Younger Dryas stadial meaning the period of significant cooling throughout NW Europe. The high pine pollen values are interpreted as being of long-distance origin during a period mainly characterized by shrubs and herbs, which allow greater influx of wind-blown pollen from longe-distance sources. This zone lasts for 4-500 years.

Around 10,500 yr B.P. the pollen assemblage changes character. When Artemisia pollen values begin to decline it is followed by increasing Betula pollen values and soon thereafter also increasing Empetrum frequencies. This zone is called Empetrum-Betula-Pinus-Gramineae pollen assemblage zone (Fig. 3). At a couple of sites on Mt. Billingen the on-set of the zone is characterized by a very minerogenic layer coupled with a *Pinus* maximum. The most dramatic change, compared with the previous zone, is the rather abrupt change from herb dominance to a more shrub and tree dominated vegetation. This could indicate a more favourable climate allowing northerly migration of birch trees. The very distinct minerogenic layer, possibly indicating a water-level change, is dated to about 10,500 yr B.P. and the beginning of the next zone, the *Juniperus-Betula-Salix* pollen zone, starts around 10,200 yr B.P. This means that it seems to have taken 3-400 years between the end of the herb zone (B2) and the beginning of something similar to the often described and discussed transition zone (cf. Berglund, 1966). In the Billingen area the supposed vegetational changes from 10,500 yr B.P. and onwards seem to reflect a rather lengthy vegetational succession. The B4 zone lasts about 400 years and after that trees dominate completely, starting with birch and followed by pine and hazel. The empiric limit of hazel is set to 9,700-9,600 yr B.P. This development suggests that the possibly most significant climatic change in the end-moraine zone occurred already in the middle of the Younger Dryas Chronozone.

The $^{18}O/^{16}O$ measurements on sediments from Sjömossen (Mt. Billingen) and Igelsjön (Valle Härad) suggest that any climatic amelioration occurred in, or even before, the early part of zone B3 (Fig. 3). In lakes situated above the marine limit the lower part of zone B3 (10,400-10,500 yr B.P.) is also characterized by rapidly increasing organic content.

MORPHOLOGY AND LITHOSTRATIGRAPHY OF THE END-MORAINE ZONE

The area west of northern Mt. Billingen (Fig. 4) is morphologically rather complex. The altitudinal difference between Lake Lången (65 m a.s.l.) and the plateau of Mt. Billingen 7 km SE of the lake is 240 m. West of Lake Lången is a very steep fault line, where the bedrock rises almost vertically to c. 130 m a.s.l. A very characteristic feature are the vast accumulations of boulders situated west of the southern Lången valley and west of the northern tip of the lake (Fig. 4).

South of the Lången valley is the Valle Härad kame complex, 10 km long and 2-5 km wide. It is a mixture of hollows, ridges, plains and valleys. It consists mainly of disturbed, incompletely sorted deposits mixed with clays, other well-sorted deposits as well as till. The material does to a major degree consist of paleozoic bedrock (mainly Cambrian sandstone, shale and limestone), which also partly forms Mt. Billingen. West of Valle Härad is a complex area (Fig. 4) of well-pronounced ridges (20-40 m high) and heights with valleys or flat, rather restricted plains in between.

In order to present a possible glacial history of this morphologically and stratigraphically very complex area, drill-hole

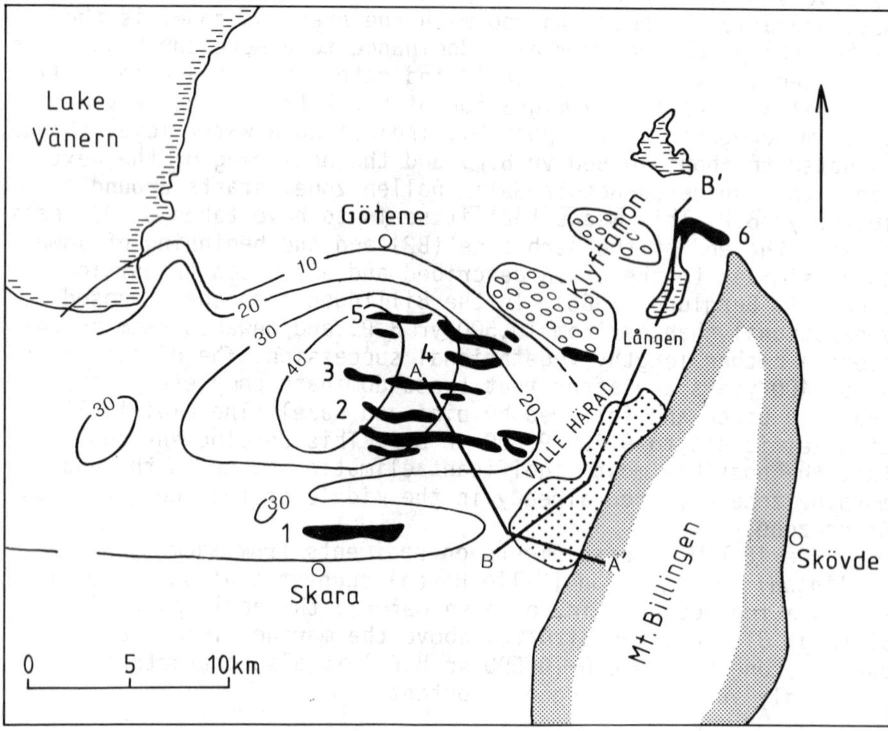

Fig. 4. The study area showing the six ridge complexes (1=Skara ridge, 2=Skånings-Åsaka ridge complex, 3=Ledsjö ridge complex, 4=Uppsala ridge, 5=Bjurum ridge complex, and 6=Timmersdala ridge), the Valle Härad complex (dotted), the two "boulder zones" on Klyftamon, the positions of profiles A-A' and B-B', and contour lines for depth of Quaternary deposits.

records from the Swedish Geological Survey were purchased and compiled, Johansson's (1943) stratigraphic descriptions were in detail studied and compiled and own investigations were carried out. The presented material is thus based on more than 100 stratigraphic descriptions with, however, some varying qualities.

The quantity of Quaternary deposits within the area is enormous (Figs. 4-5); in Valle Härad alone in the order of 1 km^3. West of Valle Härad the greatest quantities of material are found in the central and western part of the ridge area (Fig. 4) reaching depths of more than 46 m, while the Quaternary cover is very thin east and north of Götene.

As the most complete sequences are found in the ridges 1-3 (Fig. 4) the stratigraphy from those have been compiled (Table 1) in the sense that most units found outside the ridges can be

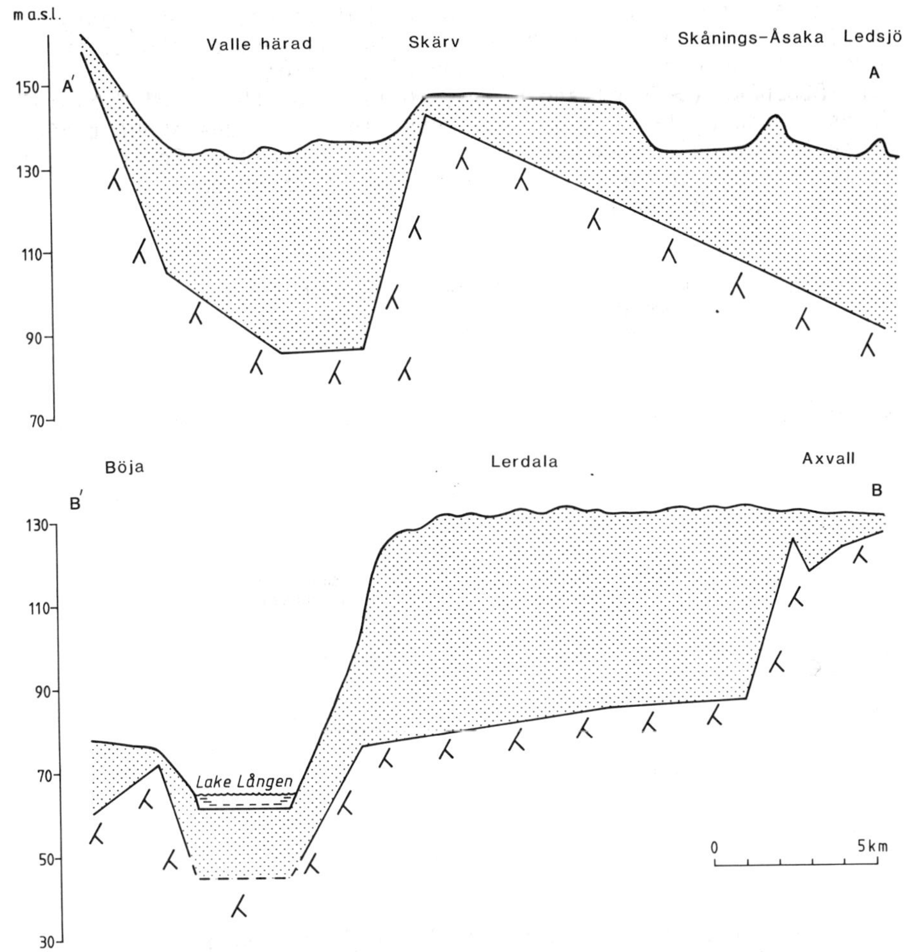

Fig. 5. Bedrock surface and Quaternary deposits (dotted) along profiles A-A' and B-B'. Note that the profiles are generalized.

correlated to this stratigraphy. In order to understand the relationship between the ridges and the valleys in between a stratigraphy for the valleys has also been set up (Table 2).

From Table 1 it is quite clear that there are two generations of glacial material with complex sequences in between. The lowermost SAND and GRAVEL (R2) seems to be especially common and thick in the area with greatest thickness where it makes up significant parts of the sequence. We think that this unit is a westerly and distal continuation of the Valle Härad deposits. We also think

Table 1. Lithostratigraphy of the ridges 1-4 based on drill-hole records and field observations. Note that if the stratigraphy should be valid also for the area north and west of those, two more units should be placed on top of the ridge stratigraphy: SILT-BOULDERS (> 30 m) and CLAY (> 6 m), i.e. the final drainage deposits and post-drainage sediments. The knowledge about their characteristics and extent is, however, incomplete. So not until it is possible to set up an independent stratigraphy for that area will they be included in a stratigraphic scheme.

Unit	Lithology	Maximum penetrated thickness (m)	Occurrence	Altitude of occurrence (m a.s.l.)	Proposed origin
R7	TILL or GRAVEL	12	Very common	137 - 118	Glacial, ice-marginal
R6	SAND (poor in shale)	11	Common	135 - 121	Glaciofluvial, redeposited?
R5	CLAY (partly deformed)	> 30	Always present	134 - 98	Glaciomarine, glacially transported and redeposited
R4	SAND	12	Rare	101 - 89	Glaciofluvial
R3	CLAY	5	Rare	89 - 84	Glaciomarine
R2	SAND and GRAVEL	> 45	Very common	145 - 100	Baltic Ice Lake drainage deposits
R1	TILL	> 4	Common	114 - 96	Glacial

that the main part of this material was deposited during a first drainage of the Baltic Ice Lake. This should have occurred in the later part of the Allerød Chronozone (Donner, 1969; Björck, 1979; Kessel and Raukas, 1980), which according to Björck (1979) at least lowered the Baltic Ice Lake 10-15 m. At this event the southern Valle Härad depression was rapidly filled in with a mixture of drainage deposits and bigger or smaller ice-bodies. Judging from the highest altitude in and west of the southern part of Valle Härad the sea level during the drainage could have been about 155 m.

Units R3-R7 (Table 1), we think, represent deglaciation, an ice readvance and final deglaciation. The CLAY (R5) building up major parts of the ridges was probably originally laid down in the area during a deglaciation phase. This phase was followed by a readvance, possibly to the Skara ridge (Fig. 4), during which the deglaciation sediments were overridden. Because of that they were usually also deformed, pushed up and transported by the readvancing glacier. An exception seems to be unit R2 and also rare finds of clay and sand (units R3-R4) in possible in situ positions. The sand was possibly deposited in front of the readvancing glacier.

The Skara ridge was formed in front of the ice as a kind of push-moraine during the maximum of the readvance. This and the other ridges contain, to a high degree, the deformed CLAY (unit R5). During standstills at the final ice-retreat additional push-moraines were formed in front of the ice (ridges 2-4, Fig. 4) often covered by till and glaciofluvial material, units R6-7. As it was transported from the north it was poor in Cambrian material.

The stratigraphy in the low-lying areas between the ridges is different (Table 2). Units V1-V3 can correspond to units R1-R3 (Table 1) representing deglaciation, sedimentation of coarse sorted deposits and a glaciomarine environment, but units V4 and V5 represent a very different depositional environment compared to units R5-R7. Unit V4, we think, was transported by meltwater from the vast quantities of dead-ice in Valle Härad. This also explains the high content of Cambrian material in unit V4. So while the ridges were formed as a kind of push-moraines during the final ice-retreat, material from the east was transported into the areas south of those. During the rapid regression later on (Björck and Digerfeldt, 1982) unit V5 was deposited as wave

Table 2. Lithostratigraphy of the valleys between ridges 1-4 based on drill-hole records and dug sections (mainly from Johansson, 1943).

Unit	Lithology	Maximum penetrated thickness (m)	Occurrence	Altitude of occurrence (m a.s.l.)	Proposed origin
V5	SAND and GRAVEL (poor in shale)	3	Very common	120 - 103	Wave reworked
V4	SAND and GRAVEL (rich in shale)	> 4	Very common	122 - 111	Glaciofluvial and/or fluvial, transported from Valle Härad
V3	CLAY	> 7	Very common	120 - 101	Glaciomarine, redeposited?
V2	SAND and GRAVEL	> 2	Common	113 - 99	Glaciofluvial or drainage deposits
V1	TILL	> 5	Common	110 - 100	Glacial

reworked material from the surrounding ridges (poor in Cambrian material).

Lundqvist (1931) made up a model for the final drainage of the Baltic Ice Lake in which the glacier was situated at the very northern tip of Mt. Billingen and the Baltic Ice Lake was lowered by 26 m. The amount fits well with Munthe's (1910, 1928b, 1940) observations (24-27 m). The same magnitude of a sudden lowering has been found in the shore displacement from Latvia (Berzin, 1967), Finland (Donner, 1969), Lithuania (Gudelis, 1969), SE Sweden (Björck, 1979), USSR (Kvasov, 1979), and Estonia (Kessel and Raukas, 1980). Furthermore, Westergård's (1926) and Bergsten's (1943) investigations east of Mt. Billingen support the idea of a suddenly lower water level in the Baltic basin at the deglaciation. Strömberg's (1977) thorough review and study of the Billingen-Baltic Ice Lake complex showed that Mt. Billingen area did not necessarily have to be completely deglaciated before the drainage could occur. We also think that the glacier's margin was south of the Timmersdala ridge (ridge 6, Fig. 4) when the drainage began. The Baltic Ice Lake water flowed partly on top of the glacier in the Lången depression and possibly partly also between Mt. Billingen and the glacier towards the south and had to change direction towards the west, 3.5 km south of Lången, where the Valle Härad deposits rise to 135-150 m a.s.l., i.e. 10-25 m above sea level at that time.

The deposits transported westwards by the drainage water possibly filled in crevasses and hollows in and between stagnant ice-bodies in front of the active ice-margin, forming ridge 5. In contrast to ridges 1-4 this irregular ridge consists mainly of coarse rather poorly sorted material, built up to max. 125 m a.s.l. According to Lundqvist (1931; Fig. 52) this zone, from Lerdala to the western part of the Bjurum ridge is rich in Cambrian sandstone in comparison to areas north and south of it. This suggests that the sandstone, which outcrops closest to Lången, was once again eroded by drainage forces. NW and W of Götene the basement rises very steeply (Mt. Kinnekulle), which forced the sediment rich drainage water towards SW and thus explaining the presence of very thick sandy-silty deposits W and SW of the ridge area.

Very soon the southern part of Klyftamon (105-130 m a.s.l.) was ice-free, but ice still occupied L. Lången (Fig. 4). The Baltic Ice Lake water washed the till and deposited cobble and gravel from Mt. Billingen (rich in Cambrian sandstone). In this way we think the southern "boulder zone" (Fig. 4), described by Lundqvist (1931), was formed.

The ice in Lake Lången and the high bedrock surface prevented the Baltic Ice Lake water from producing a boulder zone west of the lake. Not until the ice melted from the more low-lying area west of the Timmersdala ridge (Fig. 4) another "boulder zone" was formed. A zone relatively rich in Cambrian bedrock (Lundqvist, 1931) coincides with the northern "boulder zone".

At that time the remaining ice in Lången should have been cut off from the active ice-margin and the Timmersdala ridge could be formed in a gap between those. The steep slopes on both sides of the ridge are interpreted as ice-contact features.

The ice sheet kept receding north with immense water flow in front of the margin. Not until the ice-margin was situated along a line Låstad-Holmestad-N of Götene had the water flow slowed down enough to create the prerequisites for the formation or preservation of eskers and endmoraines. At that time the outlet north of Billingen was at least 2 km wide. Judging from the endmoraines between Låstad and Binneberg and those north of Holmestad the ice recession from now on was 100-200 m/yr.

STRATIGRAPHY AND DEVELOPMENT OF THE LÅNGEN BASIN

The whole development described above would not have been possible to interpret if not the litho-, bio-, and chronostratigraphy of the Lake Lången sediments had been investigated. Altogether more than 30 corings have been carried out in the lake. The lithostratigraphy has been described in detail from 22 of those and many of them have been brought to the laboratory for further analyses: pollen, diatoms and detailed lithostratigraphic descriptions. These studies have resulted in a fairly detailed division of the deposits in Lake Lången (Fig. 6). The sequence of interest for this study has been subdivided into 9 units (A-I) and the history interpreted below is based on the characteristics of each unit and its stratigraphic position, starting from the bottom of the sequence (Fig. 6).

A. Only once has it been possible to document this unit as it usually was impossible to penetrate through the covering unit B. This clay is correlated with the clay (unit R5) building up the ridges further west (Table 1) and is known, from investigations in and south of the Lake Vänern basin, (e.g. Dennegård, 1979) as being the oldest clay in the area. Geochemical analyses show that its chemical composition is very different from the composition of the younger grey clays. This is probably the reason for the colour differences. The clay is supposed to have been deposited in a glaciomarine environment, distal to the receding ice after the Baltic Ice Lake drainage during Allerød. No diatoms, but a few foraminifera were found. We think that the over-consolidated glaciomarine clay at Askeslätt, Lidköping (Dennegård, 1979) at least partly can correspond to unit A. The overconsolidation was caused by the readvancing glacier during Younger Dryas and the clay was either only overridden or transported. In any case the clay was deformed. Whether unit A in Lången occurs as occasional lumps or in its original position is not known, but it seems to be present in many parts of the basin.

B. When the glacier passed over the previously ice-free and clay-covered region, unit B was deposited as a basal till mixed

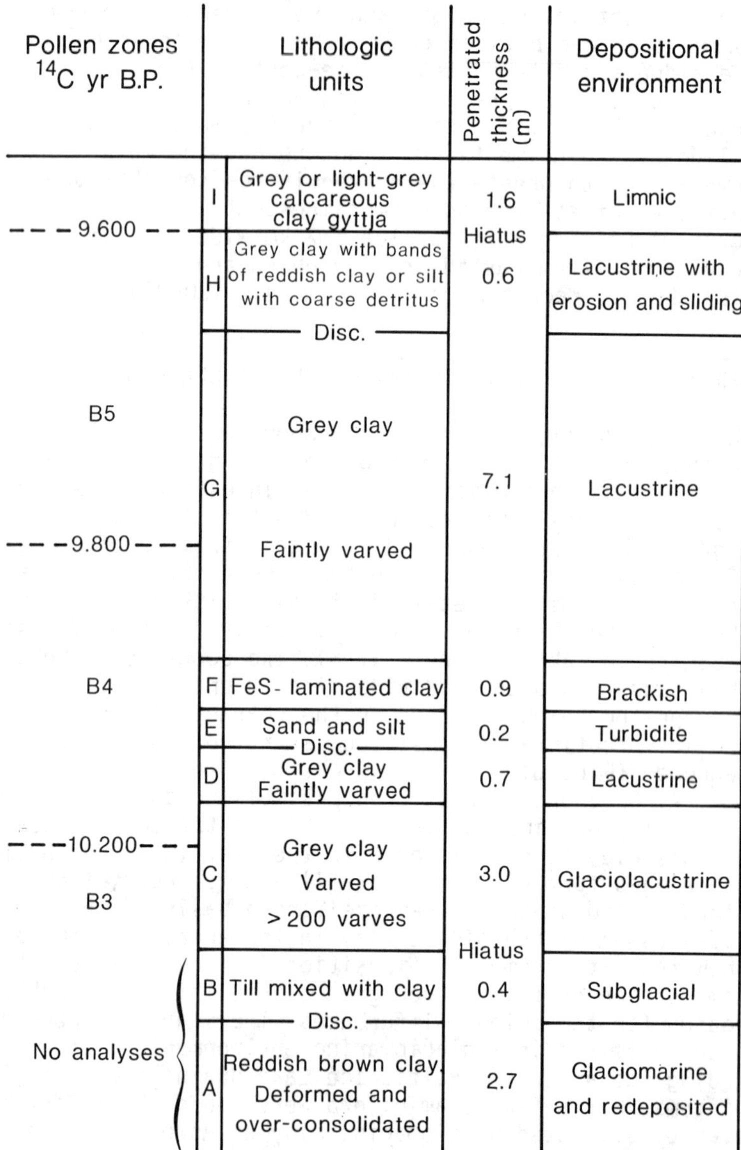

Fig. 6. Lithostratigraphy of Late Weichselian - Early Holocene sediments in Lake Lången based on 22 corings in the lake. The lithology is related to pollen zones, radiocarbon chronology and, interpreted depositional environment. Maximum penetrated thickness of each unit is also shown.

with lumps of clay. This has also been reported by Johansson (1943) from the ridge area west of Billingen. The till with lumps and bands of clay on top of the Råda delta at Lidköping (Johansson, 1943) possibly corresponds to the same readvance. Because of glacial erosion and deformation the contact between units A and B is developed as a disconformity. This unit does most likely correspond to Strömberg's (1977) "Oszillations-schicht", i.e. Caldenius' (1944) secondary drainage varve.

C. As long as the glacier covered the Lången depression no deposition, apart from unit B, took place. Therefore we can conclude that ice occupied the Lången area during the two supposed drainages from the Baltic basin. If the depression would have been free from ice at those occasions it would certainly have been filled up with drainage deposits. So a hiatus exists between units B and C. According to pollen analyses and varve counts on the grey clay of unit C the sedimentation of that unit begins around 10,400 yr B.P. It shows that the last drainage took place before that, and that the ice in Lake Lången was gone by then. It does, however, also indicate that the receding ice was not far away when sedimentation began, but in spite of that no coarse drainage deposits entered the basin. This fact together with the presence of distinct varves show that unit C was deposited in a fairly calm glaciolacustrine environment. No diatoms were found in unit C.

D. Unit D is a continuation of unit C, but the glacial influence decreases. The diatom flora is completely dominated by the oligohalobous Melosira islandica ssp. helvetica, which we think indicates a predominance of freshwater influence from the Baltic basin as its outlet was situated just north of Lången. This shows that the outflow from the east was still too great to allow marine ingression from the sea in the west.

E. This unit of sand and/or silt is not always found, but where it occurs the contact to the underlying clay is always erosive. We interpret this unit as turbidites caused by rapid regression around 10,000 yr B.P., which has not only been documented from Hunneberg (Björck and Digerfeldt, 1982) but is also indicated from our shore-displacement investigations west of Mt. Billingen. We also think that the, in Olausson (1982), often described silty layers from the Swedish West Coast can be explained by the same processes instead of regarding them as distal (>80 km) deposits from the Baltic Ice Lake drainage. This becomes even more probable if the drainage occurred a couple of hundred years earlier than previously presumed.

F. This clay unit was always found when the penetration was deep enough. Its main characteristics are the FeS-laminae and the diatom assemblage. Compared with the diatom assemblage of underlying and covering clays there are very significant increases of polyhalobous, mesohalobous and halophilous diatoms (Diploneis smithii, Nitzschia navicularis, Achnanthes hauckiana, Amphora commutata, Opephora schulzi, Synedra pulchella, Rhoicosphenia *curvata*). This increased influence from the sea water in the west is dated to c. 10,000 yr B.P., which is slightly later than the marine dominance in Lake St. Bjursjön, 2 km SW of Lake Lången (Fig. 2). This quite short-lasting phase (< 100 years) is the only period of marine influence in the Lången basin. It does probably also correspond to the opening of the Otteid and Närke straits (Fredén, 1982) allowing the entry of some sea water into the Baltic.

G. This clay is the same type of clay as in unit D. Once again the oligohalobous *Melosira islandica* ssp. *helvetica* dominates completely (72-97 %). As only scattered finds of polyhalobous or mesohalobous diatoms have been found the marine influence seems to have disappeared. If the onset of the Ancylus stage (or the end of the Yoldia Sea stage) in the Baltic is defined as the time when only freshwater from the east could pass over the Närke strait (cf. Fredén 1982) the beginning of unit G (9,800-9,900 yr B.P.) could actually correspond to that event as it in Finland has been dated to c. 9,800 yr B.P. (Glückert and Ristaniemi, 1982). So the reason for the freshwater environment is either an increased freshwater influence in the eastern Lake Vänern basin owing to uplifted and thus shallower straits allowing only one waterflow direction (from the east) or a shallowing of the Lake Lången threshold to such a degree that no salt water could enter the basin. The faintly varved character of the clay indicates a still glaciogenic influence. As the ice at that time was far north this could either originate from meltwater in the Baltic basin, or from melting dead-ice in the Lången drainage area. Finally it should be noted that the greatest thickness of unit G is found in the NW part of the lake.

H. This unit, which is very heterogenous, is a continuation of unit G mixed with bands and lumps of reddish clay and silt. Sometimes also thin layers of coarse-detritus or gyttja clay with molluscs occur. This is supposed to represent the pre-isolation phase of the lake when near-shore sediments were eroded because of the water-level lowering. The boundary between this and the underlying unit is always developed as a disconformity showing the erosive character of unit H.

I. According to pollen analyses the boundary between units H and I represent a hiatus. Unit I represents the oldest deposition in the isolated lake as indicated by the diatom analyses. From the pollen analyses the isolation of Lake Lången seems to have occurred around the rational *Corylus* limit (c. 9,600 yr B.P.). Owing to a probably distinct water level lowering in Early Holocene (Digerfeldt, 1972, 1974, 1976) the lowest part of unit I is usually strongly disturbed.

SEQUENCE OF EVENTS IN THE DEGLACIATION

Based on facts described above it is now possible to propose a

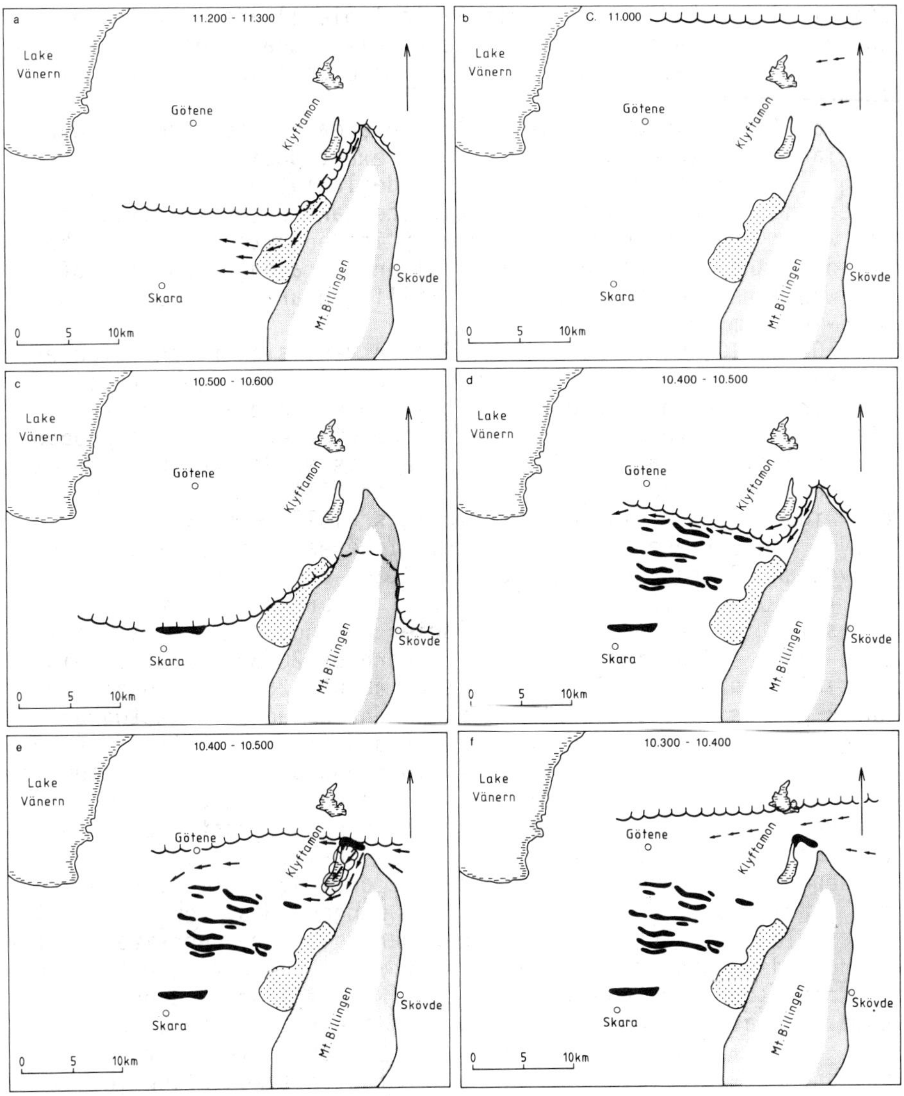

Fig. 7a-f. Supposed positions of the active ice-margin at six different occassions between c. 11,300 and 10,300 yr B.P. Arrows mark water flow from the Baltic Basin.

well dated sequence of events for the deglaciation of the area around Mt. Billingen (Fig. 7a-f).

11,200-11,300 yr B.P. (Fig. 7a): The ice-margin is approximately situated where ridge complex 2 lies today (Fig. 4) and the

Valle Härad depression was filled with drainage deposits. The distal drainage deposits (unit R2 in Table 1) are only found south of the Skånings-Åsaka ridge complex. Sea level was at about 155 m and the Baltic Ice Lake could have been lowered by 10-15 m (Björck, 1979).

c. 11,000 yr B.P. (Fig. 7b): The area becomes icefree during the late Allerød Chronozone, but the exact position of the ice-margin at that time is not known. A deglaciation rate of 100 m/year has been postulated. Baltic Ice Lake water is mixed with sea water and the reddish brown clay is deposited covering the bedrock, and glacial, glaciofluvial and, drainage deposits of the whole region. Parts of N Mt. Billingen are still occupied by dead-ice remnants.

10,500-10,600 yr B.P. (Fig. 7c): A very significant glacier readvance has reached its maximum extension. The ice covers northern Valle Härad and almost the whole ridge area and the Skara ridge is formed. In the east the glacier margin is probably in the Skövde area. The "Allerød clay" is eroded, pushed and deposited at the ice-margin. Northernmost Billingen is probably overridden by the glacier. Strömberg's (1977) > 650 varves at Skövde can correspond to a deglaciation-readvance period in the Skövde and Skara areas between c. 11,300 and 10,600.

10,400-10,500 yr B.P. (7d, e): A critical point in the ice-recession is reached when the glacier can not dam the Baltic Ice Lake water and a final lowering of perhaps 20-30 m of this big glacial lake occurs. Sea level is at about 125 m. The draining water passes east and south of Lake Lången and then westwards (Fig. 7d) depositing ridge complex 5 and finer deposits further southwest. Rapid ice-recession follows and parts of Klyftamon is heavily eroded. The glacier in Lake Lången is cut off from the active ice (Fig. 7e) and drainage water passes east, south, north and probably also on top of the Lången ice. The Timmersdala ridge is deposited as a kind of ice-contact drainage formation between the glacier and the dead-ice in Lake Lången. Heavy erosion still takes place on Klyftamon and finer particles are deposited in the bedrock-dependent depression in southwest. The water level change described earlier from some lakes on Mt. Billingen occurs at the same time as the supposed drainage of the Baltic Ice Lake.

10,300-10,400 yr B.P. (9f): The Lången valley is free from ice and calm glaciolacustrine sedimentation takes place in low-lying areas. The outlet from the Baltic is wide (> 5 km) and the ice-recession is rapid. At this time limnic sedimentation begins in the former drainage passage just south of the Lången valley and in lakes today situated 10-15 m below sea level at that time. This shows that a great deal of dead-ice was caught in the area, and on top of and between ice, small lakes were very soon formed and later submerged due to melting of underlying ice.

A PALEOCLIMATIC SYNTHESIS

From pollen analyses and lithostratigraphic evidences we can conclude that climate during Allerød in the Middle Swedish end-moraine zone as in the whole of South Sweden was favourable for tree immigration and rapid deglaciation. This changed rather dramatically around 11,000 yr B.P. when tundra vegetation replaced the former forest-tundra and the glacier moved south again redepositing Allerød sediments in front of the readvancing ice. This readvance seems to be in the order of at least 20-30 km. After about 500 years of cool climate a sudden change occurs. A plant succession takes place with immigration of e.g. Betula and Empetrum replacing the previously herb dominated vegetation. The $\delta^{18}O$ values increase and the glacier retreats quite rapidly. Therefore it seems as if the most significant climatic change around the boundary Pleistocene/Holocene in the Middle Swedish endmoraine zone occurred already in the middle of the Younger Dryas Chronozone, i.e. around 10,500 yr B.P. After that no evidences are found for any similar climatic amelioration. At around 10,200 yr B.P., when the most significant climatic change is supposed to have occurred further south (cf. Berglund, 1966; Björck, 1981), the receding glacier was since long gone from the Middle Swedish endmoraine zone and a quite lengthy and normal vegetational succession characterized the Billingen area. This difference in interpretation is either due to climatic differences within South Sweden at that time or that zone B3 (Fig. 3) corresponds to Björck's (1979, 1981) zone 6 (10,500-10,200 yr B.P.) in Blekinge, which climatic significance was not fully understood previously as glacial deposits and oxygen isotope records were never before directly related to the pollen stratigraphy of the Younger Dryas Chronozone in the Middle Swedish endmoraine zone, where the actual deglaciation took place at that time.

ACKNOWLEDGEMENTS

We are indebted to Christin Andréasson for drawing the illustrations, Benneth Dennegård for some foramenifera analyses, Ulrich Eicher for oxygen isotope analyses, Hannelore Håkansson for diatom analyses, Sören Håkansson for radiocarbon analyses, Erna Hansson for typing the manuscript, Mervi Hjelmroos-Ericsson for most of the pollen analyses and, Siv Olsson for chemical analyses.

We are also very grateful to Lena Adrielsson for discussions and comments on the text, to Benneth Dennegård for discussions concerning the Lake Vänern stratigraphy and to Bo Strömberg for very helpful discussions about the complex geology and morphology around Mt. Billingen. The work has been financially supported by the Swedish Natural Science Research Council (NFR), which we gratefully acknowledge.

REFERENCES

Berglund, B.E., 1966. Late-Quaternary vegetation in eastern Blekinge, south-eastern Sweden. I. Late-Glacial time. Opera Botanica, 12:1, 1-180.

Berglund, B.E., 1976. The deglaciation of southern Sweden. Presentation of a research project and a tentative radiocarbon chronology. University of Lund, Department of Quaternary Geology Report, 10, 1-67.

Berglund, B.E., 1979. The deglaciation of southern Sweden 13,500-10,000 B.P.Boreas, 8, 89-118.

Bergsten, K.E., 1943. Isälvsält kring norra Vättern. Meddelanden från Lunds universitets geografiska institution, Avhandling, 7, 1-245.

Berzin, L.E., 1967. On the age and sea level oscillations of early and middle Holocene transgressions in the upper part of the Gulf of Riga (in Russian with summaries in English and German). Baltica, 3, 87-104.

Björck, S., 1979. Late Weichselian stratigraphy of Blekinge, SE Sweden, and water level changes in the Baltic Ice Lake. University of Lund, Department of Quaternary Geology Thesis, 7, 1-248.

Björck, S., 1981. A stratigraphic study of Late Weichselian deglaciation, shore displacement and vegetation history in south-eastern Sweden. Fossils and Strata, 14, 1-93.

Björck, S. & Digerfeldt, G., 1982. Late Weichselian shore displacement at Hunneberg, southern Sweden, indicating complex uplift. Geologiska Föreningens i Stockholm Förhandlingar, 104, 132-155.

Caldenius, C., 1944. Baltiska issjöns sänkning till Västerhavet. En geokronologisk studie. Geologiska Föreningens i Stockholm Förhandlingar, 66, 366-382.

Dennegård, B., 1979. Kvartärpaleontologiska studier av lagerföljder från Allerød, Yngre Dryas och Preboreal tid i sydvästra Vänerområdet. Lägesrapport december 1979. Chalmers Tekniska Högskola, Göteborgs Universitet, Geologiska institutionen, Publ. B 136, 1-33.

Digerfeldt, G., 1972. The Post-Glacial development of Lake Trummen. Regional vegetation history, water-level changes and palaeolimnology. Folia Limnologica Scandinavica, 16, 1-96.

Digerfeldt, G., 1974. The Post-Glacial development of the Ranviken bay in Lake Immeln. I. The history of the regional vegetation, and II. The water-level changes. Geologiska Föreningens i Stockholm Förhandlingar, 96, 2-32.

Digerfeldt, G., 1976. A Pre-Boreal water-level change in Lake Lyngsjö, Central Halland. Geologiska Föreningens i Stockholm Förhandlingar, 98, 329-336.

Donner, J.J., 1969. Land/sea level changes in southern Finland during the formation of Salpausselkä endmoraines. Bull. Geol. Soc. Finland, 41, 135-150.
Donner, J.J., 1978. The dating of the levels of the Baltic Ice Lake and the Salpausselkä moraines in south Finland. Societas Sientiarum Fennica, Commentationes Physico-Mathematicae, 48, 1, 11-38.
Donner, J.J., 1982. Fluctuations in water level of the Baltic Ice Lake. Annales Academiae Scientiarum Fennicae A, III, 134, 13-28.
Fredén, C., 1982. An outline of the marine stage of the Väner basin. In: The Pleistocene/Holocene boundary in southwestern Sweden (ed. E. Olausson). Sveriges Geologiska Undersökning C, 794, 16-26.
Glückert, G. & Ristaniemi, O., 1982. The Ancylus transgression west of Helsinki, South Finland - a preliminary report. Annales Academiae Scientiarum Fennicae A, III, 134, 99-110.
Gudelis, V.K., 1969. Die Küstenentwicklung der Südöstlichen Ostsee während der Spät- und Nacheiszeit. Beitrage zur Meereskunde, Deutsche Akademie der Wissenschaft zu Berlin, Institut für Meereskund, 24-25, III, 10-14.
Johansson, S., 1926. Baltiska issjöns tappning. Geologiska Föreningens i Stockholm Förhandlingar, 48, 186-263.
Johansson, S., 1937. Senglaciala och interglaciala avlagringar vid ändmoränstråket i Västergötland. Geologiska Föreningens i Stockholm Förhandlingar, 59, 379-454.
Johansson, S., 1943. Jordlagren. In: Johansson, S., Sundius, N. & Westergård, A.H.: Beskrivning till kartbladet Lidköping. Sveriges Geologiska Undersökning Aa, 182, 1-185.
Kessel, H. & Raukas, A., 1980. The Quaternary history of the Baltic, Estonia. In: The Quaternary history of the Baltic (eds. V. Gudelis & L.-K. Königsson). Acta Universitatis Upsaliensis, Symposia Universitatis Upsaliensis, Annum Quingentesimum Celebrantis 1, 127-146.
Kvasov, D.D., 1979. The Late-Quaternary history of large lakes and inland seas of eastern Europe. Annales Academiae Scientiarum Fennicae A, III, 127, 1-71.
Lundqvist, G., 1921. Den baltiska issjöns tappning. Geologiska Föreningens i Stockholm Förhandlingar, 43, 381-385.
Lundqvist, G., 1931. Jordlagren. In: Lundqvist, G., Häggbom, A. & Westergård, A.H.: Beskrivning till kartbladet Lugnås. Sveriges Geologiska Undersökning Aa, 172, 1-185.
Lundqvist, G., 1961. Beskrivning till karta över landisens avsmältning och högsta kustlinjen i Sverige. Sveriges Geologiska Undersökning, Ba, 18, 1-148.
Mangerud, J., Andersen, S.T., Berglund, B.E. & Donner, J.J., 1974. Quaternary stratigraphy of Norden, a proposal for terminology and classification. Boreas, 3, 109-128.
Mörner, N.-A., 1969. The Late Quaternary history of the Kattegatt Sea and the Swedish West Coast: deglaciation, shorelevel

displacement, chronology, isostasy and eustasy. Sveriges Geologiska Undersökning C, 640, 1-487.
Mörner, N.-A., 1970. The Younger Dryas Stadial. Geologiska Föreningens i Stockholm Förhandlingar, 92, 5-20.
Munthe, H., 1910. Studies in the Late-Quaternary history of Southern Sweden. Geologiska Föreningens i Stockholm Förhandlingar, 32, 1197-1293.
Munthe, H., 1928(a). Kvartärsystemet. In: Munthe, H., Westergård, A.H. & Lundqvist, G.: Beskrivning till kartbladet Skövde, 2nd ed. Sveriges Geologiska Undersökning Aa, 121, 1-182.
Munthe, H., 1928(b). Drag ur den senglaciala utvecklingen av Billingen-Falbygden med omnejd. Geologiska Föreningens i Stockholm Förhandlingar, 50, 233-286.
Munthe, H., 1940. Om Nordens, främst Baltikums, senkvartära utveckling och stenåldersbebyggelse. Kungliga Svenska Vetenskapsakademins Handlingar, Serie 3, 19, 1-242.
Nilsson, E., 1953. Om södra Sveriges senkvartära historia. Geologiska Föreningens i Stockholm Förhandlingar, 75, 166-246.
Nilsson, E., 1968. Södra Sveriges senkvartära historia. Kungliga Svenska Vetenskapsakademins Handlingar, Serie 4, 12, 1, 1-117.
Olausson, E., (ed.) 1982. The Pleistocene/Holocene boundary in south-western Sweden. Sveriges Geologiska Undersökning C, 794, 1-284.
Ramsay, W., 1924. Strandlinjerna i södra Finland. Geologiska Föreningens i Stockholm Förhandlingar, 43, 495-497.
Sauramo, M., 1934. Zur spätquartären Geschichte der Ostsee. Comptes Rendus de la Societé Géologique de Finlande, 8, 1-60.
Sauramo, M., 1937. Das System der spätglazialen Strandlinien im südlichen Finnland. Societas Scientiarum Fennica, Commentationes Physico-Mathematicae, 9, 10, 1-23.
Sauramo, M., 1958. Die Geschichte der Ostsee. Annales Academiae Scientiarum Fennicae, Series A, III, 51, 1-522.
Strömberg, B., 1974. Billingenområdet vid slutet av den senaste istiden. Skaraborgsnatur, 11, 2-16.
Strömberg, B., 1977. Einige Bemerkungen zum Rückzug des Inlandeises am Billingen (Västergötland, Schweden) und dem Ausbruch des Baltischen Eisstausees. Zeitschrift für Geomorphologie, Suppl. - Bd., 27, 89-111.
Westergård, A.H., 1926. Kvartärsystemet. In: Westergård, A.H., Johansson, H.E. & Willén, N.: Beskrivning till kartbladet Karlsborg. Sveriges Geologiska Undersökning Aa, 162, 1-96.

CYCLIC LAMINATION IN PROXIMAL VARVES REFLECTING THE LENGTH OF SUMMERS DURING LATE WEICHSEL IN SOUTHERNMOST SWEDEN

Bertil Ringberg

Geological Survey of Sweden

Cyclic lamination in the summer layers of proximal varves has been studied in the bottomset beds of glaciofluvial deltas in the provinces of Skåne and Blekinge, southernmost Sweden. Between 43 and 56 day varves have been observed in the summer layers and the numbr of day varves are supposed to represent the length of summers during the Bölling Chronozone in Late Weichsel of southernmost Sweden.

Interseasonal lamination in annual varves has previously been discussed by Sauramo (1923) and De Geer (1933). Sauramo did not believe that day varves can be formed in varved clays, and De Geer did not find any. Andersen (1928), Jessen (1935) and Hansen (1940) think however, on the basis of observations made in ice-lake clays in Denmark and in the province of Skåne, southernmost Sweden, that day varves can be formed. Schwarzbach (1940) and Pirrus (1965 and 1968) also hold this opinion on the basis of observations of varved clay in Silesian Eulengebirge and Estonia respectively. In the works mentioned the authors discuss day rhythm in varved clays with large clay content in the winter layers as well as in the layers which are assumed to have been deposited during the night. In the proximal varves described here there generally is a low clay content in the summer layer (less than 5%).

Cyclic lamination in proximal varves has been studied at seven localities in the provinces of Skåne and Blekinge, southernmost Sweden. In all eleven of them proximal varves of this type have been found in bottomset beds of glaciofluvial delta deposits in the area where the deglaciation took place during the Bölling Chronozone in Late Weichsel.

N.-A. Mörner and W. Karlén (eds.), Climatic Changes on a Yearly to Millennial Basis, 57–62.

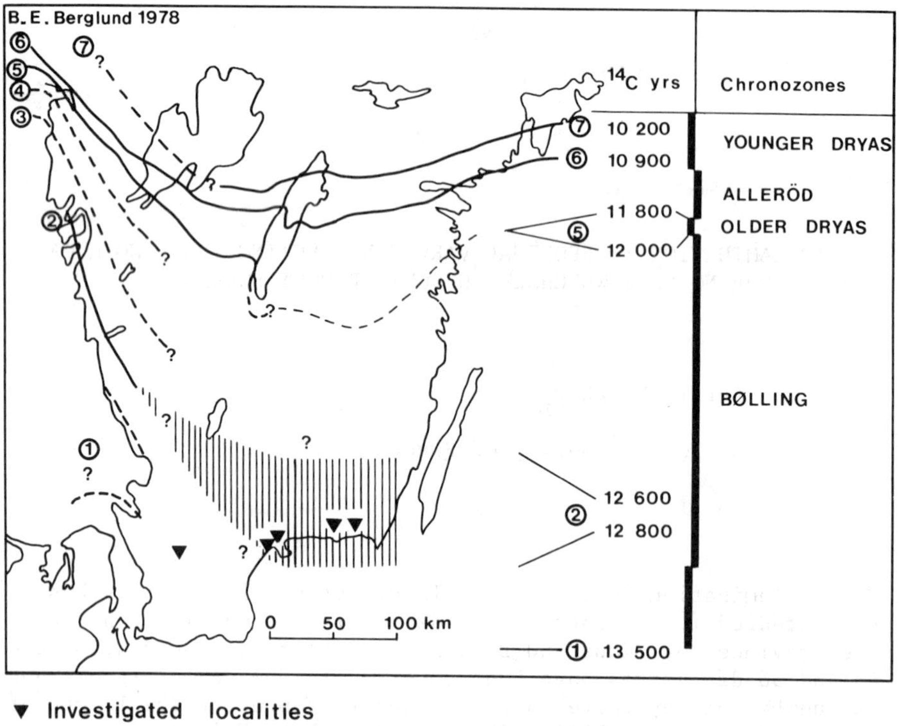

Figure 1. Map showing the investigated localities and the deglaciation chronology for southern Sweden according to Berglund (1979).

All the localities except one are situated close to the highest shoreline of the Baltic Ice Lake in Blekinge and northeastern Skåne. One of the localities, the westernmost one in Fig. 1, is situated near the marine limit of the west coast of Skåne. The highest shoreline and the marine limit have been determined in the neighbourhood of the localities (Ringberg 1971; Sandgren 1983). Therefore it is possible to establish the water depth during the sedimentation of the proximal varves to a maximum of 10—20 m.

The winter layers which limit the described proximal varves are only a few millimetres thick and consist of heavy clay whilst the summer layers, forming the principal part of the 20—75 cm thick varves, consist of sublayers of coarse silt and fine sand, alternating with sublayers of silt (Fig. 2). In certain parts of the summer layers the coarse-grained sublayers are thicker than the fine-grained which do not vary in thickness as much as the former. Both below and above the winter layers the transitions are very sharp. The lamination is very like the draped lamination

Figure 2. Cyclic lamination in a proximal varve at Vieröd in the province of Skåne. The sticks have been put in the mm-thick winter layers of clay. The coin is 25 mm in diameter.

consisting of parallell laminae which has been described by Gustavson et al. (1975). Sometimes and especially in the lowest parts of the varves ripple-drift cross-lamination occurs in the coarse-grained sublayers. Locally the fine-grained sublayers are eroded away completely or partly when ripples appear in the coarse-grained sublayers. In order to obtain complete records of sublayers in the summer layers of the proximal varves it has therefore been necessary to avoid varves where ripples occur.

From one of the localities a proximal varve has been drawn up in the diagram in Fig. 3. The diagram is constructed in the same way as a varve diagram. Each vertical line in the diagram corresponds to both one coarse-grained and one fine-grained sublayer in the summer layer of the varve. The underlying winter layer has been placed to the right in the diagram and the over-

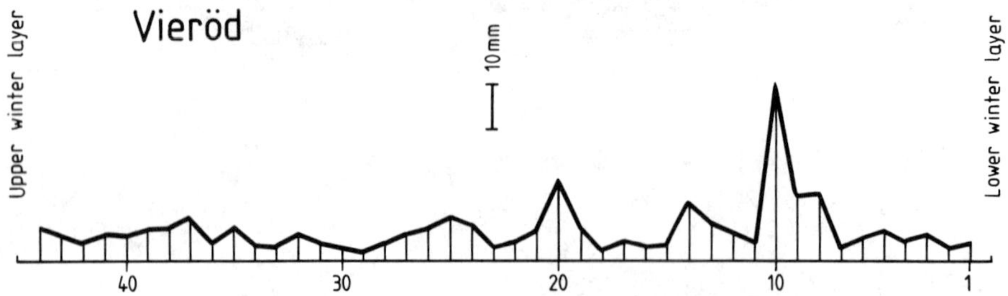

Figure 3. Diagram of the cyclic lamination in a proximal varve at Vieröd in the province of Skåne.

lying one to the left.

The observed lamination in the summer layers of the proximal varves is believed to correspond to the day and night rhythm in the transportation of material and its sedimentation in the melt-water streams. The layers of coarse silt and fine sand were deposited by day and those of silt were deposited at night.

The normal varved clay in the investigation area is not graded but has very sharp limits both inside the varves as well as at their limits (Fig. 4; Ringberg 1979). It is assumed that this is so because the water-flow increased suddenly in spring with the meltwater streams rapidly petering out after a period of intense melting. The appearance and composition of the proximal varves in the bottom beds of the deltas support this hypothesis.

The number of the day varves varies between 43 and 56 in the 11 studied proximal annual varves. In each of these annual varves there is a high and long peak after a few small day varves, a peak which coincides with the melt-water flow in spring (Fig. 3). That peak, at the beginning of the melting season, is as well-known as the day and night rhythm of the melt-water flow of the recent glaciers (see for instance Østrem 1975).

Whether later peaks of day varves in the summer layers represent variations of temperature, heavy rainfalls or both cannot be determined until the conditions of the precipitation are known.

If the described lamination in the proximal varves is formed of unmistakable day varves the duration of the intense melting period was 40—55 days during the deglaciation in Late Weichsel in southernmost Sweden. Should the same lamination be found in other regions it would then be possible to investigate the duration of the intense melting period regionally and in relation to the different periods of climate during the deglaciation.

A discussion as to the extension of the drainage area in the ice on the basis of the cyclic lamination is also possible. The length of the melt-water streams and the size of the drainage area

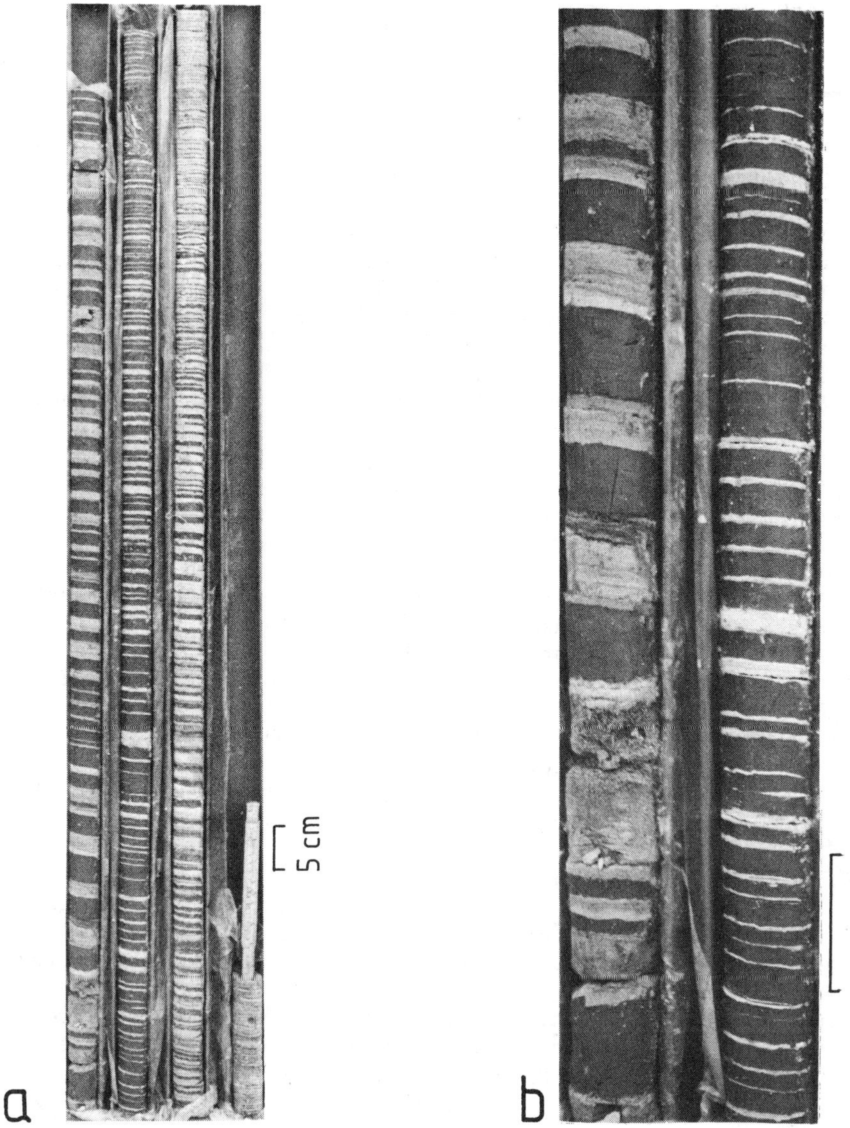

Figure 4a. Varved clay and silt in a core from Karlshamn in the province of Blekinge. The bottom varve is in the lowest left part of the core.

4b. Detail of the lowest left part of the core in Fig. 4a.

cannot have been so great during the deglaciation that the day and night rhythm of the melt-water flow was smoothed out.

References

Andersen, S.A., 1928. De danske varv. Geol.Fören. i Stockh.Förhandl., 50, 90-96.

Berglund, B.E., 1979. The deglaciation of southern Sweden 13.500-10.000 B.P. Boreas 8, 89-118.

De Geer, G., 1933. Årsvarv, äkta och oäkta, i Scanodania. Geol. Fören. i Stockh.Förhandl., 55, 655-656.

Gustavson, T.C., Ashley, G.M. & Boothroyd, J.C., 1975: Depositional sequences in glaciolacustrine deltas. In: Glaciofluvial and glaciolacustrine sedimentation (ed. A.V. Jopling & B.C. McDonald). Soc. of econ.paleont. and mineral. Spec. publ. 23, 264-280.

Hansen, S., 1940. Varvighed i danske og skaanske senglaciale aflejringer. Danm.Geol.Unders. II. Raekke, 63, 411 pp.

Jessen, A., 1935. Beskrivelse till geologiske kort over Danmark. Kortbladet Haderslev. Danm.Geol.Unders. I. Raekke, 17.

Pirrus, E., 1965. Interseasonal lamination of varved clays. In: Litologiya i stratigrafiya chetvertichnykh of Lozheniy estonii. Eesti NSV Teaduste Akadeemia Geoloogia Instituut. Tallin.

Pirrus, E., 1968. Estonian varved clays. In: Litologiya i stratigrafiya chetvertichnykh of Lozheniy estonii. Eesti NSV Teaduste Akadeemia Geoloogia Instituut. Tallin.

Ringberg, B., 1971. Glacialgeologi och isavsmältning i östra Blekinge. Sver.geol.unders. C 661, 174 pp.

Ringberg, B., 1979. Varve chronology of the glacial sediments in Blekinge and northeastern Skåne, southeastern Sweden. Boreas 8, 209-215.

Sandgren, P., 1983. The deglaciation of the Klippan area, southern Sweden. A study of glaciofluvial and glaciomarine sediments. Univ.of Lund, Dept. of Quat.Geol. Thesis 14, 99 pp.

Schwarzbach, M., 1940. Das Diluviale Klima während des Höchstandes einer Vereisung. Z. Deutschen Geol.Ges., 92.

Østrem, G., 1975. Sediment Transport in Glacial Meltwater Streams. In: Glaciofluvial and glaciolacustrine sedimentation. (ed. A.V. Jopling & B.C. McDonald). Soc. of econ.paleont. and mineral. Spec. publ. 23, 101-122.

LATE WEICHSELIAN SEA-LEVELS AND FAUNA COMMUNITIES IN NORTHERN VENDSYSSEL, JUTLAND, DENMARK

Kaj Strand Petersen

Geological Survey of Denmark

During the last decades several C-14 dates on molluscs have been obtained from Late Weichselian seabeds in Vendsyssel, northern Jutland, Denmark. On the basis of the original C-14 collection of samples it has proved possible to evaluate the fauna communities and distinguish between arctic and boreal-arctic communities at different depths. A curve of sea-level stand can be drawn for the final part of glaciation in Jutland up to the end of Allerød, showing that the highest marine shoreline was established between 14,000-13,000 B.P. and that the Zirphaea Beds, the oldest of which is dated to about 12,800 B.P., do not represent a separate transgression, but is a boreal littoral equivalent of the Upper Saxicava Sands.

INTRODUCTION AND BACKGROUND

73 years ago, in Stockholm, State geologist V. Nordmann presented his paper on climatic changes in Denmark during the Late Glacial and the Alluvial periods (Nordmann, 1910). Here Nordmann maintained that the fauna of the Zirphaea strata from Vendsyssel was a distinctly littoral fauna, and he suggested that these beds were of a more recent origin than the Younger Yoldia Clay known from northern Jutland as well. A transition between the two marine deposits was not known. This led to the definition and introduction of the so-called Zirphaea transgression.

In those days as well as today we depend on the original geological mapping of Vendsyssel conducted by State geologist A. Jessen (1899). This map shows that the Younger Yoldia Clay, ascribed to an arctic sea covering the area after the waning of the ice sheet, is found up to a present altitude of about +30 m,

N.-A. Mörner and W. Karlén (eds.), Climatic Changes on a Yearly to Millennial Basis, 63–68.

and that the highest marine limit formed in Saxicava Sands is at about +60 m. In the sixties H. Krog, S. Hansen, I. Marcussen and Erna Nordmann collected material for the later establishment of the C-14 dating record (Krog & Tauber, 1974), and part of this original material has formed the basis of the present investigations.

My own approach to the former views has been influenced by experiences obtained from work on Holocene raised marine deposits and recent bottom samples in the central and northern East Greenland Fjord Region. Here it appeared that bottom communities could be evaluated in the subfossil material (Petersen, 1978). Raised marine outcrops of deep and shallow marine deposits represent the Astarte borealis-Pectinaria zone and the Ophiocten zone known from the recent arctic Macoma calcaria bottom community from 3-14 m and 14-45 m depth of water respectively. The eulittoral boreal-subarctic element Mytilus edulis appeared in the East Greenland Fjord Region during the hypsithermal period of the Holocene. Contemporaneously and in the same region arctic elements such as Portlandia arctica persisted in deeper water.

We now have more than thirty C-14 datings from Late Weichselian marine deposits in Vendsyssel, all from the northernmost part (see insert map on Fig. 1). From most of the sites bulk samples are at hand allowing an evaluation of the macrofauna. These fauna communities have been subdivided into 1) arctic at 0-20 m depth and at more than 20 m depth, and 2) boreal-arctic at 0-20 m depth and at more than 20 m depth. This may be regarded as an oversimplification, but is, however, to a large extent justified by my investigations from East Greenland mentioned earlier. These investigations were based on works back to Jensen & Harder (1910) and followed the advices of Petersen (1910) who during the meeting in Stockholm stated that "what is most required is a study of and comparison between the fossil faunas and the molluscan faunas now living".

INVESTIGATIONS

Hiatella arctica (Saxicava arctica) is by far the most common species in the older deposits and the species which has been used for the C-14 datings of all the samples up to 13,000 B.P. Hiatella arctica has a wide bathymetric range, from 0 to 175 m (quoting Ockelmann, 1958, from East Greenland), and from the Québec region Hillaire-Marcel (1980) writes that "Hiatella arctica est thanatocénotique dans les sables". However, in the Danish Weichselian deposits this species occurs as a burrowing form as evidenced by traces of Hiatella found near the highest marine limit in the Upper Saxicava Beds as well as in the Yoldia Clay. Furthermore, Hiatella is found in the older strata, the so-called Lower Saxicava Sand which is overlain by the Yoldia Clay. On the basis of the outlined succession within the oldest dated samples a curve of sea-level stand can be drawn, but the depth must be estimated on the associated fauna elements.

The following examples are from the time about 13,000 B.P. (Fig. 1). The *Yoldia* Clay samples at hand from this time interval expose the fauna of more than 20 m water, as seen for instance at the Dybvad clay pit site (K-903, Fig. 1). A C-14 dating here gives 13,010 ± 190 B.P. The fauna consists of *Hiatella arctica*, *Mya truncata*, *Macoma torelli*, *Portlandia arctica*, *Buccinum groenlandicum* and *Cylichna occulta*. This shows that deposition occurred in deeper water - and in an arctic environment.

Foraminiferal analyses of the size-fraction 0.1-1.0 mm of the sample from Dybvad were carried out by Karen Luise Knudsen who found the following species: *Elphidium excavatum* forma *clavata*, *Cassidulina reniforme*, *Buccella frigida* var. *calida*, *Elphidium albiumbilicatum*, *Elphidium hallandense*, *Bulimina marginata*, *Islandiella helenae*, *Elphidium asklundi*, *Stainforthia schreibersiana*, *Nonion orbiculare*, *Buliminella elegantissima* and *Cibicides lobatulus*, and commented as follows: "The present foraminiferal fauna from Dybvad is mainly arctic, but a content of boreal species, as for instance *Buccella frigida* var. *calida*, *Elphidium albiumbilicatum* and *Bulimina marginata*, indicates ameliorated conditions.

The faunal composition is similar to those registered in layers representing the transition from Younger *Yoldia* Clay (*Elphidium excavatum* zone) to the *Zirfaea* Beds (*Elphidium albiumbilicatum* zone) in a boring at Skeen Møllebæk (Knudsen, in prep.). The fauna also compares with faunas from corresponding layers at Nørre Lyngby, for instance in profile 2 (Knudsen, 1978, p. 12)."

It is worth mentioning that when studying the microfauna in the *Yoldia* Clay from this region Knudsen (1978, p. 24) is able to point out sand laminae where the sand material and the foraminifera may be transported together into the area of clay deposition from more sandy shallow water deposits. Furthermore, the faunal changes *within* the sand laminae depict a gradual amelioration in climate.

So far the microfauna analyses. The oldest of the boreal-arctic molluscan littoral to sublittoral faunas has been dated to 12,770 ± 190 B.P. (K-1468, Fig. 1). The sample contains the earliest occurrence of *Zirphaea crispata*. But still younger arctic deep water faunas are found in Bindslev clay pit 12,650 ± 180 B.P. (K-891, Fig. 1), with species such as *Macoma calcaria*, *Macoma moesta* and *Portlandia arctica*.

From about 12,500 B.P. towards the end of Allerød there are many datings of the boreal-arctic community at 0-20 m depth of water. It shall be stressed that in the Late Weichselian of Vendsyssel we find arctic faunas from the deeper water deposits and contemporaneous boreal-arctic elements from the littoral and sublittoral environments. This observation forms a parallel to that obtained by my investigations in East Greenland mentioned earlier.

The boreal-arctic fauna elements are, besides *Zirphaea cris-*

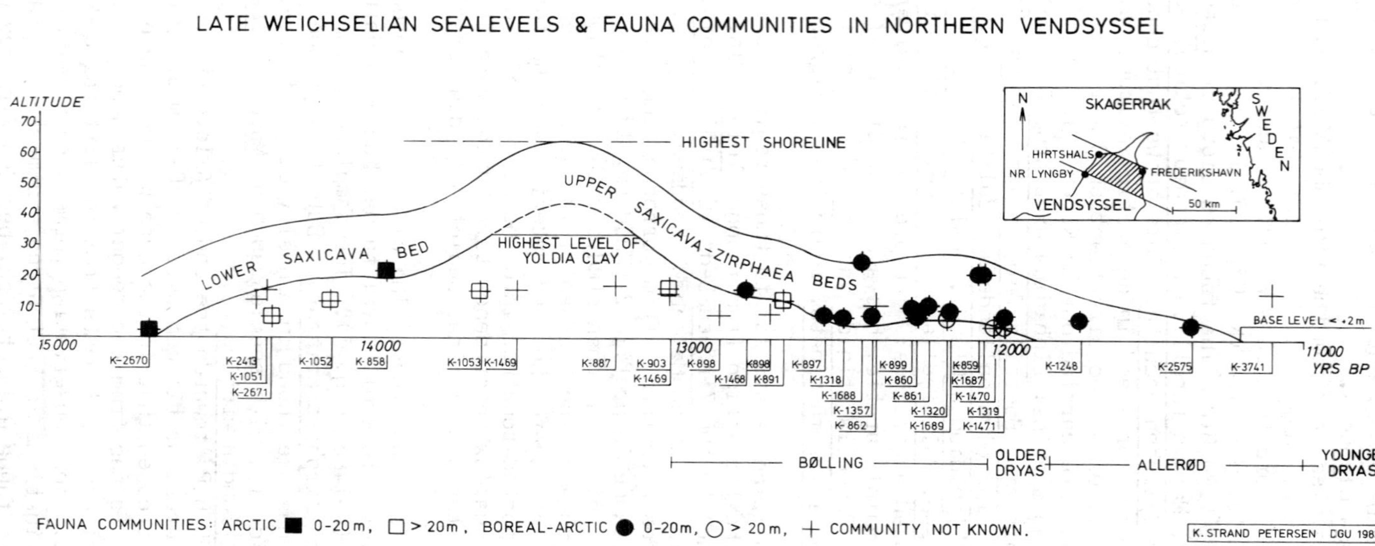

Fig. 1. The C-14 datings have all been performed by the Copenhagen Carbon-14 Dating Laboratory and published in Tauber (1966), Krog & Tauber (1974) and Aaris-Sørensen & Petersen (1984). Insert map of the investigated area.

pata which gives name to the deposit, species such as: Mytilus edulis, Macoma baltica and Cyprina islandica. There are few samples from deeper water boreal-arctic faunas, these being characterized by the presence of Cyprina islandica and the absence of the eulittoral form Mytilus edulis.

CONCLUSIONS

The present investigation shows that the highest marine shoreline of the Younger Yoldia Sea was established between 14,000-13,000 B.P. This time interval corresponds to the time of the low Baltic ice lobe, dated by the Lockarp mammoth (Berglund & Mörner, 1983), and to the time of the waning of the ice from the classical D-line in Denmark, dated to about 13,200 B.P. by a mammoth tusk embedded in the extramarginal deposits on Djursland, East Jutland (Petersen, 1983).

The deposition of Yoldia Clay containing a cold marine fauna from deeper water is now seen to continue into Bølling, but with a contemporaneous temperate fauna in the more shallow water deposits - the Zirphaea Beds. Therefore the Zirphaea Beds do not represent a separate transgression as assumed earlier, but are a boreal equivalent of the Upper Saxicava Sands. This concurs with the findings of Mörner (1969) on the Swedish west coast.

Considering the Weichselian marine record as a whole as known from Denmark, the Late Weichselian high sea-level outlined here constitutes the second of two main rises during the Weichselian.

ACKNOWLEDGEMENTS

Karen Luise Knudsen, University of Århus, is thanked for her contribution and discussions. Harald Krog, Geological Survey of Denmark, helped me organizing the samples and provided helpful information on the many sites.

REFERENCES

Berglund, B.E. & Mörner, N.-A., 1983. Late Weichselian deglaciation and chronostratigraphy of Southern Scandinavia: problems and present "state of the art". Danish Meteorol. Inst., Abstracts Second Nordic Symposium on Climatic Changes and Related Problems, Stockholm, May 16-20 1983, pp. 9-13.

Hillaire-Marcel, C., 1980. Les faunes des Mers Post-glaciaires du Québec: Quelques considerations paléoécologiques. Géogr. phys. Quat., 34, 1, pp. 3-59.

Jensen, Ad.S. & Harder, P., 1910. Post-glacial changes of climate in arctic regions as revealed by investigations on marine deposits. In: Die Veränderungen des Klimas seit dem Maximum der letzten Eiszeit, pp. 399-407. Generalstabens Litografiska Anstalt, Stockholm.

Jessen, A., 1899. Beskrivelse til Geologisk Kort over Danmark. Kortbladene Skagen, Hirtshals, Frederikshavn, Hjørring og Løkken. Danm. geol. Unders., I. række, 3, 368 p.

Knudsen, K.L., 1978. Middle and Late Weichselian marine deposits at Nørre Lyngby, northern Jutland, Denmark, and their foraminiferal faunas. Danm. geol. Unders., II. række, 112, 44 p.

Krog, H. & Tauber, H., 1974. C-14 chronology of Late- and Postglacial marine deposits in North Jutland. Danm. geol. Unders., Årbog 1973, pp. 93-105.

Mörner, N.-A., 1969. The Late Quaternary history of the Kattegat Sea and the Swedish West Coast. Sveriges Geologiska Undersökning, Ser. C., 640, 487 p.

Nordmann, V., 1910. Post-glacial climatic changes in Denmark. In: Die Veränderungen des Klimas seit dem Maximum der letzten Eiszeit, pp. 313-328. Generalstabens Litografiska Anstalt, Stockholm.

Ockelmann, W.K., 1958. Marine Lamellibranchiata. Meddr om Grønland, 122, 4, 256 p.

Petersen, C.G.Joh., 1910. Some considerations on the study of the post-Glacial climatic changes. In: Die Veränderungen des Klimas seit dem Maximum der letzten Eiszeit, pp. 329-331. Generalstabens Litografiska Anstalt, Stockholm.

Petersen, K.Strand, 1978. Kvantitative analyser af molluskfaunaen i hævede marine aflejringer ved Loch Fyne, Østgrønland. Abst. XIII. Nord. Geol. Vinterm., Jan. 1978, Copenhagen, p. 86.

Petersen, K.Strand, 1983. Stratigraphical position of Weichselian tills in Denmark. In: Ten Years of Nordic Till Research (ed. L.-K. Königsson). Stria, 20.

Tauber, H., 1966. Danske Kulstof-14 dateringsresultater II. Meddr dansk geol. Foren., 16, pp. 153-176.

Aaris-Sørensen, K. & Petersen, K.Strand, 1984. A Late Weichselian find of polar bear (Ursus maritimus Phipps) from Denmark and reflections on the paleoenvironment. Boreas, 3.

FLORA, FAUNA AND STABLE ISOTOPES IN LATE-WÜRM DEPOSITS AT LOBSIGENSEE (SWISS PLATEAU)

Brigitta Ammann[1], Louis Chaix[2], Ulrich Eicher[1], Scott A. Elias[3], Marie-José Gaillard[4], Wolfgang Hofmann[5], Ulrich Siegenthaler[1], Kazimierz Tobolski[6] and Bridget Wilkinson[7]

1 Universitiy of Berne, Switzerland
2 Museum d'histoire naturelle Genève, Switzerland
3 Instaar Boulder, Colorado (USA)
4 University of Lund (Sweden)
5 Max Planck-Institut Limnologie, Plön (FRG)
6 University of Poznań, Poland
7 University of Birmingham, England

Lobsigensee is a small lake on the western Swiss Plateau in a tectonic depression of the tertiary Molasse. The modern surface area is 2 ha and the maximum depth 2.5 m. It was chosen as one of the primary reference sites in the Swiss contribution to IGCP 158b (LANG 1983). The littoral site "150" was investigated by various methods (AMMANN & TOBOLSKI 1983, GAILLARD 1983, ELIAS & WILKINSON 1983, HOFMANN 1983, CHAIX 1983, EICHER & SIEGENTHALER 1983, AMMANN et al. 1983). The main results are summarized in Fig. 1.

RESULTS

The OLDEST DRYAS (?16,000-13,000 B.P.) includes three local pollen assemblage zones. The transition L1/L2 is comparable but not synchronous to van der HAMMEN's (1951) *Artemisia* rise at his transition from Pleniglacial to Lateglacial. During L2 the termination of in-washed sands corresponds to the commencement of the occurrence of chironomids. The transition L2/L3 is characterized in the terrestrial record by the expansion of *Betula nana*,

N.-A. Mörner and W. Karlén (eds.), Climatic Changes on a Yearly to Millennial Basis, 69–73.

in the limnic record by the first occurrence of Coleoptera, Trichoptera and Mollusca and by a change in the chironomid fauna. In the three groups of insects mentioned, cold stenothermic species are found. From the current biogeographical ranges of Coleoptera and Trichoptera a mean July temperature of 10-12°C can be deduced. Among the molluscs the proportion of the cold adapted *Pisidium* decreases, whilst that of the more temperate *Valvata piscinalis* increases.

The transition from OLDEST DRYAS to BØLLING is the most strongly marked horizon for several parameters: in the sediment it is the transition from clay to lake marl, in pollen and macrofossils a sharp juniper peak and the expansion of tree-birches indicate the reforestation of the area, among the chironomidae the cold stenothermic species disappear and among molluscs a maximum density of individuals is found.

During the BØLLING, forests of tree-birches dominated the landscape (dwarf birch had decreased). A depression in the *Betula* pollen curve, comparable to that which has often been correlated with the Older Dryas in diagrams from the northern Alpine foreland, cannot be interpreted as a cooler phase: the development of the local vegetation is proceeding, none of the minor negative fluctuations in the $\delta^{18}O$ curve coincides with this event and the first temperate beetles are found. Perhaps the increasing *Artemisia* and Gramineae indicate a drier period. From the fauna of Coleoptera and Trichoptera, mean July temperatures of 14-16°C can be deduced. According to GAILLARD(1981) and WELTEN(1982) we should omit the Older Dryas as a pollen zone and as a chronozone.

During the ALLERØD the pine forest was expanding. No significant changes in the fauna were found. The oxygen isotopes, measured in lake marl as well as in shells of *Valvata piscinalis,* show from the early Bølling and throughout the Allerød the slightly falling trend also registred in other localities.

The beginning of the YOUNGER DRYAS (L9/L10) was recorded in the pollen, in the oxygen isotopes and in the frequency of molluscs, but not in the insect fauna.

The transition from LATEGLACIAL to HOLOCENE (L10/L11) of the pollen diagram corresponds to the transition from lake marl to peat; therefore neither insects nor molluscs nor oxygen isotopes were studied in the holocene deposits.

DISCUSSION

Our data support the view that rising annual and summer temperatures favored a vegetational and faunal development reaching a major step around 13,300-13,000 B.P.. The only climatic reversal registred was the Younger Dryas.

Some of the steps of development during the Oldest Dryas (e.g. the expansion of *Betula nana* at L2/L3) and during the Bølling (e.g.

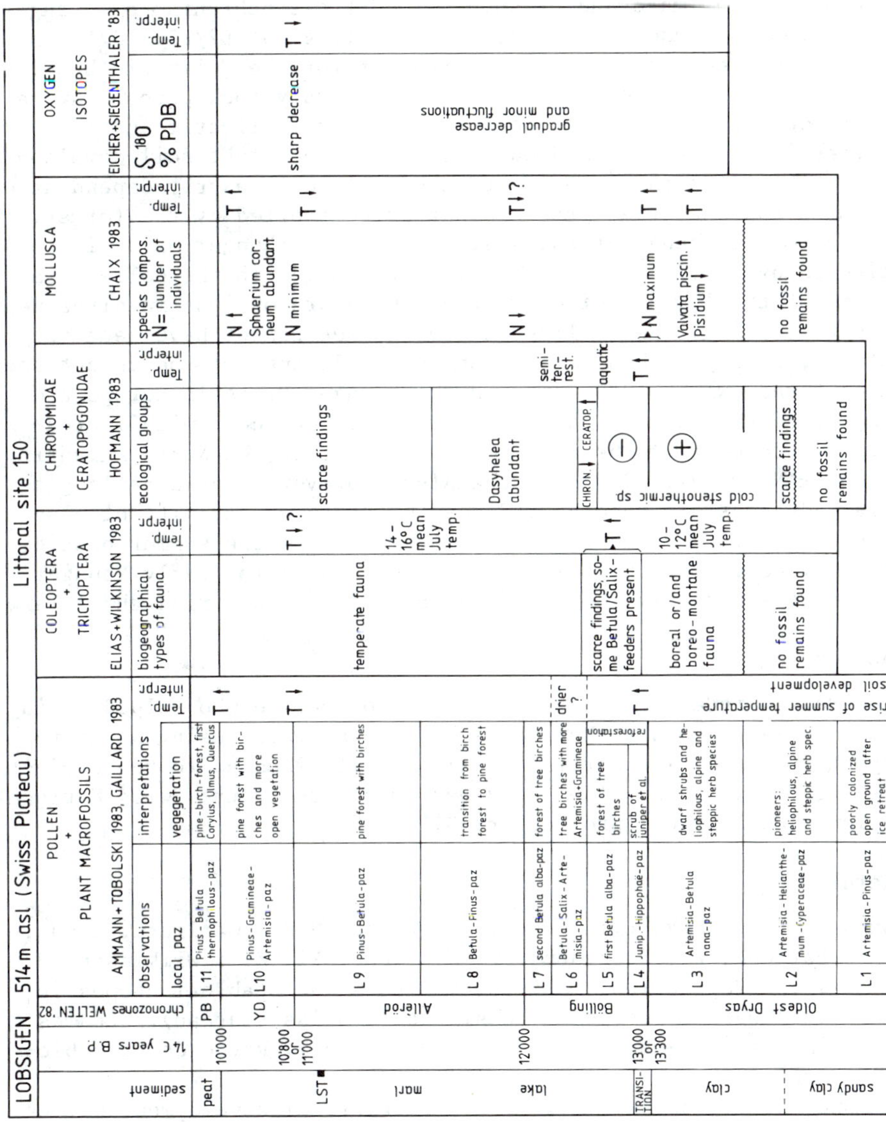

Figure 1: Principal results for vegetation, insects, molluscs and oxygen isotopes from the Late-Würm deposits at Lobsigensee (northern alpine foreland).

the shading-out of *Juniperus* and *Hippophaë* by *Betula alba* at L4/L5) could be interpreted as an indication of a warming climate but also could be due to successional phases. In the discussion concerning "climate and/or succession" the synchronism of changes in terrestrial and limnic environments (e.g. at L2/L3, L3/L4 and L9/L10) may be evaluated as an argument for the "climatic hypothesis". On the other hand successions obviously took place. Our results do not yet allow us to disentangle the climatic and the successional factor - and one should remember that soil development as well as the trophic state of a lake is partly dependent on climate and that the latter is partly controlled by the former. The climatic change at the beginning of the Younger Dryas is reflected by only slight but consistent changes in the pollen diagrams of the lowland sites, whereas these are obvious in sites near the timberline. The cooling at Lobsigensee cannot have been very strong; it lay within the temperature tolerance of the insect species found. WATTS 1980 reviewed the Younger Dryas in Europe and explained the reduced "amplitude" in the continent with increasing distance from the shifted polar front in the North Atlantic (RUDDIMAN et al. 1977). In the Alps, however, considerable glacier advances are recorded. Since the state of activity of glaciers is not only due to temperature but also the summer precipitation, we could conclude that the summers of the early Younger Dryas were not only cooler but also wetter than during the Allerød.

Acknowledgements

We are all indebted to a great number of people who helped us in various ways. Each author has acknowledged their help in his respective paper. The cordial thanks of all of us go to the project leader Prof. G.Lang. The study was supported by the Swiss National Science Foundation (project nr. 3.336-78).

References

Ammann, B., Chaix, L., Eicher, U., Elias, S.A., Gaillard, M.-J., Hofmann, W., Siegenthaler, U., Tobolski, K. & Wilkinson, B. 1983. Vegetation, insects, molluscs and stable isotopes in Late-Würm deposits at Lobsigensee (Swiss Plateau). Studies in the Late-Quaternary of Lobsigensee 7. Revue de Paléobiologie 2/2.

Ammann, B. & Tobolski, K., 1983. Vegetational development during the Late-Würm at Lobsigensee (Swiss Plateau). Studies in the Late-Quaternary of Lobsigensee 1. Revue de Paléobiologie 2/2.

Chaix, L., 1983. Malacofauna from the Lateglacial deposits of Lobsigensee (Swiss Plateau). Studies in the Late-Quaternary of Lobsigensee 5. Revue de Paléobiologie 2/2.

Eicher, U. & Siegenthaler, U., 1983. Stable isotopes in lake marl and mollusc shells from Lobsigensee (Swiss Plateau). Studies in the Late-Quaternary of Lobsigensee 6. Revue de Paléobiologie 2/2.

Elias, S.A. & Wilkinson, B., 1983. Lateglacial insect fossil assemblages from Lobsigensee (Swiss Plateau). Studies in the Late-Quaternary of Lobsigensee 3. Revue de Paléobiologie 2/2.

Gaillard, M.-J., 1981. Etude palynologique de l'évolution tardi- et postglàciaire de la végétation du Moyen-Pays romand (Suisse). Thèse Université de Lausanne 517 p, in press.

Gaillard, M.-J., 1983. On the occurrence of *Betula nana* L. pollen grains in the Late-glacial deposits of Lobsigensee (Swiss Plateau). Studies in the Late-Quaternary of Lobsigensee 2. Revue de Paléobiologie 2/2.

Hammen, T. van der, 1951. Late-glacial flora and periglacial phenomena in the Netherlands. Leidse geol. Med. 17, 71-183.

Hofmann, W., 1983. Stratigraphy of subfossil Chironomidae and Ceratopogonidae (Insecta: Diptera) in late-glacial littoral sediments from Lobsigensee (Swiss Plateau). Studies in the Late-Quaternary of Lobsigensee 4. Revue de Paléobiologie 2/2.

Lang, G., 1983. Spätquartäre See- und Moorentwicklung in der Schweiz. Stand und erste Ergebnisse eines Forschungsprogrammes des Systematisch-Geobotanischen Instituts der Universität Bern. Gesellschaft für Oekologie. Verhandlungen Band XI, 12. Jahrestagung Bern 1982 (in press).

Ruddiman, W.F., Sancetta, C.D. & McIntyre, A., 1977. Glacial/interglacial response rate of subpolar North Atlantic water to climatic change: the record in ocean sediments. Phil. Trans. R. Soc. Lond. B, 280, 119-142.

Watts, W.A., 1980. Regional Variation in the Response of Vegetation to Lateglacial Climatic events in Europe. In: Lowe, J.J., Gray, J.M. & Robinson J.E., 1980. Studies in the Lateglacial of North-West-Europe, 1-21, Oxford.

Welten, M., 1982. Vegetationsgeschichtliche Untersuchungen in den westlichen Schweizer Alpen: Bern - Wallis. Denkschr. Schweiz. natf. Ges. 95, 104 p., Basel.

LATE WEICHSELIAN FLUCTUATIONS IN THE FRENCH ALPS AND MASSIF CENTRAL FROM POLLEN ANALYSES

J.-L. de Beaulieu, J. Clerc and M. Reille

Faculté des Sciences et Techniques St-Jérôme, Marseille, France

Pollen analyses in Southern Europe mountains provided information of greatest interest concerning not only the history of mesophilous and thermophilous species refuges, as was shown by Van Campo & Aymonin (1962), Lang (1970), Beug (1975), Van der Hammen (1979), but also the variations of climate. These regions were not much affected by the Scandinavian inland ice; therefore they may have registered the first oscillations that took place at the end of the Würm.

However, in South-Eastern France, except for a study by Lang & Trautmann in Massif Central (1961), only for some ten years has light been thrown on the Late glacial and Holocene history of vegetation, mainly under the impulse of Prof. A. Pons (Reille 1975; Triat-Laval 1978; Beaulieu 1977; Wegmüller 1977; Planchais *et al.* 1977; Coûteaux 1970, 1983; Beaulieu, Pons & Reille 1982; Richard 1983).

The exploitation of our botanical records is still in progress, so that the paleoclimatic results presented here are but a provisional state of things, partly based on original results whose detailed account has not yet been published.

SOME REFERENCE SEQUENCES

In the region studied as in more northern countries, the first steps of the Late-Würmian forest development are generally marked by the flourishing of the same pioneer taxa : *Juniperus*, *Betula* and *Pinus*, to the prejudice of herbaceous formations with *Heliophytae*, *Artemisia* and *Gramineae*.

Therefore, as a support to our demonstration, simplified extracts of pollen diagrams are presented here. They show the

N.-A. Mörner and W. Karlén (eds.), Climatic Changes on a Yearly to Millennial Basis, 75–90.

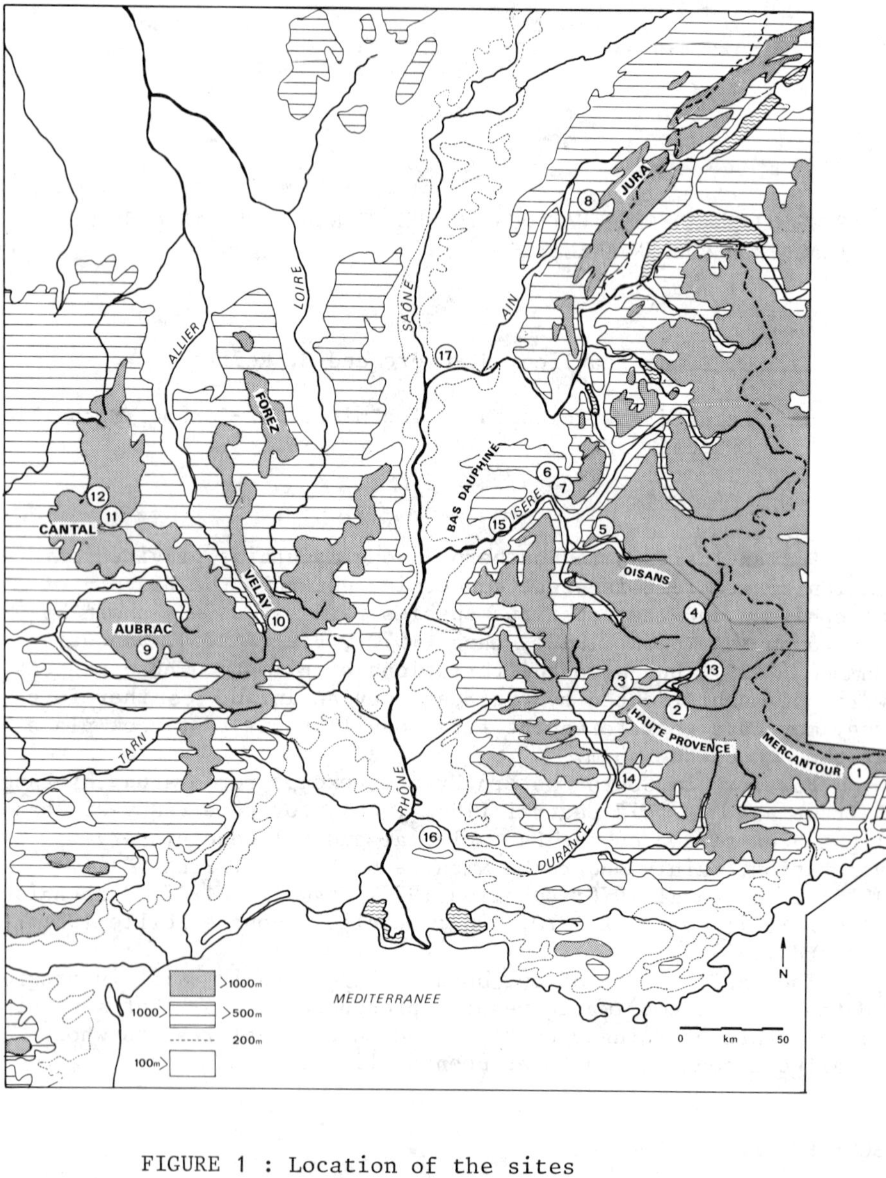

FIGURE 1 : Location of the sites

pollen variations of these five taxa as well as those of the first mesophilous trees (*Quercus* and *Corylus*) during the Late-Glacial and the Preboreal. These diagrams come from 12 reference sites (Fig. 1) : 4 in the Southern Alps (Fig. 2), 4 in the Northern Alps and the Jura (Fig. 3), 4 in the Massif Central (Fig. 4).

The French Southern Alps

Up to an elevation of about 1800 m, the Bølling *sensu lato* is characterized by a maximum of *Juniperus* which, on several occasions, has been dated around 13,300 B.P.

In the Maritime Alps (Lac Long Inférieur), *Pinus* expansion takes place very early, towards 13,500 B.P., and it is related to the climatic characteristics of the Maritime Alps (warmer and wetter) such as they still prevail to-day.

Juniperus expansion is not recorded.

In more northern sites, at a low elevation, *Betula* has an expansion phase before that of *Pinus* which occurs towards 12,500 B.P.

Therefore, it suggests that the Older Dryas took place after *Pinus* maximum. Here, the absence of an oscillation might be due to an early development of the forest.

Among all the sites of the Southern Alps, Lac Long Inférieur is the only one that shows a decrease in *Pinus* frequencies which could correspond to the Older Dryas. This decline is represented by a thin, 3 mm thick, level of grey clay embedded in gyttja.

But additional borings in the same site have shown that this level could be either absent or subdivided in three (stars in Fig. 5). In this second case, the scallopped aspect of the pollen curves cannot reflect changes in vegetation - they would not be so sudden -; they rather suggest reworkings of Oldest Dryas sediments.

Thus, the regional pollen diagrams are in the whole in agreement with the notion of "Windermere Interstadial" (Welten 1972) grouping in a same unit both the Bølling and the Allerød interstadials.

After the *Pinus* maximum, well marked up to 2000 m, the Younger Dryas is characterized by a strong decline of this tree to the benefit of steppe species and *Juniperus*. At a mean elevation, the forest does not disappear but it only becomes more open.

As a matter of fact, even at sites such as Siguret (N°13 in Fig. 1) (Beaulieu 1977; Beaulieu & Reille, unpublished), where the decrease in *Pinus* pollen values is very clear, the presence of stoma from this tree indicates its survival in the surroundings. Therefore it is difficult to precise how lower did the timberline fall. It certainly lay below the altitude of Col des Lauzes (1784 m).

Contrary to the rather insignificant Older Dryas, the Younger Dryas appears as a most important climatic event in the region.

Later, during the Preboreal, and only in the Montagne de la Blanche, a recurrence of NAP and *Artemisia* is noted. It is a

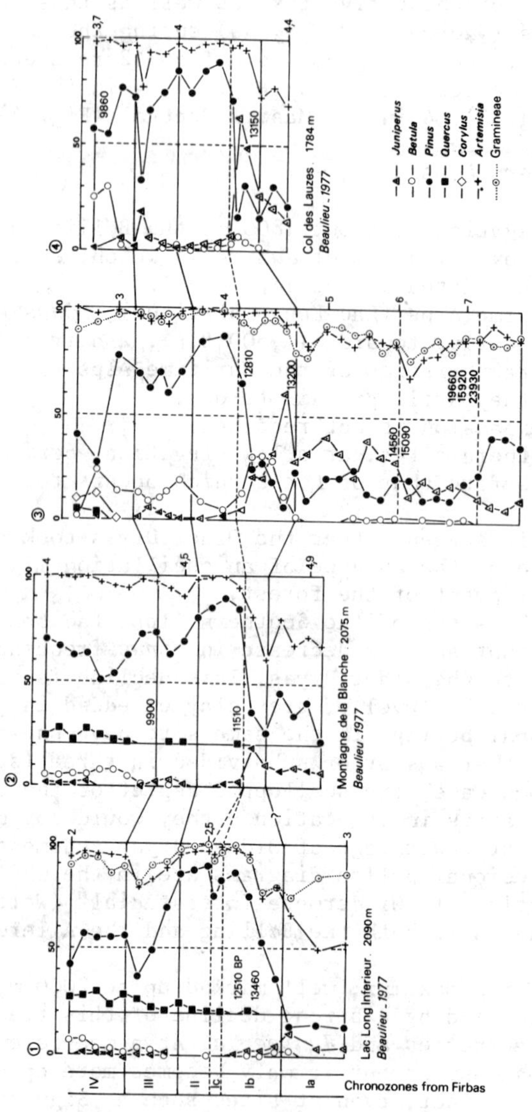

FIGURE 2 : Late-glacial vegetational evolution in the Southern Alps

locally significant event since it has also been found in two other borings in the vicinity (Beaulieu 1977). However it seems to be too isolated to have a regional climatic value.

Finally it should be mentioned that the Oldest Dryas presents a very thick sequence, older than the Bølling. Here can be recorded the first steps of the vegetational development after the retreat of the Würmian glacier which covered the place now occupied by the mire. We note :

- first, a phase with almost bare soils and with long-distance pollen transports of *Pinus*;
- second, a phase of local increase of steppe species (first *Chenopodiaceae* - not represented in our simplified diagram - then *Artemisia*) and *Juniperus* percentages, locally reflecting a thicker vegetation-cover; this phase could not be precisely dated (three contradictory 14C dates);
- lastly, a phase of early expansion of *Juniperus* from about 14,700 B.P., pointing to an improvement of climate before the Bølling.

This improvement is confirmed by the finding of *Pinus* trunks buried in sands and dated 14,000 B.P. at Salignac (N°14 in Fig. 1), 30 km below Pelléautier (Beaulieu & Tiercelin, unpublished).

The Northern Alps and the Jura

Unfortunately the Late-Glacial sequence from St-Hilaire du Rosier (Isère) is broken before the expansion of *Pinus*. However, it presents thick deposits covering the whole of the Oldest Dryas (Fig. 6; from Clerc, unpublished).

Above the blue clays without pollen, marly sediments contain two gyttja-algae levels.

The lower level is dated at about 15,000 B.P.; the lithology and a richer aquatic flora indicate a climatic improvement too weak to have changed the regional steppe vegetation but sufficient to confirm the observations made at Pelléautier.

The second date, 13,900 B.P., shows that, here, the rise of *Juniperus* curve approximately corresponds to the Bølling.

At other sites, there are obvious similarities with the Southern Alps, but the *Betula* maximum is generally higher than that of *Juniperus*. Besides, the formation of *Pinus* forest, not yet dated with precision, may have taken place later than in the Southern Alps. In this case, the slight flattening in the rise of *Pinus* could indicate the Older Dryas. But this hypothesis is not corroborated by the results of Gaillard (1981) in Switzerland. She shows that the slight decrease in A.P. relative values at the intersection of *Betula* and *Pinus* curves is not accompanied by a decrease of the regional pollen rain of trees.

The Younger Dryas is not always well marked, except in the Jura; this is probably due to the low elevation of the sites studied.

Lastly, in the region, the Preboreal deposits are very thin,

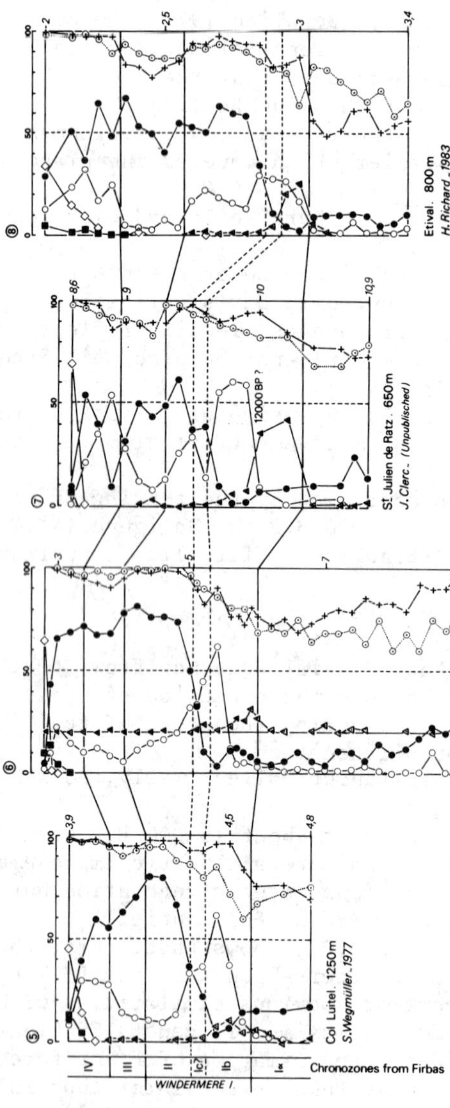

FIGURE 3 : Late-glacial vegetational evolution in the Northern Alps and the Jura

so that a detailed study of this period has been impossible up to now.

In a general manner, there is an opposition between the Southern Alps (Fig. 2), where steppe species values, especially *Artemisia*, strongly decrease as early as the beginning of the Bølling, and Bas-Dauphiné (Fig. 3) where they remain high during the first phase with *Betula* (just as at St-Hilaire) and until the beginning of the Preboreal (for example at St-Julien de Ratz).

This phenomenon can be compared with the even more surprising events appearing in the diagrams from Lower Provence (Triat-Laval 1978), where at Beauchamp-Panières (Bouches-du-Rhône; N°16 in Fig. 1), *Artemisia* frequencies hardly fall below 10% during the interstadial Bølling-Allerød and keep high during a good part of the Holocene.

Here, the persistence of the steppe is of course controlled both by the pluviometric deficit and the Mediterranean regime, but as it does not appear in the Southern Alps, it is probably also related to the existence of a "paleo Mistral", that is to say a wind sweeping down the Rhône valley and increasing its aridity.

Finally, between the Jura and the Massif Central, on the Plateau des Dombes, the ancient lake of Les Echets (267 m; n°17 in Fig. 1) which contains a sequence older than the Eemian (Beaulieu *et al.* 1980; Beaulieu & Reille 1984) shows a dry phase slightly younger than 12,000 B.P. and probably corresponding to a water deficit due to a greater increase of evaporation than of precipitation.

Before this episode, the *Artemisia* expansion, which indicates a larger steppe cover, is well dated at 15,000 B.P., that is to say somewhat later than at Pelléautier.

The Massif Central

Volcanic activity and erosion in the valleys have divided the North of Massif Central into several mountainous massifs. Some of them have been intensely prospected during the last five years : Massif du Cantal (Beaulieu, Pons & Reille 1982), Plateau de Millevaches (Denèfle *et al.* 1979), Planèze de St-Flour (Reille et Beaulieu 1979), Vivarais (Coûteaux 1978, 1984), Forez (Janssen et Van Straten 1982), Margeride (Reille & Pons 1982), Velay (Beaulieu, Pons & Reille, in press); Cézalier and Aubrac (Beaulieu, Pons & Reille, unpublished). In the context of a regional synthesis, only four sites with a well-marked Late-Glacial are presented here. A succession of vegetation stages, rather similar to those found in the Alps, is observed :

- first, a phase mainly with long-distance *Pinus* transports,
- second, a maximum phase for steppe species beginning at 15,800 B.P. (Lac du Bouchet), a date close to that obtained at Les Echets,
- third, the succession *Juniperus-Betula-Pinus*,
- lastly, a well-marked Younger Dryas.

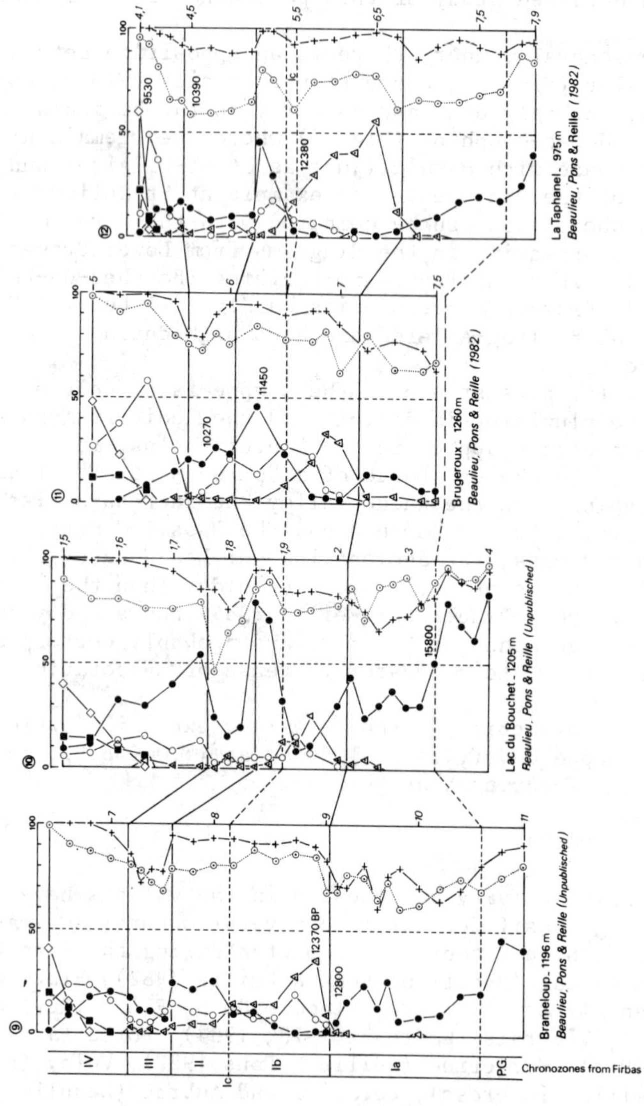

FIGURE 4 : Late-glacial vegetational evolution in the Massif Central

But herbaceous frequencies are much higher than in the Alps, and *Artemisia* values are almost always lower than those of *Gramineae*, suggesting that herbaceous formations have less affinity with steppe.

The dates found for *Juniperus* optimum, though few and unprecised, seem to indicate that it took place after 13,000 B.P., that is to say several centuries later than in the Alps.

After this maximum, *Betula* and chiefly *Pinus* frequencies suggest very clear woodstands.

These results show that during the Late-Glacial the climate was more severe in the Massif Central than in the Alps. Therefore one could expect the Older Dryas to be more strongly marked, which is not at all the case, except at La Taphanel where a layer of clay, some millimeters thick, deposited after 12,400, suggests phenomena similar to those described at Lac Long Inférieur. This could imply that the Older Dryas was marked by torrential phenomena.

Likewise, in the well-developped Preboreal, characterized by *Betula* expansion and the persistence of grass-land formations, no negative oscillation has been recorded. Moreover, several datings show that *Corylus* expansion took place before 9000 B.P., that is to say during the Preboreal.

CORRELATIONS WITH OTHER REGIONAL STUDIES

The climatic evolution such as it is suggested by pollen sequences is confirmed or precised by results obtained in other fields of research, for example the chronology of glacial advances and the oxygen isotope measurements.

Glacial advances

In the French Alps, many problems have not yet been solved. For example, the age of the "moraines internes" of the Würm maximum has not yet been precised : it might be contemporaneous either with the thermic minimum of 18,000 - 20,000 B.P. or with the first half of the Würm (Bornand *et al.* 1976).

As regards the Late-Glacial moraines, the relative chronology and the equivalences with the classical stages of Austria established by Schweitzer in the Maritime Alps (1968) have been completed by Vivian (1975) for the Northern Alps; they have been reviewed, in particular thanks to a collaboration between geomorphologists and palynologists (Beaulieu & Jorda 1977; Hannss & Wegmüller 1978; Hannss 1982; Coûteaux 1982). In spite of notable differences in details, these studies on the whole show that at the beginning of the Bølling, the glacier tongues had withdrawn in the high valleys, which suggests a sudden retreat before 13,300 B.P. probably contemporaneous with the expansion phase of steppe species dated 15,000 - 16,000 B.P. This is also in agreement with

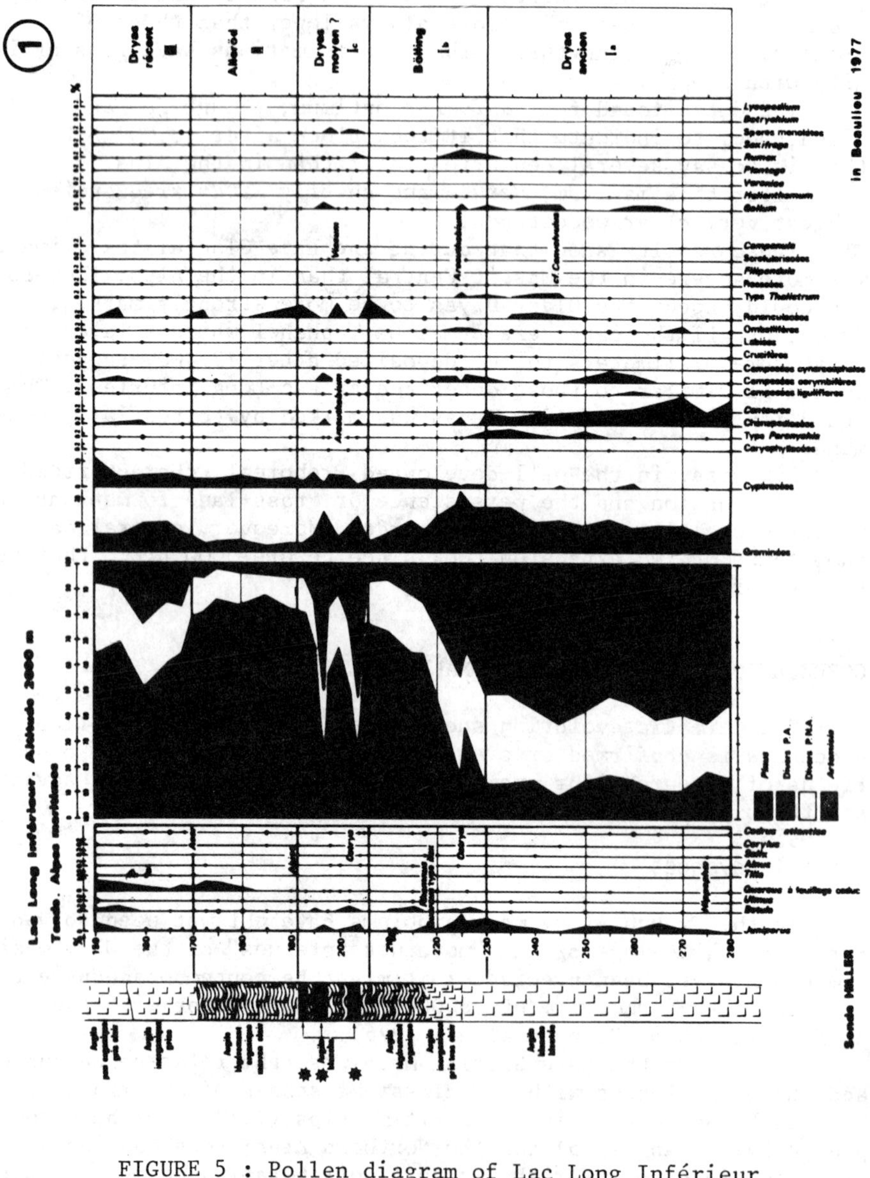

FIGURE 5 : Pollen diagram of Lac Long Inférieur
(Loc. 1 in Fig.1)

evidences of thawing recorded in sea cores (Duplessy *et al.* 1981).

According to Jorda & Delibrias (1981) an interstadial dated around 19,000-20,000 B.P. in the Southern Alps (interstadial of Bleone) took place between two glacier advances. The last glacial readvance, reaching a mean elevation of 1100 m, is dated between 19,000 and 15,000 B.P. This readvance is not reflected in the pollen diagrams from Les Echets.

In the Massif Central, no approach of this problem has been made except by Veyret *et al.* (1978) who describe moraine deposits overlying an organic layer dated 13,580 ± 250 B.P. in the Santoire valley. This implies a glacier tongue more than 30 km long. If this observation was confirmed, it would indicate an ice-cover much more important than in the Alps.

Oxygen isotope measurements

The application of oxygen isotope measurements to Swiss Late-Glacial lake sediments for paleoclimatic purposes has been developped by Siegenthaler (1971) and Eicher (1979).

In the French Alps, the study of Chirens peat bog (Eicher, Siegenthaler, Wegmüller 1981) and Siguret lake (Eicher, in press) confirms the results obtained in Switzerland : the main change in ^{18}O ratio occurs suddenly at the beginning of the Bølling, without any sign of a previous climatic improvement. This is inconsistent with the results from pollen analyses. The Older Dryas is not perceived, whereas the Younger Dryas is indicated through a strong decrease of ^{18}O ratio; the latter, more clearly than the results of pollen analyses, points to a very cold period.

CONCLUSIONS

Pollen analyses of lakes and mires in Southern France provided the following results concerning Late-Weichselian climatic fluctuations :

1. Between ca 20,000 B.P. and 15,000 B.P. there appears no change in regional vegetations that could be ascribed to climatic fluctuations.

2. Towards 15,000 B.P. (at a mean elevation), phenomena different from one region to the other take place, which can be interpreted as the first signs of the climatic improvement that marked the end of the Würm.

3. At 13,300 - 13,000 B.P., the beginning of the interval "Bølling (*sensu lato*) - Allerød" is clearly marked at any altitude by a notable increase of the pollen rain and by the expansion of *Pinus* in the Maritime Alps and of *Juniperus* in more northern sites. In the Massif Central, the optimum of *Juniperus* before ca 12,380 B.P. has not yet been dated with precision.

4. During the interval Bølling-Allerød, forest dynamics

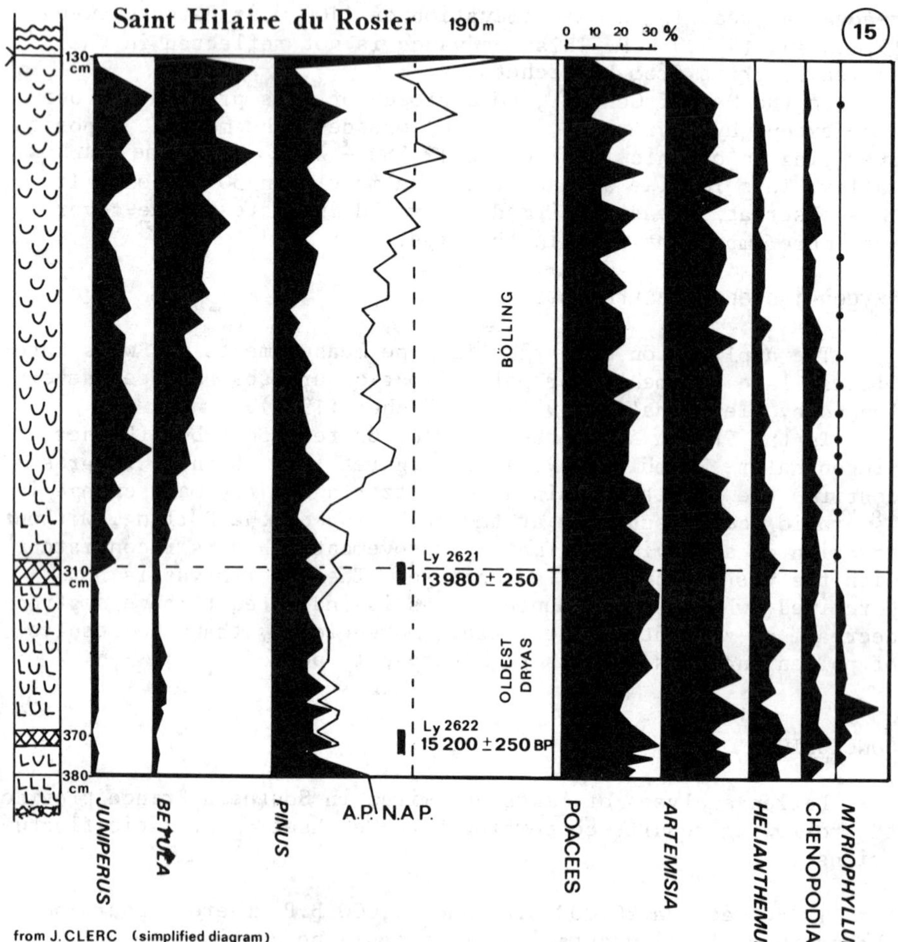

FIGURE 6 : Simplified diagram from St-Hilaire du Rosier (Loc. 15 in Fig. 1)

marked by the classical *Juniperus-Betula-Pinus* succession are generally recorded, yet with great differences : thick forest with *Pinus* covered the Southern Alps from 12,500 B.P., whereas only scattered and sparse woods are evidenced in the Massif Central; so, this region then constituted a cold center : by extrapolation from present climatic data, the abundance of snow falls could explain this situation. At the same time, in the lower Rhône valley, very arid conditions prevailed, as is shown by the presence of still abundant steppe taxa : the Mediterranean regime, a water deficit related to a low elevation and the probable existence of a "paleo-Mistral" may account for this.

5. None of our pollen analysis evidenced a cooling that could be ascribed to the Older Dryas; if it did exist, it did not affect the vegetational dynamics. At two sites (Lac Long Inférieur, Maritime Alps and La Taphanel, Auvergne) slight reworkings might indicate torrential phenomena around 12,000 B.P.

6. The cold oscillation of the Younger Dryas appears everywhere as a major event generally marked by changes in lake sediments and a reappearance of steppe taxa. The alpine sites, at a mean elevation, were less affected; *Pinus* forests, though less thick ones, were still existing.

7. Lastly, though several pollen zones may be distinguished during the Preboreal, in more than thirty sites there never appears a regressive vegetational evolution that could characterize a cold oscillation.

The hypothesis of sudden climatic changes at the transition between glacials and interglacials has recently been put forward (Flohn 1979). Though an important climatic improvement is recorded in South-Eastern France at the end of the Würm, around 13,000 B.P., there are evidences that the warming began toward 15,000 B.P., which is inconsistent with the idea of a sudden warming. The Younger Dryas (a short climatic regression without subsequent effect), however, represents a very sudden oscillation of large amplitude.

REFERENCES

Beaulieu, J.-L. de, 1977 Contribution pollenanalytique à l'histoire tardiglaciaire et holocène de la végétation des Alpes méridionales françaises. Thèse ès Sciences, Université Aix-Marseille III, 358 p., 29 fig., 39 diag. h.t.

Beaulieu, J.-L. de, Evin, J., Mandier, P., Monjuvent, G. & Reille, M., 1980. Les Echets : un marais capital pour l'histoire climatique du Quaternaire rhodanien. Mém. Mus. Nat. Hist. Nat., sér. B; Botanique, 27, 123-136.

Beaulieu, J.-L. de, & Jorda, M., 1977. Tardiglaciaire et Postglaciaire des Alpes de Haute Provence. Le glaciaire de la Moncagne de la Blanche-Trois Evêchés. Bull. A.F.E.Q.,3, 3-15.

Beaulieu, J.-L. de, Pons, A. & Reille, M., 1982. Recherches pollenanalytiques sur l'histoire de la végétation de la bordure nord du Massif du Cantal (Massif Central, France). Pollen et Spores, 24 (2), 251-300.

Beaulieu, J.-L. de, & Reille, M. A long upper Pleistocene pollen record from Les Echets near Lyon, France. Boreas, manuscrit accepté.

Beaulieu, J.-L. de & Reille, M., 1983. Paléoenvironnement des marais de Pelléautier et Siguret (Hautes-Alpes). Analyses polliniques. Ecologia Mediterranea, à paraître.

Beug, H.J., 1975. Changes of climate and vegetation belts in the mountains of Mediterranean Europe during the Holocene. Bull. Geol. Warsawa, 19, 101-110.

Bornand, M., Bourdier, F., Mandier, P. & Monjuvent, G., 1976. Les glaciers quaternaires dans les Alpes et le bassin du Rhône. In "La Préhistoire Française" , I, 1, 32-37.

Clerc, J. Première contribution à l'étude de la végétation tardiglaciaire et holocène du Piémont Dauphinois. Doc. de Cartographie écologique, à paraître.

Coope, G.R. & Pennington, N., 1977. The Windermere Interstadial of the Late-Devensian. Philos. Trans., B., 280, 337-339.

Coûteaux, M., 1970. Documents nouveaux pour l'histoire holocène de la végétation en Oisans. Doc. Cart. Végét. Alpes, 8, 115-129.

Coûteaux, M., 1978. Analyses polliniques à Peyrebeille, Mezillac et Mazan (Ardèche). Evolution holocène de la végétation et genèse des sols podzoliques à horizon noir profond. Pollen et Spores, 20 (4), 485-495.

Coûteaux, M., 1982. Fluctuations glaciaires de la fin du Würm dans les Alpes françaises, établies par des analyses polliniques. Boreas, 12, 35-56.

Coûteaux, M., 1984. Bilan des recherches pollenanalytiques en Ardèche (France). Bull. Soc. Roy. Bot. Belg., 117, sous presse).

Denèfle, M., Valadas, B. & Vilks, A., 1980. Analyses polliniques et datations au 14C sur le plateau de Millevaches (Limousin): la tourbière de Longeroux. C.R. Acad. Sc. Paris, 290, 853-856.

Duplessy, J.C., Delibrias, G., Turon, J.L., Pujol, C. & Duprat, J., 1981. Deglacial warming of the northeastern Atlantic Ocean : correlation with the paleoclimatic evolution of the European continent. Palaeogeogr., Palaeoclimatol., Palaeoecol., 35, 121-144.

Eicher, U., 1979. Die $^{18}O/^{16}O$ und $^{13}C/^{12}C$ - Isotopenverhältnisse in spätglazialen Süsswasserkarbonaten und ihr Zusammenhang mit den Ergebnissen der Pollenanalyse. Thesis, Bern, 204 p.

Eicher, U., Siegenthaler, U. & Wegmüller, S., 1981. Pollen and oxygen isotope analyses on Late- and Post-glacial sediments of the tourbière de Chirens (Dauphiné, France). Quaternary Research, 15, 160-170.

Flohn, H., 1979. On time scales and causes of abrupt paleoclimatic events. Quaternary Research, 12, 135-149.

Gaillard, M-J., 1981. Etude palynologique de l'évolution tardi- et postglaciaire de la végétation du Moyen Pays Romand (Suisse). Thèse Lausanne, 517 p., 64 Fig., 30 Tabl.

Hammen, T., van der, 1979. Changes in life conditions on earth during the past one million years. Kong. Danske Vid. Selsk. Biol. Skr., 22 (6), 1-32

Hannss, C., 1982. Spätpleistozäne bis postglaziale Talverschüttungs- und Vergletscherungphasen im Bereich des Sillon Alpin der französischen Nordalpen. Mitt. Quartärforsch. österreichischen Akad. Wiss., 4, 213 p.

Hannss, C., Wegmüller, S., 1978. Zur Alterstellung würmkaltzeitlicher lokalgletschermoränen im Dévoluy und in der Belledone (französischen Alpen). Z. f. Geltscherkde u. Glazialgeol., 12, 205-222.

Janssen, C.R. & Van Straten, R., 1982. Premiers résultats des recherches palynologiques en Forez, Plateau Central. C.R. Acad. Sc. Paris, sér. III, 294, 83-86.

Jorda, M. & Delibrias, G., 1981. Données nouvelles sur le Pleistocène supérieur des Alpes françaises du sud. Le Würm récent du Bassin de la Bléone (Alpes de Haute Provence). Bull. A.F.E.Q., 3-4, 173-182.

Lang, G., Trautman, W., 1961. Zur spät- und nacheiszeitlichen Vegetationsgeschichte der Auvergne (Französisches Zentralmassiv) Flora, 150, 11-41.

Planchais, N., Quet-Pasquier, L., Cours, P., Thommeret, J. & Thommeret, Y., 1977. Essai de palynologie côtière appliquée au remplissage flandrien de Palavas (Hérault). C.R. Acad. Sc. Paris, 284, 158-162.

Reille, M., 1975. Contribution pollenanalytique à l'histoire tardiglaciaire et holocène de la végétation de la montagne corse. Thèse, Univ. Aix-Marseille III, 206 p., 44 diag. et 5 pl. h.t.

Reille, M. & Beaulieu, J.-L. de, 1979. Analyse pollinique de l'étang de Cheylade (Massif Central, France) : histoire tardiglaciaire et holocène de la végétation de la planèze de Saint-Flour. C.R. Acad. Sc. Paris, 292, 243-246.

Reille, M. & Pons, A., 1982. L'histoire récente de *Pinus silvestris* L. en Margeride (Massif Central, France) et la signification de cette essence dans la végétation actuelle. C.R. Acad. Sc. Paris, 294, 471-474.

Richard, H., 1983. Nouvelles contributions à l'histoire de la végétation Franc-comtoise tardiglaciaire et holocène à partir des données de la palynologie. Thèse IIIe cycle, Besançon, 155 p.

Schweitzer, G., 1968. Der Formenschatz des Spät- und Postglazials in den Hohen Seealpen. Aktualgeomorphologische Studien im oberen Tinnéetal. Z. f. Geomorph., Suppl. Bd 6.

Siegenthaler, U., 1971. Sauerstoff-18, Deuterium und Tritium im Wasserkreislauf. Thesis Bern.

Triat-Laval, H., 1978. Contribution pollenanalytique à l'histoire tardi- et postglaciaire de la végétation de la basse vallée du Rhône. Thèse, Univ. Aix-Marseille III, 343 p., 32 diag. h.t.

Van Campo, M. & Aymonin, G., 1962. Le problème de l'histoire de la flore et de la végétation dans les Cévennes méridionales vu sous l'angle de l'analyse pollinique. Flora, 152, 679-688.

Veyret, Y., Brousse, R. & Delibrias, G., 1978. Première datation d'un épisode glaciaire récent dans le Massif Central français. C.R. Acad. Sc. Paris, 286, 1089-1092

Vivian, R., 1975. Les glaciers des Alpes occidentales. Impr. Allier, Grenoble, 513 p.

Wegmüller, S., 1977. Pollenanalytische Untersuchungen zur spät- und postglazialen Vegetationsgeschichte der französischen Alpen. Thseis Bern, 178 p., 10 diag.

Welten, M., 1972. Das Spätglazial im nördlichen Nordalpengebiet der Schweiz. Ber. Deutsch. Bot. Ges., 85 (1-4), 69-74.

CLIMATIC AND GLACIAL FLUCTUATIONS DURING THE LATE-GLACIAL AND THE HOLOCENE IN OISANS (ISERE, FRANCE)

M. COÛTEAUX

Faculté des Sciences et Techniques St-Jérôme, Marseille, France

ABSTRACT - During the whole Late-glacial and Holocene times, only two glacial recurrences (the Younger Dryas and the 17-19th centuries) reflect deep climatic changes. Emphasis is laid on the thick and long-lasting snow cover. Besides these two recurrence periods, the major events observed mainly evidence the influence of drought : glacial regression and formation of rock glaciers.

Several pollen diagrams (Coûteaux 1982a, 1982b, 1983a) from the alpine zone at La Muzelle (2140 m) made it possible to observe a strong glacial readvance during the Younger Dryas. A synchronization has been made with a group of moraines.

For the same period, several pollen diagrams (Coûteaux et Evin 1981, Coûteaux 1983b) from the subalpine zone at Deux-Alpes (1646 m) show an important nival phase.

In both cases, three pulsations are recorded during this recurrence phase (Fig. 1). The *Artemisia* curve reflects glacial pulsations, the *Selaginella* curve reflects nival pulsations (Coûteaux 1982c).

No other glacial or nival recurrence has been recorded during the Late-glacial, neither during the Oldest Dryas, nor during the Older Dryas. There appears a slight oscillation in *Artemisia* and *Selaginella* curves, but there is no possible correlation with moraine formations.

During the Holocene, the pollen diagrams do not show any glacial readvance, except at the end of the Subatlantic times, around the 17th-19th centuries. There are numerous and obvious indications of this ice-cover in the topography.

The interpretation of pollen diagrams is difficult, because in Oisans the glacier advance did not cover forest areas. Moreover,

N.-A. Mörner and W. Karlén (eds.), Climatic Changes on a Yearly to Millennial Basis, 91–94.

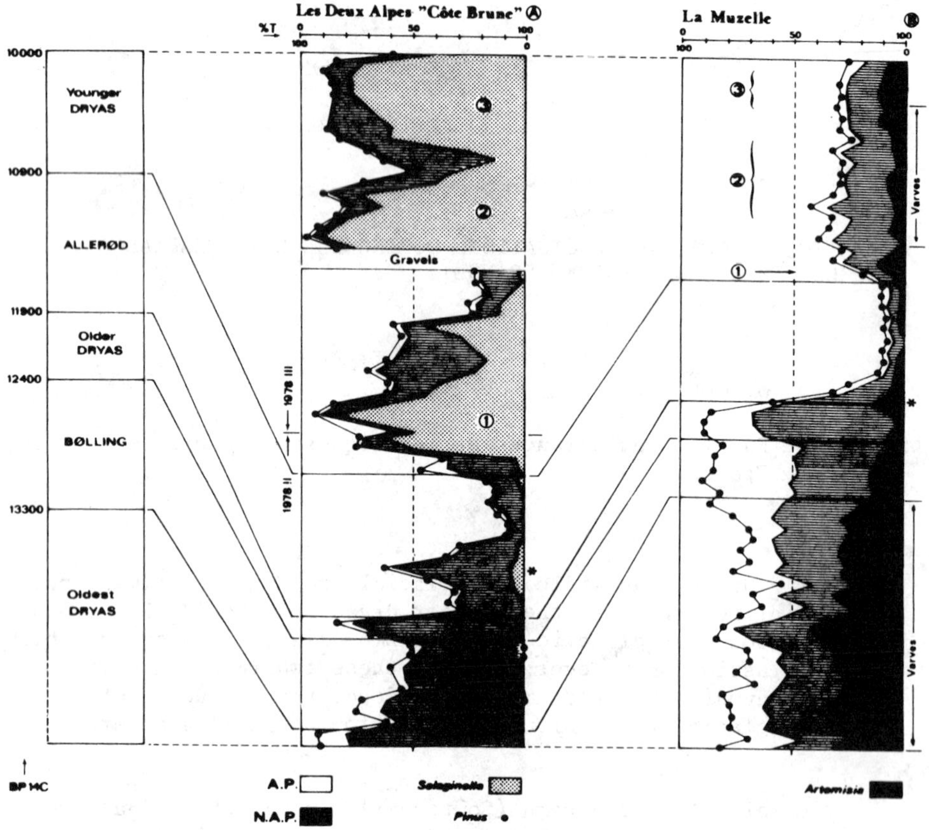

FIGURE 1 : Pollen curves of *Pinus*, *Selaginella* and *Artemisia* during the Late-glacial at Deux-Alpes and La Muzelle.

before and after the readvance period, human impact on vegetation was strong, even at a high elevation (Coûteaux 1983c).

Pollen diagrams give very different pictures of the influcence of the glacial recurrence on pollen assemblages depending on the nature of the sediment analysed and on the location of the site with regard to the glacial advance.

1. In peat bogs of the subalpine zone, beyond the area morphologically affected by the glacial advance, a decrease of the ratio AP/T may be explained by a decrease in the pollen production of scattered trees at higher altitudes.

2. In proxiglacial mineral sediments, because of the absence or spareness of a local vegetation producing pollen, only allochtonous species are recorded in the pollen diagrams. The spectra

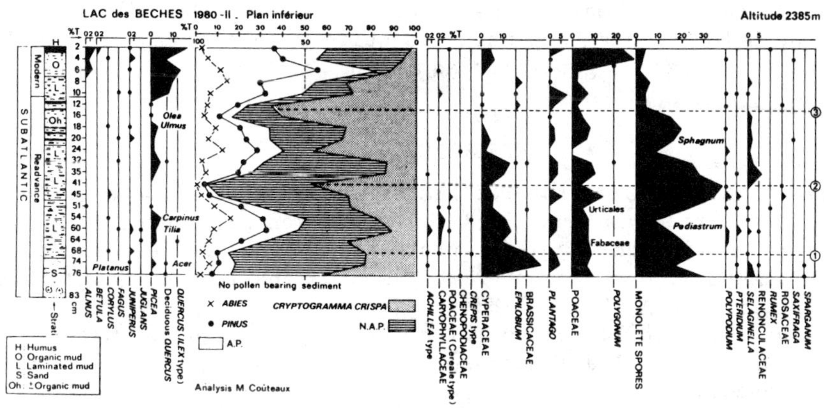

FIGURE 2 : Pollen diagram from Lac des Bèches

always contain abundant *Pinus* pollen; sometimes, they also contain *Artemisia* pollen when a proxiglacial vegetation is present in the alpine zone.

3. In the subalpine zone, below glacier areas, in mineral sediments, the variations of the snow-cover are reflected in pollen diagram through fluctuations of *Cryptogramma crispa* curve (Fig. 2). These fluctuations exhibit three glacial pulsations during the recurrence period.

Distortion phenomena in pollen spectra under the peculiar sedimentary conditions prevailing at the margin of glaciers and in snow-covered areas are difficult to interpret in terms of vegetation, but they provide valuable information on glacial episodes in the alpine zone and on nival episodes in the subalpine zone.

During the Holocene, moraines are deposited here and there, which do not delimit large ice-covered areas, but there exists remains of what could be compared to rock glaciers and to which it is difficult to assign a precise date.

In the small valley of La Lavey, the cartography of ice-covered areas during the Younger Dryas, of Holocene formations assimilable to rock glaciers, of ice-covered areas during the 17th-19th centuries and of nowadays still ice-covered areas (Coûteaux 1983d) emphasizes the strong morphogenetic consequences of two periods when climate was essentially characterized by much efficient humidity, that is to say the Younger Dryas and the 17th-19th centuries. The whole of the Holocene times and the present times are characterized by much less humidity.

In conclusion, pollenanalytical studies in Oisans enable us to propose for the main climatic factors during the Late-glacial and the Holocene a schema which is quite simple as compared with the complex systems generally admitted.

REFERENCES

Coûteaux, M., 1982a. Intérêt historique et scientifique de la tourbière de La Muzelle (commune de Venosc, Isère, France). Proposition en vue de sa protection. Actes du Colloque de Chamonix : Montagne et Aménagement. Institut de Géographie Alpine, LAMA 1982, 77-83.

Coûteaux, M., 1982b. La tourbière et le glacier de La Muzelle (Parc National des Ecrins). Rev. Parc Nat. Ecrins, 2 (sous presse).

Coûteaux, M., 1982c. Récurrences glaciaires en Oisans au Tardiglaciaire. Soc. Hydr. Fr. sect. Glaciol. Actes de la réunion du 4-5 mars 1982, 1-3.

Coûteaux, M., 1983a. Fluctuations glaciaires de la fin du Würm dans les Alpes françaises, établies par les analyses polliniques. Boreas, 12, 35-56.

Coûteaux, M., 1983b. Géomorphologie et évolution phytogéographique tardiglaciaire et holocène aux Deux-Alpes (Isère, France) : contribution pollenanalytique. Rev. Géogr. Alpine, 71 (2), 143-163.

Coûteaux, M., 1983c. Déboisements d'altitude par les charbonniers de l'Oisans : contribution pollenanalytique. 108e Congr. Soc. Sav., Grenoble (sous presse).

Coûteaux, M., 1983d. Fluctuations climatiques et fluctuations glaciaires du Tardiglaciaire et de l'Holocène en Oisans (Isère, France). Sec. Nord. Symp. "Climatic changes..." Stockholm, Mai 1983, Abstracts, 24-28.

Coûteaux, M., Evin, J., 1981. Etudes palynologiques et datages par le radiocarbone de dépôts fluvioglaciaires aux Deux-Alpes (Isère, France). C.R. Acad. Sc., Paris, 292, sér. II, 1235-1238.

THE CLASSICAL LATE WEICHSELIAN CLIMATIC FLUCTUATIONS IN MEXICO

Klaus Heine

Regensburg University, FRG

INTRODUCTION

With respect to the classical Late Weichselian climatic fluctuations (Fig. 1) of the Netherlands and Western and Central Europe many scientists tried to establish similar chronostratigraphies of climatic variations for other parts of the world (e.g. Chile, Western North America, New Zealand, South Antarctica). Recent investigations give more and more evidence that the classical West European chronostratigraphy of the Late Weichselian in many details is a *regional* stratigraphy and cannot be transferred to other areas of the world.

Therefore it is necessary to focus our attention on many different parts of the world, especially on the tropics and subtropics in order to establish a great number of seperate chronostratigraphies. I will present here the results of investigations carried out during the last two decades in the central Mexican highland (Puebla/Tlaxcala area, Fig. 2) by many scientists which were involved in the German-Mexican "Mexico Project" of the DEUTSCHE FORSCHUNGSGEMEINSCHAFT.

The conclusions concerning the Late Quaternary climatic history of the central Mexican highland are based on the results of research in geomorphology, tephrochronology, palaeopedology, and palaeoclimatology.

I. CHRONOLOGY OF GLACIAL DEPOSITS, EASTERN CORDILLERA NEOVOLVANICA, MEXICO

In many papers I have reported on the late Quaternary chronostratigraphic situation of the central Mexican volcanoes

N.-A. Mörner and W. Karlén (eds.), Climatic Changes on a Yearly to Millennial Basis, 95–115.

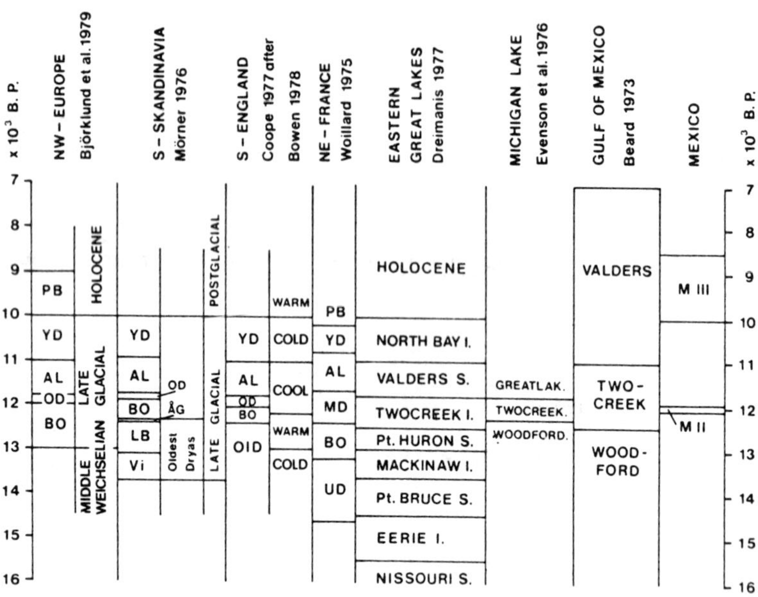

Fig. 1: Chronostratigraphy of the Late Weichselian according to various sources.
PB - Preboreal; YD - Younger Dryas; AL - Allerød;
OD - Older Dryas; MD - Middle Dryas; BO - Bølling;
I. - Interstadial; ÅG - Ågard; LB - Low Baltic;
Vi - Vintapper; OLD - Oldest Dryas; UD - Upper Dryas;
S. - Stadial

(Heine, 1975, 1978, 1980, 1983, in press). Volcanoes of the Cordillera Neovolcânica with late Quaternary glacial deposits are shown in Fig. 2. Intensive investigations concerning the glacial history of the volcanoes numbered 2 to 10 (Fig. 2) were carried out during 1971 - 1975. A summary of the palaeogeographic reconstruction of the eastern Cordillera Neovolcânica area for the time 40,000 yr BP to present is presented in Fig. 3.

The area of investigation comprises the eastern part of the volcanic belt (Cordillera Neovolcânica) that is composed of eroded late Tertiary and Quaternary volcanic and sedimentary rocks and extends for nearly 1000 kms along the 19th parallel north. The huge volcanic massifs constitute barriers which separate the valleys or basins from each other. Principal geographic features include five volcanic massifs (from west to east: Nevado de Toluca, Ajusco, Sierra Nevada with Popocatépetl, Iztaccîhuatl, Tláloc and Telapón, Malinche, and Pico de Orizaba with Cofre de Perote) and four basins (basin of Toluca, basin of Mexico, basin of Puebla/Tlaxcala, and basin of El Seco/Oriental). The elevations of the basins range from about 2200 to 2600 m, whereas the

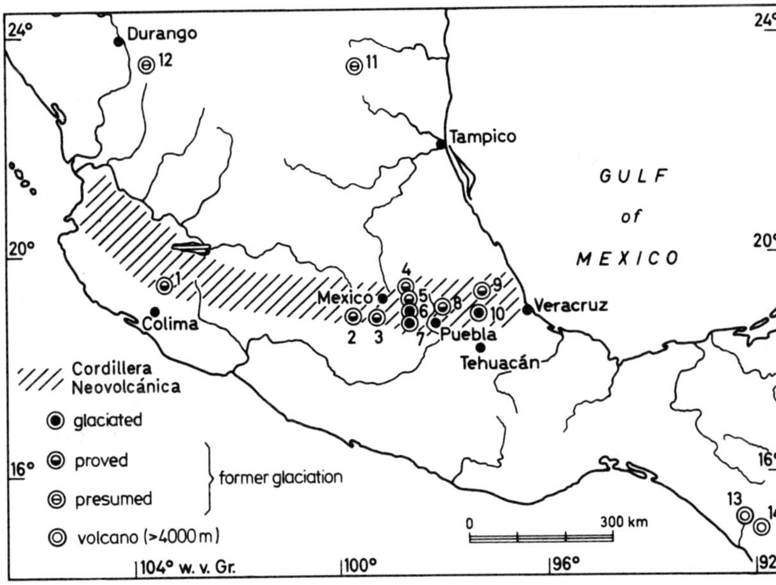

Fig. 2 : Index map of the Cordillera Neovolcânica, Mexico. Circles indicate the location of the volcanoes: 1 - Nevado de Colima, 2 - Nevado de Toluca, 3 - Ajusco, 4 - Tlâloc, 5 - Telapôn, 6 - Iztaccîhuatl, 7 - Popocatêpetl, 8 - Malinche, 9 - Cofre de Perote, 10 - Pico de Orizaba, 11 - Cerro Peña Nevada, 12 - Durango mountaneous area, 13 - Tacanâ, 14 - Tajumulco.

peaks of the volcanoes have heights between 3952 m (Ajusco) and 5700 m (Pico de Orizaba).

Different late Quaternary stratigraphic successions can be recognized within the area. The slopes of the volcanoes are dissected by barrancas (erosion gullies), radiating from the upper parts of the forest belt and descending to the basins. Thus the stratigraphic successions can easily be traced from one barranca to the other by the different layers of glacial and periglacial deposits, palaeosols, debris, fluvial gravels and sands, and loess-like so-called "toba"-sediments which are interbedded with tephra, lava flows, and ignimbrite deposits. The term "tephra" pertains to all pyroclastic fragments, such as fine and coarse ash, lapilli, volcanic bombs, and blocks.

Different tephra layers as well as some palaeosols are of great use as stratigraphic markers within the late Quaternary deposits of the volcanoes. In determining the stratigraphic succession of tephra, both field characteristics and laboratory examination were carried out. Radiocarbon dating of tephra layers has been restricted to charcoal logs and branches imbedded in the

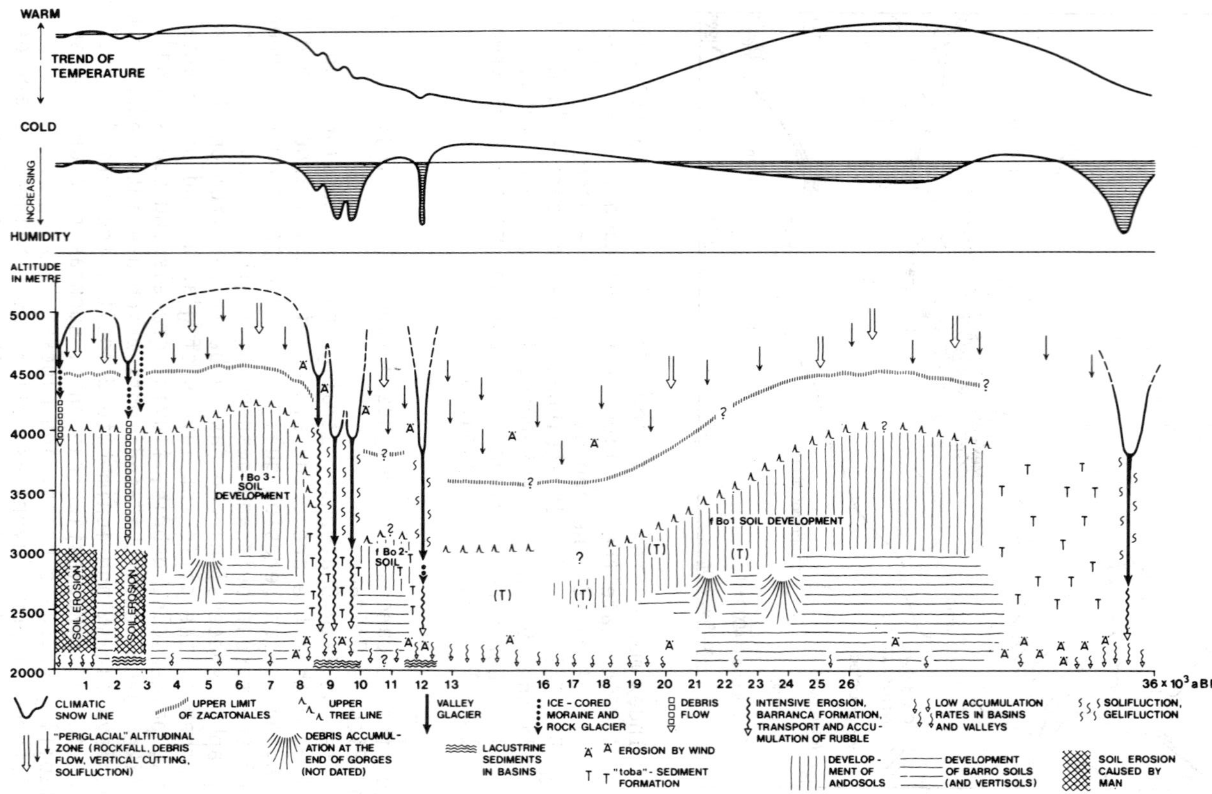

Fig. 3: Scheme of three-dimensional development of the natural environment of the central Mexican highland during the Late Quaternary.

volcanic deposits. Many samples from charcoal were dated by M.A. Geyh. In addition to these data palaeosols, wood fragments from gravels, peat, and calcrete (caliche) deposits were dated by radiocarbon. Other age determinations of the late Quaternary deposits are being attempted through studies in archaeology and prehistory. Pollen studies in cores of sediments in small volcanic craters, maars, and lakes (Ohngemach & Straka, 1983) cover the time span of approximately 35,000 to 0 yr BP. The results are compared with the chronology of the glacial deposits. Furthermore, relative age-dating methods have been used to demonstrate age differences in the till sequence; such relative dating methods include topographic position, morphologic shape of the moraines, rock-weathering parameters, thickness of aeolian "toba"-sediments, soil properties, and vegetation cover (including lichen data of Holocene glacial and periglacial deposits).

A scheme of the three-dimensional development of the natural environment of the central Mexican highland during the late Quaternary is given in Fig. 3. Periods of normal and of catastrophic processes, stability, and erosion during the past 36,000 yr BP are immediately recognizable. The processes which are controlled by the bioclimatic milieu create distinctive soils and landforms. During the past 36,000 yr BP three major periods with climatically induced high erosion intensities and glacier advances can be distinguished: (i) 36,000 to >32,000 yr BP, (ii) around 12,000 yr BP, and (iii) 10,000 to 8,500 yr BP. These periods with high erosion intensities and glacier advances coincide with climatic changes from relative aridity to greater humidity. Figure 3 shows that there is no synchroneous development of the trend of the temperature on the one hand and of the trend of the humidity on the other hand. The temperature curve for the last 36,000 yr BP is marked in the tropics of the central Mexican highland by an increase between 36,000 and 32,000 yr BP, a decrease between ca. 26,000 and ca. 16,000 yr BP (with the last glacial temperature minimum around 17,000 - 16,000 yr BP), and an increase of the temperature with minor fluctuations during the period 16,000 to 8,000 yr BP. The postglacial climatic optimum is reached 8,000 to 5,000 yr BP. During the Holocene, only between ca. 2,000 and 3,000 yr BP the climate was slightly cooler and wetter, producing a minor glacier advance.

Chronology of Late Weichselian Glacial Deposits

The chronology of the Late Weichselian glacial deposits is best known at the Malinche volcano. There, most of the late glacial tills are related to volcanic rocks and tephra as well as to organic materials that can be dated by radiocarbon. By means of tephrochronologic methods valley-to-valley correlation of tills is possible. The general distribution and dimensions of the late Quaternary glaciers of the Malinche volcano and problems associated with the identification and differentiation of each glacier advance and their age relationship were discussed by the author in previous papers (Heine, 1975, 1978, 1980, 1983). Tills with poorly preserved moraine forms are those of the M I-glacier advance between 36,000 and >32,000 yr BP and of tills deposited during the M II-glaciation about 12,000 yr BP. Lateral and end moraines deposited during the M III-glaciation between 10,000 and 8,500 yr BP show well preserved morainal forms, so do the Holocene neoglacial deposits which are divisible into two advances (M IV: 2000 - 3000 yr BP; M V: Little Ice Age). A map of the summit area of the Malinche volcano with the locations of the principal glacial and periglacial deposits is presented in Fig. 4. An outline of the local stratigraphy of the glacial deposits at the eastern slopes of the Malinche volcano (Fig. 5), based mainly on palaeosols (fBo1, fBo2, fBo3) and tephra layers (marker horizon

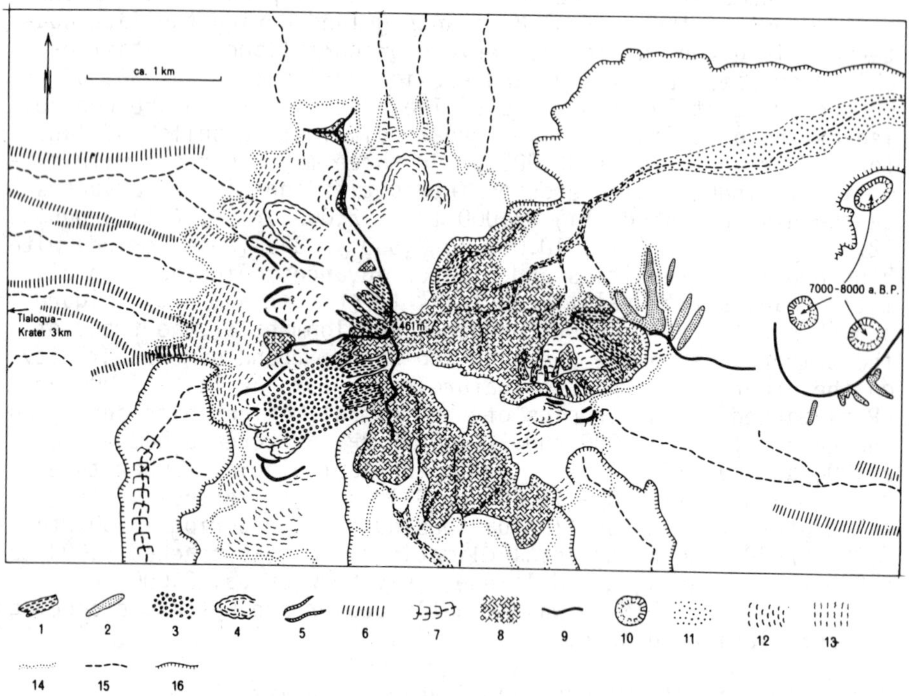

Fig. 4: Geomorphological map of the Malinche summit area. 1 - Talus debris, 2 - slope with volcanic sands, 3 - rock glacier, 4 - M IV-rock glacier, 5 - M III 3-moraine, 6 - M III 2- and M III 1-moraine, 7 - U-shaped valley, 8 - rock, 9 - crest, 10 - crater (age in ^{14}C years BP), 11 - gravel, 12 - solifluction terracette, 13 - turf exfoliation, 14 - upper timber line (ca. 3900 - 4000 m), 15 - valley, barranca, 16 - edge versus valley.

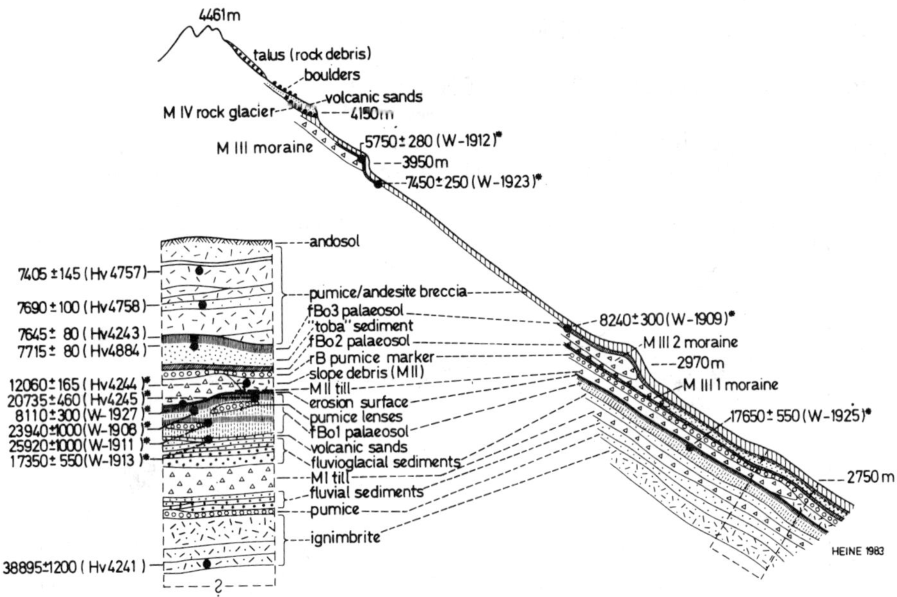

Fig. 5: Scheme of the stratigraphy of the Late Quaternary deposits at the eastern slopes of the Malinche volcano. The ^{14}C dates marked with stars refer to the western flank of the volcano.

rB), provide sufficient information of how different late Quaternary stratigraphic successions have been elaborated.

From the other volcanoes of the Cordillera Neovolcánica, deposits belonging to the major glaciations of the late Quaternary are described in detail by Heine (1975). Figures 6 to 8 give evidence of the location of the glacial and periglacial deposits on Pico de Orizaba, the southern part of Iztaccîhuatl, and the Nevado de Toluca volcano. Comparable to the late Quaternary stratigraphic successions of the Malinche volcano are the glacial de posits of these other volcanoes. Apart from the M II-glaciation that did not exist at the Nevado de Toluca volcano (see Heine, 1983), five glacier advances of the late Quaternary can be traced on the slopes of Pico de Orizaba and Iztaccîhuatl, whereas on the flanks of the less high volcanoes of Nevado de Toluca and Malinche the Holocene glacial deposits of the M IV and M V phase do not occur. Instead of the glacial deposits periglacial forms and deposits developed here (rock glaciers, ice-cored moraines, rock debris).

In the correlation diagramm (Fig. 9) most geologic-climatic unit boundaries are based either on maximum or minimum radiocarbon dates or on relative criteria. In most cases radiometric controls are very reliable and the presented boundaries will likely not be shifted on the time bar.

The Period about 16,000 to 12,000 yr BP

During the period of 16,000 to 12,000 yr BP there were no major glacier advances on the central Mexican volcanoes. The soil deve-

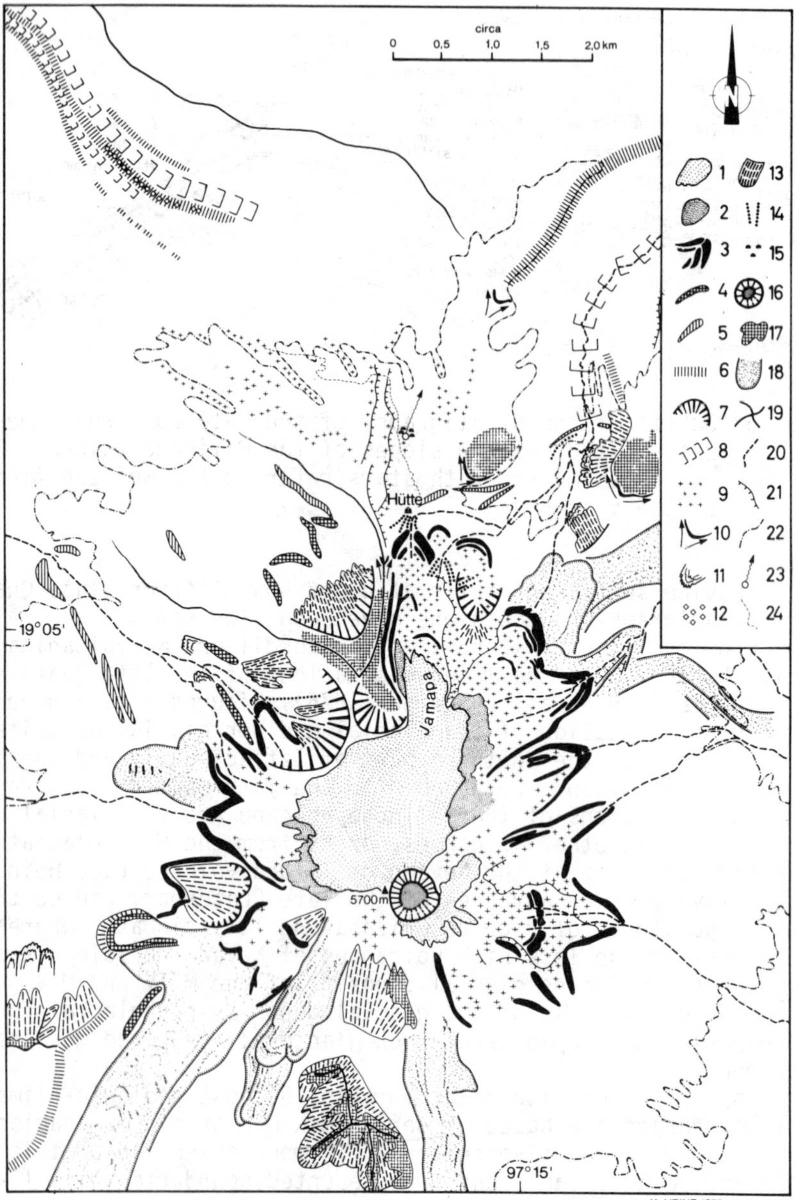

lopment of the palaeosol fBo1 terminated about 16,000 yr BP according to radiocarbon dates of charcial from pumice layers

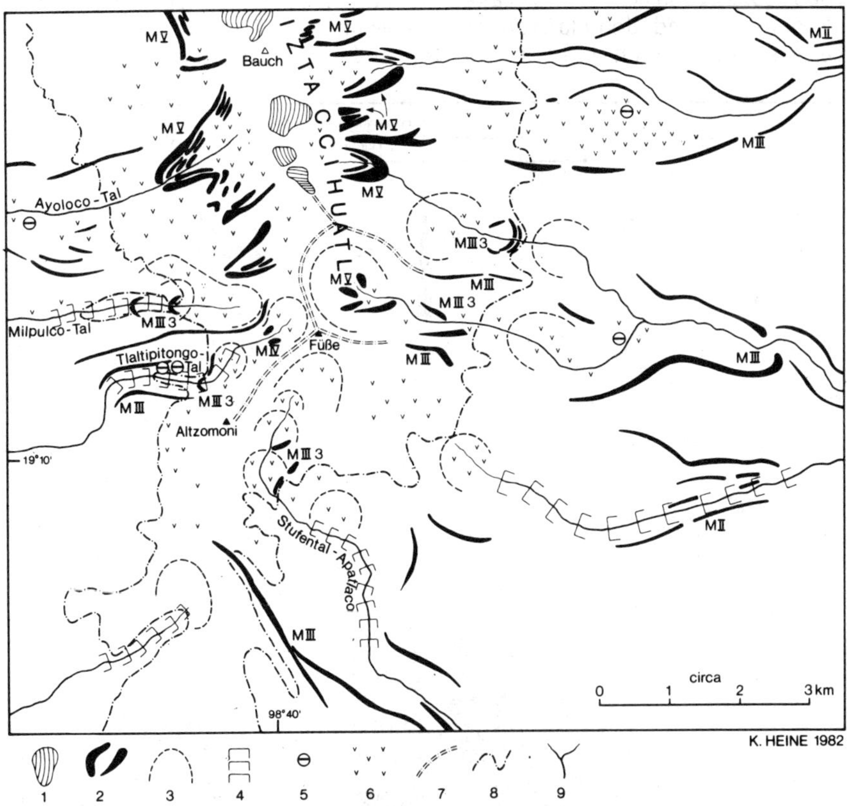

Fig. 7: Geomorphological map of the southern part of the Iztaccíhuatl massif. 1 - glacier, 2 - M V-moraine, 3 - cirque, 4 - U-shaped valley, 5 - striated ground, 6 - roche moutonée, 7 - crest, 8 - upper timber line, 9 - brook (periodic).

Fig. 6: Geomorphological map of the Pico de Orizaba. 1 - glacier, 2 - fossil ice beneath debris, 3 - M V-moraine, 4 - M IV-moraine, 5 - M III 3-moraine, 6 - M III 2- and M III 1-moraine, 7 - cirque, 8 - U-shaped valley, 9 - roche moutonée, 10 - ice-free crest during maximum glaciation, 11 - rock glacier (Little Ice Age), 12 - striated blocks on moraine, 13 - talus, 14 - debris flows, 15 - thufur, 16 - crater, 17 - steep walls, 18 - lava flow, 19 - crest, 20 - valley, barranca, 21 - edge versus valley, 22 - upper timber line (ca. 4000 m), 23 - well with brook, 24 - path.

covering the A_h-horizon of the palaeosol fBo1. No traces of erosion on the slopes of the volcanoes are visible. If the preservation of the topsoil of the palaeosol fBo1 during the period 16,000 to 12,000 yr BP is taken to be an evidence for relatively weak erosion and denudation on the mountain flanks, then the climate must have been relatively dry and cool compared with today.

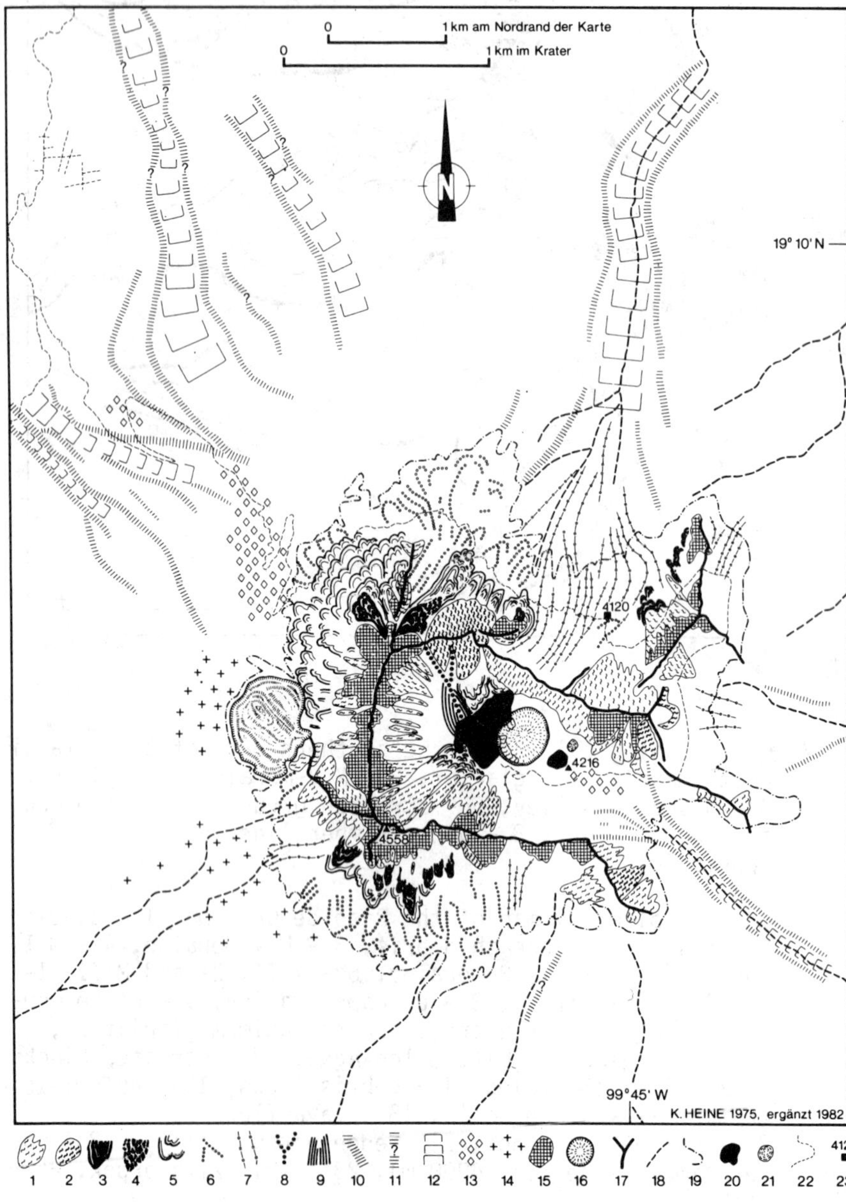

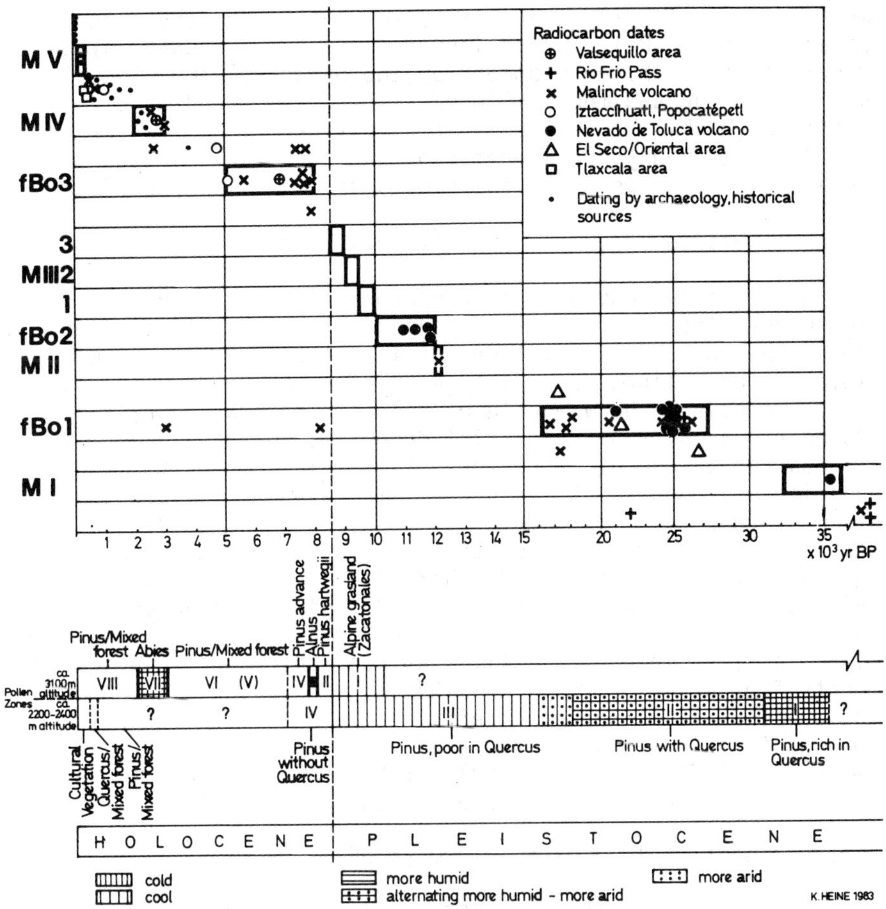

Fig. 9: Correlation diagram. Geologic-climatic unit boundaries are based on radiocarbon ages, tephrochronologic correlations, soil development, pollenanalyses, sedimentologic criteria, and topographic position. The pollen zones refer to Ohngemach & Straka (1983).

Fig. 8: Geomorphological map of the Nevado de Toluca volcano. 1 - talus, 2 - 'Glatthang' (straight smooth slope), 3 - rock glacier (phase I), 4 - rock glacier (phase I), 5 - rock glacier (phase II), 6 - rock glacier (phase III), 7 - valleys in debris, 8 - debris flow, 9 - debris flow lobe, 10 - M III-moraine, 11 - M III-moraine, not clearly identified, 12 - U-shaped valley, 13 - till, 14 - roche moutonée, 15 - rock, 16 - volcanic plug, 17 - crest, 18 - valley, barranca, 19 - upper timber line (ca. 4000 m), 20 - crater lake, 21 - dry lake, 22 - road, 23 - hut with altitude in meter.

Relative dry conditions are indicated by the preservation of the palaeosol fBo1. Relative cool conditions are indicated by the termination of the development of the palaeosol fBo1, because the soil formation and the soil properties only depend on climate and altitude (Miehlich, 1974). It is significant that the period 16,000 to 12,000 yr BP is not represented by any radiocarbon date obtained from deposits of the slopes of the volcanoes (Fig. 9).

The M II-Glacier Advance at about 12,100 yr BP

The M II-glacier advance occured around 12,100 yr BP, but only on the high volcanoes east of the basin of Mexico M II-moraines extend down to lowest elevations of less than 2750 m in huge valleys of the Malinche and Pico de Orizaba. Above 3000 to 3300 m they are covered by a mantle of volcanic tephra, debris, and younger glacial and periglacial deposits, so that the erosion of the barrancas do not reach the M II-morainic material; therefore nothing is known about the upper limit of M II-tills on the mountain slopes. The M II-glacial episode is characterized by a short duration of presumably less than 200 years according to investigations of the moraines and other glacial, periglacial, and fluvial sediments (Heine, 1983). The M II-glacier advance was caused by intense precipitation rather than by a temperature decrease. This increase in precipitation has been observed in various sediments from sites of the Sierra Nevada, the Malinche, and the Pico de Orizaba. It is not represented by any deposits east of the basin of Mexico, e.g. at the Nevado de Toluca, which I believe marks the influence of the Gulf of Mexico in respect to the short but intense increase in precipitation around 12,100 yr BP.

The Period from 12,000 to 10,000 yr BP

The period between about 12,000 and 10,000 yr BP is characterized by soil formation on the slopes of the high volcanoes. The palaeosol fBo2 was developed. According to the altitudinal belt of the formation of the palaeosol fBo2 the climate must have been relatively cool compared with today.

The M III-Glacier Advances between 10,000 and 8,500 yr BP

During the period between 10,000 and about 8,500 yr BP the complex of the M III-moraines were formed. The first and second advances of the M III-glaciation led to the formation of huge lateral and end moraines (see Fig. 4 - 8). These moraines extend from below the present tree line about 4000 m down to around 2900 m (first advance: M III 1) and 3000 m (second advance: M III 2) respectively. Recessional moraines of the M III-glaciation (M III 3) are found at about 4000 m altitude at the Malinche, whereas moraines of this advance reach farther downvalley on the flanks of the Sierra Nevada and the Pico de Orizaba, because of the greater extent of the zone of net accumulation above the firn line during the M III 3-glacier advance (Fig. 10). It is interesting to note that during the M III 3-glaciation the formation

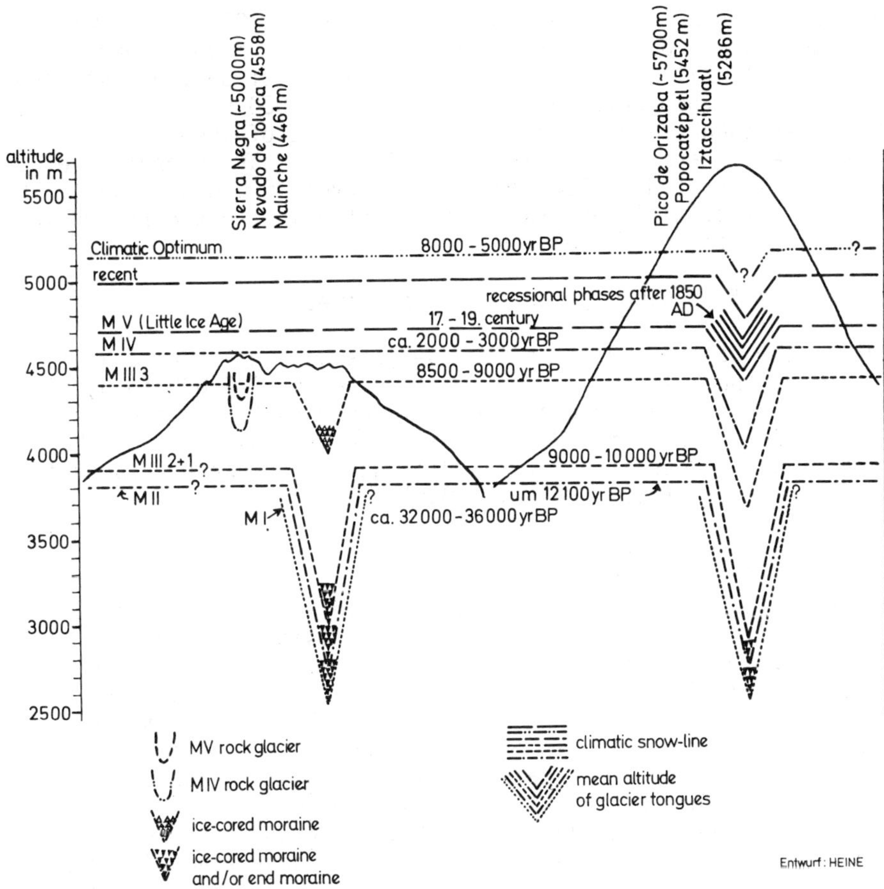

Fig. 10: Climatic snow lines reconstructed for the different Late Quaternary glacier advances.

of ice-cored moraines was very common on the Nevado de Toluca (Fig. 8) (Heine, 1976). The M III-glacier advances reflect a lowering of the climatic snowline during the period 10,000 - 9,000 yr BP of approximately 800 m in reference to the snowline of 1850 AD.

Since about 9,000 yr BP the mean annual temperatures increased so much that the relationship between temperature and precipitation led to a rapid shift of the climatic snowline to higher elevations. Only a few recessional moraines give evidence of minor glacier fluctuations during the period 9,000 to 8,500 yr BP.

The Period from 8,500 to 5,000 yr BP

The rapid deglaciation terminated ca. 8,500 yr BP. According to the radiocarbon dates of the palaeosol fBo3 that developed either on the M-III-tills and/or on volcanic tephra of post-Pleistocene age (Fig. 5) the period between 8,000 and 5,000 yr BP was slightly warmer than recent times; the development of andosols took place even in areas up to 4200 m altitude (today andosol development is restricted by low temperatures over 4000 m altitude). The postglacial climatic optimum is documented in the central Mexican highland by the palaeosol fBo3.

The Period from 5,000 yr BP to Recent Times

During the last 5,000 years, two periods with glacier advances occurred. The M IV-glaciation is dated to about 2,000 to 3,000 yr BP. A minor glaciation can be correlated with the Little Ice Age. Since about 1850 AD to 1972, the ice caps and glaciers of the Iztaccîhuatl, Popocatépetl, and Pico de Orizaba were reduced to small remnants of ice (Heine, 1983b).

Palynologic Analyses

Palynologic, sedimentary, and radiocarbon analyses of gyttja, clay, sand, ash, lake marl, and other deposits, formed in small craters and maars, provide a palaeoenvironmental and palaeoclimatic record of approximately 35,000 yr BP for the eastern Cordillera Neovolcánica (Ohngemach & Straka, 1983; Heine & Ohngemach, 1976). Of special interest is a pollen sequence of the Tlaloqua crater which is situated at an altitude of about 3100 m on the western slope of the Malinche volcano (Fig. 11).

During the M III-glaciation (lower part of the diagram, Fig. 11) the upper zone of the forest certainly consisted of pines just as today, for pine pollen mostly accounts for more than 80 % of the pollen during the time rich in NAP. *Alnus* and *Quercus* play a comparatively minor part (Ohngemach 1977). Of particular interest is the occurrence of *Picea*, a genus encountered in Mexico today only at a few relict stations. The nearest site of *Picea chihuahuana* is situated 900 km north-west of the Malinche volcano, *Picea mexicana* is found some 700 km to the north. The pollen of *Picea* occurs below about 2 m in every sample, above the 2 m level, *Picea* is no longer found. The extinction of *Picea* coincides with the rapid decrease of the NAP curve (Ohngemach, 1977). According to Heine (1975, 1980, 1983a; see Fig. 3) the climate was cooler and more humid than today during the M III-glacial advances. Such a climate may have been the precondition for the former presence of spruce at the Malinche volcano and the basin of El

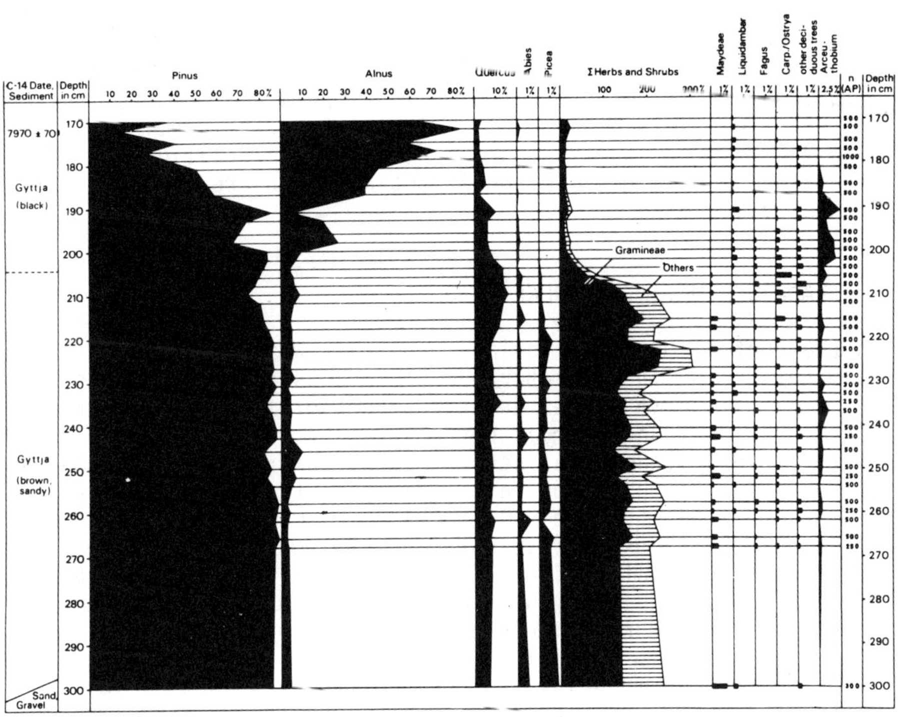

Fig. 11: Pollen diagram of the Pleistocene/Holocene boundary at the Malinche volcano, Mexico (after Heine & Ohngemach, 1976 and Ohngemach & Straka, 1983).

Seco/Oriental (Ohngemach & Straka, 1983). When the climate became drier and warmer after the M III-glaciation, *Picea* was not able to persist any longer and dies out (Ohngemach, 1977; Heine & Ohngemach, 1976). The level marked by the extinction of *Picea* is taken as the boundary between Pleistocene and Holocene vegetation. Provided that the diagram (Fig. 11) starts shortly after the beginning of M III-glaciation at 10,000 yr BP, one reaches an age of some 8,500 yr BP for this transition, if one takes further into account a radiocarbon date (ca. 8,000 yr BP) for the *Alnus* maximum at 1.7 m together with the supposition of a constant sedimentation rate (Ohngemach 1977). This agrees well with the fixation of the Pleistocene/Holocene boundary based on geomorphologic, palaeopedologic, and palaeontologic evidence (Heine & Ohngemach, 1976).

The region around the Tlaloqua crater was only a transit station to higher elevations for the *Pinus hartwegii* forest that succeeded the late Weichselian alpine grassland after about 8,500 yr BP (see also Fig. 9). Ohngemach (1977) and Ohngemach & Straka (1983) infer the presence of *Pinus hartwegii* from the occurrence of the parasite *Arceuthobium globosum* ssp. *graudicaule* by which *Pinus hartwegii* is preferably infected.

According to the pollen diagrams of Ohngemach & Straka (1983) no noticible deterioration in temperature occurred during the Late Weichselian and early Holocene trend of warming (16,000 to 8,000 yr BP). On the other hand, the M I-glaciation, the M III-glacier advances, and the M IV-glaciation are well recorded in the pollen diagrams.

II. PALAEOCLIMATIC INTERPRETATION

In the central Mexican highland, the Late Weichselian (ca. 16,000 to 8,500 yr BP) is characterized by increasing temperatures without the well-known sudden deteriorations of temperature during the Older and Younger Dryas periods. Apart from the 12,000 yr BP event (Heine, 1983a) that seems to be triggered by an enormeous increase in precipitation rates mainly in the regions facing the Gulf of Mexico-coast, only the period 10,000 to 8,500 yr BP shows climatic conditions favorable for greater glaciations on the high mountains. This Late Weichselian climatic fluctuation is caused by higher precipitation during a period with increasing temperatures between the relatively cold phase of oxygen isotope stage 2 and the warm postglacial climatic optimum of oxygen isotope stage 1. The palynologic evidence indicates, too, that the Pleistocene/Holocene boundary is dated to about 8,500 yr BP in the central Mexican Highland. Different pollen diagrams from the area prove that no abrupt climatic fluctuations in temperature comparable to the Older and Younger Dryas periods occurred. Even the 12,000 yr BP event cannot be traced in the pollen diagrams, presumably because of the short duration of the 12,000 yr BP event of less than 200 years (Heine, 1983a).

The classical Late Weichselian climatic fluctuations of the central Mexican highland are (i) with respect to the temperature; increasing temperatures without any major variations and/or abrupt fluctuations, and (ii) with respect to the precipitation: a short but great increase of precipitation about 12,000 yr BP and a period with higher precipitation between 10,000 and 8,500 yr BP. Yet, until today we cannot estimate the absolute amount of precipitation during the various Late Weichselian glaciations. The pollen studies show that the Late Weichselian fluctuations in precipitation in the basins must have been of no significant influence for the vegetation. We might conclude, because of this, and because of the shrinkage of the glaciers since the Little Ice Age maximum glacier extent in 1850 AD that took place without any significant variations in temperature and/or precipitation, that the Late Weichselian glaciations were due to only slightly altered conditions in humidity.

III. CORRELATION WITH LATE QUATERNARY EVENTS IN OTHER REGIONS

The late Quaternary environmental history of Lake Valencia/Venezuela (Bradbury et al., 1981) obtained by chemical, palaeontological, and mineralogical analyses of a 7.5 m core from the lake allows a tentative correlation with the situation in central Mexico. The data show that dry climates existed in the Lake Valencia region from at least 13,000 yr BP (= the base of the core) until about 10,000 yr BP. The Lake Valencia Basin was occupied by intermittent saline marshes at that time. During the period 13,000 to 10,000 yr BP only the pollen record gives evidence for a minor increase in moisture by a small amount of arboreal pollen (*Spondias* and *Bursera*) at a depth of 6.4 m (Bradbury et al., 1981). If one takes into account a constant sedimentation rate of the 13,000 to 10,000 yr BP deposits, the increase in moisture in what otherwise was an arid climatic period can be compared to the M II-glaciation of Mexico. About 10,000 yr BP, a permanent Lake Valencia of fluctuating salinity formed and arboreal plant communities replaced the earlier dominant xeric herbaceous vegetation and marsh plants. This zone 2 of the Lake Valencia core correlates with the M III-glaciation period of central Mexico. By 8,500 yr BP, Lake Valencia reached moderate to low salinities and discharged water; the modern vegetation became established at that time. After 8,500 yr BP, the lake twice ceased discharging as a result of reduced watershed moisture (Bradbury et al., 1981). The stratigraphic record from Lake Valencia is important because it contains a continuous, well-dated, and well-documented palaeoenvironmental history of a low-latitude, low elevation site from the late Pleistocene to the present. Table 1 is a tentative correlation of palaeoenvironmental interpretations from the Lake Valencia record with the Mexican data.

The most important conclusions from Table 1 are the coincidence of the central Mexican late Quaternary chronostratigraphy

yr BP x 10³	Central Mexico this paper	Lake Valencia Bradbury et al. 1981	
1	M V	zone 3	Climate generally similar to that of today
2	M IV		
3			
4			
5			
6			
7			
8			
9	M III	zone 2	Humid
10			
11		zone 1	More arid than today
12	M II		More moisture
13		zone 1 ? ? ?	More arid than today
14			

Table 1: Tentative correlation of palaeoenvironmental interpretations from the Lake Valencia record with the Mexican data

with the Lake Valencia core and the evidence that the Younger Dryas period did not exist, neither in Mexico nor in Venezuela.

The stratigraphic records from Mexico and Venezuela cannot be correlated with late Quaternary glaciations in Central and South America; the principal difficulty is the lack of adequate radiocarbon time control in most studies (Bradbury et al., 1981; Heine, 1975, 1983c). According to Mercer (1983), in the South American tropics, interpretations of past variations of glaciers in terms of climate is difficult; owing to the intensity of solar radiation year-round, the glaciers are very sensitive to changes in cloudiness and albedo of the glacier surface. Mercer (1983) reports that in Peru by 12,200 yr BP the Quelccaya Icecap was not much larger than it is today and that a later readvance that was in progress soon after 11,500 yr BP culminated after 11,000 yr BP. The icecap then shrank rapidly and shortly after 10,000 yr BP was little or no larger than it is today. The dating of the readvance after 11,500 yr BP suggests that it preceded the European Younger Dryas stade by a few centuries, but it does not preclude the possibility that they were simultaneous (Mercer, 1983). In southern South America no glacial geological evidence has been found anywhere for a readvance at the time of the Younger Dryas period (Mercer, 1983), although severe cooling and a rise in precipitation have been inferred from the pollen record (Heusser et. al., 1981). The analyses of fossil beetle assemblages through the Late Glacial of southern Chile implies a relative stable climate, too, and does not support, like the geomorphological evidence, the assumption of a major climatic deterioration about the same time as the Younger Dryas in Europe (Ashworth & Hoganson, 1983).

IV. CONCLUSIONS

The palaeonvironmental data from Mexico, Venezuela, and Chile support the hypothesis that Late Weichselian climates in the tropics of America were more arid than today. The classical Late Weichselian climatic fluctuations of the Older and Younger Dryas periods did not occur. There is evidence that these Late Weichselian climatic deteriorations are best documented in Scandinavia (Berglund et al., 1983; Berglund & Mörner, 1983) although even for middle Sweden Björck & Digerfeldt (1983) postulate that either many significant events occurred during Younger Dryas or hardly nothing at all, perhaps not even the 'famous' drainage of the Baltic Ice Lake at the northern point of Mt. Billingen. If we move away from Scandinavia to the Alps then the Older Dryas nearly disappers (de Beaulieu et al., 1983; Ammann et al., 1983) and the Younger Dryas cooling cannot have been very strong (Ammann et al., 1983). Observations from the Alps and Switzerland show that the Older Dryas stade can only be traced by 'reworked material' and not by pollen evidence (de Beaulieu et al., 1983) or as a somewhat 'drier phase' (Ammann et al., 1983). In the Alps, the Younger Dryas was more pronounced by palaeoenvironmental changes at greater heights, whereas on the Swiss Plateau the Younger Dryas climatic deterioration was less severe.

From the 'tropical point of view' we should not only look for evidence for the Older and Younger Dryas stades, but also for evidence for the non-existence of these climatic fluctuations. The low-latitude stratigraphies experience a Late Weichselian transition from full glacial dry and cold/cool climate to postglacial conditions *without* significant cold temperature anomalies.

ACKNOWLEDGEMENTS

The author is grateful to M. A. Geyh for radiocarbon age determinations, H. Heide-Weise for mineralogical analyses, and the DEUTSCHE FORSCHUNGSGEMEINSCHAFT for financial support.

REFERENCES

Ammann, B., Chaix, L., Eicher, U., Elias, S., Gaillard, M.-J., Hofmann, W., Siegenthaler, U., Tobolski, K. & Wilkinson, B., 1983. Vegetation, insects, molluscs and oxygen isotopes in Late-Würm deposits at Lobsigensee (Swiss Plateau). Abstracts 2nd Nordic Symp., Stockholm May 16 - 20, 1983, Danish Meteor. Inst., Kopenhagn, pp. 2-3.

Ashworth, A. C. & Hoganson, J. W., 1983. Was there a climatic change in southern Chile about the same time as the Younger Dryas in Europe? Abstracts Intern. Symp. on Late Cainozoic Palaeoclimates of the Southern Hemisphere, Swaziland 29 Aug. - 2 Sept. 1983, p. 31.

Beard, J. H., 1973. Pleistocene-Holocene Boundary and Wisconsinan Substages, Gulf of Mexico. Geol. Soc. amer. Memoir, 136, 277-316.

Berglund, B. E., Dearing, J., Lemdahl, G., Liedberg-Jönsson, B. & Persson, T., 1983. The Kullen Peninsula in SW Sweden - a palaeoecological reference site for Late Weichselian. Abstracts 2nd Nordic Symp., Stockholm May 16 - 20, 1983, Danish Meteor. Inst., Kopenhagn, pp. 14-15.

Berglund, B. E. & Mörner, N.-A., 1983. Late Weichselian deglaciation and chronostratigraphy of Southern Scandinavia: problems and present "state of art". Abstracts 2nd Nordic Symp., Stockholm May 16 - 20, 1983, Danish Meteor. Inst., Kopenhagn, pp. 9-13.

Björck, S. & Digerfeldt, G., 1983. Shore displacement and deglaciation chronology around the boundary Pleistocene-Holocene in the Middle Swedish endmoraine zone. Abstracts, 2nd Nordic Symp., Stockholm May 16 - 20, 1983, Danish Meteor. Inst., Kopenhagn, pp. 17-19.

Björklund, K. R. & Goll, R. M., 1979. Ice Age Climates of the

Norwegian-Greenland Sea. GeoJournal, 3.3, 273-286.
Bowen, D. Q., 1978. Quaternary Geology. A Stratigraphic Framework for Multidisciplinary Work. Oxford, pp. 1-221, Pergamon.
Bradbury, J. P., Leyden, B., Salgado-Labouriau, M., Lewis, W. M., Schubert, C., Binford, M. W., Frey, D. G., Whitehead, D. R. & Weibezahn, F. H., 1981. Late Quaternary Environmental History of Lake Valencia, Venezuela. Science, 214, 1299-1305.
de Beaulieu, J.-L., Clerc, J. & Reille, M., 1983. Late Weichselian fluctuations in the French Alps and Massif Central from pollenanalyses. Abstracts, 2nd Nordic Symp., Stockholm May 16 - 20, 1983, Danish Meteor. Inst., Kopenhagn, p. 8.
Dreimanis, A., 1977. Late Wisconsin Glacial Retreat in the Great Lakes Region, North America. Ann. New York Acad. Sciences, 288, 70-89.
Evenson, E. B., Farrand, W. R., Eschmann, D. F., Mickelson, D. M. & Maher, L. J., 1976. Greatlakean Substage: A Replacement for Valderan Substage in the Lake Michigan Basin. Quaternary Research, 6, 411-424.
Heine, K., 1975. Studien zur jungquartären Glazialmorphologie mexikanischer Vulkane - mit einem Ausblick auf die Klimaentwicklung. Das Mexiko-Projekt der DFG VII, Wiesbaden, pp. 1-178, Steiner.
Heine, K., 1976. Blockgletscher- und Blockzungen-Generationen am Nevado de Toluca, Mexiko. Die Erde, 107, 330-352.
Heine, K., 1978. Neue Beobachtungen zur Chronostratigraphie der mittelwisconsinzeitlichen Vergletscherungen und Böden mexikanischer Vulkane. Eiszeitalter u. Gegenwart, 28, 139-147.
Heine, K., 1980. Quartäre Pluvialzeiten und klimamorphologischer Formenwandel in den Randtropen (Mexiko, Kalahari). Arb. Geogr. Inst. Univ. Saarland (Festschr. C. Rathjens), 29, 135-157.
Heine, K., 1983a. Ein aussergewöhnlicher Gletschervorstoss in Mexiko vor 12,000 Jahren. Ein Beitrag zum Problem der spätglazialen Klimaschwankungen. Catena, 10, 1-25.
Heine, K., 1983b. Mesoformen der Periglazialstufe der semihumiden Randtropen, dargestellt an Beispielen der Cordillera Neovolcánica, Mexiko. Abh. Akad. Wiss. Göttingen, Math.-Phys. Kl. (in press).
Heine, K., 1983c. Führt die Quartärforschung zu nicht-aktualistischen Modellvorstellungen in der Geomorphologie? Colloquium Geographicum, 16, 93-121.
Heine, K., in press. Spät- und postglaziale Gletscherschwankungen in Mexiko: Befunde und paläoklimatische Deutung. In: Late- and postglacial oscillations of glaciers: Glacial and periglacial forms (ed. H. Schroeder-Lanz), in press, Balkema.
Heine, K. & Ohngemach, D., 1976. Die Pleistozän/Holozän-Grenze in Mexiko. Münster. Forsch. Geol. Paläont., 38/39, 229-251.
Heusser, C. J., Streeter, S. S. & Stuiver, M., 1981. Temperature and precipitation record in southern Chile extended to

~43,000 yr ago. Nature, 294, 65-67.
Mercer, J. H., 1983. Late Cainozoic Glacier Variations in South America South of the Equator. Abstracts Intern. Symp. on Late Cainozoic Palaeoclimates of the Southern Hemisphere, Swaziland 29 Aug. - 2 Sept. 1983, pp. 19-21.
Mörner, N.-A., 1976. Paleoclimatic records from South Scandinavia, global correlations, origin and cyclicity. Paleolimnology of Lake Biwa and the Japanese Pleistocene, 4, (ed. S. Horie), pp. 499-528.
Miehlich, G., 1974. Klima- und altersabhängige Bodenentwicklung von Vulkanascheböden der Sierra Nevada de México. Mitt. Deutsch. Bodenkundl. Gesellsch., 18, 360-369.
Ohngemach, D., 1977. Pollen Sequence of the Tlaloqua Crater (La Malinche Volcano, Tlaxcala, Mexico). Bol. Soc. botan. México, 36, 33-40.
Ohngemach, D. & Straka, H., 1983. Beiträge zur Vegetations- und Klimageschichte im Gebiet von Puebla-Tlaxcala. Pollenanalysen im Mexiko-Projekt. Das Mexiko-Projekt der DFG XVIII, Wiesbaden, pp. 1-161 + Suppl., Steiner.
Woillard, G., 1975. Recherches palynologiques sur le Pleistocène dans l'Est de la Belgique et dans les Vosges Lorraines. Acta Geographica Lovanensia, 14, 1-168.

CLIMATIC VARIATIONS IN THE LOW CONTINENTAL LATITUDES DURING THE LAST 30,000 YEARS

Michel Servant

ORSTOM, France

1. VARIATIONS IN THE WATER BUDGET

Lake fluctuations and modifications in the tropical lowlands vegetation over the last 30,000 years suggest that the climate of intertropical and subtropical zones did not evolute in the same way in the different regions of Africa and America.

In the western part of America (Bolivian Andes, latitude 18°S, southwestern part of USA, 40°N), Holocene was characterized by low lacustrine levels and Upper Pleistocene by high levels with the exception of a short period before 14,000 years BP.

In the low North latitudes (Guyana, Venezuela, Ghana) and at the Equator (Galapagos), several high lacustrine levels had been observed during Holocene and low levels before 10,000 or 12,000 years BP. In Congo (about latitude 5°S) geomorphological studies show that the rain forest was reduced during the last glacial period and then has developed gradually since 12,000 or 10,000 years BP.

In the low South latitudes of America, rain forest was degraded during the main part of Holocene in Brazil (Para area 5°S) and in Bolivia (Santa Cruz area 18°S), then it developed after 2,000 years BP.

In the Sahelian zone of Africa (Chad, East Africa), the evolution of the lakes was more complicated. Holocene was characterized by two lacustrine maxima (8,000 and 6,000 years BP.) which can be observed within Sahara up to a latitude of 25° - 30°N. The whole lake basins were gradually affected by drought after 5,000 years BP. except a minor lacustrine phase dated from about 3,000 years BP. in the southern Sahara. In Upper Pleistocene, the lakes largely developed in Chad and East Africa, except during a short period of drastic drought at about 16,000 years BP. This dry period

N.-A. Mörner and W. Karlén (eds.), Climatic Changes on a Yearly to Millennial Basis, 117–120.

seems to have lasted until 10,000 years within Sahara (NE Niger, N Ethiopia, NW Sahara). But it has been suggested that some lakes still existed until Late Pleistocene in the Saharian mountains (Tibesti).

2. WATER FLOWS

The study of fluvial terraces in Saharian massifs (Air, Tibesti) shows that water flows were much more regular than at the present time during part of the last glaciation and early Holocene, from 20,000 years to about 7,000 years BP. These regular water flows, identified by silty and well-laminated deposits, were surely induced by non-stormy and annual well-distributed precipitation. Flood flows, similar to that of today characterized the last 7,000 years. They are identified by coarse fluvial sediments and/or active erosion. Similar modifications were observed in Bolivian Andes (latitude 16-18°S). However, flood flows occurred only during short periods in Upper Quaternary in this area at about 15,000, 10,000, 7,000 years BP. and during the last 1,500 years.

3. PALEOBIOGEOGRAPHICAL ANOMALIES

The last glacial lacustrine phases are characterized by the scarcity or absence of typically tropical diatoms and by the presence of several cold eusthenotherm species in Chad and less so in northern Ethiopia. In Chad, pollen assemblages were poor in tropical taxa especially at about 25,000 years BP. but saharian and mediterranean pollens were more abundant than presently in the same area. In South America, Patagonian vertebrates were discovered in Upper Quaternary deposits of the southern tropical zone, (Argentina). These anomalies suggest that certain species of the temperate flora and fauna moved towards the Equator during the last glaciation.

4. TROPICAL GLACIERS

Moraines dated from Upper Pleistocene have been observed in all tropical mountains. Still, little is known about the evolution of Quaternary glaciers. In Bolivia and southern Peru, some fifteen C14-dates allow to plot a tentative curve of the glacial fluctuations over the last 25,000 years. The last glacial maximum occurred before 14,500 years BP. and maybe after 16,600 years BP. The minor readvances of the Late Glacial do not seem to be synchronous with those prevailing in Europe. Two maxima of glaciar retreat have been identified; the first one at about 10,000 years BP. the second one between 2,700 and 1,600 years BP.

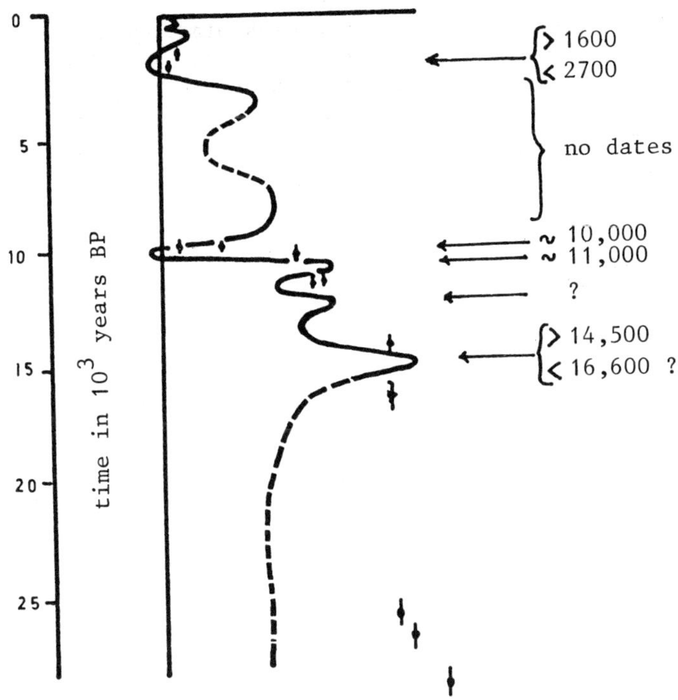

Fig. 1. Glacial fluctuations of the southern tropical Andes (Peru, Bolivia). Tentative curve.

5. PALEOCLIMATIC IMPLICATIONS

The tropical zones are actually characterized by summer and storms precipitation which are located in the east atmospheric flow. The non-seasonal and non-stormy precipitation which occurred during several periods of Upper Quaternary in the Bolivian Andes and in central saharian massifs indicate a main change in the mechanisms of precipitation and in the atmospheric circulations. Western atmospheric flows of middle latitudes, moving to the Equator during an hemispheric cooling could explain these modifications. Strong penetration of polar air masses to the low latitudes in the lower part of the atmosphere induced a drastic decrease of minimal temperatures. A more active displacement of equatorial air masses towards the middle latitudes could explain the annual well-distributed precipitation in the tropical zone especially in the tropical mountains.

The equatorial zone is actually characterized by high precipitation which are related to an active atmospheric ascendance. The reduction of rain forests and the low levels of the lakes, during the last Glacial period - and during a main part of Holocene in the south tropical regions of America - involve precipitations lower than to day and consequently a weak atmospheric ascendance. An active atmospheric subsidence seems to have occurred in the high layers of the low latitude atmosphere.

II. The Holocene

LATE-GLACIAL - HOLOCENE HIGH-FREQUENCY CLIMATIC CHANGES IN DEEP-SEA CORES FROM THE DENMARK STRAIT

Thomas B. Kellogg

SPECMAP, University of Maine at Orono, USA

ABSTRACT

Denmark Strait cores V28-14 and V29-206 contain records of high-frequency (periods of 615-784 yr) fluctuations in percent of subpolar planktonic foraminifera. The two principal species involved in the fluctuations are G. quinqueloba and G. bulloides, the former in V29-206 and the latter in V28-14. Variations in these species' relative and absolute abundance are directly related to variable dissolution and to changes in the boundary between the Irminger and East Greenland currents.

INTRODUCTION

The 275-km wide Denmark Strait separates Iceland from Greenland (Fig. 1). Water depths are generally less than 400 m, but reach maximum sill depths of 650 m (Coachman and Aagaard, 1974). Together with the Iceland-Faeroe Ridge, the sill in the Denmark Strait separates the deep basins of the North Atlantic Ocean from those in the Norwegian-Greenland Sea and Arctic Ocean, controlling bottom water flow between these regions. Despite their shallow sill depths, these straits are the most important marine connections between Arctic Seas and the remainder of the oceans.

Cores V28-14 and V29-206 were collected from critical locations for monitoring past changes in Denmark Strait water masses. Sediment, carbonate, and foraminiferal analyses of core V28-14 (64°47'N, 29°34'W from a water depth of 1855 m) showed that this 611-cm core penetrates into isotopic stage 6 or 7 (Kellogg, 1975, 1976; Kellogg et al., 1978). V29-206 (64°54.34'N, 29°17.37'W,

N.-A. Mörner and W. Karlén (eds.), Climatic Changes on a Yearly to Millennial Basis, 123–133.

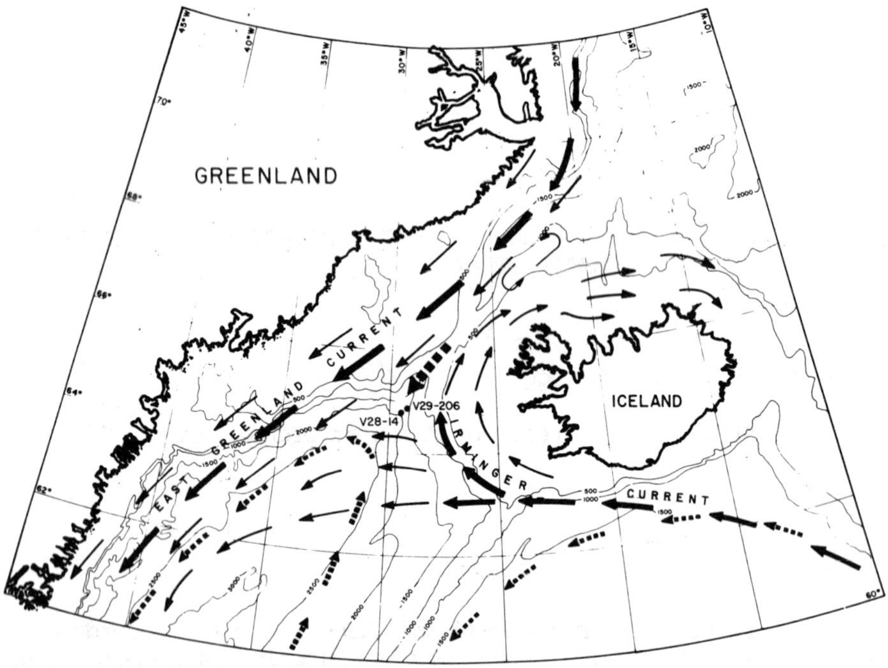

Figure 1. Map of the Denmark Strait Region showing bathymetry, surface currents (solid arrows), Norwegian Sea Overflow (dashed arrows), and locations of V28-14 and V29-206.

1080 cm) was collected at a location about 25 km from V28-14 but at a slightly shallower site (1608 m). Carbonate and sedimentologic analyses for the entire core (Kellogg, unpublished) showed that this core, like V28-14, penetrates into oxygen isotope stage 6 or 7. Both cores have nearly identical foraminiferal and sedimentologic variations above stage 5, and both have high sedimentation rates (>10 cm/10^3 yr) in stage 1.

Previous analyses of V28-14 carried out at 10-cm intervals showed that early Holocene conditions in the Denmark Strait were considerably warmer than those of today (Kellogg, 1975, Fig. 32). W. Karlén (oral communication, 1976) suggested that a closer sampling interval should be employed to clarify details of the Holocene portion of this core. Sampling was conducted at 2.5-cm intervals throughout the top 150 cm of V28-14, and at 5-cm intervals from 150 to 200 cm (below the isotope stage 2 maximum). Similar sampling intervals were used in analysis of V29-206 because it was deemed necessary to confirm that the high-frequency pattern of fluctuations of subpolar foraminifera in V28-14 was regional, and not a phenomenon confined to a single core.

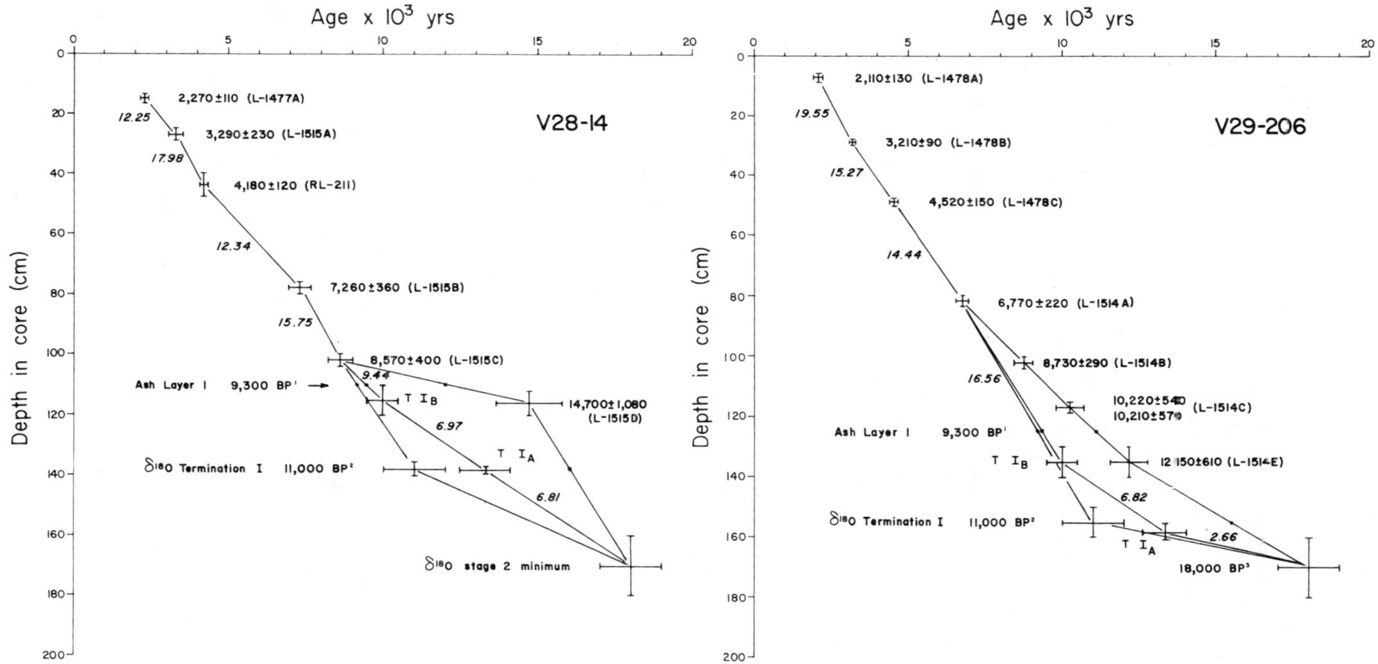

Figure 2. Age versus depth models for V28-14 and V29-206. Radiocarbon date and laboratory numbers given in boldface; sedimentation rates given in italics. Note three alternative age models below 8000 B.P.: (a) based on radiocarbon dates; (b) based on published ages of Isotope Termination I and Ash Layer 1; and (c) based on ages of Isotope Terminations I_A and I_B (see text).

STRATIGRAPHY

Stratigraphic control for cores V28-14 and V29-206 is based on carbonate curves, the positions of Ash Layer 1 (Ruddiman and Glover, 1972), oxygen-isotopic curves, and radiocarbon dates. Age versus depth plots of stratigraphic datum points for each core are shown in Fig. 2. Radiocarbon control in the top 80 to 100 cm of each core that indicates nearly linear sedimentation rates prevailed at both sites after about 8000 B.P. Prior to 8000 B.P., different stratigraphic interpretations are possible, depending on whether more reliance is placed on radiocarbon control or on the "known" ages of Termination I (11,000 B.P.: Broecker and Ku, 1969) and Ash Layer 1 (9000 B.P.: Ruddiman and Glover, 1972; 9300 B.P.: Ruddiman and McIntyre, 1973).

The three stratigraphic models considered are based on: (1) acceptance of all radiocarbon ages, including those >8000 B.P.; (2) acceptance of published ages for Isotope Termination I and Ash Layer 1; and (3) use of Isotope Terminations I_A and I_B (Duplessy *et al.*, 1981) for stratigraphic control in parts of the cores where maximum radiocarbon ages are suspected. The third model is considered most reliable, although the second cannot be ruled out at this time. Cores V28-14 and V29-206 contain considerable amounts of ice-rafted detritus below a depth of about 1 m (~8000 B.P.). This fraction of the sediment often contains detrital carbonate in other North Atlantic cores, in sufficient abundance to make radiocarbon ages (e.g., L-1514B, L1514C, and L-1514E from V29-206 and L-1515D from V28-14) unreliable. The initial age model used for spectral analyses is given by the intermediate stratigraphic interpretation shown on Fig. 2 (the third alternative discussed above).

RESULTS OF FORAMINIFERAL ANALYSES

Planktonic foraminifera encountered during analyses of V28-14 and V29-206 include sinistral and dextral morphotypes of *Globoquadrina pachyderma*, *Globigerina quinqueloba*, *G. bulloides*, and *Globigerinita glutinata*. Specimens of *Orbulina universa*, *Globorotalia inflata*, *G. hirsuta*, *G. scitula*, and *G. tumida* were seen (exclusively in the upper 100 cm of each core) but rarely comprised more than 1 to 2 percent of any sample. Plots showing variations in the percentage of dextral *G. pachyderma* (as a percentage of the total *G. pachyderma*) and of subpolar species (all species except *G. pachyderma*, expressed as a percentage of the total foraminiferal fauna) are included in Figs. 3 and 4, along with oxygen-isotope, carbonate, and foraminiferal-abundance curves, temperature estimates, and key stratigraphic marker levels.

The first-order feature of these curves is the change from glacial to interglacial conditions. The different curves show that this change is recorded at different times within each core. For example, the change to interglacial conditions is first recorded by the isotope curves at about 13,300 B.P., thus showing that disintegration of terrestrial ice sheets was the first event recorded in Denmark Strait cores. The foraminiferal fauna, which respond to changing conditions in the surface waters, did not begin to show a warming influence until nearly 8000 B.P. Both faunal indices and isotope curves for each core show rather abrupt changes from glacial to interglacial conditions. A more gradual change is seen in the carbonate and foraminiferal abundance curves.

A second-order feature, and the principal topic of discussion in this paper, is fluctuations in the percentage of subpolar foraminifera. These faunal variations are best seen in the upper 50 cm of V28-14 and above 150 cm in V29-206. Are these fluctuations real (i.e., are they representations of changing environmental conditions in Denmark Strait surface waters, or some other factor such as dissolution), or are they a local phenomenon, or an artefact of some error in sample collection, preparation, or analysis? I believe they are real for a number of reasons. First, the analytical procedure is not new but has been tested extensively in North Atlantic and Norwegian Sea sediments (Kellogg, 1975, 1976, 1980, 1984; Kellogg et al., 1978). This procedure includes calculation of possible error for each analysis (using the normal approximation to the binomial distribution [Meyer, 1965], which considers the sample size and the observed percentage, and calculates upper and lower 95% confidence limits, included in Figs. 3 and 4). This procedure shows that the fluctuations in subpolar foraminifera are statistically significant. A second test is the demonstration that these fluctuations are not restricted to a single core (they were observed first in V28-14, where their discovery led to the analysis of V29-206). The third test is that the fluctuations can be correlated from core to core. This can be seen by simply comparing Figs. 3 and 4, but a more rigorous proof is desirable. Independently dated stratigraphic models for each core (described above) were used to interpolate ages for each sample in both cores. When the subpolar foraminiferal percentages are then plotted versus time (Fig. 5), the correlation is convincing.

Temperature Estimates

Winter and summer sea-surface temperature estimates were made for each sample in V28-14 and V29-206 using Equation T3 (Kellogg, 1980). The results of this work (Figs. 3 and 4) are not completely reliable for several reasons. Although the transfer function employed is state-of-the-art for this region (Kellogg, 1980), no-analog conditions (Hutson, 1977) were encountered in V29-206. In addition, the fact that winter and summer temperature

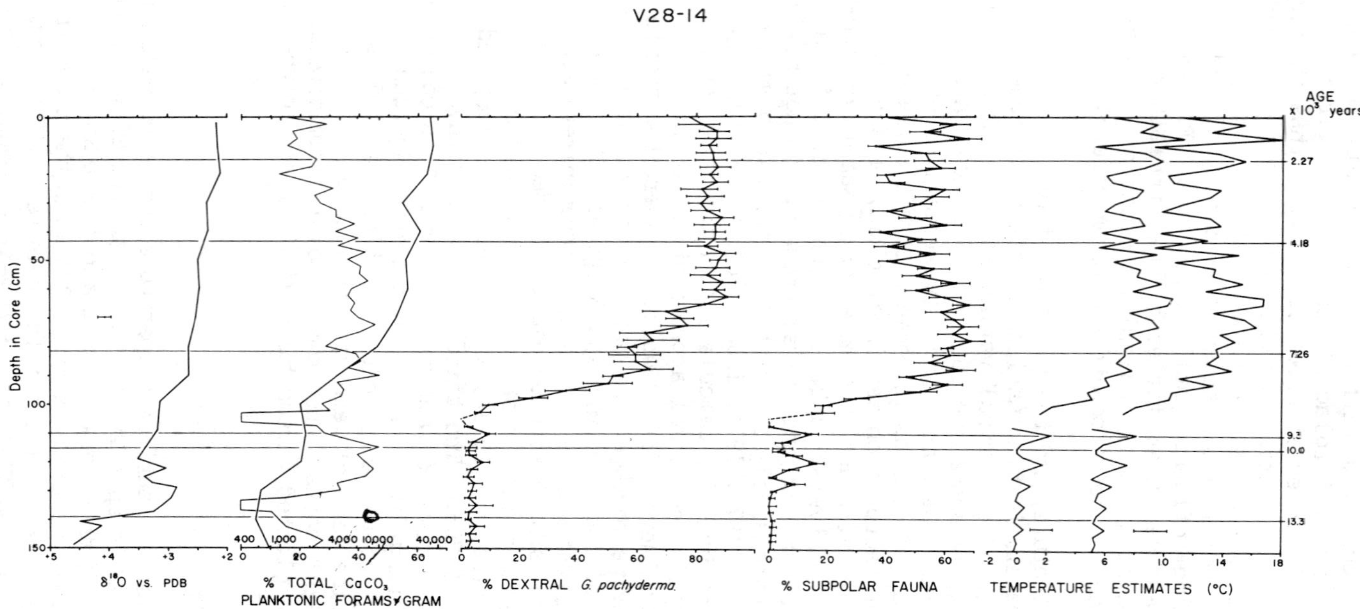

Figure 3. Oxygen isotopic, carbonate, foraminiferal abundance, faunal, and SST-estimate curves for V28-14.

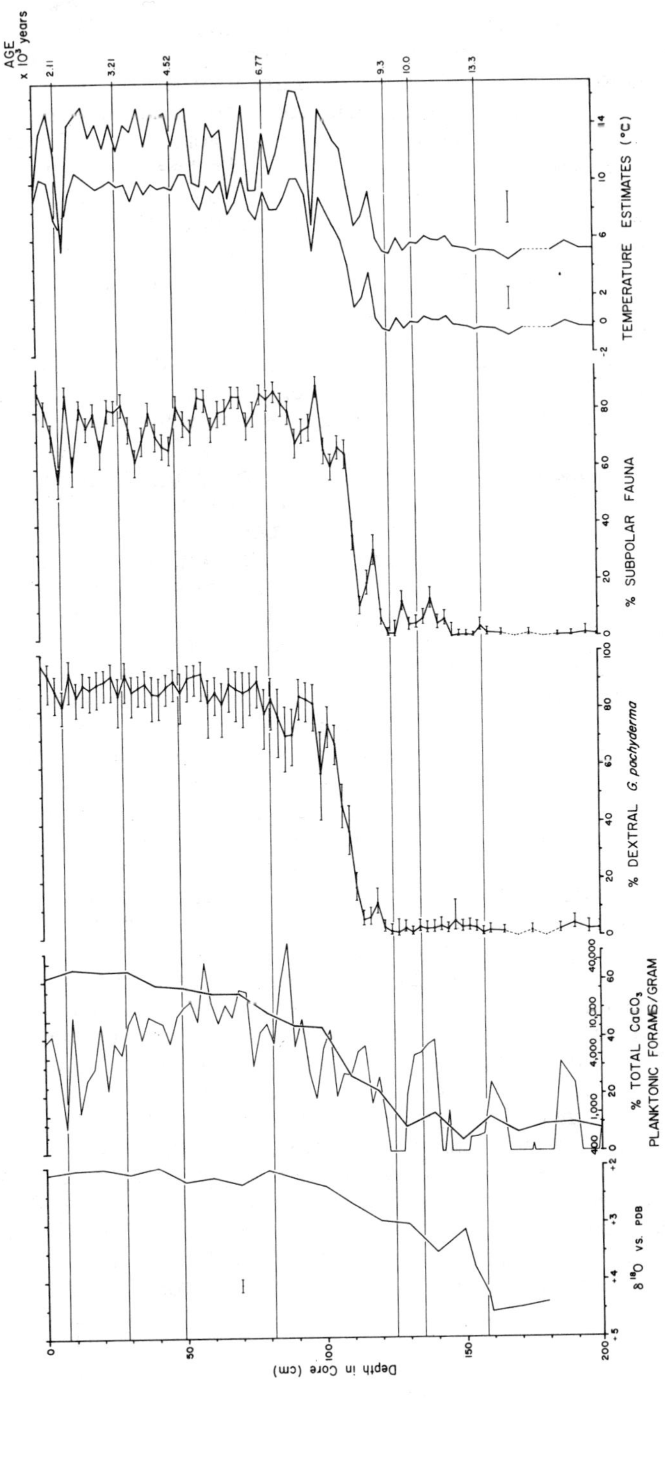

Figure 4. Oxygen isotopic, carbonate, foraminiferal abundance, faunal, and SST-estimate curves for V29-206.

estimate curves for each core display almost identical fluctuations suggests that the curves are not mutually exclusive and that perhaps a single average-annual temperature estimate would be more meaningful. Nevertheless, estimates for V28-14 are considered fairly reliable, along with those from below 100 cm in V29-206. Above 100 cm in V29-206, the estimates converge and diverge (and even cross at 10 cm), the result of the no-analog problem mentioned above. There is no easy solution to the no-analog problem described here. Reliable temperature estimates will not be produced until North Atlantic core-top material containing abundant G. quinqueloba, indicating relatively little dissolution becomes available.

Recognition of the no-analog problem in V29-206, and its absence in V28-14, led to an evaluation of the faunal makeup of the two cores to determine which subpolar species were responsible for the observed fluctuations. In V28-14, these fluctuations are caused by variations in G. bulloides; G. quinqueloba dominates the subpolar fluctuations in V29-206. G. quinqueloba is much less important (lower percentages and lower absolute numbers per gram of sediment) in V28-14 than in V29-206, indicating that this relatively dissolution-susceptible species has been largely removed from V28-14, and that low-abundance zones in V28-14 probably represent times with increased dissolution. This relationship is not the sole cause for the fluctuations in percentage of subpolar species because species fluctuations are not always in phase, because there is a gradual increase in G. glutinata toward the top of V28-14 (G. glutinata is also relatively susceptible to dissolution: Berger, 1970), and because obvious indications of dissolution (fragmented foraminiferal tests, etc.) show that dissolution is less important in V28-14 between 100 and 50 cm than it is above 50 cm. I therefore conclude that in V28-14, fluctuations in percentage of subpolar foraminifera result from an interaction between dissolution and varying environmental conditions in overlying surface waters. Specifically, the major control on species abundances, both absolute and relative, is probably directly related to the shifting boundary between the Irminger and East Greenland currents. We may infer an indirect climatic control on the temporal position of this boundary.

Invocation of changing environmental conditions as a partial cause for observed fluctuations in subpolar foraminiferal percentages in Denmark Strait cores is also supported by data on the absolute abundances of the subpolar species in V28-14 and V29-206. Numbers of specimens of each species per gram of sediment were calculated and show that, at almost every level in each core, increases or decreases in absolute abundance are consistently in the same direction. Only the magnitude of abundance variations differs from one species to another. This observation rules out any possibility that bottom currents have selectively removed or prevented deposition of foraminiferal shells. The most likely

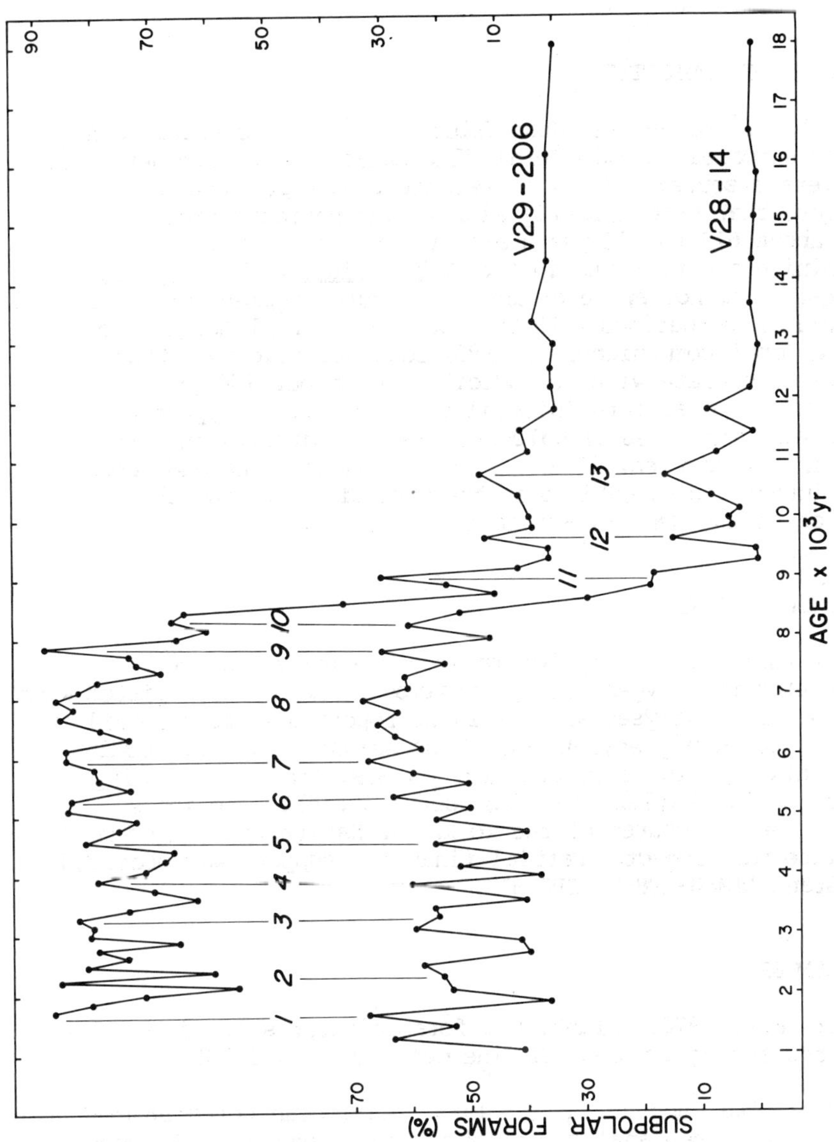

Figure 5. Subpolar foraminiferal curves from V28-14 and V29-206 plotted versus age (alternative 3 - see text). (Scale for V29-206 is on the left).

explanation for coincidental variations in absolute abundance of the three species (and of G. pachyderma) is surface productivity changes, probably related to shifts in the position of major surface currents.

A 700 YR PERIODICITY?

The author and Dr. John Imbrie have begun spectral analyses of the subpolar foraminiferal fluctuations described here, and a complete discussion is in preparation. Our preliminary results indicate that statistically significant periodicities ranging from about 615 to 784 years are present in the subpolar foraminifera curves and in the G. bulloides and G. quinqueloba curves. I am not aware of any literature documenting other climatic fluctuations with this periodicity, although Reid Bryson (oral communication, 1983) suggests that some tidal processes operate with periodicities of about 750 yr.

The subpolar foraminiferal fluctuations do appear to correlate with times of Holocene glacial advances in northern Scandinavia (Karlén, 1982). I invite suggestions and comments from interested scientists in the hope of resolving the significance of the Denmark Strait fluctuations.

ACKNOWLEDGEMENTS

I thank T.-H. Peng for radiocarbon dating and N.J. Shackleton for oxygen isotope analyses. J. Imbrie assisted with the spectral analyses which will be reported in more detail in a joint paper (in preparation). Discussions with W.F. Ruddiman, A. McIntyre, J.D. Hays, D. Kellogg, S.S. Streeter, N. Kipp, G.H. Denton, and W. Karlén provided useful insights and ideas throughout the course of the work. W. Karlén encouraged me to initiate the project. Partial financial support was provided by NSF Grant ATM80-18897 (SPECMAP).

REFERENCES

Berger, W.H., 1970. Planktonic foraminifera: selective solution and the lysocline. Marine Geology, 8, 111-138.

Broecker, W.S. and Ku, T.-L., 1969. Caribbean cores P6304-8 and P6304-9: new analysis of absolute chronology. Science, 166, 404-406.

Coachman, L.K. and Aagaard, K., 1974. Physical Oceanography of Arctic and Subarctic Seas. In: Marine Geology and Oceanography of the Arctic Seas (ed. Y. Herman), pp. 1-72. Springer-Verlag.

Duplessy, J.-C., Delibrias, G., Pujol, C. and Duprat, J., 1981. Deglacial warming of the northeastern Atlantic Ocean: correlation with the paleoclimatic evolution of the European continent. Palaeogeography, Palaeoclimatology, Palaeoecology, 35, 121-144.

Hutson, W.H., 1977. Transfer functions under no-analog conditions: experiments with Indian Ocean planktonic foraminifera. Quaternary Research, 8, 355-367.

Karlén, W., 1982. Holocene glacier fluctuations in Scandinavia. In: Holocene Glaciers (ed. W. Karlen), pp. 26-34. Striae, 18.

Kellogg, T.B., 1975. Late Quaternary climatic changes in the Norwegian and Greenland Seas. In: Climate of the Arctic (ed. G. Weller and S.A. Bowling), p. 3-36. University of Alaska, Fairbanks.

Kellogg, T.B., 1976. Late Quaternary climatic changes: evidence from deep-sea cores of Norwegian and Greenland Seas. Geological Society of America, Memoir 145, 77-110.

Kellogg, T.B., 1980. Paleoclimatology and paleo-oceanography of the Norwegian and Greenland Seas: glacial-interglacial contrasts. Boreas, 9, 115-137.

Kellogg, T.B., 1984. Paleoclimatic significance of subpolar foraminifera in high-latitude marine sediments. Canadian Journal of Earth Sciences (in press).

Kellogg, T.B., Duplessy, J.-C. and Shackleton, N.J., 1978. Planktonic foraminiferal and oxygen isotopic stratigraphy and paleoclimatology of Norwegian Sea deep-sea cores. Boreas, 7, 61-73.

Meyer, P.L., 1965. Introductory Probability and Statistical Applications. Addison-Wesley Pub. Co., Reading Mass, 339pp.

Ruddiman, W.F. and Glover, L.K., 1972. Vertical mixing of ice-rafted volcanic ash in North Atlantic sediments. Geological Society of America, Bulletin, 83, 2817-2836.

Ruddiman, W.F. and McIntyre, A., 1973. Time-transgressive deglacial retreat of polar waters from the North Atlantic. Quaternary Research, 3, 117-130.

THE REFLECTION OF ABRUPT CLIMATIC CHANGES IN THE RELIEF AND SEQUENCE OF CONTINENTAL DEPOSITS

Leszek Starkel

Polish Academy of Sciences, Poland

ABSTRACT

The abrupt climatic changes in the temperature and hydrology (precipitation), that last for only ca. one or two centuries, but are so important and continuous that they cause a disturbance and often a change in the equilibrium of the system, are reconstructed on the basis of forms and continental deposits in the temperature zone. The following abrupt changes are discussed for the last 11000 years; 10.9-10.6 Ka, 8.4-8.0 Ka, 5.0-4.5 Ka, 2.8-2.4 Ka BP and the Little Ice Age. The following abrupt climatic changes have been distinguished: a/abrupt warmings and coolings and b/short waves of increased humidity and increased flood frequency.

INTRODUCTION

By an abrupt climatic change, I understand a change in temperature or precipitation lasting either for one or two centuries or shorter, that is so important and continuous that it causes a disturbance or a change in the metastable equilibrium of the geo-ecosystem.

These abrupt climatic changes are reflected in continental deposits, and will be discussed using examples of forms and deposits found in the present day temperate zone. Four types of abrupt climatic changes may be distinguished and discussed:

a) Abrupt warmings, causing rapid melting and recession of ice sheets, mountain glaciers (temporary increase in river discharge and sediment yield) and permafrost (decrease in river discharge) and a rise in vegetation density and invasion of forest communi-

N.-A. Mörner and W. Karlén (eds.), Climatic Changes on a Yearly to Millennial Basis, 135–146.

ties (where solution dominates over suspension).

b) Abrupt coolings, causing readvance of glaciers, creation of new permafrost layers, retreat of forest vegetation and increases in fluvial activity (tendensy to braiding), solifluction and eolian processes.

c) Abrupt rise in precipitation and/or humidity, coinciding with heavy rainfalls, causing increased vegetation density, increased solution, higher frequence of floods and gravitation processes and readvance of glaciers.

d) Abrupt decrease in precipitation and/or humidity, causing stabilization of channels due to lower flood frequence or slope and channel instability due to sparse vegetation or rare but catastrophic floods in semi-arid zones.

In most cases two of these tendencies coincide; temperature and precipitation change due to variation in the air circulation pattern (cf. Flohn 1979). In order to avoid drawing incorrect conclusions when looking for climatic factors, a study of forms and deposits should exclude from consideration areas of eustatic or tectonic activity (Schumm 1979, Frenzel 1983).

THE ROLE OF EXTREME EVENTS AND THE THRESHOLD CONCEPT

The role that extreme climatic events plays must not be overlooked. In the fluvial environment the changes in river discharge (Qw) and sediment load (Qs) are closely connected and each rapid turn in one of these two components brings about a change of the channel parameters (to braiding or to meanding) with a tendency to downcutting or to aggradation (Schumm 1977, Starkel 1983). It is also reflected in the deposition and in the grain size composition (Baker 1983). One of the most exact indicators is the change in the rate of deposition as well as breaks in it, in the process of soil formation. In semi-arid climates rare events play a leading role even though they do not change the slopes of river channels. On the other hand, in humid temperate climate slackwater deposits are not as readily preserved due to the rapid recovery by vegetation and the lack of long term stability of alluvial channel cross-sections (Baker 1983).

These rare extreme events can coincide with abrupt changes of climate. Most frequently such change can be brought about by the change in the frequency of extreme events (Starkel 1983). If the extreme event does not coincide with the change in climate, the river or slope tries to reach its former equilibrium again (e.g. Cimarron River, Schumm 1977, Starkel 1976). The fluvial system, as well as other systems, tend to reach in their evoulution a threshold value (Schumm 1977). This can be of external

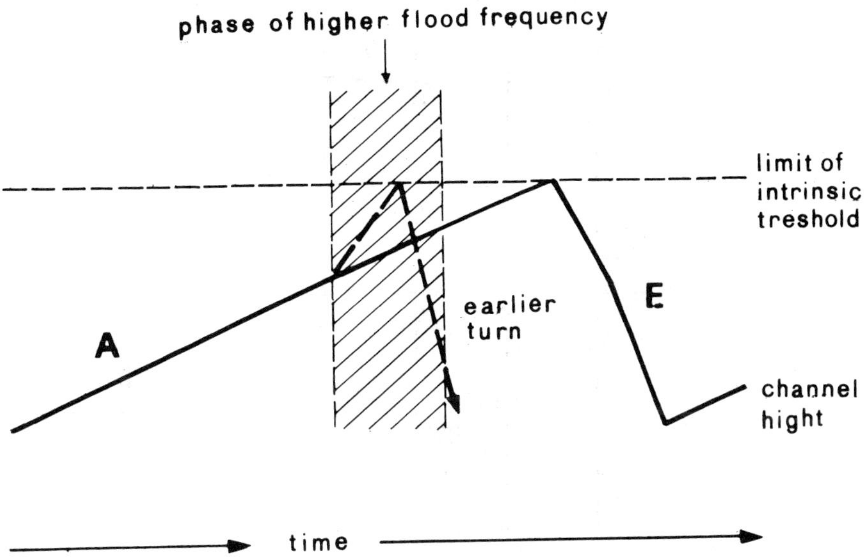

Fig. 1. The importance of the overlapping of the higher flood frequency phase with the general trend for reaching the limit of intrinsic threshold by the river channel (after Starkel 1983).

origin (climatic change) or intrinsic, connected with self-regulation of the system. The reaching of the limit of the intrinsic threshold, which may cause change from aggradation to erosion (downcutting) can be much earlier in time due to change towards higher frequencies of floods (Starkel 1983, Fig. 1). In such cases we observe the overlapping of two kinds of geomorphic thresholds, which is very common.

In the boundary zones of ecotones a single extreme event can cause a general change (Frenzel 1977). Even a slow trend can be reflected in abrupt change of forms or sediment faces. Therefore, we must be very careful in the reconstruction of abrupt climatic changes, and look more for evidence in the global, zonal or regional scale, trying to find some substantial changes in the central parts of the vegetation-climatic zones. The rapid downcutting of rivers after deglaciation (St. Onge 1972, Koutaniemi 1979) although regionally significant, do not indicate any simultaneous abrupt climatic event. In the fluvial environment the reflection of a particular event is much greater in a small basin than in a large basin, and the event can more easily cause a transformation of the system (Church 1981). In the annual laminated deposits of lakes, shorter or longer phases reflecting higher or lower delivery of matter and thus indicating variations in the hydrological regime can usually be found, but they do not reflect abrupt climatic changes. This is illustrated by the diagramme, elaborated

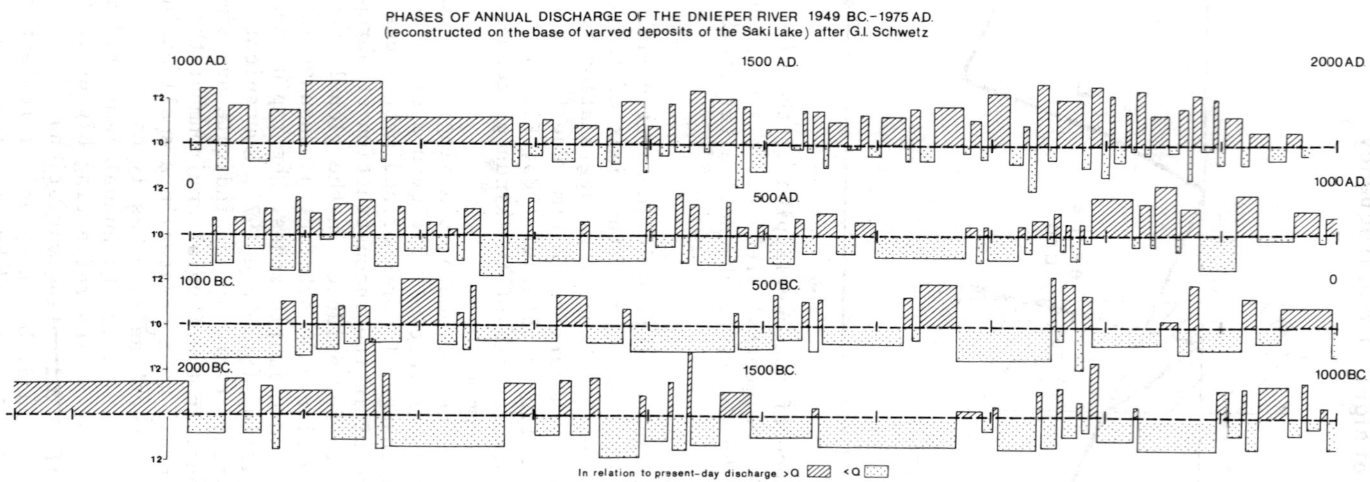

Fig. 2. Fluctuations of the mean annual discharge of the Dniepr river 2149 BC-1975 AD, reconstructed on the bases of varved deposits of the Saki lake (compiled after G.J. Schwetz 1978).

for the annual variations in the discharge of the Dnieper river, based on the lacustrine sediments of Saki lake (Schwetz 1978, Fig. 2). Therefore, in order to distinguish abrupt climatic changes from other factors, we should study larger areas and find characteristic evidence in all types of environments (e.g. fluvial, slope, lake, bog, dunes etc).

EVIDENCE OF ABRUPT CLIMATIC CHANGES DURING THE LATE GLACIAL AND HOLOCENE

Among abrupt changes documented by radiocarbon dating the following are discussed: The cooling at the beginning of the Younger Dryas (ca. 10.9-10.6 ka BP), the warming at the Late Glacial/Holocene Boundary (ca. 10.3-10.0 Ka BP), the rise in rainfall and probable cooling at the Boreal/Atlantic break (8.5-8.0.Ka BP) at the Atlantic/Subboreal break (5.0-4.5 Ka BP), at the Subboreal/Subatlantic break (2.8-2.4 Ka BP), and in the XVII-XIX centuries (Little Ice Age). The phases of increased fluvial activity (with a higher flood frequencey) in the temperate zone are presented in Fig. 3. The coincidence of the beginning of most of these phases with the beginning of glacial advances and high lake levels is remarkable. These simultaneous changes can only be interpreted as indicators of abrupt climatic changes.

At the end of the Alleröd, ca. 10.8-10.6 Ka BP, there occurred a rapid cooling, reflected in the abrupt extension of the Scottish ice cap (Sissons 1977, Lamb 1977). A rise in activity in fluvial and slope processes was noted for southern Britain (Rose et al. 1980), and in other parts of western Europe (Munault and Paulissen 1973, Starkel 1977). At the archeological site Całowanie, in the middle of the Vistula river, a rise in the river activity, followed by the building up of the former bar by a dune, was proved (Schild 1982). Ths aeolian phase started after 10.660±100 BP and ended before 10.435±90 BP. Cooling and a decrease in precipitation caused the blowing of sand from the braided river channel and the formation of the dune.

The next rapid warming took place at the beginning of the Holocene (10.3-10.0 Ka BP). A rapid retreat of the Scandinavian ice sheet reaching up to 250 m/yr (Mörner et al. 1973) coincides well with the decrease in fluvial activity expressed in the change from channel bars to swampy overbank deposits in southern Poland (Starkel et al. 1981), in Belgium (Munault and Paulisson 1973) or in the rapid change in the size of meanders in the San valley (Szumański 1982). This warming was not accompanied by a rise in precipitation.

About 8.5-8.0 Ka BP a rise in flood frequence has been registered, well known from southern Germany (dated by dendrochronological methods - Becker and Schirmer 1977), in Norway (alluvial fans with highest rate of deposition ca. 8400 BP - Sonstegaard and Mangerud 1977) and from the Carpathian valleys. In the last region

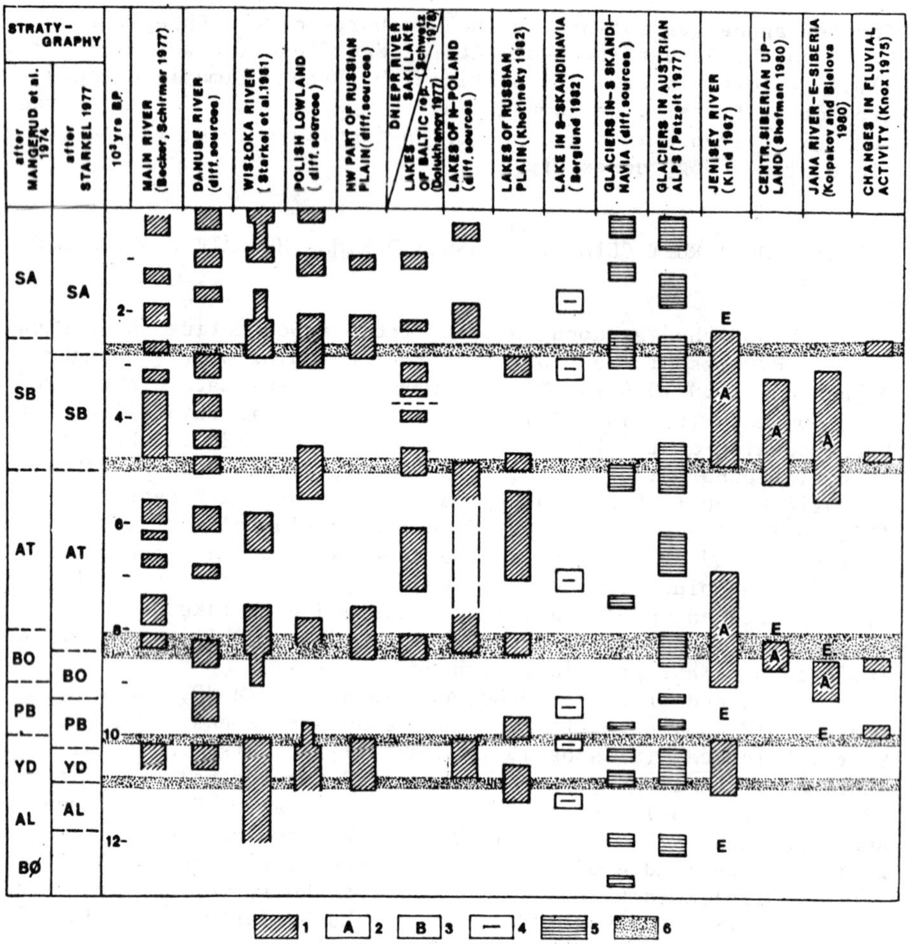

Fig. 3. The position of abrupt changes of hydrological regime in the temperate zone in the Late glacial and Holocene. Compilation based on information about river activity compared with changes of lake levels and fluctuations in mountain glaciers.
Key: 1) increased fluvial activity or high lake water level, 2) aggradation phase (after author), 3) downcutting phase (after author), 4) low lake water level, 5) phase of glacial transgression, 6) phases of abrupt climatic changes.

the oxbow lakes mires or overbank deposits at many localities have been rapidly covered by sandy bars or by thick alluvial fans of tributary valleys (Starkel 1977, Fig. 4). In the Carpathians this was followed by the activization of landslides (Gil et al 1974) and, in the limestone areas, of intensive leaching and pre-

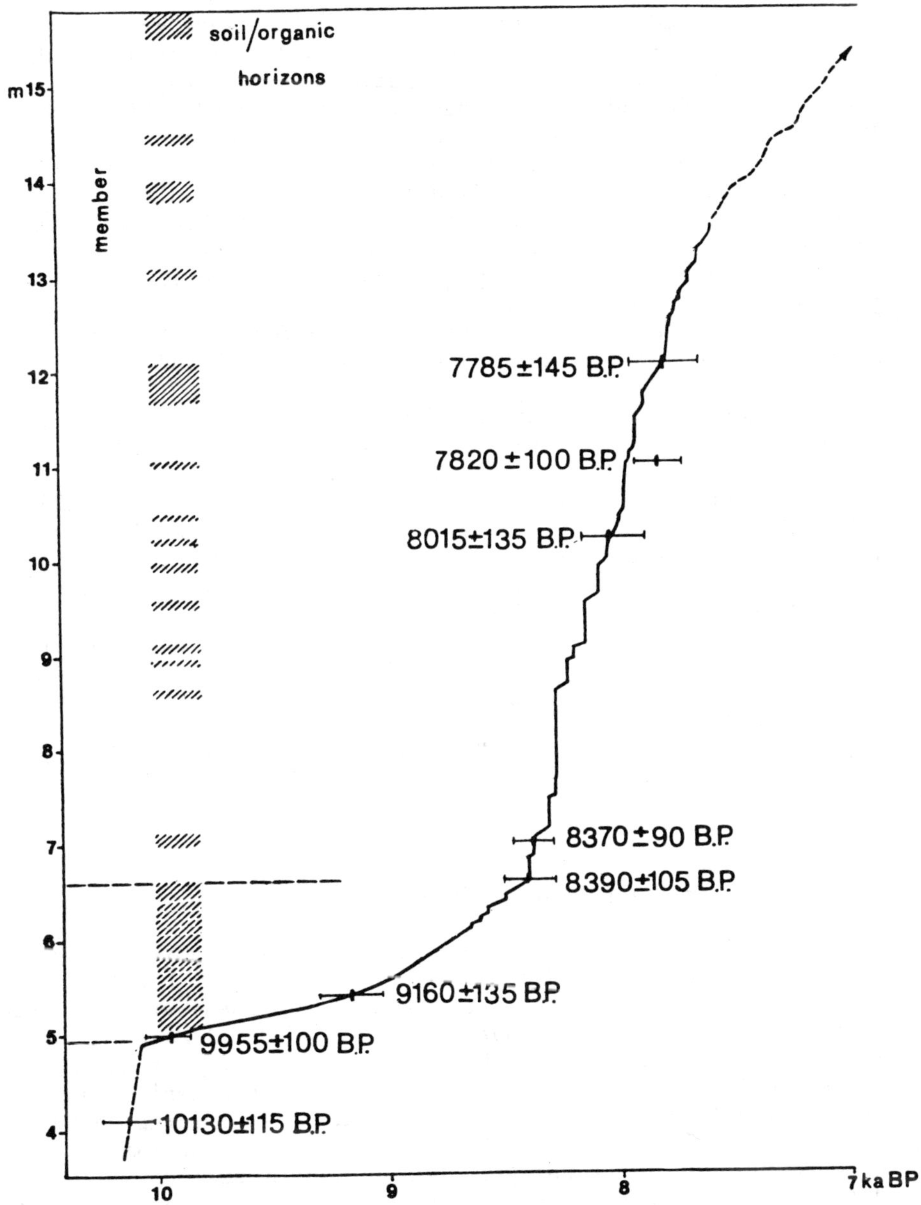

Fig. 4. Changes of rate of deposition of oxbow lake and alluvial fan deposits at Podgrodzie in Wisloka valley (with rapid break ca. 8400 BP) - after Niedzialkowska et al. (1977) and Starkel et al. (1981).

cipitation of $CaCo_3$. In Zelse cave stalagmites were growing up to 100 mm per century in the period 8350-8080 BP (Franke and Geyh 1971). At the Calowanie locality mentioned above, the last sandy intercalation in the paleochannel fill occurred 8360±75 BP (Schild 1982). It can be explained as the beginning of downcutting of an existing channel. This was also a phase of distinct advance for mountain glaciers (Patzelt 1977 and others). It should be mentioned that a supply of minerogenic material has been observed in many lakes in Eastern and Central Europe (Yakovleva et al. 1981, Kurotskina 1981). This phase was dated in Woryty lake to just before 8440±100 BP (Pawlikowski et al. 1982). All these events coincide with the phase of drastic changes found by the use of statistical methods by Wendland and Bryson (1977) and later in fluvial deposits by Knox (1975).

Younger episodes of abrupt changes are not so well expressed due to the overlapping with phases of human activity. However, Fig. 3. shows a relatively good coincidence of three cooler and humid phases: ca. 5-0-4.5 Ka BP, 2.8-2.4 Ka BP and in the XVII-XIX centuries. During all these time there was a rise in lake levels (cf. Dolukhanov 1977, Niewiarowski 1978, Berglund 1983) an advance of mountain glaciers (Patzelt 1977, Röthisberger el al. 1980, Denton and Karlén 1973) and a rise in fluvial activity with frequent avulsions of the river channels (Kozarski and Tobolski 1981, Starkel et al. 1981 and others). Especially well documented is the Little Ice Age, using glaciers, dated avalanches, landslides and floods (Grove 1972, Lamb 1977, Pfister 1979). A careful analysis shows that each "abrupt change" towards a rise in humidity is composed of a sequence of extreme events - it is the clustering of events in a short time.

TWO TYPES OF ABRUPT CLIMATIC CHANGES

Two types of abrupt climatic changes can be distinguished. One type consists of directed changes in temperature with step-like rapid warmings and coolings (cf. Flohn 1979). These are expressed in the general change of tendency in slope and valley bottom evolution, such as those at the beginning of the Late Glacial and at the beginning of the Holocene. The warmings cause rapid change from braided to meandering river channels or from large to small meanders. These are typical shiftings of the equilibrium caused by external factors. A delay can exist in some areas, e.g. the deforestation of dunes a few centuries after the main change. It should not be forgotten that after diminishing the role of some of the leading factors there can follow a relaxation phase with a slowly decreasing amplitude of fluctuations (Rose et al. 1980), which can only be indirectly connected with a proceeding abrupt climatic change.

The second type of abrupt climatic change consist of "short term waves" of mainly increased humidity or of a high frequency of heavy rains (100-400 yr long) superimposed on the longer fluctuations. The repetition of these ca. once every 2000 years during the Holocene is interpreted as some kind of cyclicity in the air circulation pattern and precipitation regime (Lamb 1977, Starkel 1983). These short humid waves are reflected in the existence of parallel fills with paleochannels (Troll 1957, Becker and Schirmer 1977, Starkel 1977, Brakenridge 1981) or in superposition of slope covers related to episodes with disturbed slope stability (landslides - Gil et al. 1974); solifluction layers - Furrer and Bachmann 1972). During these waves advances of glaciers and high water levels in lakes occur (Fig. 3). In some cases, the destruction of the geo-eco-system equilibrium coincide with these waves. More frequently, however, we observe only a rise in amplitude of the intrinsic oscillations.

REFERENCES

Baker, V.R., 1983. Paleoflood hydrological analysis from slackwater deposits. Proceedings of the Holocene Symposium. Quaternary Studies in Poland, 4, Poznań, 19-26.

Becker, B., Schirmer, W., 1977. Palaeoecologic study of the Holocene valley development of the River Main, Southern Germany. Boreas 6:4, 303-321.

Berglund, B., 1983. Palaeoclimatic changes in Scandinavia and on Greenland - a tentative correlation based on lake- and bog stratigraphical studies. Quaternary Studies in Poland 4, Poznań, 27-44.

Brakenridge, G.R., 1981. Late Quaternary floodplain sedimentation along the Pomme de Terre river, Southern Missouri. Quaternary Research 15, 62-76.

Chotinsky, N.A., Starkel, L., 1982. Naturalne i antropogeniczne poziomy graniczne w osadach holoceńskich Polski i centralnej cześci Niziny Rosyjskiej. Przegl.Geogr. 54:2.

Church, M., 1981. Reconstruction of the hydrological and climatic conditions of the past fluvial environments. In: L. Starkel and J.G. Thornes. Palaeohydrology of river basine. IGCP Project, No 158. Technical Bulletin BGRG. Geoabstracts, Norwich, 50-79.

Denton, G.H., Karlén, W., 1973. Holocene climatic variations, their pattern and possible cause. Quaternary Research 3:2, 155-205.

Dolukhanov, P.M., 1977. The Holocene history of the Baltic sea and ecology of prehistoric settlement. Baltica 6, 227-244.

Flohn, H., 1979. On time scales and causes of abrupt paleoclimatic events. Quaternary Research 12, 135-149.

Franke, H.W., Geyh, M.A., 1971. ^{14}C-Datierungen von Kalksinter aus slovenischen Höhlen. Der Aufschluss 22, 7/8, 235-237.

Fenzel, B., 1977. Postglaciale Klimaschwankungen im südwestlichen Mitteleuropa. In: Dendrochronologie und postglaziale

Klimaschwankungen in Europa. Erdwis. Forschung 13, Wiesbaden, 297-322.
Frenzel, B., 1983. On the Central-European water-budget during the last 15 000 years. Quaternary Studies in Poland. Poznań 4, 45-59.
Furrer, G., Bachmann, F., 1972. Solifluctionsdecken in Schweizerichen Hochgebirge als Spiegel der postglacialen Landschaftsentwicklung. Zeitschrift für Geomorph., Suppl. 13, 63-172.
Gil, E., Gilot, E., Kotarba, A., Starkel, L., Szczepanek, K., 1974. An early Holocene landslide in the Niski Beskid and its significance for paleogeographical reconstructions. Studia Geomorphologica Carpatho-Balcanica 8, 69-83.
Grove, J.M., 1972. The incidence of landslides, avalanches and floods in eastern Norway during the Little Ice age. Arctic and Alpine Research 4, 131-138.
Knox, J.C., 1975. Concept of the graded stream. Theories of Landform Development. Eds. W.N. Melhorn and R.C. Flemal. Binghamton, New York, 169-198.
Koutaniemi, L., 1979. Outline of the development of relief in the Oulanka river valley, North-eastern Finland. Acta Universitatis Ouluensis A, 82. Geol. 3, 29-38.
Kozarski, S., Tobolski, K. (eds.), 1981. Guide-book of excursion. Symposium "Paleohydrology of the temperate zone". Poznań. Adam Mikhiewicz Univ., 120.
Kurotchkina, A.A., 1981. Paleolimnology of the Woshe and Kubenskoya lakes and reflection of the human impact (in Russian). In: G.G. Martinson, N.N. Dawydova (eds.) Paleolimnologitcheskij podchod k izutcheniu antropogennogo wozdieystwia na oziera. Geogr. Obsch. SSSR, Leningrad, 62-72.
Lamb, H.H., 1977. Climate: Present, past and future, Vol II. Climatic history and the future. Methuen, London, 835.
Mörner, N.A. (ed.), 1973. The Pleistocene/Holocene boundary. A proposed boundary - stratotype in Gothenburg, Sweden. Symposium INQUA 1973, 163.
Munaut, A.V., Paulissen, E., 1973. Evoluation et paléo-écologie de la vallée de la petite Néthe au cours du post-würm (Belgique). Annales de la sociéte Géologique de Belgique 96:2, 301-348.
Niedzialkowska, E., Skubisz, A., Starkel, L., 1977. Lithology of the Eo- and Meso-holocene alluvid in Podgrodzie upon Wisloka river. Studia Geomorphologica Carpatho-Balcanica 11, 89-100.
Patzelt, G., 1977. Die zeitliche Ablauf und das Ausmass postglacialer Klimaschwankungen in den Alpen. In. Dendrochronologie und postglaciale Klimaschwankungen in Europa. Erdwiss. Forschung 13, Wiesbaden, 249-259.
Pawlikowski, M., Ralska-Jasiewczowa, M., Schönborn, W., Stupnicka, E., Szeroczyńska, K., Woryty near Gietrzwald, Olsztyn Lake District. NE Poland-Vegetational history and lake development during the last 12 000 years. Acta Palaeobotanica 22:1 85-116.

Pfister, Ch., 1979. The reconstruction of past climate: the example of the Swiss historical weather documentation project (16th to early 19th century). Review papers. International Conference on Climate and History, University of East Anglia, 128-147.

Rotnicki, K., 1970. Glówne problemy wydm śródladowych w Polsce s świetle badań w Weglewicach. Prace Kom. Geogr. Geol. Poznań Tow. Przyj, Nauk 11:2, 1-146.

Röthlisberger, F., Haas, P., Holzhauser, H., Keller, W., Bircher, W., Renner, R., 1980. Holocene climatic fluctuations radiocarbon dating of fossil soils (fAh) and woods from moraines and glaciers in the Alps. In: Geography in Switzerland, Geographica Helvetica 35;5, 21-52.

Rose, J., Turner, C., Coope, G.R., Bryan, M.D., 1980. Channel changes in a lowland river catchment over the last 13 000 years. Timescales in Geomorphology (eds. R.A. Cullingford et al.), J. Wiley and Sons, London, 160-175.

Schild, R., 1982. Stratygrafia archeologiczna wydm śródladowych widziana z Mazovsza, Reczniki Gleboznawcze PTG 33:3-4, Warszawa, 59-79.

Schumm, S., 1977. The fluvial system. Wiley-Interscience Publ., New York, 338.

Schwetz, G.I., 1978. Multi-centennial changes of the run-off of the Dniepr river (in Russian). Gidreometeoidat, Leningrad, 82.

Sissons, J.B., 1977. The Loch Lomond readvance in the Northern Mainland of Scotland. In: Studies in the Scottish Lateglai cial Environment. Pergamon Press, 45-60.

Sønstegaard, E., Mangerud, J., 1977. Stratigraphy and dating of Holocene gully sediments in Os. western Norway. Norsk Geologisk Tidsskrift 57, 313-346.

Starkel, L., 1976. The role of extreme (catastrophic) meteorological events in the contemporaneous evolution of slopes. Geomorphology and Climate, J. Wiley and Sons, London, 203-246.

Starkel, L., 1977. Paleogeografia holocenu. PWN, Warszawa.

Starkel, L., 1983. The reflection of hydrologic changes in the fluvial environment of the temperate zone during the last 15 000 years. In: Background to Palaeohydrology: A Perspective (ed. K.J. Gregory), J. Wiley and Sons, London, 213-235.

Starkel, L., (ed.), Alexandrowicz, S.W., Limek, K., Kowalkowski, A., Mamakowa, K., Niedzialkowska, W., Pazdur, M., 1981. The evolution of the Wisloka valley near Debica during the Lateglacial and Holocene. Folia Quaternaria 53, Kraków, 91.

St-Onge, D.A., 1972. Sequence of glacial lakes in North-central Alberta. Geological Survey of Canada, Bulletin 213, 16.

Szumański, A., 1982. Teh evolution of the lower San river valley during the Lateglacial and Holocene. Prace Geograficzne IGiPZ PAN, spec. issue 1, Warszawa, 57-78.

Troll, C., 1957. Tiefenerosion. Seitenerosion und Akkumulation der Flüsse im fluvioglazialen und periglazialen Bereich. Geomorphologische Studien. Machatsche-Festschrift Gotha, 213-226.

Yakovleva, L.W., Chonutova, W.I., Siergyeva, L.W., Drabkowa, W.G., 1981. Postglacial deposition in the Latviian lakes and its change under the human influence (in Russian). In: G.G. Martinson, N.N. Dawydova (eds.). Paleolimnologitcheskij podchod k izutcheniu antropogennogo wozdieystwia na oziera. Geogr. Obsch. SSSR, Leningrad, 36-51.

Wendland, W.W., Bryson, R.A., 1974. Dating Climatic Episodes of the Holocene. Quaternary Research 4:1, 9-24.

WATER-LEVEL CHANGES, CLIMATE AND HUMAN IMPACT: A PALAEOHYDROLOGICAL STUDY OF KRAGEHOLM LAKE (SCANIA, SOUTHERN SWEDEN)

Marie-José Gaillard

Lund University, Sweden

A palaeohydrological study of Krageholm lake (Krageholmssjön, Scania, Southern Sweden) is presented. The expansion stages of human influence are described on the basis of a pollen diagram. A reconstruction of the postglacial water-level changes in Krageholmssjön is proposed and compared to the results obtained earlier in Southern Sweden. The possible connections between human activity, climate and water-level changes are discussed.

1. INTRODUCTION

Krageholmssjön (Fig. 1) has been chosen as a palaeoecological site to describe the history of regional and local vegetation, palaeohydrological changes and soil erosion during the Holocene. The main purpose of the palaeohydrological study is reconstruction of postglacial water-level changes in the lake, including their dating and the analysis of causes (climate, human impact).

The palaeohydrology of lakes in South Sweden has been studied in great detail by Digerfeldt (1972, 1974, 1975, 1976). Apart from the ancient lake of Torreberga (Digerfeldt, 1971) Krageholmssjön is the first eutrophic lowland lake of South Sweden investigated in relation to its palaeohydrology (Fig. 1).

The methods suggested by Digerfeldt (1972) and applied in many oligotrophic lakes of southern Swedish uplands (Digerfeldt, 1974, 1975, 1976) were used in the same way for Krageholmssjön.

N.-A. Mörner and W. Karlén (eds.), Climatic Changes on a Yearly to Millennial Basis, 147–154.

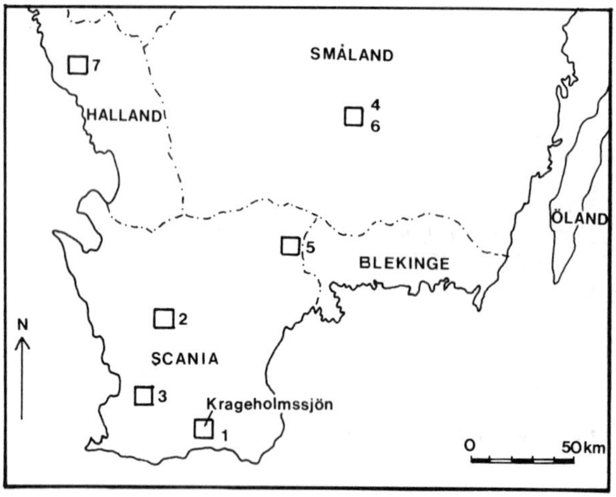

(1) Bjärsjöholmssjön (Nilsson 1961)
(2) Agerödsmosse (Nilsson 1964)
(3) Torreberga (Digerfeldt 1971)
(4) Trummen (Digerfeldt 1972)
(5) Immeln (Digerfeldt 1974)
(6) Växjösjön (Digerfeldt 1975)
(7) Lyngsjö (Digerfeldt 1976)

Figure 1. Sites investigated for palaeohydrology.

2. RESULTS

The results of pollen-macrofossil analysis and of the stratigraphical study will be presented in detail in a future publication (Gaillard, 1984). The aim of this paper is to present briefly the first results of our investigation, to propose a preliminary interpretation of the water level changes in Krageholmssjön and to compare it with results obtained in Southern Sweden, and finally to discuss the possible causes of palaeohydrological changes (climate, human impact).

2.1. Human impact in the lake environment

The pollen-analysis of a core taken at 180 m from the shore (in the SW bay of the lake, Fig. 2) gives good evidences for human activity in the surroundings since Early Neolithic. Since dating Krageholmssjön sediments by ^{14}C is questionable, the pollen-zones defined and dated by Nilsson (1961, 1964) in southern Scania have been used. The four human expansion stages described by Berglund (1969) in South Scandinavia are present in Krageholmssjön pollen diagram (Fig. 4):
(1) The Early Neolithic"landnam"period (around 5200 BP), however weaker than it is shown in Bjärsjöholmssjön (Nilsson, 1961).
(2) The Middle Neolithic (around 4000 BP) and the Late Neolithic (around 4500 BP). After this very pronounced latter expansion phase, there is a continuous human influence up to about 2000 BP. Nevertheless, there are indications for a slightly decreased human activity at the end of Late Subboreal (Bronze Age).

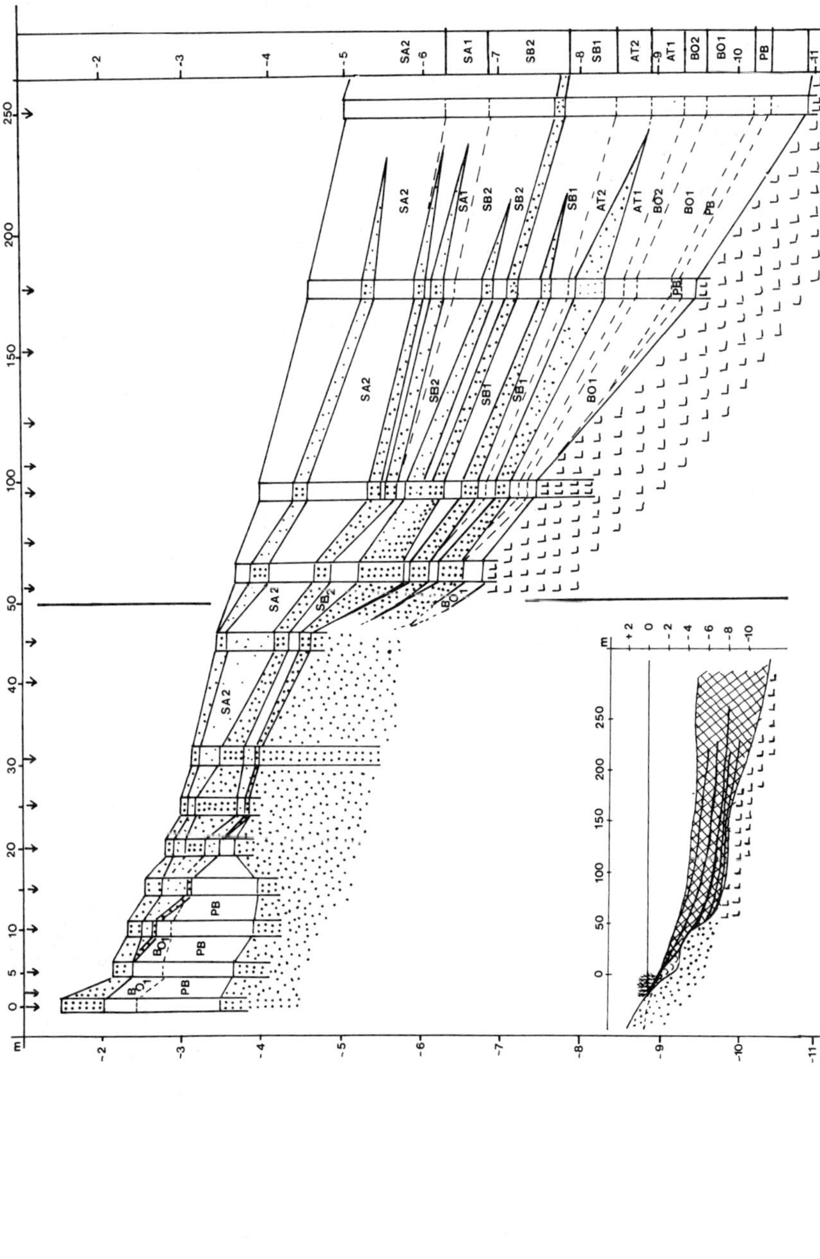

Figure 2. Section of the southwestern bay of Krageholmssjön. The sandy layers were correlated by physical and pollen analyses.

(3) The Pre-Roman Iron Age (around 2000 BP). During the end of Early Subatlantic, there is evidences for a reduction of the human influence around the lake during the Iron Age.
(4) The Viking Time (around 1000 BP). After this expansion, there is a continuous and increasing human influence during the Middle Ages.

2.2. Water-level changes in Krageholmssjön

The stratigraphical study was carried out in the south-western bay of Krageholmssjön. The simplified section given in Fig. 2 illustrates the preliminary interpretation based on physical analyses of sediments (amounts of coarse minerogenic and coarse organic matter), pollen-analysis and macrofossil analysis. The stratigraphy is characterized by a certain number of sandy-silty layers correlated and dated by pollen-analysis. Most of them have been interpreted, in a first attempt, in terms of water-level lowerings. Nevertheless, it is important to mention that a sandy layer can be due to many different causes as shown in Fig. 3.

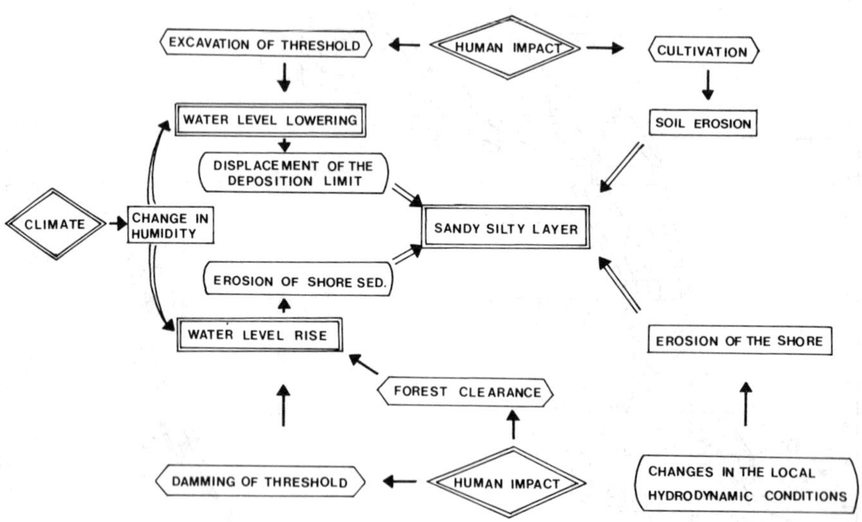

Figure 3. Schematic presentation of factors which should be considered when interpreting sandy-silty layers in terms of lake water-level changes.

The uppermost limits of pollen-zones in the section (Fig. 2) give evidences for (1) a very drastic water-level lowering which occurred at the end of Pre-Boreal and beginning of Early Boreal, causing the erosion of most of the Pre-Boreal sediments (the Pre-boreal and first Early Boreal sediments were preserved close to the shore, where the reed vegetation had a sheltering effect), and (2) a rise in water-level during Early Boreal.

Water-level lowerings are indicated by increases in the contents of fruits and seeds from the reed vegetation, coarse organic and minerogenic matter, (1) at the transition Early Atlantic/Late Atlantic, (2) during Early Sub-Boreal, (3) at the beginning of Late Sub-Boreal, (4) during Early Sub-Atlantic, and (5) during Late Sub-Atlantic. Higher sediment limits, as well as decrease in fruits and seeds, coarse minerogenic and organic matter indicate a pronounced rise in water-level during the end of Late Sub-Boreal and at the beginning of Late Sub-Atlantic.

3. DISCUSSION AND CONCLUSIONS

The six different water-level lowerings described in Krageholmssjön are compared with the results obtained in Southern Sweden in Fig. 4. The first two evident lowerings in Krageholmssjön most probably have climatic causes and are due to a general decrease in average humidity, which could be a combination of a decrease in precipitation plus an increase in temperature. The lowering which occurs at the transition between Pre-Boreal and Early Boreal has been recorded in all the lakes studied by Digerfeldt (1971, 1972, 1974, 1975, 1976). From these results, we can assume that the lowering began at the end of Pre-Boreal and culminated at the very beginning of Early Boreal. The second lowering has been registered in at least two cases: in Växjösjön, where it is indicated among other things by a sandy layer, and in Torreberga, where there is evidence of a rise in water-level from the beginning of Late-Atlantic and of a generally low water-level during Boreal and Early Atlantic.

In both Lake Immeln and Lake Trummen, there are indications for low water-level from Late-Atlantic to Early Sub-Boreal, and good evidence of a culmination of the water-level lowering at the transition between Early Sub-Boreal and Late Sub-Boreal. The possible lowering registered in Krageholmssjön during Early Sub-Boreal could be connected with the general low level defined in Lake Trummen and Lake Immeln.

From the beginning of Early Sub-Boreal, the influence of man must be considered when interpreting the causes of sandy layers and water-level changes (Fig. 3). However, since the water-level lowering at the beginning of Late Sub-Boreal is recorded in many lakes and in regions where human impact was small, it can be assumed that this event has a climatic cause.

An Early Sub-Atlantic water-level lowering is also indicated by a sandy layer in Växjösjön. In both Lake Trummen and Lake Immeln, the poor representation of the Early Sub-Atlantic pollen-zone was associated with a water-level lowering.In this

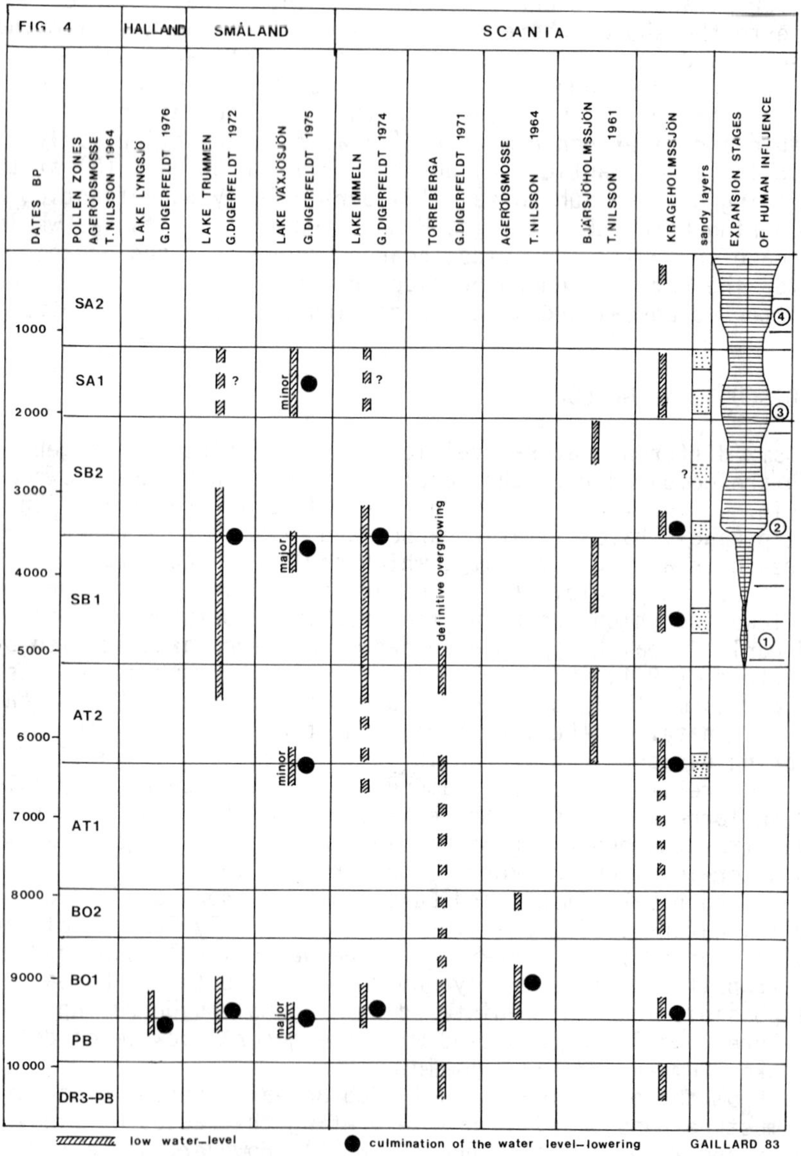

Figure 4. Water-level fluctuations in Krageholmssjön and in South-Swedish lakes. The expansion stages of human influence recognised in Krageholmssjön pollen diagram are presented on the right.

case again, it is not possible to exclude the intervention of man. Nevertheless, the possible water-level lowering in these four lakes are apparently contemporary; this fact could indicate a climatic cause.

The recent water-level lowering in Krageholmssjön is obviously the consequence of human activity and is most probably due to the excavation of the threshold.

If the preliminary interpretation of this palaeohydrological study is correct, it is interesting to note that the two rather strong human interference periods of Late Neolithic and Pre-Roman Iron Age correspond to low water-level periods. On the other hand, during the end of Sub-Boreal, a period evidently characterised by higher water-level, there are indications of slightly decreased human activity around the lake (Bronze Age). During the beginning of Late Sub-Atlantic, the rise in water-level is again connected with a period of reduced human activity (Late Iron Age). From the Viking Time, in spite of a rather high water-level, the human expansion is very important and continuous throughout Middle Age. If the assumption is made that water-level lowerings during Sub-Boreal and Early Sub-Atlantic have climatic causes, it follows that human activity around the lake was directly connected with water-level changes and climate (temperature and precipitation changes) until the Viking Time. Since that time, the fluctuations of human activity have probably been much more a function of political-economical than climatic-ecological factors (Berglund, 1969).

ACKNOWLEDGEMENTS

This study is part of the "Ystad project" (the human landscape during the last 6000 years) headed by B.E. Berglund. I am grateful to Professor Björn E. Berglund and Docent Gunnar Digerfeldt who offered me excellent working conditions in the Department of Quaternary Geology (University of Lund) and gave valuable criticism of the manuscript. The study was supported by a scholarship from the Swiss National Science Foundation.

REFERENCES

Berglund, B.E., 1969. Vegetation and human influence in South Scandinavia during Prehistoric time. Oikos Suppl, 12, 9-28.

Digerfeldt, G., 1971. The Post-Glacial development of the ancient lake of Torreberga, Scania, South Sweden. Geol. Fören. Stockh. Förh., 93, 601-624.

Digerfeldt, G., 1972. The Post-Glacial development of Lake Trummen. Regional vegetation history, water-level changes and palaeolimnology. Folia Limnol. Scand., 16.

Digerfeldt, G., 1974. The Post-Glacial development of the Ranviken bay in Lake Immeln. I. The history of the regional vegetation and II. The water-level changes. Geol. Fören. Stockh. Förh., 96, 3-32.
Digerfeldt, G., 1975. Post-Glacial water-level changes in Lake Växjösjön, central southern Sweden. Geol. Fören. Stockh. Förh., 97, 167-173.
Digerfeldt, G., 1976. A Pre-Boreal water-level change in lake Lyngsjö, central Halland. Geol. Fören. Stockh. Förh., 98, 329-336.
Gaillard, M.J., 1984. A palaeohydrological study of Krageholm lake (Scania, southern Sweden). I. The regional vegetation history. II. The water-level changes. LUNDQUA Report, 25.
Nilsson, T., 1961. Ein neues Standardpollendiagramm aus Bjärsjöholmssjön in Schonen. Lunds univ. årsskr. N.F. 2, 56:18.
Nilsson, T., 1964. Standardpollendiagramme und C14-Datierungen aus dem Ageröds Mosse in Mittleren Schonen. Lunds univ. årsskr. N.F. 2, 59:7.

TREES AND SHRUBS IN THE SHETLAND ISLANDS : EVIDENCE FOR A POSTGLACIAL CLIMATIC OPTIMUM?

Jacqueline Birnie

Aberdeen University, Scotland.

The Shetland Islands, by virtue of their location, offer considerable potential for the linking of the terrestrial climatic records of Britain and Scandinavia. Also, due to their small land area and remoteness from the continent of Europe, the islands should have responded sensitively to changes in ocean temperature. Minimal isostatic control has led to sea level rising throughout the Postglacial, impeding drainage and impounding sediment in most low-lying valleys, hence providing well-preserved records of environmental change. Included is pollen and macrofossil evidence for a limited phase of tree and shrub growth throughout Shetland within the period 8000 to 4500 BP. In this paper the evidence for this phase of vegetation history is critically assessed, and possible climatic implications are discussed.

The Shetland group consists of over 100 islands with coastlines deeply indented by voes. No part is further than 5 km from the sea, and the environment is not only humid, cool and windy, but also characterised by the highest recorded atmospheric salt levels in Europe (Johnston, 1974). The growing season is short, equivalent to that at about 300 m asl. on the Scottish mainland (Spence, 1979). At the present time the islands are virtually treeless, and the landscape appears bleak and windswept. Much of the area is peat-covered, and heath and bog communities of blanket peat and peaty podsols predominate, with small areas of improved pasture and some cultivation in lowland areas. Those few trees that do exist are in sheltered sites behind walls or around croft houses, and are severely stunted and pruned where they are exposed to the wind. Their prolific branching is characteristic of salt

N.-A. Mörner and W. Karlén (eds.), Climatic Changes on a Yearly to Millennial Basis, 155–161.

damage, and it would appear that salt spray, coupled with mechanical damage caused by high winds, is the major cause of growth limitation. Whilst grazing must also affect growth, even plantations excluding animals suffer from very slow growth and heavy losses during winter gales, and regeneration is not taking place. It is therefore reasonable to conclude that Shetland is beyond the natural tree line under present climatic conditions.

In contrast to the scarcity of living trees and shrubs, the remains of branches, trunks and root systems are widespread in the peat of both upland areas and valley basins, extensively exposed where peat is cut for fuel. In upland areas the wood consists of remains restricted to a single 10 to 20 cm basal layer underlying true blanket peat (Lewis, 1907 and 1911). In lowland basins wood fragments are distributed over a greater depth. Birch and willow are the most widespread types, with hazel locally abundant. Pine, alder and mixed-oak elements are thought to be absent. This wood forms the most unequivocal evidence for the existence of some form of woodland in the islands. There are five radiocarbon dates for woody remains in Shetland, ranging from 7870 to 4586 radiocarbon years BP (Fig.1). The regional picture, however, has to be provided by the pollen record of lake sediments and peat.

Relative pollen representation in postglacial sediments in Shetland does show a peak in arboreal pollen. At peak, tree and shrub pollen contributes between 40 and 50 per cent of the pollen total in contrast to the 15 per cent maximum of the present and early postglacial (Johansen, 1975; Birnie, 1981). However, there is a major problem in the interpretation of pollen records from small islands with restricted capacity for local pollen production, as variations in long-distance input may overwhelm local effects. The modern contribution of long-distance pollen, as represented in moss polsters from a range of sites, gives some indication of its past status. All arboreal pollen at the present time originates outside Shetland, and comprises a maximum of 15 per cent of the pollen total at lowland sites, increasing in upland areas (Birnie, 1981). Measurements over the open sea (Erdtman, 1943) show that around one-fifth of the long-distance component may be non-arboreal, giving an estimated total (arboreal plus non-arboreal) of 18 per cent of the lowland pollen spectra. Thus a past increase of tree pollen to 40 per cent is most unlikely to result from changes within the long-distance input.

The ratio of pine pollen to birch pollen provides an additional index of past long-distance input. There is no macrofossil evidence for postglacial growth of pine in Shetland, and its presence in Orkney is uncertain (Moar, 1969). Pine pollen is therefore almost certainly of long-distance origin, and probably dominated the long-distance spectrum due to its status in

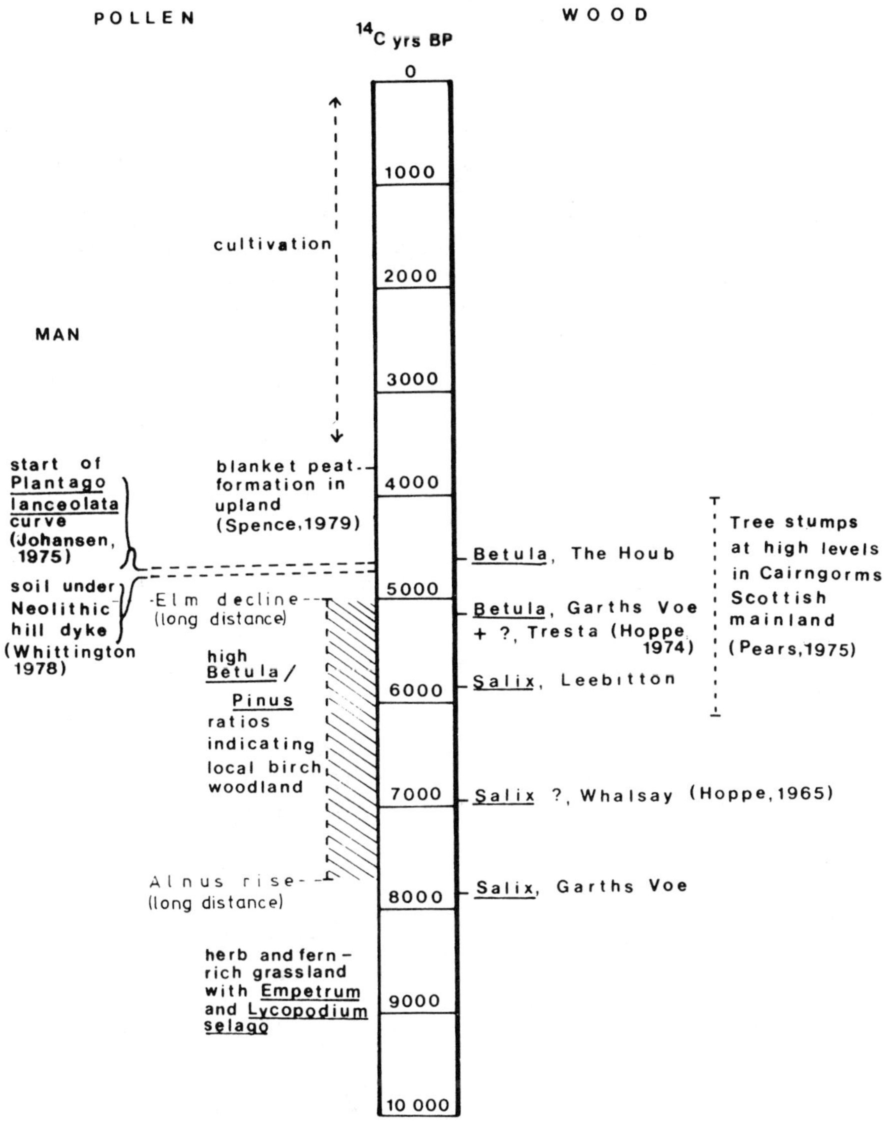

Fig.1 Dated evidence for environmental change in Shetland in the Postglacial period.

the source areas of Britain and Scandinavia, and its suitability for distant transport. In modern Shetland spectra the pine: birch ratio is between 1:1 and 1:3, but at the postglacial tree pollen peak the average ratio was 1:23. Pre-peak ratios are between 1:3 and 1:6. The peak is thus clearly defined as representing tree and shrub pollen of local origin.

Having established that the pollen record is not an illusion, it is of interest to date the deposits recording the duration of tree and shrub growth. Unfortunately, radiocarbon analyses of these sediments are not available. However, the clear representation of long-distance pollen in Shetland means that two major dated horizons from the British pollen record are identifiable, even though the trees represented were not growing in the islands. These horizons are the alder rise and the elm decline. The latter is one of the least time-transgressive vegetation changes, and so where a single decline in elm pollen representation is recorded, a date of around 5000 BP is indicated (Smith and Pilcher, 1973). The alder rise is more variable, but at least is unlikely to predate 7800 BP. The period of local tree and shrub growth in Shetland lies within, or very close to, these pollen boundaries at every site so far investigated. This is in accordance with the radiocarbon dates on the wood macrofossils, with the earliest date at 7870 ± 50 BP and the latest 4586 ± 40 BP.

Accepting the existence of an exceptional vegetation type in Shetland for a limited period of the Postglacial, possible climatic implications may be discussed. In general terms, the presence of trees and shrubs suggests that the climate must have been different from now, and some aspect of the degree of windiness, affecting both mechanical damage and quantities of airborne salt, is the most obvious explanation. This is also consistent with the model for Postglacial atmospheric circulation in the North Atlantic described by Lamb *et al* (1966) in which circulation weakened as the contrast in temperature between North America and Scandinavia lessened after 8500 BP. The consequent decrease in storminess would not only improve the environment of the outer islands of Scotland, but also that of the mountains, where Pears (1975) has obtained a cluster of radiocarbon dates ranging between 6000 and 4000 BP for tree stumps well above the present natural tree line. However, in upland areas changes in precipitation must also have been a controlling factor on growth, and dates for blanket peat formation, for example 3733 BP from Shetland (Spence, 1979, p.100), coincide with the end of the period of tree and shrub vegetation.

More detailed examination of the rest of the Shetland vegetation record, both before and after the wooded phase, and of changes that took place whilst trees and shrubs were relatively widespread, may shed further light on this circumstantial evidence for climatic control.

Prior to the arrival and spread of birch and hazel in Shetland, the vegetation is represented by a pollen record dominated by grasses, with some crowberry, and spores of the fir clubmoss characteristically present. Vegetation cover was closed, with some evidence for peat formation and the beginning of acidification of lakes, indicating that frost disturbance of soils had ceased. Such a vegetation could have been similar to a montane grass-heath of the present, where the length of snow lie and/or the shortness of the growing season preclude woody growth. But were trees and shrubs excluded or had they simply not had time to reach Shetland during this period? The arrival and spread of birch in the northern mainland of Scotland is dated to around 9250 BP (Birks, 1977). The birch pollen expansion in Shetland appears to postdate 7800 BP. Is a dispersal lag of 1450 years, in a rapid colonist, reasonable?

Once shrubs did begin to grow in Shetland, a gradual change in the composition of the vegetation is detectable in the pollen record at several sites. Where willow was the initial colonist, with a fern- and herb-rich understorey, birch replaced the willow later, and the understorey changed to species-poor heath. Whereas willow shrub development coincided with an improvement in the base-status of lakes, the birch was characteristically accompanied by acidification. These changes can be explained in terms of natural soil development, but leaching and podsolization may equally have been accelerated by increased precipitation. An alternative consideration is that grazing animals, and possibly man, may have arrived on the islands during this period. Direct evidence of occupation post-dates this period, yet several plant types which disappear from the record during this phase of shrub and tree growth, for example willow and polypody, continue to grow in inaccessible ungrazed and unburnt localities at the present time (Spence, 1979). Animals and man as agents of these changes cannot be dismissed.

Finally, the decline in trees and shrubs is the vegetation change in which climatic control is least clear, due to the number of other factors which may have been involved. It is certainly possible that there was an increase in storminess at or after 4500 BP, and that this in itself, or through increased precipitation, waterlogging and blanket peat development, prevented growth and regeneration of trees and shrubs. However, the earliest dated evidence for human settlement in Shetland is around 4700 BP (Whittington, 1978; Johansen, 1976), coinciding with the end of tree and shrub growth, and this neolithic settlement and cultivation, which was apparently widespread in upland areas, would have been likely to result in the destruction of any such vegetation. At some sites the pollen record shows that clearance and cultivation immediately followed the disappearance of trees and shrubs, but at others the association is not so clear, with a birch decline in some cases where there is no evidence for human interference. Therefore, the question of climatic controls must remain open until there is accurate dating of the decline at a wide range of sites.

The Shetland record is introduced here because it has been neglected as a source of information on climatic history, and is potentially very valuable due to the location of the Islands. The evidence for tree and shrub growth in the mid-Postglacial may indicate climatic change: an optimum (for Shetland) of decreased wind speeds. However, the problems of interpretation from this one record serve to remind those who reconstruct climatic history and argue over details of periodicities, just how insubstantial some of the footings still are. There remains much that may be done to improve climatic information retrieval from the sedimentary record, a process which must continue if theories of climatic change are to be well founded.

References

Birks, H.J.B., 1977. 'The Flandrian Forest History of Scotland: a preliminary synthesis' in 'British Quaternary Studies, Recent Advances' (ed. F.W. Shotton), pp.119-135. Oxford, Clarendon Press.

Birnie, J.F., 1981. 'Environmental changes in Shetland since the end of the last glaciation'. Unpubl. Ph.D thesis. University of Aberdeen, 542 pp.

Erdtman, G., 1943. 'An Introduction to Pollen Analysis'. Waltham, Mass. (*Chronica botanica*).

Hoppe, G., 1965. 'Submarine peat in the Shetland Islands'. Geogr. Annlr. 47A, 195-203.

Hoppe, G., 1974. 'The glacial history of the Shetland Islands'. Trans. Inst. Br. Geog. Special publication No.7, 197-210.

Johansen, J., 1975. 'Pollen diagrams from the Shetland and Faeroe Islands'. New Phytol. 74, 369-387.

Johansen, J., 1976. 'The age of the introduction of *Plantago lanceolata* to the Shetland Islands'. Dansk. Geol. Unders. 1976, 45-48.

Johnston, J.L., 1974. 'Shetland habitats, an outline ecological framework' in 'The Natural Environment of Shetland' (ed. R. Goodier), pp.33-49. Edinburgh : Natural Conservancy Council.

Lamb, H.H., Lewis, R.P.W. and Woodroffe, A., 1966. 'Atmospheric circulation and the main climatic variables between 8000 and 0 BC : meteorological evidence'. Proc. Int. Sym. on World Climate 8000 to 0 BC. Proc. Roy. Met. Soc. 99.

Lewis, F.J., 1907. 'The plant remains in the Scottish peat mosses, III : The Scottish Highlands and the Shetland Islands'. Trans. Roy. Soc. Edinb. 46, 33-70.

Lewis, F.J., 1911. 'The plant remains in the Scottish peat mosses, IV : The Scottish Highlands and Shetland'. Trans. Roy. Soc. Edinb. 47, 793-833.
Moar, N.T., 1969. 'Two pollen diagrams from the Mainland, Orkney Islands'. New. Phytol. 68, 201-208.
Pears, N.V., 1975. 'Radiocarbon dating of Peat Macrofossils in the Cairngorm Mountains, Scotland'. Trans. Bot. Soc. Edinb. 42, 255-260.
Smith, A.G. and Pilcher, J.R., 1973. 'Radiocarbon dates and the vegetational history of the British Isles'. New. Phytol. 72, 903-914.
Spence, D., 1979. 'Shetlands Living Landscape'. Thule Press, Shetland. 152 pp.
Whittington, G., 1978. 'A sub-peat dyke on Shurton Hill, Mainland, Shetland'. Proc. Soc. Antiq. Scot. 109, 30-35.

HOLOCENE PALAEO-WINDS AND CLIMATIC CHANGES IN WEST GREENLAND AS INDICATED BY LONG-DISTANCE TRANSPORTED AND LOCAL POLLEN IN LAKE SEDIMENTS.

Bent Fredskild

Botanical Museum, Copenhagen University, Denmark

Introduction

Long-distance transported exotic pollen have repeatedly been used in correlating arctic pollen zones with more southerly ones, e.g. by Hyvärinen (1972) on Spitzbergen, and they have also been used in an attempt to reconstruct palaeo-winds reaching Baffin Island (Nichols et al. 1978; Andrews et al. 1979; see also discussion in Barry et al. 1981). Ten years ago I tried to trace source-areas of exotic pollen in S. Greenland lake diagrams (Fredskild 1973), but as a great deal of further material is now at hand (Fig. 1), it seems reasonable to make an up-to-date review.

Quite many species of exotic pollen are found in S. Greenland samples, like Oak (Quercus), Hemlock (Tsuga), Elm (Ulmus), Beech (Fagus) and Ash (Fraxinus), but only a few in statistically sufficient numbers: Pine (Pinus), Spruce (Picea), Alder (Alnus), Wormwood (Artemisia) and Ragweed (Ambrosia). Among these the last mentioned is most rare, but even so, its occurrence in Holocene Greenland sediments may give information as to palaeo-winds.

Ragweed (Ambrosia)

The occurrence of 150 Ragweed pollen in four S. Greenland lakes (60-61° N) and four Godthåbsfjord lakes (64° N) varies with time (Fig. 2), best visualized by the mean number of pollen per sample for each millennium of calibrated dates. The picture seems clear: summer winds from the Ragweed area in the midwest prairie region were frequent in early Holocene, peaking in the period 7-6,000 uncalibrated radiocarbon years B.P., and then decreasing until the

N.-A. Mörner and W. Karlén (eds.), Climatic Changes on a Yearly to Millennial Basis, 163–171.

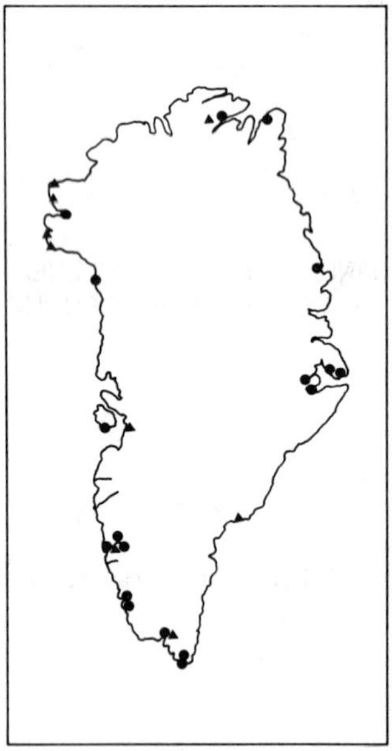

Figure 1. Map of Greenland showing position of lakes (dots) and peat deposits (triangles), from where pollen diagrams covering the whole or part of Holocene are at hand.

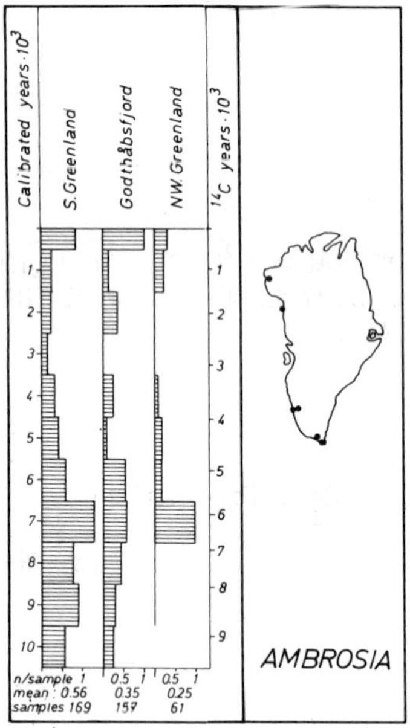

Figure 2. Mean number of Ragweed pollen per sample per millennium of radiocarbon dates, calibrated according to Damon et al. (1973) and Stuiver (1971).

period c. 3,200-2,400 B.P., followed by a slight increase in the coming millennium. The increase in the past five hundred years is a reflection of the recent forest clearance in N. America, and in fact the most pollen are found in the uppermost sample from the very lake bottom. As for the two NW. Greenland lakes (75-78° N) the picture is similar, yet the total lack of Ragweed pollen in the first two millennia might indicate that not until 7-6,000 B.P. did the pollen-carrying southwesterly winds to a greater extent penetrate that far north. This is confirmed by other exotics.

Pine (Pinus) and Spruce (Picea)

Pollen from coniferous trees are found in virtually all S. and W. Greenland pollen samples in a number, sufficiently high to allow the calculation of influx, which is the number of pollen deposited per cm^2 per year. There is, however, a marked difference in Pine and Spruce pollen influx.

Pine pollen are fairly frequent throughout (Fig. 3), especial-

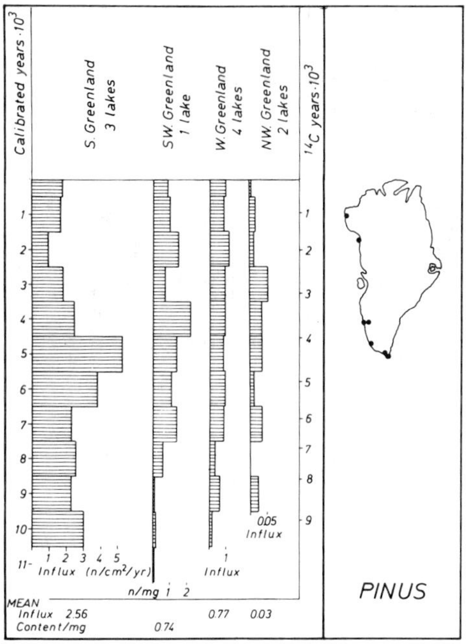

Figure 3. Mean influx or content of Pine pollen per millennium of calibrated dates.

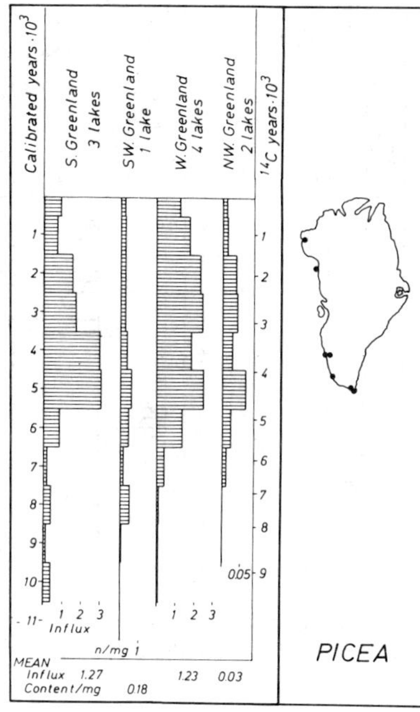

Figure 4. Mean influx or content of Spruce pollen per millennium of calibrated dates.

ly in the south, where the influx peaks 5-4,000 B.P., followed by a decrease to roughly 1/5 around 2,000 B.P. In the SW. and W. Greenland lakes the influx is fairly low until c. 7,000 B.P., but from then constant at a higher level. The Pine pollen in the first millennium in the histogram from the NW. Greenland lakes (note change in unit) are from the very deepest sample in only one of the lakes. This sample was rich in rebedded interglacial pollen, and the Pine pollen therefore are considered "old". Later on, these lakes differ from the more southerly ones in the markedly lower influx in the recent 2½ millennia.

Spruce pollen influx (Fig. 4) shows quite another picture: it is very low - or even missing - up till 7,000 B.P., peaking in the S. Greenland diagrams 5-3,000 B.P. and then decreasing, whereas the decline is not as pronounced in SW. and W. Greenland. In the NW. Greenland lakes no Spruce pollen is found in the first two millennia, and like with Pine, the influx decreases upwards. The great difference in Pine and Spruce pollen influx is perhaps best illustrated when looking at the ratio, which in many lakes changes from up to ten Pine to one Spruce in early Holocene to one Pine to two Spruce around 5,000 B.P.

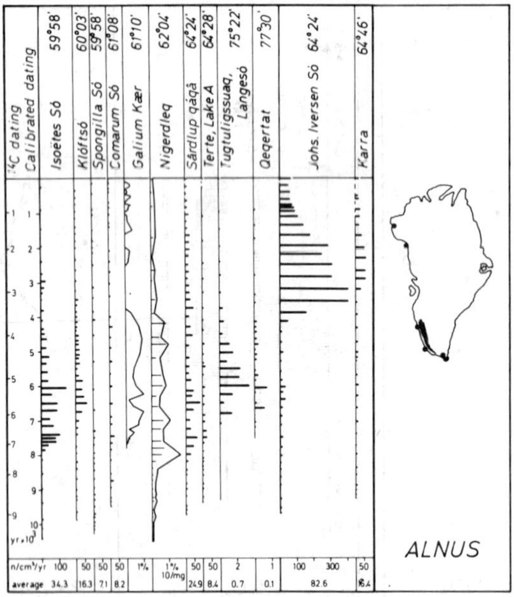

Figure 5. Influx (thick bars) or percentages (curves) of Alder pollen in twelve lakes, of which the ten (left) are outside the present area (shaded) of Green Alder (Alnus crispa).

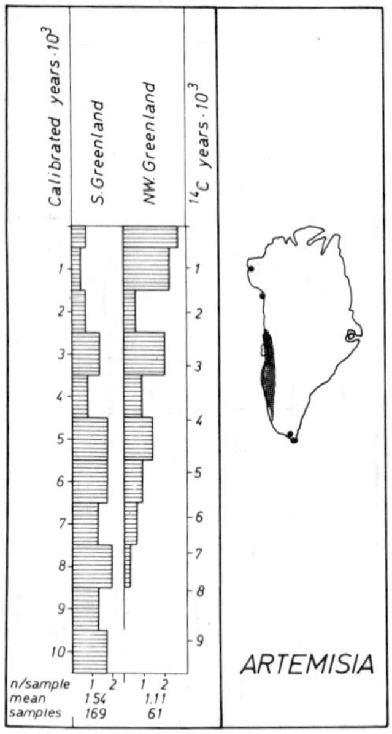

Figure 6. Mean number of Wormwood pollen per sample per millennium of calibrated dates. The present area of the Greenland representative of Wormwood (Artemisia borealis) is shaded.

A reasonable explanation would be to assume palaeo-winds in spring and early summer to S. and W. Greenland up till c. 7,000 B.P. to come mainly from the area south of the Laurentide Ice shield, where the Spruce-covered area was a narrow rim between the wider Pine area to the south and the ice to the north. Then the gradually deglaciated Labrador became the main source area. The change in Pine: Spruce ratio around 5,000 B.P. indicates that from that time winds from the Canadian taiga became more frequent.

Alder (Alnus)

Contrary to the plants mentioned, Alder - in this case the Green Alder - is growing in a restricted area in W. Greenland today (Fig. 5). The eagerly discussed question, when Alder spread to Greenland (Iversen 1954; Kelly & Funder 1974) will not be dealt with, but recent investigations (Fredskild 1983) indicate a late immigration. The ten lakes to the left in Fig. 5 are all outside the present area. The histograms with thick bars show pollen influx, the curves percentages, but the overall picture for the S. and W. Greenland lakes is very similar: around 7,000 B.P. there is a marked increase in influx, lasting till around 3,500 B.P. A similar maximum, of course with much lower influxes - note the reduction in unit to 1:100 - is seen in the NW. Greenland lakes, but as was the case with Ragweed, Spruce and supposedly also Pine, pollen-carrying winds did not reach that far north until between 6,500 and 6,000 B.P. The decline in all lakes around 3,500 B.P. seems to reflect conditions in Labrador, where Alder spread along the coast around 7,200 B.P. (Jordan 1975), showing metachronous maxima (Morrison 1970; Lamb 1980; Short & Nichols 1977), until a synchronous decline is registered west of Ungava Bay c. 3,500 B.P. (Richard 1981). Winds from this area must have been frequent during springtime, at least in the said period. Whether the Alder decline in the Greenland diagrams simply reflects the Labrador decline, or whether changing wind directions are included, cannot be decided on this basis alone.

Accidently Alnus spread to W. Greenland about the same time, almost immediately obtaining its maximum, as seen especially in the histogram from Johs. Iversen Sø, situated right in the heart of the Alder area.

Wormwood (Artemisia)

The present distribution in Greenland of Wormwood (Artemisia borealis), the pollen of which cannot be distinguished from other Artemisia pollen, is restricted to sunny slopes, especially in the interior between 62° and 72° N on the west coast. The histograms (Fig. 6) show the mean pollen per sample for four S. and two NW. Greenland lakes, well outside this area. The content of Artemisia pollen is fairly constant in the S. Greenland diagrams up till c. 4,000 B.P., then decreasing slightly. The picture in the NW. Greenland lakes is quite different: from low values in early Holocene, the frequency increases to reach a maximum around 3,000 B.P., contemporary with the minimum in Ragweed, and another one in the past 1½ millennium. This seems to indicate that the major part of Wormwood pollen comes from the arctic tundra regions of N. America, where the southward retreat of the timberline gave place to extended open communities with i.a. Artemisia frigida on xeric sites (Ritchie & Hare 1971; Nichols 1975).

Discussion

Summing up the evidences of palaeo-winds to the western half of Greenland, it can be stated that in early Holocene an all-year anticyclone over the Laurentide Ice caused winds from south of the ice cap to reach S. Greenland, but not NW. Greenland. In Mid-Holocene the gradually disintegrating Laurentide Ice reduced

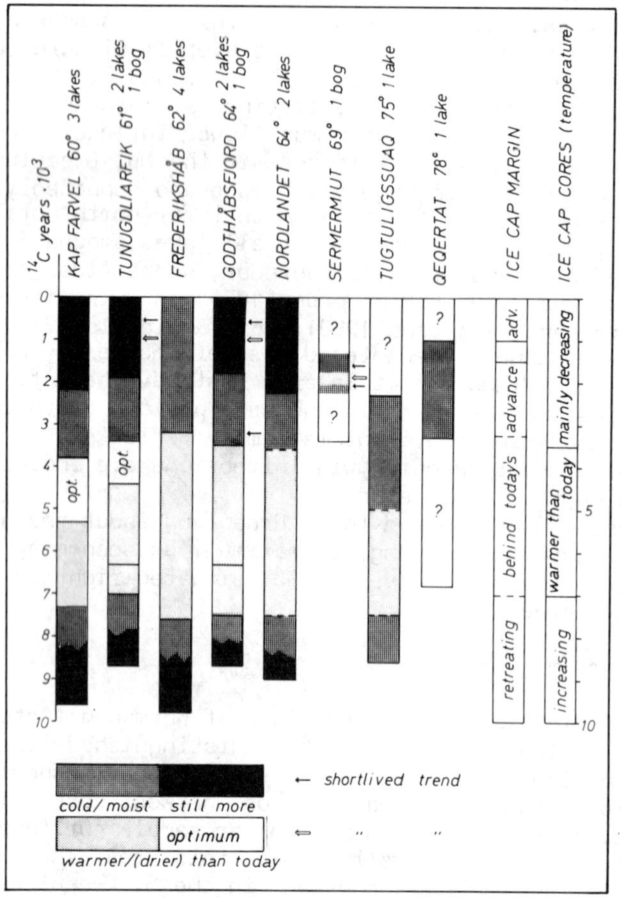

Figure 7. Inferred Holocene climatic conditions in the western half of Greenland. Sources: see text. The radiocarbon datings are uncalibrated.

the influence of the continental anticyclone (Macpherson 1982), permitting the development of more northerly cyclone tracks and the penetrating of warm, moist air passing Labrador and reaching all of W. Greenland. Later on, the southward movement of the polar-front caused winds from the Canadian taiga and tundra to reach NW. Greenland, whereas the Labrador region, now with wider areas covered with forest-tundra, remained the main source area of exotic pollen reaching the southwestern part of Greenland.

Consequently a question quite naturally arises: do these indices of changing winds in spring and early summer to the western half of Greenland match other inferred evidences of Holocene climatic changes in this part of the world? As to the western half of Greenland the evidences are summarized in Fig. 7. The "climate bars" begin at the time, when the sediment in the lakes changes from clay and sand to gyttja, supposedly some centuries after deglaciation. At first widespread arctic-alpine pioneer plants of open soil and snowbeds dominate, and two-three centuries later, plants especially ericaceous dwarf-shrubs from snow-protected heaths emerge. Another proof of a cool climate is made by the high-arctic Crustacea Lepidurus arcticus, which was living in most S. Greenland lakes, more than 500 km to the south of its present south limit in the lowland. But southwesterly winds were frequent, the temperature increased rapidly, and at a certain time, usually around 7,500 B.P., it reached and passed today's level.

During this warm period, minor changes, not caused by competition from new immigrants, are registered in some of the lake pollen diagrams. At what time the temperature peaked during the Hypsithermal, is not fully clear, but it may well have been fairly late. In most diagrams changes in the vegetation show a marked temperature decline and/or increase in humidity between 4,000 and 3,500 B.P., contemporary with the change in exotic pollen that indicated more frequent winds from the taiga and tundra. At a later date, c. 2,000 B.P., the climate became even colder and/or more humid. Minor fluctuations (marked by arrows in Fig. 7) in the past two millennia are mainly registered in peat deposits.

The high arctic diagrams look different. In the extremely maritime Melville Bugt lakes on Tugtuligssuaq, warmer and less humid conditions seem to have been prevailing in a period up to around 5,000 B.P., succeeded by a more humid period. On Carey Øer, west of Thule, the formation of a 2.6 m thick peat deposit began c. 6,500 B.P. and ceased 4,500-4,000 B.P. coinciding with the Hypsithermal (Brassard & Blake 1978). In the northernmost lake on Qeqertat, unambiguous evidences of an increased activity in frost boils surrounding the lake are seen at 3,300 B.P., initiating a colder period, which may have lasted until c. 1,000 B.P. (Fredskild, in print).

Kelly (1980) has summarized all evidences of glacier fluctuations and of now extinct thermophilous molluscs along the W. Greenland coast, concluding that Hypsithermal started c. 8,000 B.P. and lasted until between 3,500 and 3,000 B.P., followed by major glacier advances around 2,000 B.P. This trend is also reflected in the ice cap core (Dansgaard et al. 1973).

But also the west side of Davis Strait-Baffin Bay shows a similar Holocene climatic development. A diagram (Macpherson 1982, Fig. 9), showing the inferred Holocene climatic conditions in northeastern North America, matches Fig. 7 very well. Also, further north, a Mid-holocene climatic optimum melted the fiord-ice on Ellesmere Land to such an extent that driftwood was washed ashore, with greatest abundance in the period c. 6,000-4,200 B.P.

(Stewart & England 1983).

So summing up: the inferred palaeo-winds are likely to have reflected the Holocene climatic history of this part of the world - it would be most peculiar if they did not.

REFERENCES

Andrews, J.T., Webber, P.J. & Nichols, H., 1979. A late Holocene pollen diagram from Pangnirtung Pass, Baffin Island, N.W.T., Canada. Rev. Palaeobot. Palynol., 27, 1-28.

Barry, R.G., Elliott, D.L. & Crane, R.G., 1981. The palaeoclimatic interpretation of exotic pollen peaks in Holocene records from the eastern Canadian Arctic: a discussion. Rev. Palaeobot. Palynol., 33, 153-167.

Brassard, G.R. & Blake, W. Jr., 1978. An extensive subfossil deposit of the arctic moss Aplodon wormskioldii. Can. J. Bot., 56, 1852-1859.

Damon, P.E., Long, A. & Wallick, E.I., 1973. Dendrochronologic calibration of the carbon-14 time scale. Proceedings of the 8th International Conference on radio carbon dating, Wellington, New Zealand October 1972. Contribution No. 57, Departm. of Geoscience, University of Arizona.

Dansgaard, W., Johnsen, S.J., Clausen, H.B. & Gundestrup, N., 1973. Stable isotope glaciology. Meddr Grønland, 197(2), 1-53.

Fredskild, B., 1973. Studies in the vegetational history of Greenland. Palaeobotanical investigations of some Holocene lake and bog deposits. Meddr Grønland, 198(4), 1-245.

Fredskild, B., 1983. The Holocene vegetational development of the Godthåbsfjord area, West Greenland. Meddr Grønland, Geoscience, 10, 1-28.

Fredskild, B., in print. Holocene pollen records from West Greenland. In: Quaternary studies on Baffin Island, Baffin Bay, and West Greenland (ed. J. Andrews). Allen & Unwin.

Hyvärinen, H., 1972. Pollen-analytic evidence for Flandrian climatic change in Svalbard. Acta Univ. Ouluensis, A 3, Geologica, 1, 225-236.

Iversen, J., 1954. Origin of the flora of Western Greenland in the light of pollen analysis. Oikos, 4, 85-103.

Jordan, R., 1975. Pollen diagrams from Hamilton Inlet, Central Labrador, and their environmental implications for the northern Maritime Archaic. Arctic Anthropology, XII-2, 92-116.

Kelly, M., 1980. The status of the Neoglacial in western Greenland. Rapp. Grønlands geol. Unders., 96, 1-24.

Kelly, M. & Funder, S., 1974. The pollen stratigraphy of late Quaternary lake sediments of South-West Greenland. Rapp. Grønlands geol. Unders., 64, 1-26.

Lamb, H.F., 1980. Late Quaternary vegetational history of southeastern Labrador. Arctic and Alpine Research, 12,2, 117-135.

Macpherson, J.B., 1982. Postglacial vegetational history of the eastern Avalon Peninsula, Newfoundland, and Holocene climatic change along the eastern Canadian seaboard. Géographie physique et Quaternaire, XXXVI, 175-196.

Morrison, A., 1970. Pollen diagrams from interior Labrador. Can. J. Bot., 48, 1957-1975.

Nichols, H., 1975. Palynological and paleoclimatic study of the Late Quaternary displacement of the boreal forest-tundra ecotone in Keewatin and Mackenzie, N.W.T., Canada. Institute of Arctic and Alpine Research, Occasional Paper, 15, 1-87.

Nichols, H., Kelly, P.M. & Andrews, J.T., 1978. Holocene paleowind evidence from palynology in Baffin Island. Nature, 273, 5658, 140-142.

Richard, P.J.H., 1981. Paléophytogéographie postglaciaire en Ungava par l'analyse pollinique. Paléo-Québec, 13, 1-153.

Ritchie, J.C. & Hare, F.K., 1971. Late-Quaternary vegetation and climate near the Arctic tree line of Northwestern North America. Quaternary Research, 1,3, 331-342.

Short, S.K. & Nichols, H., 1977. Holocene pollen diagrams from subarctic Labrador-Ungava: Vegetational history and climatic change. Arctic and Alpine Research, 9,3, 265-290.

Stewart, T.G. & England, J., 1983. Holocene sea-ice variations and paleoenvironmental change, northernmost Ellesmere Island, N.W.T., Canada. Arctic and Alpine Research, 15,1, 1-17.

Stuiver, M., 1971. Evidence for the variation of atmospheric C^{14} content in the Late Quaternary. In: Late Cenozoic Glacial Ages (ed. K.K. Turekian), pp. 57-70. Yale University Press.

PEAT INCEPTION AND CLIMATIC CHANGE IN NORTHERN QUEBEC

Serge Payette

Université Laval, Québec, Canada

Minerotrophic and ombrotrophic peatlands are widespread all over the boreal regions and represent one of the most characteristic features of glacial landscapes both in North America and in northern Eurasia (Moore & Bellamy, 1974). Peatlands develop in sites where organic deposition exceeds decomposition. Peat accumulates in wet depressions by the filling-in process, and on slopes of particularly humid regions by the building-up process (Tarnocai, 1978).

In eastern Canada, outside the maritime regions where blanket bogs are found (Damman, 1979; Pollett, 1972), peatlands are mainly of the raised-bog and aapamire types (Damman, 1979). Along a latitudinal gradient, the convex raised bogs are progressively replaced northward by more flattened raised bogs because of abundant water supply (Damman, 1979), and finally by minerogenic patterned peatlands in the northernmost boreal regions, and form peatland complexes in large flat areas (Allington, 1961; Damman, 1979; Grondin & Ouzilleau, 1980; Sjors, 1963). In the forest tundra zone, peat plateaus and palsa bogs evolving under permafrost influence are the dominant peatland types (Payette et al., 1976; Zoltai & Tarnocai, 1975).

The present study is based on an analysis of the onset of peat accumulation in different parts of subarctic Québec, and their timing within the Holocene climatic framework. Peat inception in northern Québec varied throughout the Holocene, depending on deglaciation, catchment characteristics and climatic conditions. 14C-dates of basal peat in many areas are used to evaluate the minimum age of deglaciation. However, peat inception may be a response to climatic change instead of being only related to landscape release after deglaciation. To be sure of any geological or paleoecological interpretation, one must be a-

N.-A. Mörner and W. Karlén (eds.), Climatic Changes on a Yearly to Millennial Basis, 173–179.

ware of the particular conditions that gave rise to peat development at each site and, of course, of the peat material formed. In this connection, the timing of ombrotrophic peat inception in northern Québec around some subarctic lakes is of interest regarding paleoclimatic conditions, because Sphagnum peat initiation on well-drained deposits is directly related to atmospheric humidity rather than to site drainage conditions. It is thought that the climate of large lakes in subarctic Québec is conducive to ombrotrophic peat accumulation and peatland formation. I present here data on ombrogenic bogs developed on well-drained substrates, most likely of the blanket-bog type, around large lakes in the continental parts of subarctic Québec.

INCEPTION OF PEATLAND AND OMBROTROPHIC PEAT

The chronology of peatland and ombrotrophic peat inception in northern Québec (Fig. 1) was constructed using the method employed by Ochietti & Hillaire-Marcel (1977), Gagnon & Payette (1981), and Filion (1983). The time-spectrum of peatland inception spans a 6000-year period (Fig. 1a). Peatland inception includes here various wetland types (from fens to bogs) found in humid sites, and ombrotrophic peat deposits developed on well-drained substrates. As suggested by the cumulative statistical weight distribution of 14C-dates, peatland development was particularly active around 5500, 5000, 4500-4400, 4000-3800, 3000-2800, 2100, and 1600 BP. The building-up of ombrotrophic peat (Fig. 1b) around Clearwater Lake (56°N, 74-75°W) in northern Québec extended at least from 5000 to 1600 BP. Ombrotrophic peat inception was somewhat stimulated along lake shorelines and on small islands facing large fetch (5 to 15 km) around 4000-3800, 3000, 2100, and 1800-1600 BP.

These azonal blanket bogs and peat ramparts show distinct transverse profiles (Fig. 2). They developed either on rock outcrops and on weathered deposits bearing a brunisolic (or podzol-like) soil (Canada, 1978). The peat material is generally composed of Sphagnum and black spruce wood (Picea mariana (Mill.) BSP.), i.e. the remains of a vegetation similar to that presently growing on the surface. However, two isolated blanket bogs found at the centre of Clearwater Lake are made of Polytrichum and Dicranum remains (Fig. 2). Charcoal occurs frequently at the contact between the buried soil and the peat rampart.

The development of the small insular blanket bogs was rather rapid at ca 5000 and 4100-4000 BP, since basal peat on those sites covers non-pedogenetically weathered deposits (white dots in Fig. 1b). Elsewhere the deposits underlying the basal peats are all pedogenetically weathered, an indication that peat was accumulating at a slower rate. Available 14C-dated basal peats from an insular blanket bog, located at the centre of Clearwater Lake, indicate that ombrotrophic peat expansion was

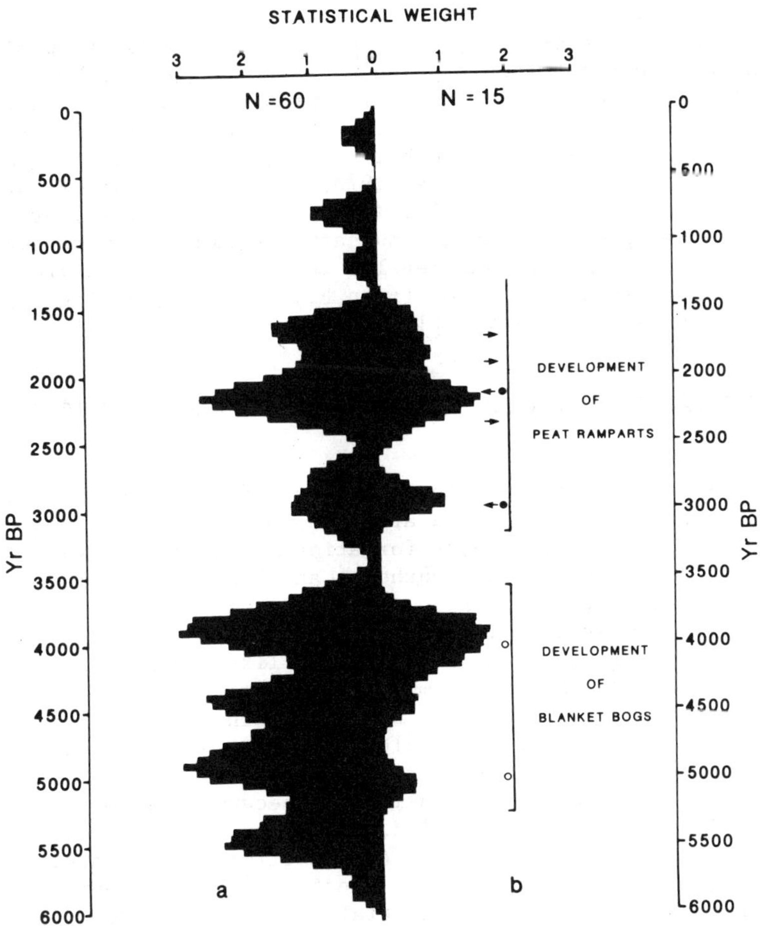

Figure 1: Distribution of 14C-dates of basal peats in northern Québec
a. Available dates from various peatlands (fens, ombrotrophic peats, palsa bogs, peat plateaus)
b. Inception dates of ombrotrophic peat in Clearwater Lake.

rapid between 4450 and 4000 BP, and rather slow after 4000 and 3000 BP (Fig. 2).

Along the lakeshores, some small blanket bogs and peat ramparts show peat outcrops exposed by wave action. The basal peat usually rests on boulders under the present lake level, suggesting that they developed at a time when the lake level was lower (→ on Fig. 1b). On the other hand, peat ramparts formed on brunisols and podzols (Canada, 1978) containing charcoal at the contact with the peat material (black dots in Fig. 1b) are associated with climatic change, indicating cooler and more humid conditions and higher lake levels. These buried soils dip gently under the present lake level (← on Fig. 1b), suggesting minor but significant lake-level fluctuations.

PALEOCLIMATIC SIGNIFICANCE OF PEATLAND AND OMBROTROPHIC PEAT INCEPTION

Peatland development in subarctic Québec dates back to the mid-Holocene, at the end and after the Hypsithermal period (Fig. 1a). The onset of peat formation in the region was delayed compared to northwest and southern Canada because of late deglaciation (Richard, 1977; Tarnocai, 1978). Considering the long ca 1500-2000-year time-lag between deglaciation and peat inception in most of the study sites, it is likely that peatland formation after 6000 BP was stimulated by humid climates, as for example between 6000 and 3800 BP; subsequent favourable peat growth periods were at 3000, 2100, and 1600 BP (Fig. 1a). The most important peat growth period was between 5500 and 3800 BP; afterwards peatland formation lowered because most of the peat-prone sites were already filled.

More specifically related to the climate, the onset of ombrotrophic peat accumulation in Clearwater Lake occurred around 5000 BP, and was somewhat accelerated around 4000-3800 BP when the climate became probably more humid (Fig. 1b). After 3100-3000 BP, ombrotrophic peat accumulated on near-shore well-drained soils in response to lake-level fluctuations probably induced by climatic change. The period between 5000 and 3800 BP was characterized by blanket-bog initiation and rapid development, and the period after 3100-3000 BP was primarily conducive to peat-rampart building (Figs. 1 & 2).

Because of the delicate balance between temperature and air humidity in subarctic climates, it seems likely that large lakes create locally more humid conditions, favourable for the formation of blanket bogs and peat ramparts. These subarctic blanket bogs, described here for the first time, differe from the oceanic blanket bogs (Damman, 1979; Moore & Bellamy, 1974; Pollett, 1972) in their botanical composition and in the climatic conditions causing their initiation. It is interesting to observe the relative synchroneity in their development with northwest Euro-

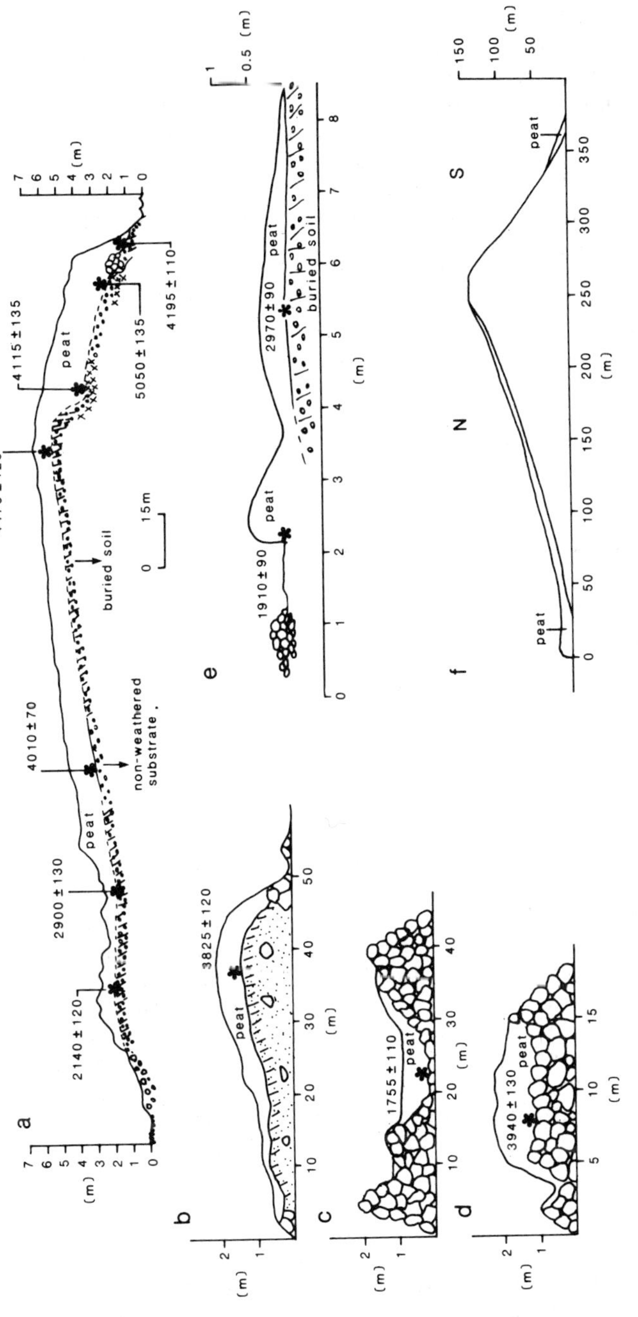

Figure 2: Insular blanket bogs (a,b,c,d,f) and peat ramparts (e) in Clearwater Lake area.
a, b, c, f: Sphagnum, black spruce and ericaceous peat
d: Dicranum and Polytrichum peat
e: Sphagnum and black spruce peat

pean blanket bogs developed since the end of the Hypsithermal, 5000 to 4000 BP ago (Moore, 1975; Pears, 1975). In northwestern Europe, the inception of blanket bogs coincides with human activity (Moore, 1979) and it is not certain whether the humid climate prevailing at that time is the sole factor in ombrotrophic peatland initiation; prehistoric forest clearance could have favoured the hydrological conditions conducive to peat formation (Moore, 1979). In subarctic Québec, human interference has been very scattered in time and space (Samson, 1981), and it is unlikely that it had any influence on ombrotrophic peat building-up. Subarctic blanket-bog development is thus directly related to the regional lake climate that prevailed since deglaciation, and more particularly between 5000 and 3800 BP when the climate was very humid and somewhat milder than during the following millenia. In that connection, it is noteworthy that many parts of the subarctic blanket bogs are now fossil because of the cessation of peat accumulation.

The lakeshore peat ramparts are much less extensive than the blanket bogs, and they expanded later, after ca 3000 BP, in response to climatic changes causing small lake-level fluctuations. A cold/dry climate with relative low lake-level possibly existed around 2300 BP and between ca 1900 and 1500 BP, and a cold/humid climate with relative high lake-level around 3000 and 2200-2100 BP. These preliminary data are consistent with Filion's (1983) paleoclimatic reconstruction from dune ecosystems in norther Québec.

ACKNOWLEDGEMENTS

Prof. Antoni W. H. Damman from the University of Connecticut made very useful comments on an earlier draft of the manuscript. Technical assistance in the field by the late François Gariépy and by Line Gauthier was greatly appreciated. I extend my appreciation to Michel Allard (Université Laval), to Réjean Gagnon (Université du Québec in Chicoutimi), and to Carole Mathieu (Université de Montréal), for providing unpublished radiocarbon dates from northern Québec peatlands. This work has been supported from operating grants of the Natural Sciences and Engineering Research Council of Canada and of the Ministère de l'Education du Québec (FCAC Programme).

REFERENCES

Allington, K.R., 1961. The bogs of central Labrador-Ungava: an examination of their physical characteristics. Geogr. Ann., 43: 407-417.

Damman, A.W.H., 1979. Geographic patterns in peatland development in eastern North America. Proceedings of the International Sympo-

sium on Classification of Peat and Peatlands. Finland. Intern. Peat Soc.: 42-57.
Canada, 1978. Le système canadien de classification des sols. Publ. 1646, Agriculture Canada, Ottawa, 170 p.
Filion, L., 1983. Dynamique holocène des systèmes éoliens et signification paléoclimatique (Québec nordique). PH.D. Thesis, Université Laval, Québec, Canada. 123 p.
Gagnon, R., & Payette, S., 1981. Fluctuations holocènes de la limite des forêts de mélèze, Rivière aux Feuilles, Nouveau-Québec: une analyse macrofossile en milieu tourbeux. Géogr. phys. Quat., 35: 57-72.
Grondin, P. & Ouzilleau, J., 1980. Les tourbières du sud de la Jamésie, Québec. Géogr. phys. Quat., 34: 267-299.
Moore, P.D., 1975. Origin of blanket mires. Nature, 256: 267-269.
Moore, P.D., 1979. Hydrological and Biological Interactions in Blanket Mire Inception and Development. Acta Univ. Oul., A.82. Geol., 3: 57-63.
Moore, P.D. & Bellamy, D.J., 1974. Peatlands. Springer-Verlag, New York. 221 p.
Ochietti, S. & Hillaire-Marcel, C., 1977. Chronologie 14-C des événements paléogéographiques du Québec depuis 14000 ans. Géogr. phys. Quat., 31: 123-133.
Payette, S., Samson, H. & Lagarec, D., 1976. The evolution of permafrost in the taiga and in the forest-tundra, western Québec-Labrador Peninsula. Can. J. For. Res., 6: 203-220.
Pears, N., 1975. The growth of hill peats in Scotland. Geol. For. Stockh. Forhandl., 97: 265-270.
Pollett, F.C., 1972. Classification of peatlands in Newfoundland. Proc. 4th Intern. Peat Congress, Helsinki. vol. 1: 101-110.
Richard, P., 1977. Histoire post-wisconsinienne de la végétation du Québec méridional par l'analyse pollinique. Service de la recherche, Direction générale des forêts, Ministère des Terres et Forêts du Québec. Vol. 1, 312 p. and Vol. 2, 142 p.
Samson, G., 1981. Préhistoire du Mushuau Nipi, Nouveau-Québec. Etude du mode d'adaptation à l'intérieur des terres hémi-arctiques. Ministère des Affaires culturelles du Québec. 2 Vol., 808 p.
Sjors, H., 1963. Bogs and fens on Attawapiskat River, Northern Ontario. Nat. Mus. Canada, Bull. 186: 45-133.
Tarnocai, C., 1978. Genesis of organic soils in Manitoba and the Northwest Territories. in Quaternary Soils (ed. W.C. Mahaney), p. 453-470. GeoAbstracts, England.
Zoltai, S.C. & Tarnocai, C., 1975. Perennially frozen peatlands in the Western Arctic and Subarctic of Canada. Can. J. Earth Sci. 12: 28-43.

THE NILE FLOODS AS A GLOBAL CLIMATIC/SOLAR PROXY

Rhodes W. Fairbridge

Columbia University, New York 10027

ABSTRACT

Flood levels of the Nile are available since the year AD 622, with some gaps (during periods of political instability). The maxima represent the summer monsoonal rains of Ethiopia and Sudan, furnished largely by the Blue Nile, whereas the minima in winter reflect the equatorial rains of East Africa supplied more sluggishly by the White Nile. Fourier and Maximum Entropy analyses have disclosed what appear to be the lunar nodal nutation cycle of 18.6 yr and the Gleissberg solar cycle of 78 yr, and other longterm periodicities, but only a weak representation of the 11-22 yr solar cycles. Statistical smoothing provides a chronology of trends, and a third technique, a search for climatic "singularities" (exceptional seasons) in comparison with computed celestial events, each provides further insight.

INTRODUCTION

Some 2400 years ago Herodotus observed the Nile delta and concluded that it was deposited by progressive increments of sediments carried down by the summer floods. Somewhat later, I was invited to study the Nile sediments in the area that was to be flooded by the Aswan High Dam (Fairbridge, 1962, 1964). Thanks to radiocarbon-dating, supplemented by stratigraphic and archeological surveys (Wendorf and Schild, 1976; Said, 1981; Adamson et al, 1980; Williams and Faure, 1980; Williams and Adamson, 1982) our early conclusion is now abundantly confirmed, that over the last 20,000 yr the Nile flow has oscillated from gigantic floods, up to four-times greater discharge than present

N.-A. Mörner and W. Karlén (eds.), Climatic Changes on a Yearly to Millennial Basis, 181–190.

maxima, to years of total failure. In the driest time, corresponding to the Pleniglacial peak, C. 19,000 to 17,000 BP, siltation had raised the flood plain of the middle Nile (above the First and Second Cataracts) to 40 m above the present surface. At times, the discharge "must have been reduced to the merest trickle" (Fairbridge, 1963). This corresponded to the time of the Sebilian culture, when Sandford and Arkell (1933) concluded that rainfall in Nubia must have dwindled until it finally ceased altogether. Subsequent workers recognized several aggradation stages in the late Pleistocene to early Holocene in the middle Nile valley, but unfortunately mistook aggradation to reflect high discharge. The silts of the middle Nile were extensively dissected and carried out to sea during the high-discharge phases subsequent to about 13,500 BP, as shown by deep-sea cores in the eastern Mediterranean.

Entering the Pharaonic era, both documents (papyri) and inscriptions demonstrate further dramatic fluctuations in Nile flow. A Middle Kingdom floodmark at Semna (75 km S of Wadi Halfa), attributed to Amenemhat III about 3750 BP is no less than 8 m above the present highest flood level (Ball, 1939). Downstream, in Lower Egypt, abormally high floods were recorded repeatedly during the epoch 3790 to 3725 BP (Bell, 1975), in contrast to catastrophically low levels around 4130 to 4080 BP, one of the cultural "dark ages". Exceptionally high levels are recognized again around 2450 BP (Tousson, 1923). Schove (1977) suggested that the periods of very low levels in AD 53 to 63, 151 to 170 and 254 to 265 were responsible, in part at least, for economic crises in the Roman world.

Comparison of the tentative curve of Nile discharge with the evidence of late Pleistocene to Holocene climate fluctuations (pollen records, ice-core oxygen isotopes) and with sea level show a close chronologic parallelism, "implying a common cause, though, of course, with differing amplitudes. We feel that this common factor can only be solar radiation" (Fairbridge, 1962, p. 110).

To test this hypothesis against actualistic data we need two time series: year-by-year discharge data, and year-by-year meteorological data. These can then be compared with astronomical evidence.

RECENT NILE DISCHARGE

The course of the Nile is divided roughly into three parts: that below the First Cataract (Aswan), the middle Nile (marked by six cataracts, extending up to Khartoum where it divides in two), and the two upper branches (the Blue Nile, from Ethiopia; the White Nile, from Uganda). At present the Blue Nile furnishes on average 68% of the flood discharge and 72% of the sediment load; it is fed by summer monsoonal rains that fall in Ethiopia and southeastern Sudan. The White Nile is supplied by equatorial rains, but its flow is strongly retarded by the swamps of the Sudd (which means "barrage" in Arabic) and its discharge at Cairo is measured by the winter, "minimum" level. It would really be better to refer to the two extremes as "summer" and "winter" discharges rather than maxima and minima, because they represent quite different watershed areas.

From the sedimentological and stratigraphic point of view, the middle Nile, between the cataracts, represents an ideal area of study, because local rainfall is minimal and the cataracts serve as natural barrages behind which sediments become ponded unless flushed out during high-discharge episodes. From Khartoum to Aswan the middle Nile drops 295 m in 1810 km, but from Aswan to the mouth the river drops only 76 m in 970 km.

Prior to the building of the Aswan High Dam ("Sudd-al-Ali"), the mean maximum summer discharge (in September) at Wadi Halfa was 712 million m^3/day. In contrast, the main winter discharge (in May), with 83% coming from the White Nile, was 45 million m^3/day. The annual average was 84 billion m^3/yr, but the range (during the recent recording period) was 41 to 140 billion m^3/yr (Hurst, 1954).

An analysis of total Nile discharge, considered since the year 1700, by Riehl et al (1979), disclosed a long-term interval of positive cumulative deviations from the average, running from a zero at 1725 and ending at 1840 (115 yr), reaching a peak about 1780. A second positive swing, from 1840 up till 1977 (possibly continuing) had its peak about 1897 (suggesting a longer cycle).

Prior to the above, in the period of the "Little Ice Age" of the 17th century (the Maunder Minimum of the sunspot cycle), snow fell in the high mountains of Ethiopia, above 3000 m, where such events are not experienced today (Hövermann, 1954). According to Gribben and Lamb (1978), at that time the summer westerlies weakened by about 30% or shifted south by 5° or so, while the subtropical monsoonal rains decreased and also receded southwards.

To compare the Nile flood behavior with authenticated meteorological data, one is essentially restricted to the present century. During this 80-yr period there have been five episodes of extremely "low Nile" (Quélennec, 1974; Flohn, 1981). They correspond to the great Sahelian droughts from the western Sahara to Somalia (Faure and Gac, 1981) and to low levels of Lake Chad and the East African lakes, thus:

(a) 1911-1915
(b) 1918-1928 (minor)
(c) 1941-1944
(d) 1969-1972 (minor)
(e) 1973-1976

At the first minimum the total discharge of the Nile at Aswan was only 45% of the average flow.

As pointed out by Flohn (1981, p. 399), the climatic perturbations were not limited merely to the North African region. Widespread droughts occurred at those times also in India and N.E. Brazil, although in the southern hemisphere tropics of Australia at those same times there were catastrophic rains. On the coast of Peru, instead of the usual aridity and upwelling, the warm El Nino current also generates heavy rains, in contrast to the Altiplano (Lake Titicaca) which has droughts. In 1972 in particular intense cold was registered from the Canadian Arctic to Greenland.

As pointed out by Faure and Gac (1981) in their analysis of the Senegal R. discharge, these dry periods are related in a complex way to the sunspot cycle but more to the 22 yr heliomagnetic cycle. The 1941/44 episode corresponds to a sunspot co-minimum with Sun/ Barycenter orbit, coinciding with the Saturn/Jupiter Lap (19.857 yr). A beat-frequency effect between the heliomagnetic cycle and the 18.6 yr lunar-tidal/nutation cycle is proposed for the U.S. droughts (Bell, 1981), of c 20.5 yr, also suggested in temperature series (Mock and Hibler, 1976), in Greenland ice cores and other proxies.

ANALYSIS OF LONG-TERM NILE DATA

The documentary evidence of the summer and winter discharge levels of the Nile is remarkable, and is complete on a year-by-year basis, save for certain regrettable gaps which correspond to loss of records or political upheaval. The records extend back to the year AD 622 (Year 1 of Mohammedan calendar). Earlier attempts at their analysis have been made by many scientists including Brooks (1949), Jarvis (1936), Popper (1951), Hurst (1952, 1954), Hövermann (1954), Herring (1979), Riehl et al (1979), and others. The most recent work was presented by Hassan (1981), who revised the source material and following the Ghaleb calibration, converted the data into meters above sea level. He very kindly furnished me with the complete printout, which has been provided a basis for Fourier analyses by my colleague Sultan Hameed and MESA (maximum entropy spectral analysis) by another colleague, Robert Currie. Both their work will be published elsewhere, but I have their preliminary results for testing by yet another technique.

The method employed by Hassan was mainly by preparing running means at various intervals; Hameed used Fast Fourier Transforms, and Currie used MESA. We are thus aware that both the lunar 18.6-yr cycle and various solar cycles are represented in the record.

My approach to the material has been twofold. First is by using a method recommended by Schove (1971, 1979) involving the search for singularities and patterns. This technique has been successfully applied since nearly a century ago when de Geer used it in obtaining teleconnections between late glacial and post-glacial varves. A "singularity" is a single day or a unique year in any meteorological time-series that is altogether unusual. Such "events", by definition, are unusual in the context of the selected time frame. The events become exceptionally useful when they cross over several watersheds or environmental boundaries. Some may even be global in effect, so that they constitute "silver spikes" or "silver threads" in the correlated time series. Often they are triggered by the Quasibiennial Cycle/ Southern Oscillation/El Niño effects (Landsberg, 1962; Ebdon, 1975; Sugiura and Poros, 1977; Fairbridge, 1983).

In the Nile records, over long periods of relatively normal solar radiation, the flood levels vary year to year within a range of about 0.5 m. But for a single year, exceptional lows may involve a fall of 2 m. Notably, during the disturbed (sunspot dearth) periods, the regular year-to-year fluctuations may be as much as 2 m, and exceptional lows can be 6 to 7 m below exceptional highs. In the framework of the Nile hydrologic regime, a 2 m drop in flood level can reflect a 50% drop in total discharge.

Notable lows (more than 2 m below average) have occurred as follows (with "silver spikes", i.e. extremes, marked with a star):

A.D. 650, 735, 764, 830, 832, 833, 841-2, 903, 927, 967(*), 1007, 1144, 1200-02(*), 1373(*),1404, 1450, 1715-16, 1766, 1788, 1809, 1824-25, 1833, 1835-59, 1845, 1855, 1864, 1868, 1873, 1877, 1899, 1913.

Since A.D. 1360 the very high years (mostly above 19 m; after 1800, above 20 m) have been as follows:

A.D. 136, 1426-43, 1512-14, 1594-96, 1603, 1611-12, 1625, 1698, 1724, 1727, 1738-47, 1751, 1754-58, 1763-64, 1767-78, 1770-71, 1775, 1770-8-, 1787, 1798, 1800, 1851, 1853, 1856, 1861, 1863, 1866, 1869-70, 1872, 1874(*), 1876, 1878(*), 1887, 1892, 1894-95, 1908, 1916-17.

Notable gaps in the record occur during the Turkish period 1520-1700, and following the Napoleonic occupation 1800-1825. The "average" flood height is a moving one because of secular siltation (0.8 to 1.0 mm/yr, or 0.8 to 1.0 m per millennium), but accelerations occurred from time to time because of political events and mismanagement (Hassan, 1981).

My second approach has been to compare the Nile data with the known lunar and solar cycles, and then with historical data on droughts, using the 20th-century evidence (noted earlier) that the major drought cycles are to be followed right across the Sahel belt of North Africa to the Nile headwaters. Naturally, there are also political and social factors, as well as metero-logical stochastic factors that make one drought area worse than another, but I have assumed that if there is any North African drought of historical (documentary) importance, it is worth noting against the Nile discharge record. My sources for this

North African drought information come mainly from Schove (1977) and Herring (1979), with some Ethiopian data from Hövermann (1954) and Wood (1977).

Inasmuch as the drought periods are controversial, because many social scientists tend to ignore natural causes and attribute them entirely to social events and stochastic fluctuations, it would be helpful to subject the data to rigorous statistical treatment. This is not at present feasible, however, because the record is extremely scattered and of variable reliability; furthermore, as one goes back in time the recorded events become progressively more sparse.

Also controversial are the planetary and lunar cycles discussed in a separate article (Fairbridge, this volume).

NILE FLOODS, DROUGHT AND PLANETARY PERIODICITIES

Brooks (1949) saw an 11 to 22-yr sunspot signal in the Nile record, but this signal is weak statistically. A comparison between sunspot data and Nile floods back to AD 622 discloses that summer and winter flood levels do not consistently correlate with either high or low sunspot conditions. Nevertheless, at times there is an alignment with various planetary signals. The relationship is quite complex, and although some data gaps exist in the time series, the evidence is as follows (I used a $\pm$ 2 yr precision for the sunspot data prior to the telescopic era):

1. Nile summer-flood minima correlate with sunspot minima at the following Saturn/Jupiter Lap (S/JL) years ($\pm$2 yr):

A.D. 633, 688, 730, 966, 989, 1065, 1086, 1167, 1212, 1389, 1666, 1723, 1766, 1784, 1823, 1844, 1866, 1901. (Through 1340 yr, there were 120 SS cycles, and 17 exact coincidences, despite many data gaps.)

2. Nile summer-flood maxima correlate with sunspot maxima at the following S/JL years ($\pm$2 yr):

A.D. 672, 760, 771, 852, 925, 949, 1125, 1227, 1249, 1307, 1328, 1405, 1426, 1607, 1625, 1883.

3. Over an interval of 600 yr (1360-1960), 24 high flood levels correlate with SS maxima, out of a possible 54. Over some 800 yr (622-1422) 63 high flood levels correlate with SS maxima ($\pm$2 yr) out of a possible 72 cycles.

4. Infrequently, the flood maximum matches a sunspot minimum that correlates with a Saturn/Jupiter Lap year, e.g.

A.D. 1345, 1368, 1587, 1644, 1744, 1923.

5. As concerns sunspot magnitudes, there are repeated periods of very low activity (or maxima R_m = < 80) with an average duration of 79 years (Stuiver and Quay, 1980), although the length variability is quite large. The most striking "dearth" episodes appear to correspond to times when the S/B_R (Sun/Barycenter Orbital Radius) showed maximum fluctuations (> 1.5 million km), accompanied by minor intermediate (IF) orbital

fluctuations. These are marked by stars (*) on the following list:

- o Dark Ages Minimum c. 660-740
- o Medieval Minimum c. 1010-1090
- o Wolf Minimum (*) 1282-1342
- o Spörer Minimum (*) 1416-1534
- o Maunder Minimum (*) 1645-1714
- o 19th C. Minimum (*) 1796-1833

Comparison with Nile-flood data shows that in general, the sunspot dearths correspond to times of extreme fluctuations or to protracted lows (Hassan, 1981). Mean cumulative discharge during the last 250 yr (Riehl et al, 1979) seems to match more closely the uneven (39-39-100) Gleissberg-Landscheidt periodicities, (Gleissberg, 1965; Landscheidt 1981), but have not yet been examined in detail.

6. According to both Hameed and Currie (personal communications) the lunar cycle of 18.6 yr provides a powerful signal in both summer and winter Nile-discharge data. Hameed et al (1983) found it in the 500-yr rainfall data from China. Currie (1981, 1982, and 1984) has identified the 18.6-yr period in MSL, air-pressure, and precipitation records from many different areas, and has extended the meteorologic base line with dendrochronological series in great number. In the 600 yr interval AD 1360 to 1960, I found 12 high flood levels on the Nile matching lunar declination or tidal maxima out of a total possible of 32 (although the data have many gaps in this period). Many low Niles match North African droughts back to AD 1340 (the limit of seemingly reliable data: Schove, 1977; Herring, 1979). The recent Sahel disasters were also among the most persistent (1974-1981: Lamb, 1982), but not unique.

A very important aspect of this lunar role has emerged from a recent thesis (Campbell, 1983; see also Campbell et al 1983), that considered the lunar tidal forcing in convection which is strongly influenced by the double sunspot (Hale) cycle (Bhalme and Mooley, 1980). When the Northern Hemisphere is warm and the zonal circulation strong, the lunar tides have only a weak effect; but in contrast, during the cool (blocking) epochs, the atmosphere is more "susceptible" to tidal forcing. The forcing was 2 to 3 times stronger in 1895 (N. Hemisphere cold) than in 1930 (N. Hemisphere warm). The year 1894 was an SS maximum and lunar-declination maximum. The lunar cycle must therefore be examined side by side with the solar record and with the long-term climatic proxies.

In studying the North African drought years, I compared them with both sunspot and lunar cycles. One may note an approximately 4-yr difference between maximal lunar declination and tidal effect: a periodic change in phase between correlations with one or the other needs careful study. Also curious is the way the droughts in certain epochs match sunspot maxima (e.g. 1860-1894, 1693-1705, 1405-1558) whereas at other times (e.g. 1901-1944,

1712-1733, 1587-1620), they consistently align with sunspot minima. Referring to Ethiopian droughts, cited by Wood (1977), all except one coincide (±2 yr) with sunspot minima. In many cases examined, the droughts came at or close to the maximum lunar tidal periodicities, (18.6 yr, 31.01 yr: see Wood, 1978) so this would also seem to be a promising line of future research.

CONCLUSIONS

1. Although my results are still extremely tentative and urgently require rigorous analysis, the messsage seems to be clear: the Nile-flood levels provides the world's longest and best-quantified climatic proxy, providing year-by-year values spanning an interval of 13 centuries.
2. A similar period covered by a NASA-JPL planetary ephemeris record permitted an analysis of orbital motions and alignments that appear to trigger solar emissions. Lunar (tidal) cycles are found to be interlocked with the same periodicities.
3. Nile floods and North (and East) African droughts both disclose interesting chronological relationships with both solar and lunar periodicities. More intense study should make prediction feasible in a way that can be of inestimable value for mankind.

REFERENCES

Adamson, D.A.,Gasse, F., Street, F.A. & Williams, M.A.J. 1980. Late Quaternary History of the Nile. Nature, 288, 50-55.

Ball, J., 1939. Contributions to the geography of Egypt. Cairo, Surv. and Mines Dept. 308 p.

Bell, B., 1975. Climate and the history of Egypt. Amer. Jour. Archaeology, 79, 223-269.

Bell, B., 1981. The combined solar and tidal influence in climate. In: Sofia, S. (ed.), Variations in the Solar Constant. Washington, NASA publ.241-254.

Bhalme, H.N., and Mooley, D.A., 1981. Cyclic fluctuations in the flood area and relationship with the double (Hale) sunspot cycle. Jour.Appl. Meteorol. 20, 1041-1048.

Brooks, C.E.P., 1949. Climate through the Ages, London, Benn (2nd ed.), 395 p.

Campbell, W.H., 1983. Possible Tidal Modulation of the Indian Monsoon Onset. University of Wisconsin-Madison, Dept. Meteorol., Thesis. 139.

Campbell, W.H., Blechman, J.B., 1983. Long-period Tidal Forcing of Indian Monsoonal Rainfall: an Hypothesis. Journal Climate & Appl. Met., 22, 287-296.

Currie, R. G., 1981. Amplitude and phase of the 11-year term in sea level - Europe, Journal Roy. Astron. Soc., 67, 547-556.

Currie, R. G., 1982. Evidence for 18.6 year term in air pressure in Japan and geophysical implications, Geophys. Jour. 321-327 Roy Astron. Soc., 69.

Currie, R. G., 1984. Evidence of 18.6 year nodal droughts in western North America during the last millennium, Journal Geophys. Res. (green), in press.

Ebdon, R.A., 1975. The quasi-biennial oscillation and its association with tropospheric circulation patterns, Meteorol. Magaz., 104, 282-297.

Fairbridge, R.W., 1962. New radiocarbon dates of Nile sediments, Nature, 196, 108-110.

Fairbridge, R.W., 1963. Nile sedimentation above Wadi Halfa during the last 20,000 years. Kush (Khartoum), 11, 96-107.

Fairbridge, R.W., 1964. "African Ice-Age Aridity." in Problems in Palaeoclimatology, ed. A.E.Nairn, NY., Interscience Publ. (J. Wiley & Sons), 356-363.

Fairbridge, R.W., 1983. The Pleistocene-Holocene boundary Quaternary Science Rev., 1 (3), 215-244.

Faure, H., and Gac, J.Y., 1981. Sahelian drought to end in 1985? Nature, 291, 475-478. Also: Ibid., 293, p. 414.

Flohn, H., 1981. Tropical climate variations during late Pleistocene and early Holocene, In: Berger, A. (ed). Climatic Variations and Variability: Facts and Theories, Dordrecht: D. Reidel, 233-242.

Gleissberg, W., 1965. The eighty-year cycle in auroral frequency numbers, Jour.Brit. Astron. Assoc., 75, 227-231.

Gribben, J. and Lamb, H.H., 1978. Climatic change in historical times, In: Gribben, J. (ed.). Climatic Change, Cambridge Univ. Press, 62-82.

Hameed, S., Yeh, W.M., L I, M.T., Cess, R.D. and Wang, W.C., 1983. An analysis of periodicities in the 1470 to 1974 Beijing precipitation record. Geophys. Res. Lett., 10 (6), 436-439.

Hassan, F.A., 1981. Historical Nile floods and their implications for climate change, Science, 212, 1142-1145.

Herring, R.S., 1979. Hydrology and chronology - the Rodah Nilometer as an aid in dating interlacustrine history. In: Webster, J.B. (ed.). Chronology, Migration and Drought in Interlacustrine Africa. Longman and Dalhousie University Press, 39-86.

Hövermann, J., 1954. Ueber die Hohenlage der Schneegrenze in Aethiopien und ihre Schwankungen in historischer Zeit. Nachr. Akad. Wiss Göttingen, 11a (6), 111-137.

Hurst, H.E., 1952. The Nile, London: Constable, 331.

Hurst, H.E., 1954. Measurement and utilization of the water resources of the Nile basin, Proc. Inst. Civil Eng., 3 (3), 1-50.

Jarvis, C.S., 1936. Flood-stage records of the river Nile, Trans. Amer. Soc. Civil Eng., 101, 1012-1071.

Lamb, P.J., 1982. Persistence of Subsaharan drough, Nature, 299, 46-48.

Landsberg, H.E., 1962. Biennial pulses in the atmosphere, Beitr. Physik Atmosph., 35, 184-194.

Landscheidt, T., 1981. Swinging sun, 79-year cycle, and climatic change, Jour.Interdiscipl. Cycle Research, 12, 3-19.

Mock, S.J. and Hibler, D.J. III, 1976. The twenty-year oscillation in eastern North America temperature records. Nature, 261, 484-486.

Popper, W., 1951. The Cairo Nilometer, Berkeley, University California Press.

Quélennec, R.E., 1974. Variations hydro-climatiques dans le bassin du Nil. Soc. Hydrotechnique France, 13th Journées d'Hydraulique, Q. 1 (Rpt 11), 7p.

Riehl, H., El-Bakry, M., & Meitin, J., 1979. Nile River discharge, Monthly Weather Review, 107, 1546-1553.

Said, R., 1981. The Geological Evolution of the River Nile, New York - Heidelberg - Berlin: Springer-Verl., 151 p.

Sandford, K.S. and Arkell, W.J., 1933. Paleolithic Man and the Nile Valley in Nubia and Upper Egypt, Chicago Univ. Oriental Inst., Publ. 17, 1-92. (Also: Ibid. 36, 1939: 1-105).

Schove, D.J., 1971. Varve teleconnection across the Baltic, Geogr. Annaler, 53 A, 214-234.

Schove, D.J., 1977./ African droughts and the Spectrum of Time, In: Dalby, D., Harrison Church, R.J., and Bezzaz, F. (eds.) Drought in Africa, London, Internat. Afr. Inst., Vol 2, 37-53.

Schove, D.J., 1979. Varve-chronologies and their teleconnections, 14,000-750 B.C., In: Schlüchter, C., (ed.) Moraines and Varves, Balkema, Rotterdam, 319-325.

Stuiver, M. & Quay, P.D., 1980. Changes in atmospheric carbon-14 attributed to a variable sun. Science, 207, 11-19.

Sugiura, M. & Poros, D.J., 1977. Solar generated quasi-biennial geomagnetic variation, J.Geophys. Res., 82, 5621-5628.

Tousson, O., 1923. Anciennes branches du Nil, Inst. Egypte Mem., 4, 212.

Wendorf, F. and Schild, R. 1976. Prehistory of the Nile valley, New York: Academic Press, 404 p.

Williams, M.A.J., and Faure, H. (eds.), 1980. The Sahara and the Nile, Rotterdam: A.A. Balkema, 634 p.

Williams, M.A.J. and Adamson, D.A., 1982. A Land between Two Niles, Rotterdam: A.A. Balkema, 246 p.

Wood, C.A. 1977. Preliminary chronology of Ethiopia droughts, In: Dalby, D., Harrison-Church, R.J., and Bezzaz, F. (eds.), Drought in Africa, London: International African Institute.

Wood, F. J., 1978. The Strategic Role of Perigean Spring Tides, In: Nautical History and North American Coastal Flooding, 1635-1976, NOAA, U. S. Dept. Commerce, Washington, 538 p.

RAINFALL PATTERNS AND LAKE LEVEL VARIATIONS AT LAKE CHAD

Dieter Jäkel

Free University of Berlin, Germany

Investigations of climatic fluctuations in the Sahelian zone frequently correlate the level of Lake Chad with the discharge data of the Senegal and Niger rivers and indeed the curves show remarkable similarity. However, the validity of this method should first be examined since, according to a personal comment by Monsieur Durand in the Office de la Recherche Scientifique et Technique d'Outre-Mer (ORSTOM), N'Djamena, 1978, the Bol station diagram of Lake Chad's water level during the period 1870-1953 is based mainly on values derived from Nile discharge data. Actual measurements were recorded only in 1907, 1908 and 1912-1919 and continuous observation first began in 1954 (Fig. 1). Individual measurements (marked with a cross in Fig. 1) and the curve of recorded data (solid line) show good agreemont with the reconstructed curve (broken line), which justifies the method chosen. How-

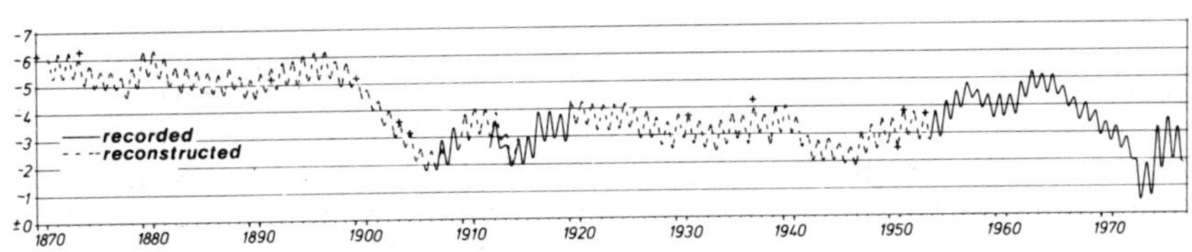

Figure 1. Level of Lake Chad according to ORSTOM (N'Djamena, 1978). Base line: 277.87 m (IGN, 1956) at Bol. Cross: depth marker.

N.-A. Mörner and W. Karlén (eds.), Climatic Changes on a Yearly to Millennial Basis, 191–200.

ever, a comparison with other records shows that data from Bol (14°43′E; 13°28′N) have an exaggerated amplitude from 1973 to 1977, i.e. they are not representative for the entire lake. A. Chouret (1977, Fig. 9) demonstrates this with great clarity. As the diagram shows, a reduction in lake level below 279 m a.s.l. results in considerable differences in water level in various parts of the lake. This is due to the barrier between Baga Kawa and Baga Sola (Fig. 2), which prevents the through-flow of water. Thus, the lake is subdivided into isolated basins. At the extremely low levels of 1973-1977 it was possible to cross the lake dryshod and the northwestern half of the lake was sometimes completely dry. The years 1905 and 1908 (cf. the maps by Freydenberg and Tilho in Chouret, 1977, Fig. 7a and 7b), 1914 and 1915 probably had similar conditions, but this is not directly visible from the diagram (Fig. 1). According to a village elder on the island of Haikoulou in 1973, this situation occurred in 1905 and 1907 at least.

The author's recent evaluation of monthly rainfall data from all climate stations in the Central African Republic, Cameroun, Niger, Chad and 120 stations in Nigeria has shown that direct correlations exist between rainfall in the lake basin and gauge height at Bol between 1936 and 1973. A computer program was developed to obtain year-by-year rainfall maps on the basis of equidistant grids. The mean annual rainfall distribution was then determined (Fig. 3). The extreme variability of precipitation sometimes reveals interesting ecological parameters. For example, the Sahelian zone corresponds to the range of the 300 mm isohyet, so that the Sahel may be defined as a zone of average rainfall between 200 and 500 mm (Fig. 3). Figure 3 shows the 300 mm isohyet in the years 1972 (southern limit of the hatched Sahelian zone) and 1950 (northern limit). Interannual climatic conditions may vary from one extreme to the other, as the years 1949 and 1950 show. The 300 mm isohyet for the year 1949 corresponds more or less to that of 1972.

This extreme variability is well illustrated by Figure 4, a profile ranging from tropical rainforest to desert, from Bangui in the Central African Republic to Faya Largeau in northern Chad. The rainfall of selected years is plotted, together with the recorded extremes, so that possible spatial shifts may be calculated. These may amount to 300 km a year, i.e. desert-like conditions may advance this far within one year. This demonstrates the shortcomings of studies on the extent of the Sahara that compare vegetation cover in the rainier fifties

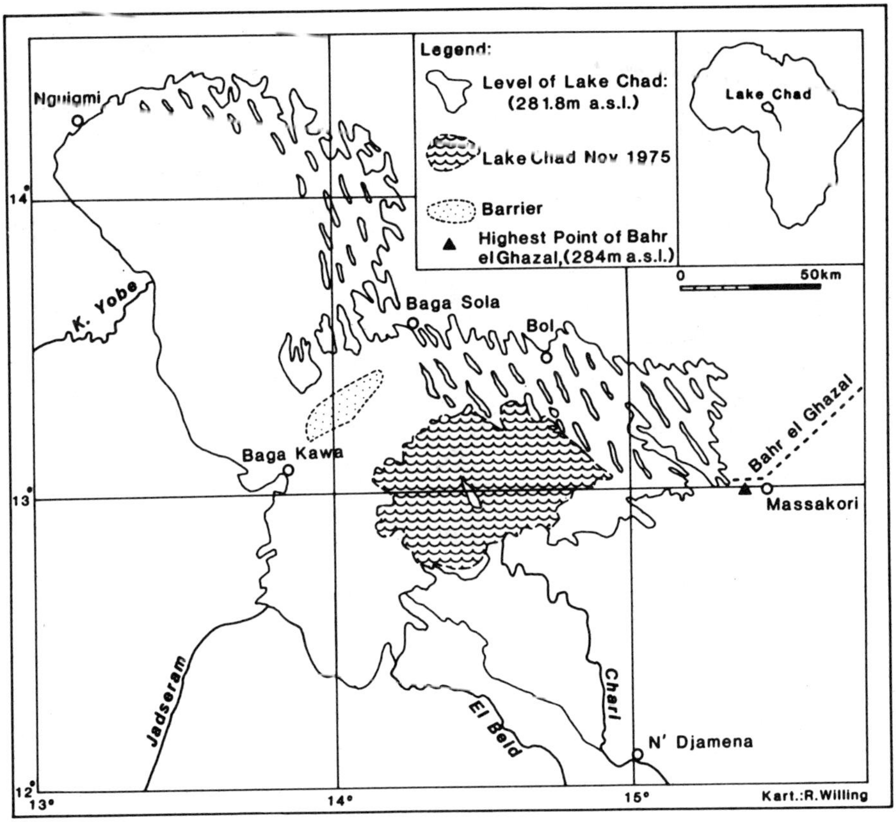

Figure 2. Extent of Lake Chad at its normal level of 281.8 m a.s.l. and Lake Chad in November 1975 (after A. Chouret, 1977, Fig. 1 and 4b).

with that in single years of the present period. Values between 30 and 120 km are quoted (U. Helldén, 1978, p. 32), giving a desertification rate of 2-7 km/yr if these 30 and 120 km respectively are divided by the number of years between each gauging. In my opinion, however, these spatial shifts remain within the normal range of present-day climatic fluctuations. In fact, the results described here indicate that the shifts may be even greater. (The environmental stress caused by human impact is not considered here.)

To compare rainfall in the catchment area of Lake Chad's affluents with the lake level sub-programs were developed in order to détermine the significance of the

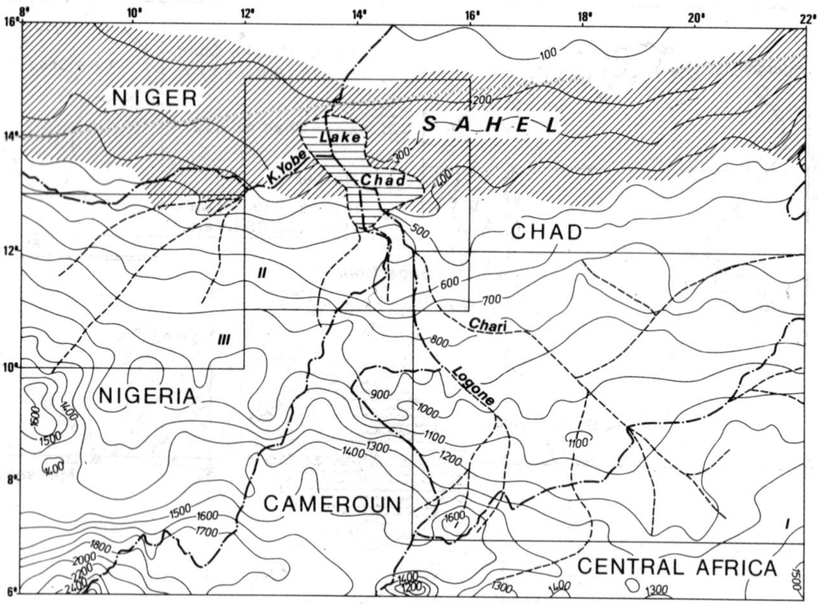

Figure 3. Mean rainfall in the catchments of Lake Chad affluents. Region I: Chari and Logone. Region II: Lake Chad and margins. Region III: catchment of Kamadougué Yobe.
Northern limit of the Sahel (hatched area): the 300 mm isohyet in 1950. Southern limit: the 300 mm isohyet in 1972.

individual affluents. Region I comprises the main tributaries Chari and Logone, Region II the actual lake, its immediate margins, and its affluents Jadseram and El Beid, and Region III the Kamadougué Yobe catchment. A rainfall diagram was constructed for each region for the period 1936-1973 (Fig. 5, I, II, III). Diagram IV is a combination of I, II and III and covers the entire Lake Chad catchment. We operated with the same number of matrix points in each case to exclude the risk of distortion. The diagrams of the three sub-regions show a fairly close resemblance, indicating that the entire catchment of Lake Chad's affluents represents a coherent climatic region. No contrary trends were established, hence diagram IV may be justifiably considered representative of the entire catchment. From this, an integ-

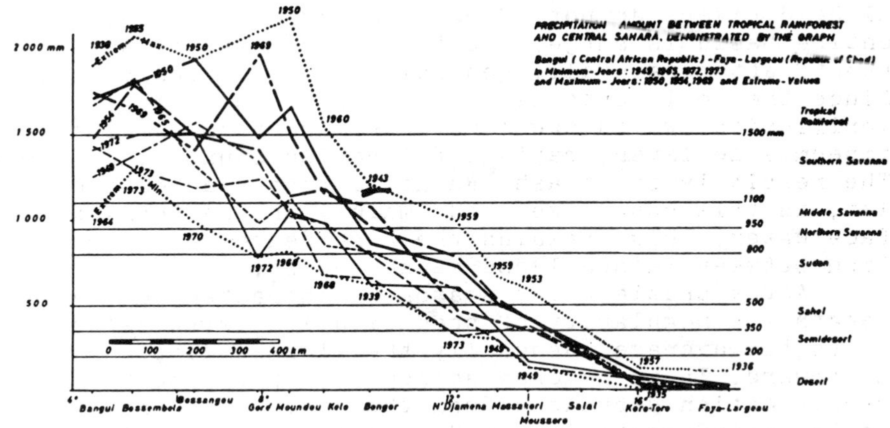

Figure 4. Precipitation along a profile from Bangui (Central African Republic) to Faya Largeau (northern Chad).

rative mean value of yearly rainfall was calculated, which corresponds more or less to the 1970 value.

An additional diagram on the same basis was produced for the area between 2°30′ W-5°30′ E and 6°-14°N, i.e. parts of western Nigeria, Niger, Upper Volta, Ghana, Togo and Benin (Fig. 5, diagram V). This curve differs considerably, suggesting that here the trend of climatic variation deviates from that of the Lake Chad area. This raises the question whether the wave-like path of the zonal "Walker Circulation" discussed by Flohn, 1981, and Fleer, 1971, also occurs in the Sahel.

Finally, diagram IV of Lake Chad basin was compared with the curve of gauge height at Bol (Fig. 5, VI). The majority of years demonstrate a direct correlation between the total precipitation of the catchment and the rise or fall in gauge height. Exceptions to the rule that an increase in rainfall over the previous year corresponds to a rise in gauge height, and a decrease to a fall, are due to the distribution of rainfall in the individual sub-regions and to the length of the dry season. Values in the catchment of the affluents Chari and Logone are the decisive factors here.

The dominant influence of Chari and Logone is also visible in the geomorphology of the area as a result of sediment transport. They create a rapidly increasing alluvial fan, whereas the fan at the mouth of Kamadougué

Yobe is comparatively underdeveloped. The Jadseram and El Beid rivers do not affect the lake shoreline. The entire basin is subject to the rapid onset of surface runoff after rainfall, generally short intensive storms. Since the soil contains clay minerals, especially montmorillonite, which swell when wet and form an almost impermeable layer, rapid infiltration cannot take place. The result is sheetwash. Rainfall thus reaches the outlet, in this case Lake Chad, mainly in the form of surface water. This explains the high degree of correlation between rainfall and gauge height.

A favourable rainy season is characterized by its length and regularity of precipitation, with values above the average. Generally the ITCZ is displaced northward. In an adverse season, it shifts south and precipitation decreases from the desert marginal regions to the rainforest; in the Sahelian zone the rainy season may be short and below the average, or long with extensive dry intervals between rainfalls. The water levels at Lake Chad demonstrate the following pattern:

1) The lake level sinks in spite of an increase in rainfall over the preceding year when the dry season in the upper catchment of Chari and Logone (Fig. 3, Region I) lasts 4 months.
2) It rises in spite of a decrease in rainfall over the previous year when the dry season in the above area lasts only 2 months.
3) It remains stable in spite of a slight decrease in rainfall when the dry season in the above area lasts 3 months.
4) It rises when rainfall increases and the dry season lasts about 3 months.
5) A short, but intense rainy season in the Kamadougué Yobe catchment (Region III, Fig. 3) with above-average values is able to compensate in the overall budget for the effects of a lengthy dry season in Region I.

From this analysis we may conclude that under present climatic conditions the Kamadougué Yobe alone would not be able to maintain a persistent outlet at Lake Chad. In successive seasons with deficient rainfall, such as

Figure 5. Rainfall histograms of regions I, II, III (Fig. 3), histogram of rainfall in the Lake Chad basin (IV) and in the area between 2°30′ W-5°30′ E and 6°-14°N (Ghana, Upper Volta, Togo, Benin, W. Niger, W. Nigeria (V), and curve of gauge height at Bol - extract from Fig. 1 - (VI).

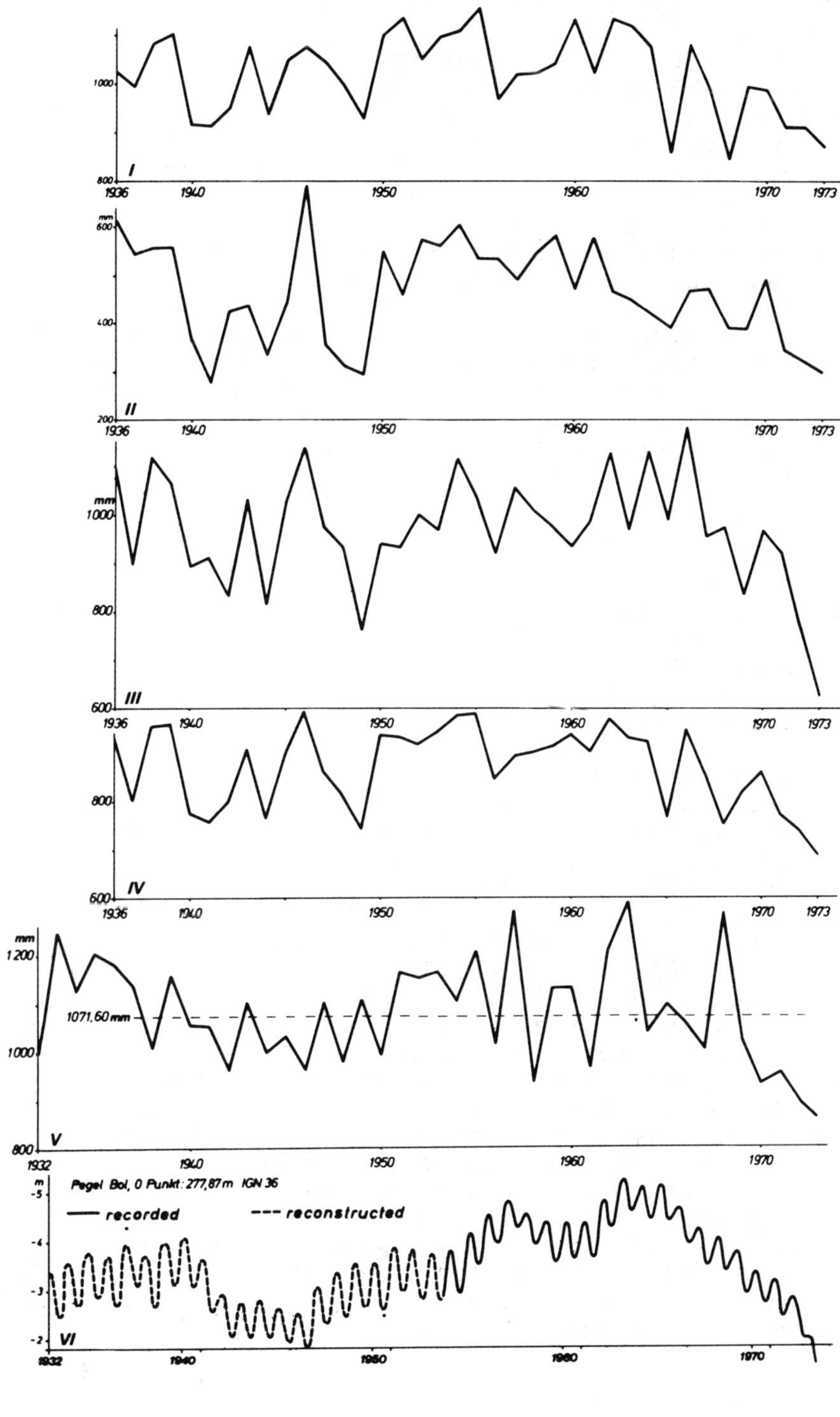
1000
800
I
1936
1940
1950
1960
1970
1973
mm
600
400
200
II
III
1200
IV
V
1071,60 mm
1932
Pegel Bol, 0 Punkt: 277,87 m IGN 36
recorded
reconstructed
m
-5
-4
-3
-2
VI

1972, 1973, 1975, the lake decreases in length by up to 150 km (cf. Fig. 2).

It may be assumed that if the climatic pattern of 1972 and 1973 lasted longer than 30 years, the Sahara would encroach about 300 km southward and, vice versa, would retreat by the same amount if rainfall occurred at the 1950 level for about 30 years running. Within 8 to 10 years, the lake would rise from its normal level of 281.8 m (cf. Fig. 2) to above 284 m a.s.l. and flow northeastward into Bahr el Ghazal, as must have been the case between 1870 and 1875 (cf. Fig. 1) judging from the topographic situation and the markers in Gustav Nachtigall's day. From this we may conclude that a considerable shift of the southern limit of the Sahara, in response to climatic fluctuations, is possible within the normal circulation pattern evidenced by available records and is not necessarily due to fundamental climatic change.

References

Cabot, J. & Bouget, C., 1972. Atlas pratique du Tchad. IGN, Paris.

Carmouze, J.P. & Dupont, B., 1970. Nouvelles Approximations sur la Bathymetrie et la Superficie du Lac Tchad. ORSTOM, Centre de Fort Lamy, Oct. 1970, 5 p.

Carmouze, J.P., Chouret, A. & Durand, J.R., 1973. Données récentes sur l'Evolution du Lac Tchad en 1972-73. ORSTOM, Centre de Fort Lamy, May 1973, 11 p.

Chouret, A.,1977. La Persistance des Effets de la Sécheresse sur le Lac Tchad. - ORSTOM, Centre de N'Djamena, Oct. 1977, 26 p.

Durand, A.,1978. Etude du Quaternaire récent dans la région des deltas du Chari (bassin du Tchad). Rapport préliminaire. ORSTOM, Centre de N'Djamena, Université de Niamey, Jan. 1978, 123 p.

Fleer, H., 1981. Large-scale tropical rainfall anomalies. Bonner Meteorologische Abhandlungen, 26, 114 p.Bonn.

Flohn, H., 1971. Tropical circulation pattern. Bonner Meteorologische Abhandlungen, 15, 55 p., Bonn.

Helldén, U., 1978. Evaluation of LANDSAT-2 Imagery for desertification studies in Northern Kordofan, Sudan. Lunds Universitets Naturgeografiska Institution, Rapporter och Notiser 38, Lund.

Jäkel, D., 1977. Preliminary account of studies on the development and distribution of precipitation in the Sahel and adjoining areas. Applied Sciences and Development, Vol. 10, p. 81-93, Tübingen.

Jäkel, D., 1979. Die Klimaoszillationen des 20. Jahrhunderts und ihre Auswirkungen im Bereich des Sahels. Verhandlungen zum 42. Deutschen Geographentag 1979, Göttingen, pp. 298-300, Steiner, Wiesbaden.

Maley, J., 1981. Etudes palynologiques dans le bassin du Tchad et paléoclimatologie de l'Afrique nord-tropicale de 30 000 ans à l'époque actuelle. Travaux et Documents de l'ORSTOM, No. 129, 586 p., Paris.

Nicholson, S.E., 1978. Climatic variations in the Sahel and other African regions during the past five centuries. Journal of Arid Environments, 1, pp. 3-24.

Nicholson, S.E. , 1979. Climatic fluctuations in the Sahel. Verhandlungen zum 42. Deutschen Geographentag 1979, Göttingen, pp. 297-98, Steiner, Wiesbaden.

Servant, M. & Servant, S. 1970. Les formations lacustres et les diatomées du quaternaire récent du fond de la Cuvette Tchadienne. Revue de Géogr. Phys. et de Géol. Dynamique, Vol. 12, 2. Fasc., 1, pp. 63-76.

Tetzlaff, G. & Adams, L.J. 1983. Present-day and Early Holocene evaporation of Lake Chad. In: Variations in the Global Water Budget (eds. A. Street-Perrott et al.), pp. 347-360, Reidel, Dordrecht/Boston.

ASECNA: Direction de la Météorologie Nationale, République Unie du Cameroun. Fiches de la Division Climatologique, jusqu'au 31 Décembre 1974. Archive Climatologique Douala.

ASECNA: Représentation au Niger, Service Météorologique. Pluviometries Mensuelles et nombre de jours de pluie, 1961-1974, Niamey.

ASECNA: Service Météorologique du Tchad. Fiches de la Division Climatologiques 1932-1973. Archive Climatologique, N'Djamena.

ASECNA: Représentation en République Centafricains, Service Météorologique. Fiches Pluviometriques Journalières, jusqu'au 31 Dec. 1974, Bangui.

Agrometeorological Bulletin, Nigerian Meteorological Department. January 1974-August 1974, Lagos.

Nigerian Meteorological Services. Archive of Climatology. Data of monthly rainfall, Lagos.

ORSTOM. Relèves Pluviometriques Journalières, jusqu'au 31 Dec. 1960, Tome I, II, République du Niger, Ministre des Travaux Publies des Mines et de l'Urbanisme, ASECNA, Service Météorologique.

Carte de l'Afrique Centrale au 1:200 000, Feuille ND 33, Publ. 1959, IGN Paris, Brazzaville.

Carte de l'Afrique au 1:1000 000, Feuille ND 33, Publ. 1960, 1969, IGN Paris.

ASECNA: Représentation au Dahomey, Service Météorologique. Bulletin Climatologique. Janvier 1972 jusqu'au 1er Octobre 1973, Cotonou.

ASECNA: Représentation au Dahomey, Service Météorologique, Fiches Pluviométriques, jusqu'au 31 Décembre 1974, Cotonou.

ASECNA: Représentation en République Haute-Volta, Service Météorologique. Fiches Pluviométriques, jusqu' au 31 Décembre 1974. Quagadougou.

Ghana Meteorological Services. Archive of Climatology. Data of monthly rainfall, Accra.

Résumé Mensuel du Temps. République Togolaise. Service de la Météorologie Nationale. Janvier jusqu'au Décembre 1972.

ASECNA = Agence pour la Sécurité de la Navigation Aérienne en Afrique et à Madagascar.(Agency for Air Navigation Safety in Africa and Madagascar).

ORSTOM = Office de la Recherche Scientifique et Technique Outre-Mer. (Overseas Scientific and Technical Research Department).

TROPICAL PEATS: HYDROGEOLOGIC AND CLIMATIC CONTROL

Faure, H., Gac, J.Y., Hillaire-Marcel, C., Lezine, A.M., Monteillet, J., Ngom, P.M., Pezeril, G. & Saos, J.L.

CNRS; France and ORSTOM, Université de Dakar, Direction des Mines et de la Géologie; Senegal

Along the flat sandy coasts, estuaries and deltas of West Africa, from South Mauritania to Nigeria, Holocene sediments fill wide areas of low-lying countries. The thickest deposits (over 30 m) are found in ancient river valleys or in deep depressions between Upper Pleistocene eolian sand dunes (Niayes).

The infilling of these ancient depressions is made of very fine grain material (silt, clays) and reworked sand that are very rich in organic vegetal components and peat layers.

Most of these sediments, sand and fine organic silt hold at various dept ground water. Its salinity is ranging from sea water along the coast and estuaries, to fresh water inland, along rivers and in lenses floating above the coastal salted bevelled water.

The present disposition of the depressions shows various environments ranging from mangrove estuaries and lagoons to evaporated dry salt pans (sebbhas) or permanent fresh water intradune lakes. The extreme variety of hydrological situations is recorded in the chemistry of soil and water and in the vegetation.

The different types of depression are essentially controlled by the water table relation to the topographic surface. In each particular place the local annual hydrologic budget determines the geochemical and botanical environment. When the water table changes in response to climatic crisis, the environment of each depression tends to change and goes through different types in a logical sequence.

During the Holocene and Upper Pleistocene a drastical change of the water table position and environment was a consequence of postglacial changes. The water table position underwent a general rise favourable to rapid accumulation for three major causes: Sea level, climate and local paleogeography.

1. The Sea level rose (at a mean rate of 1 cm per year) from

N.-A. Mörner and W. Karlén (eds.), Climatic Changes on a Yearly to Millennial Basis, 201–203.

17,000 to 7,000 B.P. and forced a rise of the water table along the coast and estuaries. From 7,000 to present 2 or 3 minor sea level fluctuations produced environmental changes competing with climatic water table changes (e.g. Faure, 1980; Faure et al., 1980).

2. The climatic rainfall increase at early Holocene induced a water table rise, that is to say fresh water lens more or less thick floating above the sea level controlled water table. A more important feeding of the underground waters by rains and rivers occurred during periods of increased run-off at Holocene (e.g. Faure & Gac, 1981; Gac et al., 1983).

3. The most effective paleogeographic change of water table is the construction of successive littoral sand barriers that favoured accumulations of fine material and mangrove peats in shaded areas.

In deltas, progradation and meandering cause rapid changes in local rates of sediment accumulation as well as in the situation of favourable depressions.

The thick peat accumulations are thus the result both of post-glacial sea level rise that controlled deep water anoxic depressions and Holocene wet climate that favoured vegetation. The oldest radiocarbon dates from Benin and Ivory Coast show propitious conditions for peat formation around 23,000 and 19,000 years B.P. In the Guinea Golf, shelf peats of 12,000 to 11,000 years B.P. are found at 60 m water depth. In Senegal, dates between 9,000 and 8,000 B.P. are obtained at a depth between 28 m and 23 m from the thickest infilling of valley and from continental shelf mangrove. The deepest paleo-valley and Niaye show a nearly continuous peat accumulation locally of several meters covering the end of the Holocene up to the present.

The "Direction des Mines et de la Géologie" of Senegal has estimated a volume of above 50 10^6 m^3 of peat in the Niayes zone. Its exploitation would save a precious amount of fuel and preserve the Sahelian zone from deforestation for fire wood production.

A better understanding of the detailed processes of the tropical peat genesis could help to reconstruct the proper environmental conditions so as to renew the peat formation after its exploitation. In this perspective man could make the peat a semi-renewable energy ressource growing again for the future.

REFERENCES

Faure, H., 1980. Late Cenozoic vertical movements in Africa. In: Earth Rheology, Isostasy and Eustasy (N.-A. Mörner, Ed.), Wiley, 465-469.

Faure, H., Fontes, J.C., Hébrard, L., Monteillet, J. & Pirazzoli, P.A., 1980. Geoidal changes and shore level tilt along the Holocene estuaries: Sénégal River area, West Africa. Science 210, 421 - 423.

Faure, H. & Gac, J.-Y., 1981. Will the Sahelian drought end in 1985? Nature 291, 475-478.
Gac, J.-Y., Monteillet, J. & Faure, H., 1983. Marine shorelines in estuaries as paleoprecipitation indicators. In: Variations in the global water budget (A. Street-Perrott, M. Beran & R. Ratcliffe, Eds.) Reidel, 361-370.

CAUSES OF A LATE-ATLANTIC SWAMP-FOREST SUBMERGENCE IN THE CENTRAL RHINE/MEUSE DELTA

Orson van de Plassche

Institute of Earth Sciences, Free University, Amsterdam

It is argued that the late-Atlantic phase of swamp-forest submergence in the central part of the Rhine/Meuse delta (Van der Woude, 1981) has no direct paleo-climatic significance. The principle cause was adjustment of the river gradient induced by the seaward displacement of the zone of maximum tidal range reduction within the intracoastal area.

INTRODUCTION

Over the past 25 years tens of 14C dates have been published for the base of peats that originated on the slopes of Late-Glacial riverdunes in the Rhine/Meuse delta in response to the postglacial sea-level rise (Jelgersma, 1961; Louwe Kooijmans, 1974, 1976; Van der Woude, 1981; Van de Plassche, 1982). These compaction-free samples provide excellent time/depth data from which not only water-level changes can be derived but also spatial and temporal changes in river gradient. The aim of this paper is to draw attention to the importance of gradient adjustment when interpreting evidence of rapid water-level rise in deltaic sequences.

GRADIENT LINES AT 6700, 6050, 5300 AND 4450 BP

The position of six riverdunes in the Rhine/Meuse delta for which time/depth data are available is indicated in Fig. 1. Figure 2 gives a cross-sectional view of the (projected) position of the dunes. The dots on the vertical lines (symbolizing the dunes) indicate the position of radiocarbon-dated basal peat samples. All samples, except for those from the Hillegersberg

N.-A. Mörner and W. Karlén (eds.), Climatic Changes on a Yearly to Millennial Basis, 205–214.

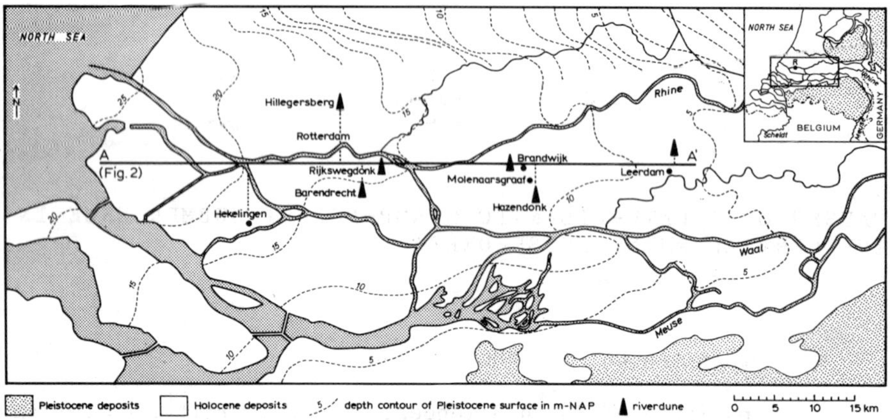

Figure 1. Map of the Rhine/Meuse delta showing the location of sites mentioned in the text.

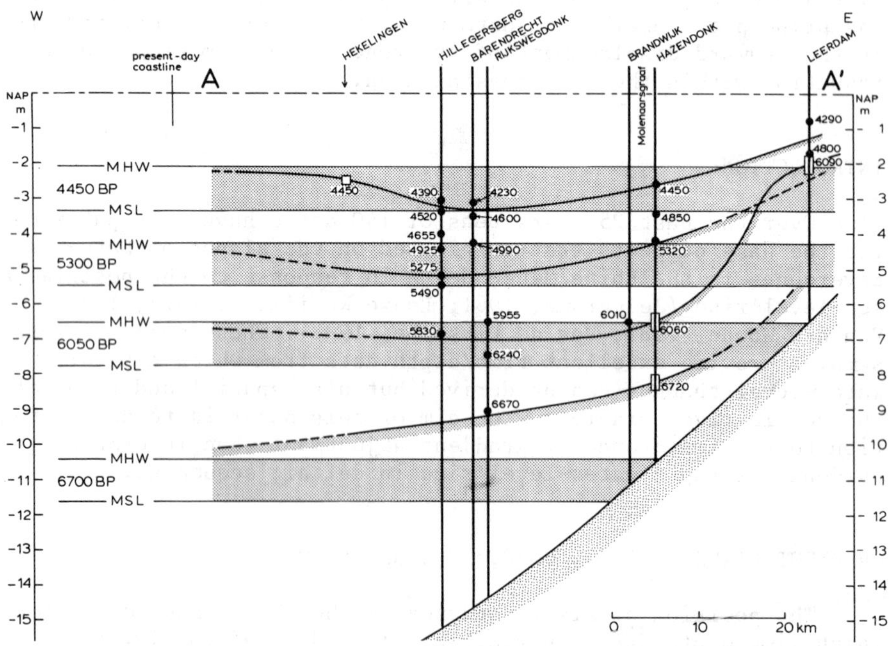

Figure 2. Gradient lines at 6700, 6050, 5300, and 4450 BP derived from published time/depth data (see text for references).

riverdune, consisted of wood peat. This may be assumed to have been formed at or very close to the local mean high water (MHW) level or, where tidal influence was absent, the average groundwater table. The Hillegersberg samples consisted of *Phragmites/Carex* peat, which may have been formed at a water depth of a few or several decimeters. As for the 4450-BP datum at Hekelingen, this represents local MHW derived from the compaction-corrected height of tidal creek levees on which remains of the Vlaardingen-Ib Culture were found (Louwe Kooijmans, 1976). The 'ages' of the open boxes on the Hazendonk riverdune refer to the radiocarbon age of samples taken at the base of two intercalated peat layers situated *c.* 1 km north of the riverdune. The original altitude of the lower sample was obtained by estimating the amount of compaction. For the 6060 BP sample the altitude ranges between *c.* -6.80 and -6.25 m NAP (Dutch Ordnance Datum), the lower value being the compaction-free position of the clay below the sampled peat bed on the dune and the higher value being the highest observed position of sandy channel facies of the same clastic unit. The height range of the open box on the riverdune near Leerdam similarly refers to the height of highest observed, compaction-free, sandy channel deposits and the compaction-free position of correlative clay on this dune. The age of 6090$\pm$70 BP refers to the base of the peat that was formed on top of the clay (Van der Woude, 1981). The sandy channel deposits emerged some time after *c.* 6100 BP as shown by altitudinal-stratigraphical, pollen analytical and radiocarbon data (see Van der Woude, 1981 for details). The water level attained at about 6100 BP was not reached again until about 4800 BP. This observation is of importance when reconstructing the former gradient lines.

From the data depicted in Fig. 2 and those mentioned above the approximate gradient lines as existed at about 6700, 6050, 5300 and 4450 BP have been derived. As discussed below, the changes in configuration of these gradient lines can be largely understood in terms of sea-level change and spatial variation in tidal range reduction.

SEA-LEVEL CHANGE AND STORAGE-BASIN EFFECT

The left part of Fig. 2 shows the approximate position of MSL (and assumed coastal MHW) at 6700, 6050, 5300 and 4450 BP (Van de Plassche, 1982). As is to be expected, the river gradient decreased as sea level increased. An exception, however, is the 6050 BP gradient line which, despite an overall sea-level rise of almost 4 m since 6700 BP, seems to have undergone a temporary gradient increase with respect to the situation at 6700 BP. This deviation can be satisfactorily explained by the so-called storage-basin effect: the lowering of the MHW level within smaller or larger parts of the intracoastal area due to frictional dissipation of energy as the tidal wave moves into the estuaries, lagoons and/or tidal gullies and creeks, and/or by the fact that (parts of) the intracoastal area may act as a storage basin (Van Veen, 1950; Zonneveld, 1960; Van de Plassche, 1980). The principle is illustrated schematically in Fig. 3, which shows that a given gradient line is a function of the gradient effect (the upstream increase in height of tidal levels solely due to the river gradient) and the storage-basin effect.

From the smooth seaward inclined 6700 BP gradient line in Fig. 2, it seems safe to conclude that no storage basin effect was present in the area east of Hekelingen around 6700 BP. Apparently, the river gradient and the distance to open sea were too great for any tidal influence to reach so far inland.

Between 6700 and 6050 BP a storage-basin effect entered the area under discussion. For the present argument it is important to note that the 6050 BP gradient line remains sub-horizontal up to the Molenaarsgraaf area. This indicates that the tidal amplitude reduction reached its maximum there (*cf.* Fig. 3).

Around 5300 BP peat growth in the area of Rotterdam (Fig. 1) took place close to MSL (Fig. 2), implying near absence of tidal influence there. This indicates that, compared to the situation at *c.* 6000 BP, the zone of maximum tidal range reduction had migrated seaward. This observation ties in well with the conclusion reached by the author in an earlier paper (Van de Plassche, 1980), that "... a seaward shift of the fluvial zone" must have taken place between 6000 and 5500 BP. A plausible explanation for this "seaward shift" is that it resulted from an adjustment of the river gradient in response to the westward migration of the zone of maximum tidal range reduction mentioned above.

The 4450 BP gradient line (Fig. 2) shows the storage-basin effect very clearly.

LATE-ATLANTIC SWAMP-FOREST SUBMERGENCE IN THE AREA NEAR MOLENAARSGRAAF

Around 6050 BP, following several centuries of fluvial clastic deposition, an extensive swamp forest came into existence in the area near Molenaarsgraaf (Van der Woude, 1981). From the beginning small lakes were present in this swamp. At a certain stage, however, these lakes began to expand and new open water spaces originated causing the death of large parts of the swamp forest. Apparently a rapid water-level rise occurred. The timing of this event is not sharply defined. Given the fact, however, that locally the organic lake deposits rest directly on the clay deposited previously, or on only a thin layer of swamp-forest peat (see

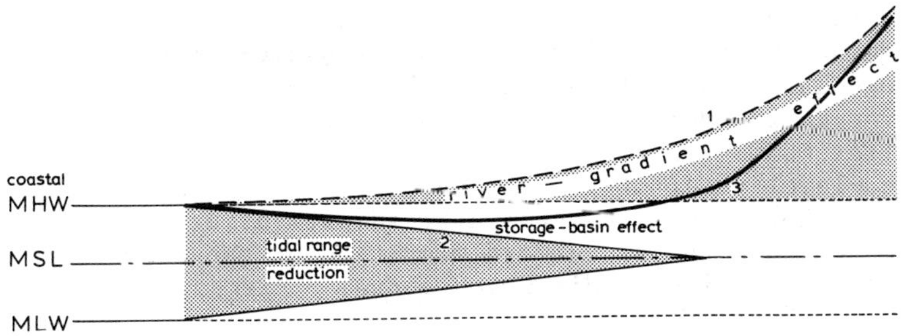

Figure 3. Schematic diagram illustrating how the gradient of a river entering a tidal basin is a function of the river-gradient effect and the storage-basin effect.

Van der Woude, 1981, Figs. 6 and 7), it would seem that the water-level rise began rather soon after 6000 BP. A 14C date for the top of the peat below organic lake sediment gave an age of 5590± 70 BP. In view, however, of the thickness of the peat at the sampling site and the diachronous nature of the submergence process, this date cannot be considered representative of the onset of the water-level rise. According to Van der Woude (1981, pers. comm.) this date might be more representative of the phase of maximum lake expansion. As to the cause(s) of the swamp-forest submergence in the Molenaarsgraaf area, Van der Woude (1981) attributed this to an "... increased eustatic component in the local water-level rise" and to compaction of peat and clay in the subsoil.

Compaction

Gravitational compaction of deeper-lying deposits may under favourable conditions add considerably to a relative water-level increase. However, in attempting to understand the late-Atlantic rise at Molenaarsgraaf it cannot be assigned too much significance because a consistent relationship between gyttja occurrence and compaction-susceptibility of underlying deposits appears to be missing (see Figs. 6 and 7 in Van der Woude, 1981).

Increased see-level rise

No direct evidence of an increase in the rate of MSL rise between 6000 and 5500 BP is available. However, as shown in Fig. 5 the time/depth data from the Hillegersberg and Rijksweg-donk riverdunes may be interpreted to indicate that a rapid water-level rise occurred prior to about 6000 BP. This was followed by a retardation up to about 5700 BP, after which the rate of rise increased again until 5500 or 5400 BP. If these variations in the rate of local water-level rise reflect synchronous, though possibly less pronounced, changes in rate of MSL rise, it can be concluded that from about 5700 BP onward an "increased eustatic component" was indeed involved in the swamp-forest submergence near Molenaarsgraaf. It was noted above, however, that the water-level rise in the Molenaarsgraaf area probably began relatively soon after 6000 BP and this cannot be explained in terms of increased sea-level rise. It is proposed here that this earlier part of the water-level rise was caused by adjustment of the river gradient in response to the seaward migration of the zone of maximum tidal range reduction as discussed above.

In summary, at least three factors may be held responsible for the origin and enlargement of open water spaces in the swamp forest that came into existence near Molenaarsgraaf around 6050 BP: differential compaction, river-gradient adjustment and increased sea-level rise.

DISCUSSION

The status of the triple-factor explanation for the late-Atlantic swamp-forest submergence given above is, of course, hypothetical: the local influence of differential compaction is likely but cannot be proved; the rate and timing of the seaward migration of the maximum storage-basin effect and the related river-gradient adjustment cannot be determined, and the sea-level increase between 5700 and 5500 or 5400 BP is based on the assumption that the water-level increase suggested by the Hillegersberg riverdune data directly reflects an increase in the rate of sea-level rise. Moreover, the factor 'river discharge' has so far been left out of consideration. However, as discussed below, several arguments and facts can be put forward in favour of the proposed explanation.

River discharge

From the evidence presented by Van der Woude (1981) it can be concluded with confidence that in the first three centuries after a swamp forest was established in the Molenaarsgraaf and Leerdam study areas (between 6100 and 6000 BP) little or no clay deposition took place there. It is improbable, therefore, that the water-level rise responsible for the early stages of swamp-forest submergence in the Molenaarsgraaf area was caused by increased river discharge in the period 6000-5700 BP. Since the water level in the Hillegersberg area was stable or rose only slowly during this period, a hydrodynamical adjustment of the river gradient seems to be the only possibility left to explain the onset of the swamp-forest submergence (unless a sudden diversion of a river course was involved).

Distribution of organic lake deposits

Van der Woude did not find any organic lake deposits within the swamp-forest peat that developed in the Leerdam area after about 6100 BP. According to Van Dijk (pers. comm.) the easternmost occurrence of the gyttja deposits is about 3 km east of Molenaarsgraaf (*cf.* the 6050 BP gradient line in Fig. 2). Again, these observations plead against increased river discharge as a causative factor, since this would also have affected the Leerdam area. On the other hand, they do support an explanation of the water-level rise in terms of gradient adjustment and/or increased sea-level rise.

Spatial variation of the storage-basin effect as a function of rate of sea-level change

The explanation for the earlier stage of swamp-forest submergence proposed above (gradient adjustment) suggests a connection between slowing down of sea-level rise and seaward migration of the storage-basin effect. Indeed, one may expect a causal relationship between both factors to the extent that a retardation in sea-level rise or a stillstand favour narrowing of coastal inlets, shallowing of (parts of) the intracoastal area, and peat formation. Such changes tend to increase the rate of tidal amplitude reduction and will therefore lead to development of a storage-basin effect closer behind the coastline. On the other hand, when the rate of sea-level rise accellerates, coastal inlets may be widened and water depth increases. As a result the tidal wave can penetrate deeper into the intracoastal area and the zone of maximum tidal amplitude reduction will occur at a greater distance from the coastline. The first case will lead to an increase in water level in that part of the intracoastal area where river-gradient adjustment occurs. The opposite case (rapid sea-level rise) can, in this model, be expected to lead to a water-level lowering or at least a slowing down of the water-level rise in the zone of gradient adjustment. Figure 4 is an attempt to illustrate this principle schematically.

The simple model sketched above predicts phase differences between rates of water-level rise in the seaward and landward part of the intracoastal area, which offers a possibility for testing the model. As yet, insufficient data are available for rigorously verifying the hypothesis. However, it appears that what little data there are do not contradict it.

Figure 5 compares, for the period 5500-4300 BP, water-level changes in the Hillegersberg-Rijkswegdonk riverdune area and in the Molenaarsgraaf area. It should be noted that the youngest part of curve 1 (4600-4250 BP) can be regarded as approximating the rise of MSL (Van de Plassche, 1982). The older part of the

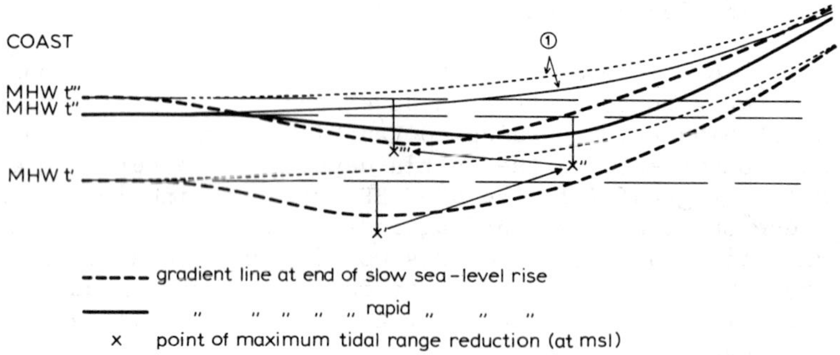

Figure 4. Model illustrating the assumed effect of rapid and slow sea-level rise on the position of the zone of maximum tidal range reduction and related water-level changes due to river-gradient adjustment. (1): gradient line in case of no tidal range reduction.

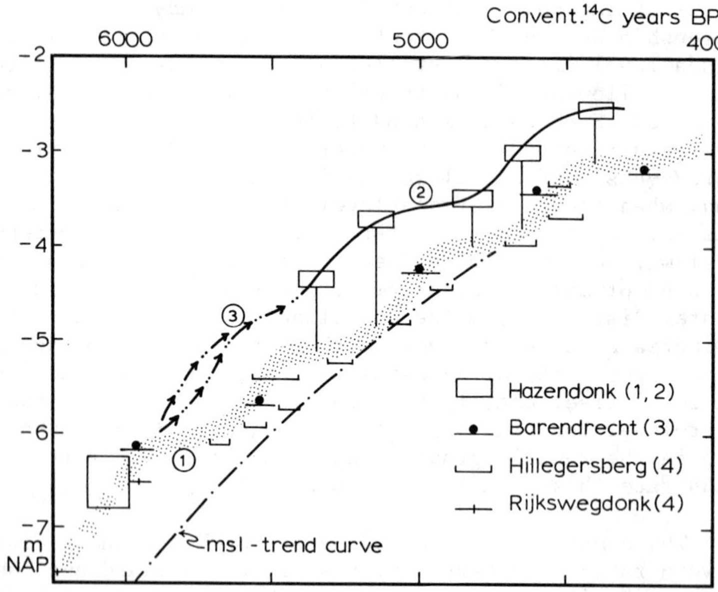

Figure 5. Time/depth diagram comparing water-level changes in the vicinity of Rotterdam (Hillegersberg, Barendrecht, Bolnes) (curve 1) and the area near Molenaarsgraaf (curve 2) to test the river-gradient adjustment model which predicts that trends in rate of change in the first-mentioned area can be out of phase with those in the latter. Curve 3 is a prediction, based on that model, of the approximate course of the water-level rise in the Molenaarsgraaf area between 6000 and 5400 BP.

water-level curve must be considered a possible interpretation of the time/depth data. Its function is to stress probable or possible changes in rates of water-level rise in the seaward part of the intracoastal area. It is assumed that these changes were caused by synchronous changes in rate of sea-level rise. Although curve 2 (Fig. 5) is based on much fewer data, it shows that changes in rate of water-level rise in the Molenaarsgraaf area are, as predicted by the model, out of phase with changes further west. This is particularly clear for the period 5400-4700 BP; for the period 4700-4300 BP a similar tendency might be discerned, but the differences are too small to be convincing with respect to error margins.

CONCLUSIONS

The above considerations lead to the conclusion that river-gradient adjustment and the related water-level rise was the principle cause for the swamp-forest submergence in the Molenaarsgraaf area after 6000 BP. On the basis of the model developed above it is to be expected that as the rate of sea-level rise began to increase (after 5700 BP) the water-level rise in the eastern part of the intracoastal area began to slow down. Thus one may predict that the course of the water-level rise in the Molenaarsgraaf area between 6000 and 5400 BP was more or less similar to that indicated by curve 3 in Fig. 5. Additional time/depth data from the Hazendonk riverdune will be collected to check the predicted curve.

In the context of paleo-climate studies a general conclusion must be that one cannot ignore the influence of river-gradient adjustment when interpreting and correlating evidence for (rapid) water-level changes in deposits such as found in the Rhine/Meuse delta. This applies in particular to that period of coastal plain evolution during which conditions were favourable for the zone of maximum tidal range reduction to migrate considerably within the intracoastal area.

FINAL REMARKS

A factor which has not been considered above is tidal amplitude. Changes in coastal tidal range can also affect the position of the maximum storage-basin effect in the intracoastal area without a change in MSL position. The influence of this factor is very difficult to evaluate, but it cannot be ignored since the configuration of the North Sea during the period 7000-5500 BP was subject to changes which may have resulted in displacement of the amphidromic points.

Finally, the fact that in many places in the Molenaarsgraaf

area the swamp forest persisted throughout the submergence phase indicates that the water-level rise was not, in an absolute sense, too rapid for peat accumulation. Apparently, ecological factors prohibiting peat growth in and around the borders of the expanding open water spaces were also involved. The influence of wave action, mentioned by Van der Woude (1981), may not have been the only factor.

ACKNOWLEDGEMENTS

The paper benefited much from discussions with Geo van Dijk. I thank Piet Cleveringa and Wim Roeleveld for comments and Truus Snijder for the typing.

REFERENCES

Jelgersma, S., 1961. Holocene sea-level changes in The Netherlands. - Mededelingen Geologische Stichting, C-IV (7), 100 pp.
Louwe Kooijmans, L.P., 1974. The Rhine/Meuse delta; four studies on its prehistoric occupation and Holocene geology. - Thesis, Leiden, 421 pp.
Louwe Kooijmans, L.P., 1976. Local developments in a borderland: a survey of the Neolithic at the Lower Rhine. - Oudheidkundige mededelingen LVII, 227-297.
Van de Plassche, O., 1980. Holocene water-level changes in the Rhine/Meuse delta as a function of changes in relative sea level, local tidal range, and river gradient. - Geologie en Mijnbouw, 59 (4), 343-351.
Van de Plassche, O., 1982. Sea-level change and water-level movements in The Netherlands during the Holocene. - Mededelingen Rijks Geologische Dienst, 36-1, 1-93.
Van Veen, J., 1950. Eb- en vloedschaar systemen in de Nederlandse getijwateren. - Tijdschrift Koninklijk Nederlands Aardrijkskundig Genootschap, Tweede Reeks 67, 303-350.
Van der Woude, J.D., 1981. Holocene paleoenvironmental evolution of a perimarine fluviatile area. - Thesis, Amsterdam, 112 pp.
Zonneveld, I.S., 1960. De Brabantsche Biesbosch. - Thesis, Wageningen. Also in: Verslagen Landbouwkundige Onderzoekingen, 65-20, 210 pp.

LATE POSTGLACIAL BEACH RIDGES ON THE ISLAND OF GOTLAND, SWEDEN, AS INDICATORS OF A PROBABLE PERIODICITY IN THE CLIMATE

Arne Philip

Dept. of Phys. Geogr., Uppsala University, Sweden

Ten localities with well-developed late Postglacial beaches are discussed. After photo-documentation from the air the localities were levelled in the field. By means of archaeological data and graphical correlation in a shoreline diagram the age of the ridges is calculated. It is shown that a period of at least 20 years per beach ridge is common both where the land uplift is great and where it is smaller. At some localities the period may be even shorter. Obviously no tectonic movements can be responsible for this periodicity. Hence the explanation seems to be cyclic variations in climate.

The island of Gotland (Fig. 1) is situated in the Baltic Sea, 90 km east of the Swedish mainland. There is no tide in the Baltic but the sea-level fluctuates on average 1.5 m between low and high levels during the year.

The bedrock of Gotland chiefly consists of easily eroded Silurian limestone, which provides material for impressive series of beach ridge accumulation.

Because of the land uplift, which is slow and almost constant, the beach ridges have been raised above the present sea-level. The oldest - from the Baltic ice-lakes - are now 40 to 80 m above the sea-level and the younger closer to the present coastline and on low elevations.

The relative land uplift of Gotland is not known in any detail. The official values - c. 2.5 mm/yr in the northern part of the island and c. 0.6 mm in the southern - are in fact extrapolated from the mainland. From the shore-displacement curves of Gotland it is obvious that the absolute uplift amounts to c. 4.0 mm/yr in the northern and c. 1.1 mm/yr in the southern part of the island.

N.-A. Mörner and W. Karlén (eds.), Climatic Changes on a Yearly to Millennial Basis, 215–218.

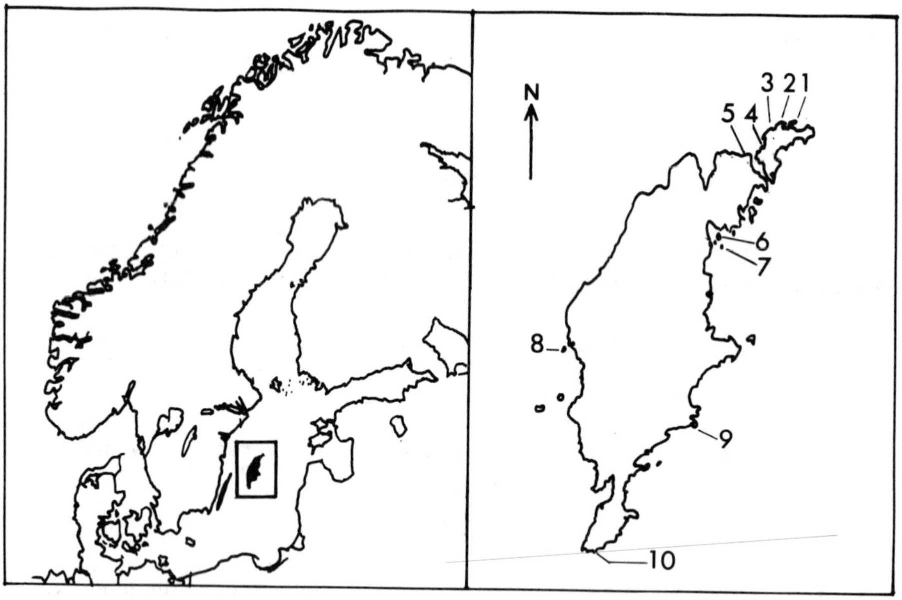

Fig. 1. Maps showing the Baltic Sea, Gotland and the localities discussed. Locality No. 1: Norsholmen, Fårö parish, 2: Langhammars, Fårö, 3: Digerhuvud, Fårö, 4: Lauters, Fårö, 5: Trälgars, Bunge, 6: Asunden, Othem, 7: Majgu, Othem, 8: Utholmen, Västergarn, 9: Närsholm, Närs, 10: Barshage, Sundre.

As a part of the former IGCP-project 61, Holocene sea-levels, the author has documented and levelled a large number of localities with well developed beach ridges on Gotland. The work is still in progress and data may be revised.

The ten localities here discussed have been plotted in a shoreline diagram where the shorelines are dated by graphical calculations. The age of the lowest shorelines is controlled by archaeological data. The age of higher and older shorelines is calculated from C14-datings on Gotland and by sea-level data from Denmark, Finland and various places within Sweden.

In Fig. 2 the beach ridge profiles from localities No. 5 and 10 are correlated tentatively. Both profiles have the same estimated age (Table 1), and the difference in altitude illustrates the slow land uplift at the southernmost edge of Gotland and the rapid uplift in the northern. The numbers below the profiles refer to synchronous regressions.

A numerical analysis of the beach ridge systems is given in Table 1. The calculations give periodicities (for the ridge/spur formation) ranging from 18 to 30 years. The mean cycle is 21 years.

As tectonic movements cannot be responsible for so short periodicities, the origin must be eustatic, meteorologic or oceanographic, i.e. "climate" in a broader sense.

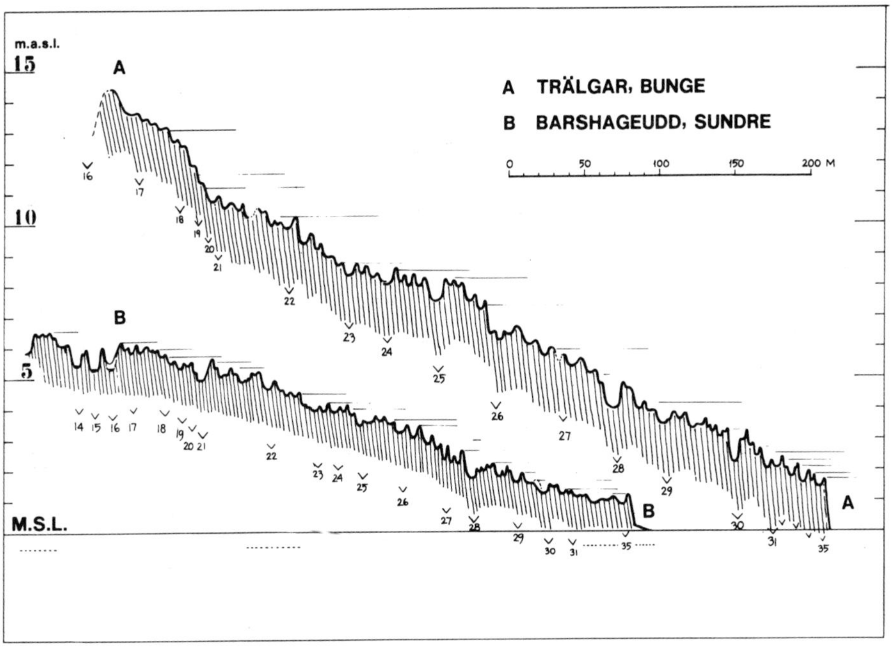

Fig. 2. Levelled profiles of localities No. 5 and 10. The numbers below the profiles refer to tentatively correlated regression marks (furrows). For the correlation the altitude a.s.l. of the last ridge before the regression is used.

Figure 3 shows one of the authors aerial photographs typical of the beach ridge systems in the localities discussed. These photographs were used (in addition to field studies) to decipher the beach ridges and leeside spurs (see Table 1).
Aerial photographs of the localities No. 6, 9 and 10 have previously been published in the Yearbook of the Swedish Natural Science Research Council 1975, pp. 127-129.

The aerial photograph is published by permission of the Swedish Defence Staff and is copyright of the author.
The investigation has been financed partly by grants from the national IGCP-61 project. The aerial documentation on photographs has been financed by grants from IGCP Project Ecostratigraphy.

Table 1. Numerical analyses of the beach ridge/spur systems in the ten localities discussed.

Locality No.	Place and parish as in Fig. 1.	Length of profile in meter	Max. number of ridges/m of altitude	Mean number of spurs/m (counted)	Altitude in m.a.s.l.	Estimated age in years BP (appr.)	Total number of ridges/spurs (est.)	Periodicity years/ridge-furrow
1.	Norsholmen, Fårö	290	10	14	5.6	1720	80	21.5
2.	Langhammars, -"-	485	10-12	12	9.7	2280	114	20
3.	Digerhuvud, -"-	1500	12	12	20.2	4700	240	19.6
4.	Lauters, -"-	205	9-10	11	7.0	1800	77	23.4
5.	Trälgars, Bunge	1480	11	11.6	12.7	2950	148	19.9
6.	Asunden, Othem	440	9	15	9.4	2640	140	18.8
7.	Majgu, -"-	250	16	16	9.0	2620	144	18.2
8.	Utholmen, Västergarn	660	19-22	22	5.9	2140	112	19.1
9.	Närsholm, När	290	25	16	5.6	2680	90	29.8
10.	Barshage, Sundre	740	19-23	27	5.4	2950	146	20.2

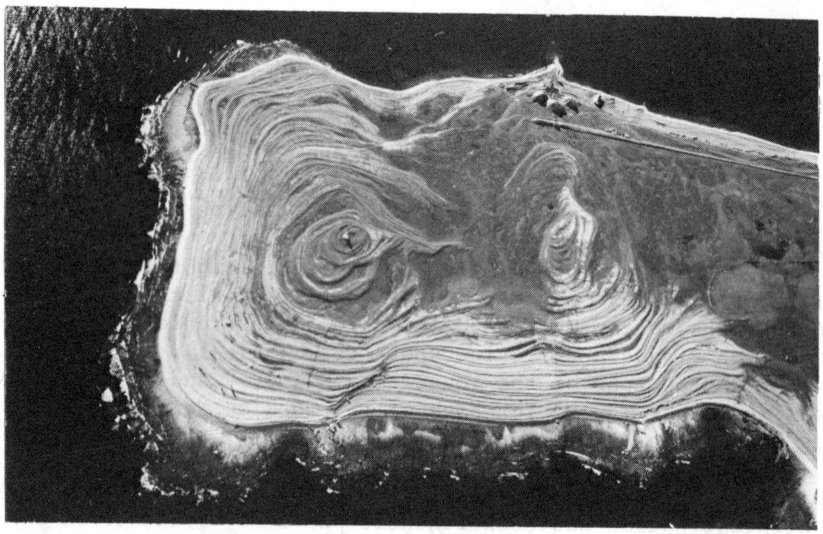

Fig. 3. Loc. No. 1, Norsholmen, Fårö parish. 80 spurs have been counted on the photography taken 6/10 1972.

SIZE VARIATIONS IN ATLANTIC, SUBBOREAL AND MODERN MOOSE TEETH

Elisabeth Iregren

Museum of National Antiquities
Box 5405, S-114 84 Stockholm

It is well known that several mammalian species have changed in size since the Pleistocene. Degerbøl (1933) showed that carnivores in subfossil and archaeological materials in Denmark usually have decreased in size to the present time. Hooijer (1950) discusses the decrease in terms of a general trend during the Quaternary.

Kurtén has also considered this question in several publications (1953, 1959, 1965). Regarding bears, Kurtén (1959) reports an "oscillation seemingly in rapport with the climatic history: a tendency to increase during glaciation, and to decrease during interglacials". The rate of change is faster during the Quaternary than during the Tertiary. "The trend is a phyletic dwarfing at a progressively faster rate." The term "postglacial dwarfing" is used by Kurtén.

The evolutionary trends seem to vary from species to species, and it appears difficult to establish a general trend. It seems possible that not only climate,but also biotope and factors such as e.g. winter habit of a species may influence size (cf. the badger in Degerbøl, 1933).

As a part of the "Early Norrland" research project, bone material from 174 dwelling sites in northern Sweden was studied (Ekman and Iregren, 1983). The moose (Alces alces L.) was by far the dominant prey species, but at most sites the bones had been accidentaly burnt and the teeth destroyed. At three sites, however, cultural layers had been formed and had thus preserved the teeth of moose hunted by early man. These sites are:

(A) Site Tåsjö 101 (Lat. 64°5'N., Long. 16°11'E.) dating at about

N.-A. Mörner and W. Karlén (eds.), Climatic Changes on a Yearly to Millennial Basis, 219–221.

Site	Tåsjö 101	Åsele 1023-24 B	Åsele 1023-24 S	Anundsjö Agnsjön	Jämtland recent
lengths					
N	20	5	18	9	12
$\bar{M}$	41.40	40.20	41.47	40.00	37.17
S.D.	2.09	1.52	2.27	2.08	1.01
breadths					
N	20	3	20	9	-
M	20.58	21.50	21.23	21.28	-
S.D.	0.59	0.50	0.98	0.91	-

Table 1. Mandibular teeth (M_3) of the moose (Alces alces L.) from dwelling sites and of modern material. The mean and standard deviation are given for lengths and breadths.

5700-4800 ^{14}C years BP (i.e. the AT zone) (T1/2 5568 y.).

(B) Site Åsele 1023-24 (Lat. 63°57'N., Long. 16°17'E.) dating at about 4300-3300 ^{14}C years BP (i.e. the SB zone).

(C) Site Anundsjö, Agnsjön (Lat. 63°33'N., Long. 17°55'E.) dating at about 3700 ^{14}C years BP (i.e. the SB zone).

The last molar in the mandible (M_3) was measured in fully grown animals. All material derives from periods of climatic optimum in Scandinavia - c. 2.5 °C above the present mean temperature.

Table 1 shows that the mean values of the site materials are greater than those of modern moose. Tooth size has evidently decreased from Atlantic and Subboreal times to the present. The material will be more fully presented elsewhere (Iregren, in prep.).

Table 2 presents the results of the comparisons. Student's t-test was applied. Comparisons between sites show no significant differences. However, differences were found when the archaeological material was tested against the modern sample.

The present study of the variation of the tooth size of moose displays some weaknesses. The decline in tooth size that was found would be more securely established if the recent material had come from the same area as the samples from the dwelling sites, as several subunits of moose populations are known (Reuterwall, 1980). It would also have been preferable if the modern material had been somewhat older, i.e. prior to the eradication of the moose during previous centuries (cf. Ekman, 1918, p. 259-69), and therewith correlated genetic effects. The results would also have been more reliable if the comparative material had come from a period before man's ecological disturbances in historic time.

Sites	t	DF	p
A x B	0.36	28	N.S.
A x C	1.44	20	N.S.
A x B+C	0.92	36	N.S.
A x D	7.52	22	<0.001
B+C x D	4.53	32	<0.001
A+B+C x D	5.13	46	<0.001

Table 2. Comparison (tooth length) between Atlantic (A), Subboreal (B,C) and modern (D) moose populations.

Despite the above-mentioned weaknesses in the material, I hypothesize that there is an evident decline in tooth size of the moose since the climatic optimum (the AT-SB zones), and that this is part of a general trend in mammalian evolution. The results encourage further studies of this nature on cervids.

References:

Dergerbøl, M. 1933. Danmarks pattedyr i fortiden i sammenligning med recente former I. In: Viddenskabelige Meddelelser 96:II, pp. 357-641. København.

Ekman, J. & Iregren, E. 1983. Archaeo-Zoological investigations in northern Sweden. (Early Norrland 8). Stockholm:KVHAA.

Ekman, S. 1918. Några jaktbara djurarters historia i Sverige under senare tid. Svenska jägarförbundets tidskrift 56, pp. 259-69.

Hooijer, D.A. 1950. The study of subspecific advance in the Quaternary, Evolution 4, pp. 360-1.

Iregren, E. in prep. Size of the moose (Alces alces L.) in Sweden in relation to climate. A study of teeth from refuse layers at dwelling sites.

Kurtén, B. 1953. On the variation and population dynamics of fossil and recent mammal populations. Acta Zoologica Fennica 76, pp. 1-122.

Kurtén, B. 1959. Rates of evolution in fossil mammals. In: Genetics and Twentieth Century Darwinism: Cold Spring Harbor Symposia on Quantitative Biology 24, pp. 205-15.

Kurtén, B. 1965. On the evolution of the European wild cat. Felis silvestris Schreber. Acta Zoologica Fennica 111, pp. 1-26.

Reuterwall, C. 1980. Genetic variation in a large game species, the moose (Alces alces): Patterns of differentiation and some management implications. Diss. Ser. 2. Universitatis Regia Stockholmiensis.

III. The Last 1,500 Years

CLIMATE AND HISTORY IN NORTHERN EUROPE AND ELSEWHERE

Hubert H. Lamb

University of East Anglia, Norwich, England.

ABSTRACT

The first part of the paper reviews the long record of climate since the last ice age and more particularly over the last few thousand years, including something of how that record has been reconstructed, its reliability, and some of the patterns of variability from year to year and over longer periods.

Effects on the landscape and on human economies are then considered. When we view human history against this background, there are important coincidences which argue for a strong underlying influence of the climatic fluctuations on human affairs - especially near the poleward and desert margins of civilization. But it is not difficult to find instances of anomalous effects and some of these cases are considered briefly.

An understanding of the connexions between climate and human history may have important lessons for today's world and for the future.

INTRODUCTION

Historians in most countries have generally been reluctant to believe that changes of climate affect the subject of their concern and need to be considered as part of the story. The situation is rather different here in northern Europe, where Professor Andreas Holmsen and the Nordic Research Project on Deserted Farms and Villages 1300-1600, the Ødegardsprosjekt, have pioneered in the field. In Iceland the late Vilhjalmur Stefansson (e.g. 1943) and Sigurdur Thorarinsson (1956), and others, opened up the subject already some decades ago. Yet,

N.-A. Mörner and W. Karlén (eds.), Climatic Changes on a Yearly to Millennial Basis, 225–240.

even in the case of Greenland, it has been argued by some historians - by historians whose own experience is of life in middle latitudes - that the demise of the Old Norse Colony during the later Middle Ages was not really due to the worsening of the climate and increase of sea ice so much as to the colonists' failure to adapt themselves to an eskimo way of life.

We shall look at this Greenland question further in what follows. But we may do well to recognize that there seems to be a general resistance in mankind since the modern scientific and technological revolution to the idea that our way of life may be drastically affected by natural changes and fluctuations in the physical environment. This attitude may hold dangers for the future. Our forefathers had a very different view, based on bitter experience of disasters from famine and disease. But even in our own day climatic fluctuations have imposed similar distresses on the poorer peoples of Africa - particularly in the Sahel zone and in Ethiopia and Somaliland - and have led to unfulfilled plans in the agricultural development of no less a country than the Soviet Union.

It is important to arrive at a sound and well informed understanding.

THE CLIMATIC RECORD

The necessary background to our study is the history of the climate itself since the last ice age. Our first outline knowledge of this came from the work of Scandinavian botanists, from Blytt (1876) and Sernander (1910, 1926) to Faegri (1964) and Iversen (1973). This knowledge has in recent decades been greatly improved by a host of new lines of evidence that provide corroboration and by new techniques, particularly radiocarbon dating (e.g. Olsson, 1970) and its refinement by calibration against the American bristlecone-pine tree-ring chronology (which goes back many thousand years) and against other objects of known age. The contribution of botany to improving the detail of the record has continued to be big, and quite recently Hafsten (1981) has reported finds of tree remains well above the present limits of trees of comparable size on Dovrefjell and on Hardangervidda in southern Norway, shown by radiocarbon measurements to have grown in the warmest postglacial times and in the Roman Iron Age 0-400 AD respectively. Stumps of full grown trees occur widely in Scotland 30 to 70 metres above the present tree line, generally dating from 2000 to 5000 BC (Pears, 1972): in a few cases the heights are up to 200m above the present natural forest limit. In central Europe it is reported that the upper forest limit descended by 200 metres between 1300 and 1600 AD, after the medieval warm period (Firbas and Losert, 1949).

Assessments of past temperatures over the last thousand years or rather more have also been derived from statistical studies of the details in historical documents (Lamb 1965, 1977, 1982). And these estimates can be controlled by reference to field evidence of various kinds, from pollen-analytical studies in peat bogs, from cultivation records, from glaciers, and so on.

Here we shall base our considerations on the reconstructed temperature record for central England and a reconstruction of the variation over recent centuries in the sea surface temperatures prevailing between Iceland, Scotland and south Norway (Lamb, 1979). The English temperature record is chosen because it has been worked out in most detail, covering the period AD 800 to the present[1], and the sea area mentioned is so far the only one for which an estimate of conditions at the climax of the Little Ice Age is available. Figure 1 shows the prevailing temperatures in England from AD 800 to 1950 (as successive 50-year averages from 1100) and gives an indication of the margins of possible error. The changes in prevailing temperature level over the last 20,000 years range from the final climax of the last Ice Age to the warmest postglacial times some 4,000 to 8,000 years ago and a wavering decline since.

The salient features which may have affected human history seem to be:

(i) the temperature levels established in the warmest and coldest times, including the medieval warm epoch around AD 900-1300 and the Little Ice Age in more recent centuries, and how these may have affected the extent of land available for human activities - cultivation and settlement, and so on.

(ii) the rapidity of the changes, and the still largely unknown detail of year to year variations during the periods of rapid change between one more stable regime and another. The year to year variability must, it seems have been abnormally great at such times and there are indications in some tree-ring sequences that this was so (see Lamb, 1977 and data in Fürst 1963). Disappointing harvests, sometimes leading to famine, and disasters of quite various kinds (droughts, landslides, great snows and frosts, storm floods and incursions of the sea) are most likely to have occurred at these times.

(iii) changes in the ocean are also implied, which presumably affected fish stocks, spawning grounds etc., and within the last 400 to 500 years are known to have done so and continue to do so today.

(iv) changes in the penetration of polar water southwards from the East Greenland Current into the Atlantic, passing both sides of Iceland (i.e. along the coast of Greenland, also round the east coast of Iceland and at times near to the Faeroe Islands) are observed to occur on all time scales

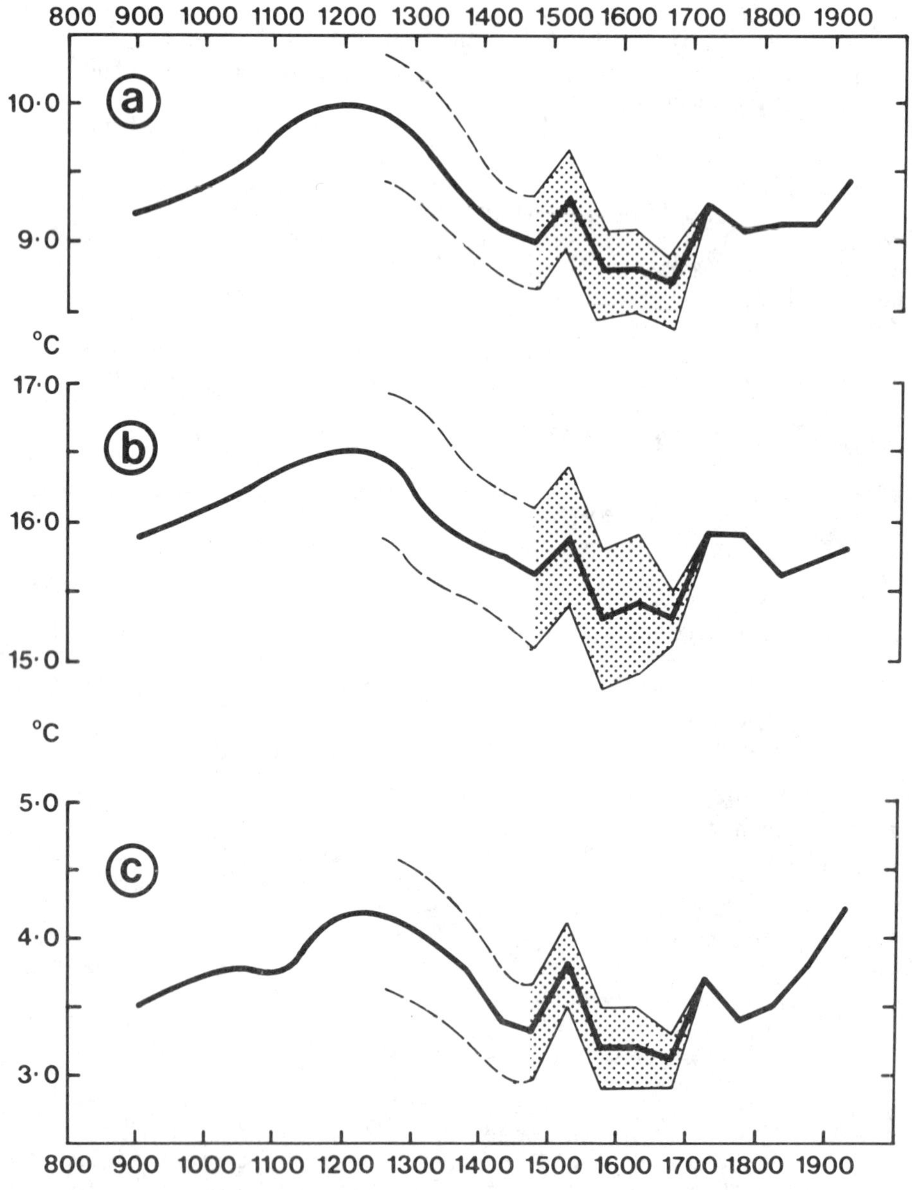

Fig. 1: Temperatures in central England since AD 800. Probable averages for
(a) the whole year
(b) July and August
(c) December, January and February.
The shaded area indicates the range of uncertainty of the values (approximately 3 times the standard-error of the estimates).

(Lamb 1977, 1979, 1982; McIntyre et al. 1972) from days and weeks at the present time up to many thousands of years. Although in the present epoch the Faeroe Islands (with 100-year average sea surface temperature 7.7°C) are normally in the midst of the main drift of warm, saline water of Gulf Stream origin flowing into the Norwegian Sea and the Arctic, at times the polar water approaches from the north and lowers the water surface temperature by several degrees. For thirty years between 1675 and 1704, at the climax of the Little Ice Age development, the ocean surface between Iceland and the Faeroe Islands seems to have been about 5°C colder than the 1870 to 1970 average.

Over and above these variations, indications have been found in recent years that an oscillation of prevailing temperature level somewhat paralleling the change from the warmth around AD 1100 to the colder regime around 1700 may have occurred in each of the previous four millennia. A parallel series of events in the last millennium before Christ has long been known in outline (e.g. Brooks 1926, Gams 1937).

If we could also establish a rainfall history in corresponding detail to the temperature history summarized above, we would again find variations. Over northern and central latitudes of Europe generally the long-term averages may have been 10 to 20% greater in the warmest postglacial epochs and 5 to 15% less than now in the Little Ice Age. Experience within the present century indicates that significant further variations of rainfall may have been superimposed over more localized regions, as the rainfall difference between the western and eastern sides of mountains changed with changes in the prevailing winds. (The frequency of westerly winds covering the British Isles as a whole seems to vary between about 20 and 40% over periods lasting up to a century.)

The dryness or wetness of the ground is, however, a different question, since it depends on the amount of evaporation as well as the rainfall. It seems that the colder periods of our climate and much of the periods of cooling were characterized by generally wetter ground, especially in the higher latitudes of Europe including northern parts of the British Isles. The greater rainfall in the warmer periods seems rather to have been associated with heavier thunderstorms in summer, and the greater frequency of clear skies, warmth and sunshine meant that the ground dried more readily. The wetness of the Little Ice Age, particularly in the sixteenth and seventeenth centuries and in some of the earlier stages of its onset around AD 1300 and 1450 is well demonstrated by studies of the peat bogs of northern and western Europe (see for example Fig. 2) as well as from documentary sources and historical accounts. - e.g. the frequency of landslides in west Norway (Grove 1972) and accounts of the lowlands and the glens of Scotland in the sixteenth to nineteenth centuries as "dissected ... by immense watery tracts Leven is now" (Fraser 1970). By contrast, in the thirteenth century some of the greatest buildings of the time seem to have suffered subsidences and collapse with shrinkage due to drying out of the soil about their foundations.

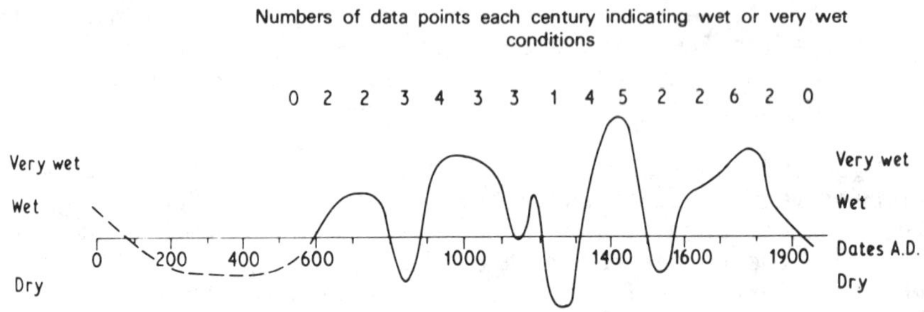

Fig. 2: Variations of the surface wetness of Bolton Fell Moss, a peat bog northeast of Carlisle near the English-Scottish border, derived by Dr K Barber from pollen and macrofossil analysis, soil chemistry measurements and records of land-use history.

Another important matter is the long-term variations in storminess in different latitudes, discussed by me elsewhere in this volume and shown for instance by the record of great sea flood disasters around the coasts of the North Sea (Fig. 3). The existence of real variations is plain, however incomplete the reports from early times and despite some contribution from minor variations of the prevailing sea level (Mörner 1980, Tooley 1974), e.g. those due to melting of the world's glaciers in warm epochs (Lamb 1981, see also Lamb 1957 for an instance of variations of storminess affecting Iceland and the Norwegian Sea.)

The main effects of the changes of the overall average temperature level, referred to under (i) above, were probably the changes in the average length of the growing season, accumulated summer warmth, incidence of frost (especially prolonged frosts), and the average number of days with snow cover. In England the growing season in the warmest decades of the present century, between 1930 and 1949, was about 10 to 20 days longer than in the late nineteenth century and probably three to five weeks longer than in the late seventeenth century. In the high Middle Ages, with a still higher temperature level, the average growing season was probably five to seven weeks longer than in parts of the fifteenth, sixteenth and seventeenth centuries. Thus, the late medieval climatic decline between AD 1300 and 1600 can hardly have failed to have serious effects

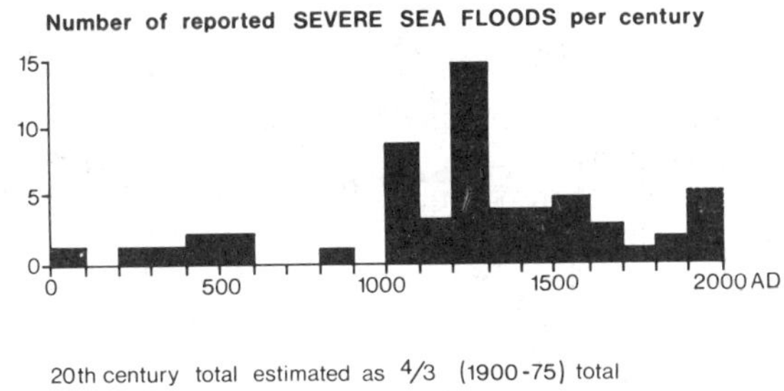

Fig. 3: Numbers of reported severe sea flood disasters causing much loss of life on land century by century. (The record is necessarily limited to cases of which reports have survived to our own times and cannot be complete in the earliest centuries.)

on European agriculture. In more continental climates the shortening of the growing season may have been as great as this only within 200 to 300 metres of the upper limit of cultivation and on the upland pastures (saeters and alps), but there the effects seem to have been sufficient to cause widespread abandonments. In the case of the duration of snow cover, however, the change was greatest on the low ground. Even in southern England there are a few reports in the seventeenth century of winters with from 70 to 102 days with snow lying, and the average was probably about 20 to 30 days, compared with twentieth century averages of 5 to 15 days and an extreme about 60 days (in 1962-3).

As regards accumulated summer warmth, Parry (1975, 1978) has shown by close examination of meteorological data and historical farm records how this must have affected the issue in the Lammermuir Hills in southeast Scotland, nor far from Edinburgh. (cf. also Sandnes 1971 for an example in Norway.) In Fig. 4

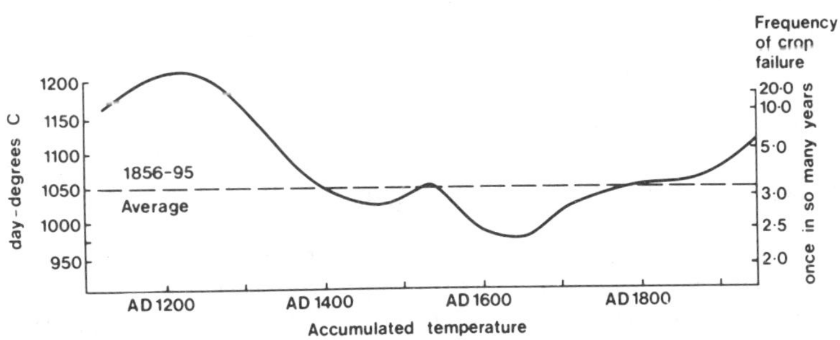

Fig. 4: Estimated average accumulated warmth of the growing seasons at different times past and probable frequency of failure of the oat harvest on high-level farms in the Lammermuir Hills in southeast Scotland at the height of the late nineteenth century upper limit of cultivation. (From Parry 1978)

Parry has taken for the period 1856-1895 the average accumulated day degrees above a base level of 4.4°C for the season in which oats would grow at the known upper limit of their cultivation in this area, and has applied the historical variation shown by our reconstruction of the temperature variations since AD 1100 in central England to show how the accumulated warmth of the growing season presumably varied. The right-hand scale shows the frequency with which the oat crop would be expected to fail at the various levels of summer warmth. This scale was derived by consideration of the best and worst years within the period for which we have full records of both temperature and crop results. It appeared that other aspects of the climate such as raininess and windiness of the summers were sufficiently correlated with the temperature that the latter could be treated as the one basic indicator. We see that in the present climate failures may be expected about one year in seven at the upper limit of cultivation: in the thirteenth century this may have been reduced to one year in twenty and in fact cultivation was extended to levels almost 100 metres higher. But in the late seventeenth century the failure rate approached one year in two and many examples of failures in successive years would be expected. This would assuredly lead to eating the next year's seed corn and then to famine. And, of course, the upper limit of cultivation had to come down to lower levels in consequence.

An aspect of climatic behaviour which may be very awkward for the human economy and a potential cause of disaster, particularly famine, e.g. through drought or plant diseases, is the observed - but not yet explained - tendency for several years with the same markedly unusual character to occur in clusters. Examples of such close successions of years - in this case with anomalous frequency of E or NW winds or calms at Copenhagen - are seen in Fig. 5. The succession of cold years in the 1690s in which the harvests failed in all the upland parishes of Scotland in seven years out of eight is a notorious example of what the effects can be.

EFFECTS ON HUMAN HISTORY

When we come to consider how the various fluctuations and changes of climate may have affected human history, we must treat as separate issues the case of communities directly hit by a climatic event and the more difficult problem of tracing the influence on communities which were much less severely affected or not directly affected at all. It is logical to look for most effect near the margins of the inhabited world, in the North, on the uplands as in the Alps, and near the deserts of Africa and Asia.

To take extreme cases first: when the steady rise of sea level with the melting of the ice-sheets and glaciers of the last

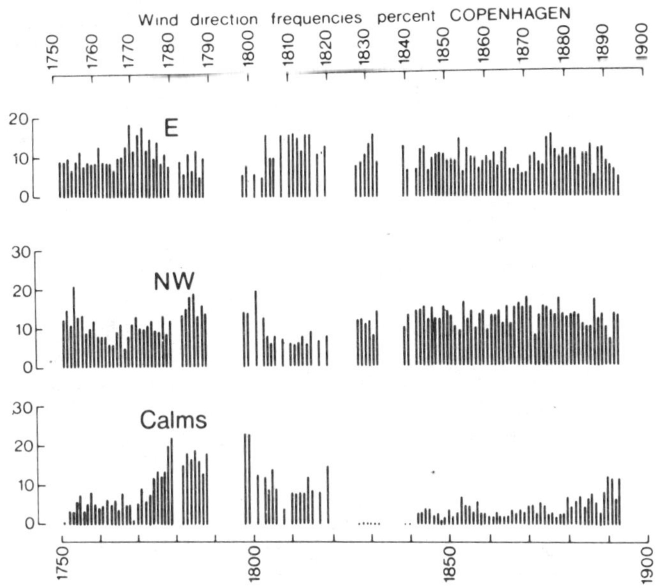

Fig. 5: Frequency of surface winds from E and from NW, and frequency of no wind, reported at Copenhagen in different years from 1752 to 1893.

glaciation flooded the great plain in the North Sea basin, the people that were living there had no choice but to go. And this process continued with a long succession of disasters, involving sometimes hundreds of thousands of deaths, as great storm surges of the North Sea took more and more of the low-lying coastlands (see Lamb 1977, pp. 120-128; 1982, pp. 109, 155) until the late Middle Ages and after. A similar compulsion affected the peoples living in North Africa when the Sahara and much of Arabia turned to desert from the Savanna-like terrain where they had herded cattle and hunted wild game before 3000 BC (Lamb 1982, p. 117). In some valleys in southern Arabia also cultivation of land which subsequently became desert had been practised.

Was the situation really less compelling for the Old Norse Greenlanders when the sea ice increased over the period between AD 1190 and about 1500, ultimately blocking their coasts? This and the increasing storminess of the northern seas cut them off from Europe. No doubt they could have tried an eskimo way of life, but the Old Norse Greenlanders - as G.H. Petersen of the Zoologisk Museum, Copenhagen has pointed out to me at this symposium - were themselves skilled hunters and fishers and may by the fifteenth century have had little to learn from the eskimos. Archaeology indicates that their health and stature was declining; and a point would surely be reached at which a people depressed in mind by their situation, and probably by watching an increased death rate among the frail children and old people, would lack the morale to undertake bold changes. And what of the eskimos that were encroaching more and more into the area as they abandoned their former hunting grounds farther north, presumably because of increased freeze-up in the Arctic north and movement of the fish and seal populations? It may be that the resources of the area would no longer have yielded sustenance for the human populations of both races crowded together in a limited part of southwestern Greenland. A likelier suggestion, also by Petersen is that the final blow (the coup de grace) to the Norse population - and probably to many of the eskimos - was the ill chance that a visiting ship, or shipwrecked fishermen arriving from England or Germany, may have brought an infection (the plague?) that flared up as an epidemic among the Greenland population, whose run-down state and long isolation left them with no resistance to it. This would be no more than a repeat in Greenland of the consequences which afflicted the American Indian populations of North and South America after the arrival of the Europeans about the same time.

In the Sahel zone of western Africa within the last decade an analogous situation developed. After six years of drought between 1968 and 1973, the rainfall has never returned to its former average level. And in the extreme phase of the drought around 1972-1973, with hundreds of thousands of cattle dying and multitudes of human deaths besides, the population fled southwards, crossing the national frontiers that were an awkward relic of the old colonial administrations. Some complete national territories looked like becoming uninhabitable, and would have become so if the climatic situation had developed further in the same direction. In fact, there has been little further change.

In all these cases the compelling power of a change of the prevailing climate has virtually dictated the action taken by the peoples affected. But, when we look farther afield, there are hints of a much farther reaching influence of the climatic trends and, on the other side, no great difficulty in finding examples of apparently contrary effects.

Students of history can hardly fail to be struck by the apparent coincidence of the high points of cultural achievement in the late Stone Age and Bronze Age development of trade and communications across Europe and of sea-going communications all along the Atlantic coasts and island chains, and again later in Roman times, and thirdly in the high Middle Ages, with the crests of the temperature curve. And the parallelism seems to go into some closer detail too.

In central Norway between AD 800 and 1000 the area of settlement, forest clearance and farming, which had been more or less unchanged since early Iron Age times, and had retreated in places since AD 400, rather quickly spread 100 to 200 metres farther up the valleys and onto the heights (Holmsen 1961). It retreated again just as decisively between AD 1300 and 1500, not only due to the Black Death (mannedauen) which did not strike until after the population decline and abandonments had begun. This seems to be the finding of investigations of the contemporaneous retreat and social unrest in Scotland, England, Denmark (Christensen 1938, Gissel 1973) and central Europe, as well as Holmsen's (1978) studies in Norway, Salvesen (1979) in Jämtland Sweden and of the Ødegaardsprosjekt in general. The parallelism with the temperature curve in the course of the medieval warm-up and later decline is close. In the case of abandonments of many villages, where the dates are known, they appear to be concentrated in the severest decades (see Lamb 1977, 1982). Examples include Hoset in eastern Trøndelag in the 1430s and again in the 1960s, numerous villages in central and eastern England between 1430 and 1480, and one in the Southern Uplands of Scotland in the 1690s. And it was during the long climatic decline that the cultural capital, as one may describe it, of northern Europe moved southward from Trondheim successively to Bergen and Oslo, and later farther from the ocean to Copenhagen and Stockholm. In 1536 Norway lost its existence as a separate country. So too, in Scotland in 1436, after the murder of the king, the capital was moved for security to Edinburgh Castle from the former northern seats in Dunfermline and Scone (near Perth); after the famine and other disasters of the 1690s and later, in 1707 the parliament and administration in Edinburgh were moved to London with a deeply felt loss of the identity of Scotland.

Was the parallelism with the temperature curve just coincidence? Or was there a remorseless pressure exerted by the harvest failures and increasing dislocation of life in the most exposed places, however well some of the more fortunate communities and individuals elsewhere fared? One prominent member of the old Scottish Parliament in Edinburgh in 1698, Andrew Fletcher of Saltoun, berated the well-to-do people of the prosperous, corn-growing lowland districts near Edinburgh for their unconcern over the disaster in all the upland parishes where the people were reduced to gathering nettles to eat and were dying in huge numbers of starvation and disease.

There seems little doubt that when, in the sixteenth century, the Baltic herring shifted to the North Sea, England and the Netherlands prospered while people in the northern countries suffered. And for a long time during the climatic deterioration after AD 1300 the cod fishery seems to have been improving up the Norwegian coasts, as if the increased flow of ice southward in the East Greenland Current was being compensated by an increased flow of Atlantic water northwards off Norway. However that may be, the improvement in the coastal fishery led to more human settlement along the coast, partly an actual movement northward of fisher folk from the districts near Bergen (precisely during the time of general climatic decline) and partly a drift to the coast of those who were abandoning farms in the interior in north Norway. In Denmark and Sweden, too, there are complicating details, with serious effects of the climatic worsening (storm losses of coastlands and abandonment of farms) in Jutland but much less effect, if any, in the more eastern islands. In southern Sweden there were abandonments also, but much less in the north (Osterberg 1977) perhaps because settlement had been so sparse there that only the best sites were occupied.

Most remarkable of all the apparently contradictory developments is the great age of expansion of Norway's shipping and sea-going trade between 1680 and 1710, just coinciding with what appear to have been the bitterest years of the climatic development. There is no doubt, of course, that that was the time when the fisheries finally went into decline (and at one stage in the 1690s disappeared) along the whole Norwegian coast. It is recorded however, that it was in the years of poor harvests in the late seventeenth century that the farmers who were near enough to the coast of south Norway and had good timber on their land - some of them had oak - took to ship-building and trading their timber abroad in their own ships. The development was also helped by the immigration of experienced mariners and shipwrights from Holland, seeking the advantages of a neutral land while the powerful nations of Europe further south were at war with each other.

At the same epoch in Iceland life was at a low ebb. The advancing glaciers were overrunning farms at the same time as the fisheries were failing, and there were notable outbreaks of disease probably partly due to undernourishment. Gunnar Gunnarsson (1980) has interestingly stressed how the situation in that country was made worse by its social conditions and the customs of the time. The landowners used their power in what was still a largely feudal society to restrict the drift of workers from the farms to try their luck with fishing on the coast. And there were regulations restricting the fishing, allowing no more than one hook on a line and forbidding the use of worms as bait. Custom also dictated the use of open boats. And even a hundred years later, when the government in Copenhagen tried to stimulate recovery in Iceland by introducing decked vessels and the use of many hooks on the lines, it met with prolonged and general opposition. This is surely an example of lack of resilience and more than ever stubborn clinging to old ways in peoples whose morale has been sufficiently lowered.

It seems to the writer undeniable that pressures due to climatic change were present underlying all the historical events we have discussed. The impact is clear in the case of the most devastating climatic events upon the societies that were directly hit. These we may describe as impacts of the first order. But, as we turn our attention to societies which were less and less affected, we find their reactions more and more diverse, depending on the range of choices open to them and, in some cases, the opportunities to profit at the expense of the communities whose former activities declined. Among those who suffer the greatest restriction of their former activities, depression, undernourishment and ill health seem liable to undermine their capacity to adapt to different patterns of life; and so they miss such opportunities of self-help as exist. This characteristic has been noted by the Brandt Commission Report (1980) as operating in the Third World today. Perhaps we should also note a lesson from the exodus in recent years of the so-called "boat people" from south-east Asia, and of Africans from Uganda and the Ogaden, in years of both climatic (monsoon irregularities) and political stress, that political prejudice may also enter in and dictate which groups shall be the main victims of a natural disaster.

[1] The oxygen isotope proxy temperature record from northwest Greenland produced by Dansgaard and his co-workers (1975) is even more detailed but is considered too distant from the north European regions with which we are here concerned, since there is some clear evidence of a different history there (see Lamb, 1977, pp. 429-430).

REFERENCES

Blytt, A. 1876. Essay on the immigration of the Norwegian flora during alternating rainy and dry periods. Christiana (A. Cammermeyer). 89 pp.

Brandt, W. and others 1980. North-south: a programme for survival. The Report of an Independent Commission on International Development Issues under the Chairmanship of Dr. Willy Brandt. London and Sydney (Pan Books).

Brooks, C.E.P. 1926. Climate through the ages. London (Benn).

Christensen, A.E. 1938. Danmarks befolkning og bebyggelse i middelalderen. Article in: Nordisk Kultur II: Befolkning i Middelalderen, 57 pp. Copenhagen (Adolf Schück).

Dansgaard, W., Johnsen, S.J., Reeh, N., Gundestrup, M., Clausen, H.B. and Hammer, C.U. 1975. Climatic changes, Norsemen and modern man. Nature, 255, 24-8. London.

Faegri, K. and Iversen, J. 1964. Textbook of pollen analysis. Copenhagen (Munksgaard), New York (Hafner) and Oxford (Blackwell). 237 pp.

Firbas, F. and Losert, H. 1949. Untersuchungen über die Entstehung der heutigen Waldstufen in den Sudeten. Planta, 36, 478-506. Berlin.

Fraser, Lady Antonia, 1970. Mary Queen of Scots. London.

Fürst, O. 1963. Vergleichende Untersuchungen über räumliche und zeitliche Unterschiede interannueller Jahrringbreitenschwankungen und ihre klimatologischer Auswertung. Flora, 153, 469-508. Jena.

Gams, H. 1937. Aus der Geschichte der Alpenwälder. Zeitschrift des deutschen und Österreichischen Alpenvereins, 68, 157-170. Stuttgart.

Gissel, Sv. 1973. Nogle kronologiske synspunkter vedrørende agrarkrisen i senmiddelalderen. Arbejdspapir 1972-73 nr.47 (København 28.3 1973). Copenhagen (Det Nordiske Ødegardsprojekt (Dansk Afdeling)).

Grove, J.M. 1972. The incidence of landslides, avalanches and floods in western Norway during the Little Ice Age. Arctic and Alpine Research, 4 (2), 131-138, Boulder, Colorado.

Gunnarsson, G. 1980. A study of causal relations in climate and history, with an emphasis on Icelandic experience. Meddelande fran Ekonomisk-Historiska Institutionen, nr. 17, Lund University. Lund.

Hafsten, U. 1981. An 8000 years old pine trunk from Dovre, south Norway. Norsk Geografisk Tidsskrift, 35, 161-165.

Hafsten, U. 1981. Palaeo-ecological evidence of a climatic shift at the end of the Roman Iron Age. Striae, 14, 58-61. Uppsala.

Holmsen, A. 1961. Norges Historie. Oslo, Bergen (Universitetsforlaget). 480 pp.

Holmsen, A. 1978. Hva kan vi vite om agrar katastrofen i Norge i middelalderen? Oslo, Bergen, Tromsø (Universitetsforlaget). 128 pp.

Iversen, J. 1973. The development of Denmark's nature since the last glacial. Geol. Survey of Denmark, V Series, No. 7c. Copenhagen (Reitzel). 126 pp.

Lamb, H.H. 1957. On the frequency of gales in the Arctic and Antarctic. Geographical Journal, 123, 287-297. London.

Lamb, H.H. 1965. The early medieval warm epoch and its sequel. Palaeogeography, Palaeoclimatology, Palaeoecology, 1, 13-37. Amsterdam (Elsevier).

Lamb, H.H. 1977. Climate: present, past and future, vol. 2: Climatic history and the future. London (Methuen) 835 pp.

Lamb, H.H. 1979. Climatic variation and changes in the wind and ocean circulation: the Little Ice Age in the northeast Atlantic. Quaternary Research, 11, 1-20.

Lamb, H.H. 1982. Climate, history and the modern world. London (Methuen). 387 pp.

McIntyre, A., Ruddiman, W.F. and Jantzen, R. 1972. Southward penetrations of the North Atlantic polar front: faunal and floristic evidence of large-scale surface water-mass movements over the last 225,000 years. Deep-Sea Research, 19, 64-77. London (Pergamon).

Mörner, N.-A. 1980. The northwest European "sea-level Laboratory" and regional Holocene eustasy. Palaeogeogr., Palaeoclimatol., Palaeoecol., 29, 281-300.

Ogilvie, A.E.J. 1984. The past climate and sea-ice record from Iceland, Part 1: data to AD 1780. Climatic Change 6, 131.

Olsson, I.U. (editor) 1970. Radiocarbon variations and absolute chronology (Nobel Symposium 12). Stockholm (Almqvist & Wiksell) and New York (Wiley). 657 pp.

Østerberg, E. 1977. Kolonisation og kriser: bebyggelse, skattetryck, odling och agrarstruktur i västra Varmland ca. 1300-1600. Lund (Det Nordiska Ødegardsprojektet, Publikation Nr. 3) (Gleerup). 308 pp.

Parry, M.L. 1975, Secular climatic change and marginal agriculture. Pp. 1-3 in Publication No. 64, Transactions of the Institute of British Geographers.

Parry, M.L. 1978. Climatic change, agriculture and settlement. Folkestone (Dawson) and Hamden, Connecticut (Archon Books-Studies in Historical Geography). 214 pp.

Pears, N.V. 1972. Interpretation problems in the study of tree-line fluctuations. In: Research papers in forest meteorology (ed. J.A. Taylor). pp. 31-45. Aberystwyth (Cambrian News). 196 pp.

Salvesen, H. 1979. Jord i Jemtland. Østersund (A/B Wisenska bokhandelens förlag). 187 pp.

Sandnes, J. 1971. Ødetid og gjenreisning. Oslo, Bergen, Tromsø (Universitetsforlaget). 385 pp.

Sernander, R. 1910. Die schwedischen Torfmoore als Zeugen postglazialer Klimaschwankungen. 11th International Geological Congress, Stockholm.

Sernander, R. 1926. Postglaziale Klimaverschlechterung. In Ebert's Reallexikon der Vorgeschichte. Berlin.

Stefansson, V. 1943. Greenland. London (Harrap). 240 pp.

Thorarinsson, S. 1956. The thousand years struggle against ice and fire. Museum of Natural History, Miscellaneous Papers No. 14. Reykjavik.

Tooley, M.J. 1974. Sea level changes during the last 9000 years in northwest England. Geogr. J., 140, 18-42. London.

SETTLEMENT EXPANSION - SETTLEMENT CONTRACTION: A QUESTION OF WAR, PLAGUE, ECOLOGY OR CLIMATE?

Björn Ambrosiani

Statens Historiska Museum
Stockholm, Sweden

Abstract

For a long time discussion has concentrated on war and plague as explanation for settlement contraction during Scandinavian prehistory and middle ages. In this paper, the author proposes that the reasons could be more complicated. Settlement changes should be studied against the background of population pressure, land usage, climate and technological innovations. The land upheaval diagram for the Mälaren area is one way to study the background. The ancient monuments survey and the medieval documents give information about the settlement. Together, they make it possible to study the changes over a long period of time in ecology and settlement.

The question of settlement expansion - settlement contraction is often discussed archaeologically using a more or less adequate data base of variations in the quantities of finds and ancient monuments. There are many examples of such changes as these, not least in northern Europe during the period known here as the Iron Age and as the Roman and post-Roman periods in the rest of Europe.

On the Baltic islands, in south Sweden and south-west Norway there was an enormous increase in the number of graves and settlements during the Roman Iron Age and up to A.D. 500 when a significant change seems to have occurred; this is reflected in the quantity of hoards and deserted settlements, and in a decreasing number of graves. There was a new phase of expansion during the

N.-A. Mörner and W. Karlén (eds.), Climatic Changes on a Yearly to Millennial Basis, 241–247.

Viking period (A.D. 800-1100) and the middle ages up to c. A.D. 1300. The Viking raids can be seen as an aspect of this expansion and have been regarded by many as an effect of overpopulation combined with good ship-building techniques and a desire for adventure.

A later population contraction is indicated by the many late medieval deserted farms, a subject which is being studied in depth by the nordic research project "Ödegårdsprojekt" (e.g. Gissel et.al. 1981).

Many explanations for these happenings have been looked for, most attention being paid to the periods of decline. Wars and plunder have been put forward as the main reasons for the deposition of hoards ever since Sture Bolin's epoch-making dissertation of 1926 on hoards, in which he suggested that warfare made their retrieval impossible.

The quantity of gold finds and deserted buildings were very soon brought together in discussions about warlike activities and migrations in the Baltic region, one facet of the folk-movements of the Migration period in Europe as a whole. There were also attempts to identify some of the actors in these events with the Svear who were thought to have extended their sphere of influence at that time from the Mälaren area to Götaland (Stjerna 1905; Nerman 1923, 1925). Mårten Stenberger also discussed this question in his dissertation Öland under äldre järnåldern and in the Vallhagar publication (1955, p. 1169 ff.). He largely agreed with the theory that warlike activities were the reasons for the concealment of gold and for the ruined farms. But Vallhagar also contains a chapter which points more towards ecological facts such as soil exhaustion and deforestation (Arrhenius 1955, p. 1053 ff.). Arrhenius thought that the soil was damaged through over-grazing and desiccation, followed by a rainy period which waterlogged the ground. This possible catastrophe would have been the reason why the settlement was not re-established later (cf. Almgren 1957).

Subsequent discussion on settlement contraction has concentrated on a number of questions. Was there really a contraction of settlement or only a movement to new, previously unutilized, settlement sites? It has, however, often been emphasized that many existing settlements stretch as far back, or even further, in time as the deserted ones, so we must be dealing with a true contraction. Also, the deserted farms have often been shown to lie on marginal agricultural land with a limited agrarian capacity which therefore more rapidly became uncultivable. What we are seeing, then, is an alternation between expansion towards the margins and contraction/concentration towards the central areas of settlement. A recent synthesis of the discussions can be found in Widgren 1983, p. 8 ff.

The late medieval contraction has long been attributed to the effects of the Black Death, starting with the catastrophic inroads into the population in 1349-50. Lars Walløe has recently (1982) studied the way in which the plague affected parts of the Norwegian population and how long the affects lasted before the period of plague died out during the 16th century and population growth began again. But at the same time many historians have suggested that the crisis in agrarian society was already evident before the 1350s through declining rents and an increase in waste land. The historians, therefore, wish to devalue the importance of the plague for the fluctuations in population.

So it is all the more interesting that from an archaeological point of view plague has been given as one of the explanations for the migration-period decline in Europe (e.g. Gräslund 1973); for instance, the great Justinian plague of the 540s which emptied the Mediterranean towns and villages, and aggravated the economic and political crises of the Roman world. It is not impossible that the infection was carried over the borders to Germania by the many trading contacts and returning legionary soldiers whose pay is reflected in the gold hoards, hidden then, only to be recovered in modern times.

As a background to all these theories there is always the question of how the population increased and what part this increase played in the historical pattern of settlement. The 18th-century Malthusian theory, which maintains that population is limited by economic conditions and plague, has lately been supplanted by that of Ester Boserup who postulated a more or less steady population growth and suggested that when the ecological ceiling is reached population growth can continue because the ceiling is raised by technological and agricultural innovations. She developed and synthesized these ideas in 1981, devoting some chapters to prehistoric and medieval developments.

It is of course difficult to pass judgement on a discussion of this type and to give a simple answer such as that the cause is one or other factor and that all other factors are of no importance. In fact, everything should be brought together in a systems theory which can produce a complicated historical reality out of more or less disparate facts. Settlement changes should actually be studied from a completely different standpoint from earlier ones: as a combination of population pressure, land usage and climate, questions which have hitherto only been touched on, at least in Scandinavian discussions. One of these factors has become my particular field of research: the climate and its influence on shoreline displacement in an area of land upheaval such as Scandinavia.

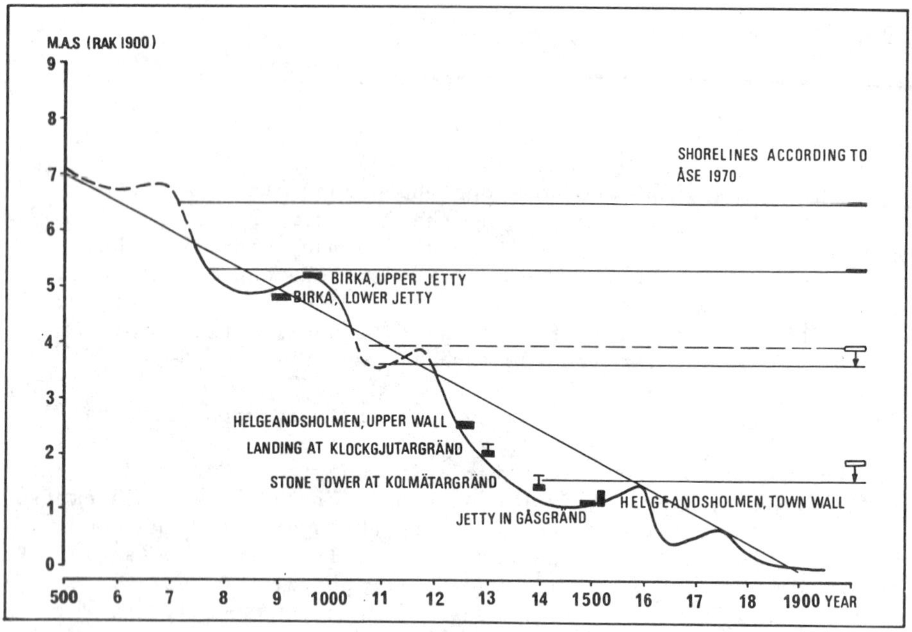

Figure 1. The Viking Age and later land upheaval curve for the Mälaren area, redrawn after B. Ambrosiani 1982, Fig. 5.

Scandinavia has long been recognized as an area of land upheaval with extremely variable land and sea levels during post-glacial times. These changes have usually been assumed to form a linear or almost linear progression. There are, however, some archaeological land upheaval values from the Mälaren area which are very inconsistent, for example the jetties in Viking-age Birka and late medieval Stockholm. The former suggests an average land upheaval of more than 50 cm per century whereas the latter indicates less than 35 cm per century. The level for the newly discovered town wall by Riksdagshuset in Stockholm is still more remarkable for there the level averages only c. 25 cm per century.

These variable values can, however, be seen to be climate-associated in that they mainly reflect eustatic changes of world sea level.

Ambrosiani (1981, 1982) and Mörner (1980, 1983) have shown that the shorelevel displacement in the Stockholm area is characterized by "steps" caused by the interaction between eustatic transgressions and regressions and a crustal uplift of about 0.5 m per century. During the last 1000 years such steps are recorded during the Viking Age and the 13th and 16th centuries.

The results can also be compared with the dendro-chronological and glacialogical analyses which Wibjörn Karlén and Thomas Bartholin (1983) have made for the northern Scandinavian mountain region. The widths of annual rings vary according to a pattern which agrees with the assumed warm and cold periods which may be seen in the shoreline displacement diagram. This implies that the Viking age was a warm period compared with the last 1000 years, followed by a cooler phase during the middle ages and a renewed warm period in the 16th century, after which the variations seem to have been more limited.

It is obvious that in the Viking period there was better crop production and population survival. A greater number of surviving children resulted in increased population pressure which led to the foundation of many new farms in the Scandinavian woodland region. An expansion of this kind can be established in east Sweden where 1000 - 2000 farms in the 8th century increased to c. 4000 - 5000 farms in c. 1100 and to three or four times that number c. A.D. 1300. In south Scandinavia much of the population expansion is also reflected in a pronounced growth of villages.

In the 12th and 13th centuries the eastern Swedish expansion spilled over onto infertile climate-dependent soils with poor production capabilities. The climate deterioration which can be established from the shoreline displacement diagram and dendro-climatological curves must have had disastrous consequence on the chances of survival in agrarian communities. The well known climatic change at the beginning of the 14th century must, through the poverty of the crops, have rapidly diminished resistence in a population which had quickly outgrown the potential for further inner colonization. Resistence against disease would also have declined, and when the Black Death struck in 1349-50 a large proportion of the population was wiped out. Certainly, it has been proved that some areas were not apparently affected by any noteworthy population decline, but other areas were almost wiped out. These differences have led to the diminution of plague as an influential factor in settlement.

In fact, a secondary concentration effect needs to be considered; after the plague the surviving population was concentrated on the best land and the less suitable land was virtually totally abandoned, the climate being one of the reasons.

Put in this perspective access to land, population pressure and climate all influence both the expansion and contraction periods of settlement, so we have to analyse this complex historical picture through inter-disciplinary studies by various branches of research such as archaeology, history, technical history, quarternary geology, dendrochronology and glaciology. No single explanation is true; the answer lies in a combination of different factors.

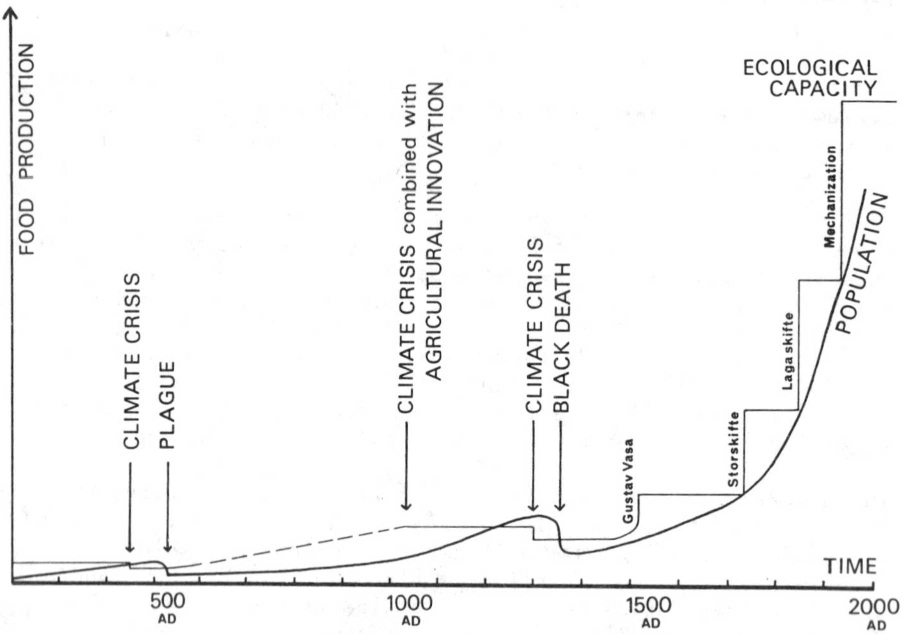

Figure 2. Factors, important for settlement growth and contraction during the Iron Age and later periods.

The development can be shown in a diagram (fig. 2) of the same type as that constructed by Stig Welinder in 1975 on the basis of Boserup's theory, but taking into account corrections such as climate and so on. The complex interaction between population and settlement obviously has its roots in the climate and the ecological capacity of different times. It is obvious that the Malthusian theory is relevant to the earlier part of the diagram and that of Boserup is applicable to the last 200 or 300 years.

Translated by Helen Clarke

References

Almgren, B., 1957. Vallhagar, Byn och Boken. Fornvännen 52, 227-246.

Ambrosiani, B., 1981. Changing Waterlevels and Settlement in the Mälar District since AD 700. Striae, 14, 140-143.

Ambrosiani, B., 1982. Mälarstäderna och landhöjningen. Bebyggelsehistorisk Tidskrift 3, 71-81.

Arrhenius, O., 1955. The Iron Age Settlements on Gotland and the Nature of the Soil. Vallhagar II, 1053-1064. Copenhagen.

Åse, L.-E., 1970. Shore-displacement in Eastern Svealand and Åland during the last 4000 years. Diss. Medd. från Naturgeografiska Institutionen vid Stockholms Universitet Nr A31.

Bartholin, Th., Karlén, W., 1983, Dendrokronologi i Lappland. Dendrokronologiska sällskapet. Meddelande 5.

Bolin, S., 1926. Fynden av romerska mynt i det fria Germanien. Lund.

Boserup, E., 1981. Population and Technological Change. University of Chicago Press, Chicago.

Gissel, S., Jutikkala, E., Österberg, E., Sandnes, J., Teitsson, B., 1981. Desertation and land colonization in the Nordic countries c 1300 - 1600. Stockholm.

Gräslund, B., 1973. Äring, näring, pest och salt. TOR, 274-293.

Mörner, N.-A., 1980, Late Quaternary sea level changes in northwest Europe: A synthesis. Geol. Fören. Stockh. Förh., 100, 381-400.

Mörner, N.-A., 1983. Stockholm excursion guide, Second Nordic Symp. Climatic Changes and Related Problems, Stockholm, May 1983, 16 p.

Nerman, B., 1923. En utvandring från Gotland och öns införlivande med Sveaväldet. KVHAA handlingar, 34, III F. 1:4.

Nerman, B., 1925. Det svenska rikets uppkomst. Stockholm.

Stenberger, M., 1933. Öland under äldre järnåldern. KVHAA.

Stenberger, M., 1955. The Reasons for the Abandonment of Vallhagar. Vallhagar II, 1161-1185. Copenhagen.

Stjerna, K., 1905, Svear och Götar under folkvandringstiden. Svenska Fornminnesföreningens Tidskrift XII, 339-360.

Walløe,L., 1982, Pest og folketall 1350 - 1750. Norsk Historisk Tidskrift 1/1982, 1-45.

Welinder, S., 1975. Prehistoric agriculture in Eastern Middle Sweden. Acta Archaeologica Lundensia, Series in 8° Minore No 4.

Widgren, M., 1983, Settlement and farming systems in the Early Iron Age. Acta Universitatis Stockholmiensis. Stocholm Studies in Human Geography 3.

CLIMATIC REFLECTION IN VARVED LAKE SEDIMENTS

Renberg, I., Segerström, U. and Wallin, J.-E.

Umeå University, Sweden

ABSTRACT

The individual thicknesses of about 1300 varves (230 A.D. - 1552 A.D.) of the varved lake sediment from Lake Judesjön, N. Sweden, were measured to study short-term climatic changes. The results indicate a cyclic variation in varve thickness, with a period length of 30-40 years, dependent on variation in the amount of organic material deposited annually. This variation is most likely to be due to changes in the primary production of the lake caused by variation in the amount of insolation during the summer of different years.

INTRODUCTION

Varved sediments are present in several of the lakes in northern Sweden (Renberg 1981a, 1983). It is not unusual to find unbroken varve series covering the last five to six thousand years in lakes situated along the Bothnian coastland.

Each varve consists of the organic and inorganic material deposited during the course of one single year. The sediment deposited during different seasons of the year is also more or less clearly visible in each varve (Fig. 1).

Varved lake sediments are potentially useful for investigating both long- and short-term changes of the climate in the past, because any climatic changes will have affected the varves, in several ways, for example: The seasonal and annual deposition rate of organic matter and mineral grains, the composition of the subfossil assemblages of plant and animal remains, and other phenomena, e.g. the chemical composition of the material deposited

N.-A. Mörner and W. Karlén (eds.), Climatic Changes on a Yearly to Millennial Basis, 249–256.

during the different seasons of the year.

During the past few years we have counted the number of varves in sediment cores from several different lakes, in the course of which we have often gained the visual impression that varve thickness, and other characteristics such as the mineral grain content, vary rhythmically both on a long-term (centennial) and a short-term (annual and decadal) basis. Varve thickness is closely related to the annual sediment deposition rate. Varve thickness measurements therefore provide a simple method of following the year-to-year changes in sediment accumulation, which it is difficult to measure by any other technique.

This paper deals with some of the results obtained from measuring the individual thickness of about 1300 consecutive varves (230 A.D. to 1552 A.D.) in a sediment core taken from Lake Judesjön. Indications of a 30-40 year cyclic variation were found. A possible explanation for this variation in varve thickness is discussed.

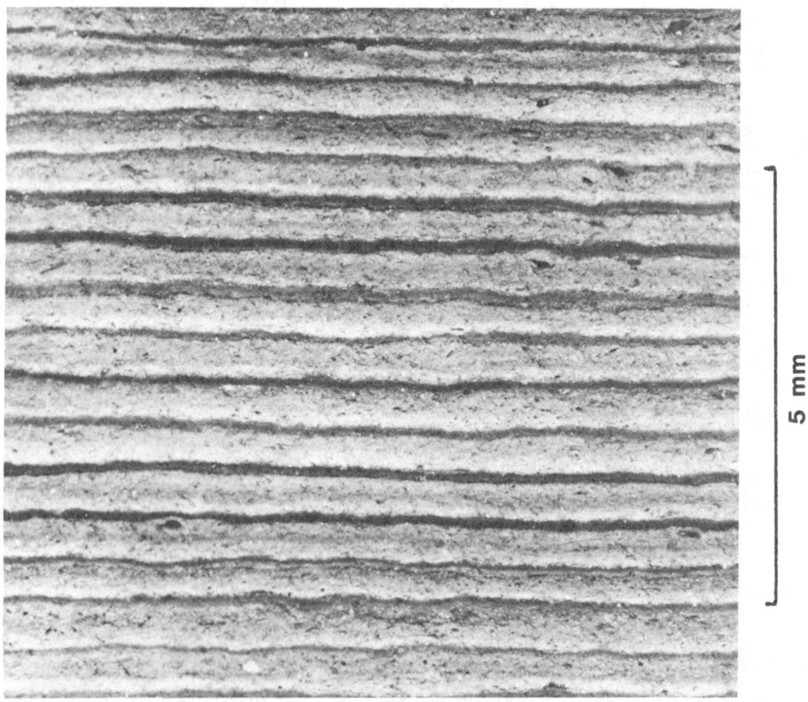

Fig. 1. A sample of varved lake sediment. Each varve consists of a more or less well-developed spring layer of mineral grains, a mainly organogenic summer layer, and a thin winter layer with a high organic content. The dark winter layer is the most distinct one in every varve.

THE INVESTIGATED LAKE AND THE ANALYSIS METHOD USED

Lake Judesjön is situated in Ångermanland province, N. Sweden (Fig. 2). The lake lies 78 m above sea-level and was isolated from the sea, by the landupheaval about 6000 years ago, according to the varve counts. The lake has a surface area of 15 ha, and a drainage basin of 200 ha and a maximum depth of 15.6 m. It is a dimictic lake.

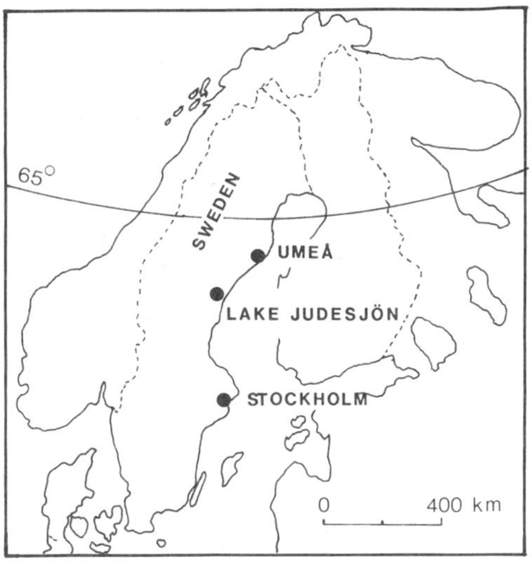

Fig. 2. Map showing the location of Lake Judesjön.

Several replicate cores were taken with a Russian peat corer (chamber length 1 m, diameter 8 cm) in winter 1983 from the deepest part of the lake. The cores were placed on short planks in the field and carefully wrapped in plastic and aluminium foil to prevent drying of the sediment.

A few days later we made the varve measurements, using equipment designed for making tree-ring measurements at the Swedish University of Agricultural Sciences (SLU), Umeå. Before counting and measuring the varves the cores were cleaned and covered with strips of semi-pliable, transparent, plastic (see the description given by Renberg 1981b).

A time-series analysis (SPSS, Subprogram SPECTRAL - Spectral Analysis of Time Series) was made, using the individual varve thickness values for the period 230 A.D. to 1552 A.D. The reason for starting the analysis 230 A.D. was that the previously deposited sediment contains more mineral matter. The mineral matter masked the variation in varve thickness due to the variation in deposition of organic matter, which was what we wished to study. After the 16th century A.D. varve thickness can be expected to be influenced by the increase in lake productivity and in inflow of allochthonous material due to agricultural activities, which commenced in the lake catchment area sometime during the 16th century (Segerström, Renberg & Wallin 1984).

The values for the annual deposition or organic and of mineral matter, were estimated in the following manner: Samples representing 4 cm^2 of the lake bottom and including a certain number of varves (2-35) were cut out from the sediment cores. The samples were dried, weighed and then combusted (at +550°C, 4 hours). The deposition values were then obtained by dividing the values for ignition loss and ash weight by the total varve area of each sample (4 cm^2 x the number of years in the sample).

The degree of error involved in making the varve counts does not exceed 2-3 %. The degree of error in measuring the varve thickness is more difficult to estimate. Varve measurements were made on two replicates, but only the results from one series were used for the time-series analysis. The results for the two separate series were very similar, however, indicating that only a minor degree of error will be involved.

RESULTS AND DISCUSSION

The time-series analysis indicates that there is a cyclic variation in the varve thickness, with a period length of 30-40 years (Fig. 4). The cyclic variation can also be descerned from the graphs for individual varve thickness and the same values expressed as 21-year running means (Fig. 3).

The thickness of each varve is mainly a function of the contents of organic matter, mineral grains and water. The water content is the most important factor in this lake sediment, because it accounts for more than 70 % of the fresh weight of the sediment. The water content of each varve depends on deposition of both the organic matter and the mineral matter. The organic matter is the more crucial element in a sediment with a rather low mineral grain content such as the one studied. This means that the changes in varve thickness recorded in this lake sediment are primarily due to the annual variation in the deposition of organic material.

The organic sediment fraction consists of both autochthonous and allochthonous materials. We consider that the autochthonous proportion is the greater, because the drainage basin is rather small, there are no peatlands situated close to the lake, and the runoff from the drainage basis is low, except during the snowmelt period. The close correlation between the deposition of organic matter and mineral grains that would be expected if the major part of the organic influx was washed into the lake during the spring flood, was not observed. The influx of windblown allochthonous material into most lakes is small, obviously even for light particles such as pollen grains (Peck 1973). Detailed investigations

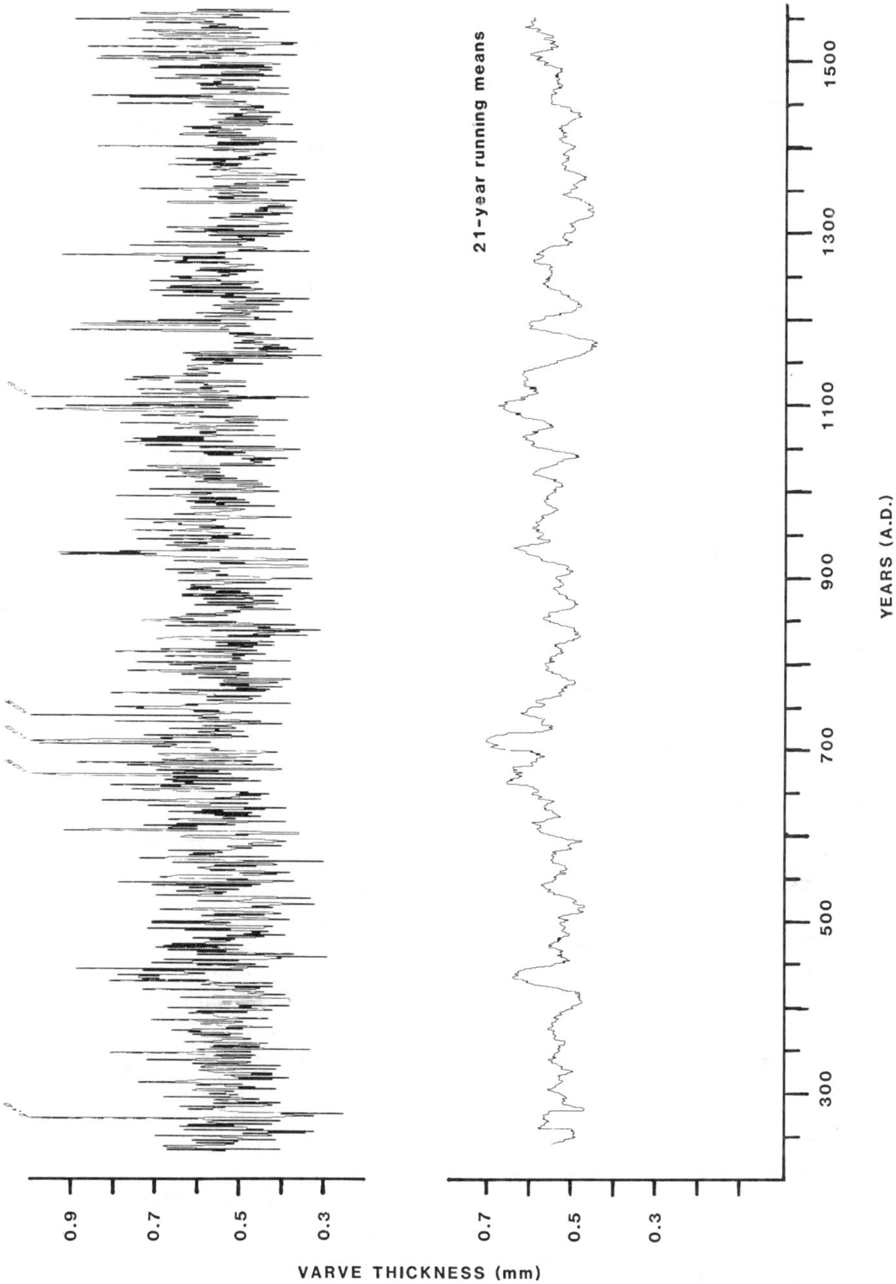

Fig. 3. Graph showing the individual thicknesses of about 1300 consecutive varves from Lake Judesjön and the same values expressed as 21-year running means (below).

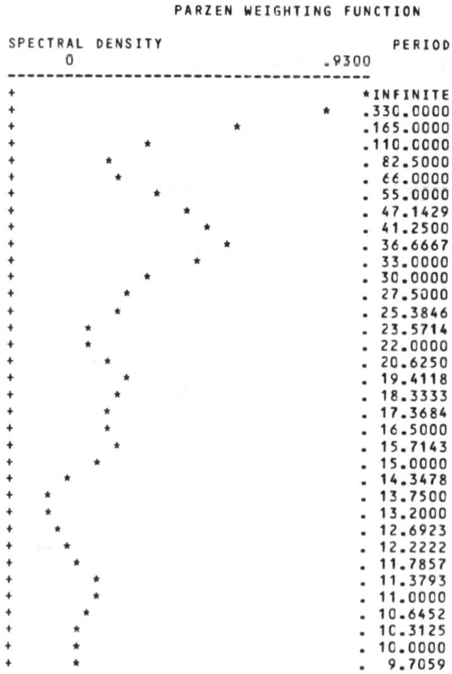

Fig. 4. Results of the time-series analysis.

of the carbon cycle in lakes (see the review by Wetzel 1975) also point to the importance of the autochthonous primary production, by the phytoplankton and the littoral flora, as being the major contributors of the particulate organic carbon of natural lake systems.

What may have caused the rhytmic changes in lake productivity? The weather situation during the vegetation period is the most reasonable answer. Climatic factors are of fundamental importance for the productivity of any lake. Although changes in precipitation amounts may influence the nutrient balance of a lake, and hence primary productivity, light and temperature (two closely-correlated factors) are probably the most important factors responsible for the fluctuation in the productivity of natural lakes. The rate of photosynthesis and of algal growth are closely related to the amount of solar insolation. The incoming rate of solar radiation reaching the lake surface is dependent on the prevailing weather conditions. Heavy cloud cover can reduce the value of incoming radiation to less than 50 % of that on a sunny day (Strahler 1973).

We would like to summarize the foregoing discussion of the results of the measurements of varve thickness in the sediment cores from Lake Judesjön by putting forward the following hypothesis: In northern Sweden the weather situation in summer during the past two millenia has (still does?) fluctuated cyclically between being sunnier/warmer and cloudier/colder (June - August). The period length has been about 30-40 years. We hope to be able to confirm this hypothesis by carrying out further investigations, including detailed analyses of the varved sediments from other lakes.

It is tempting to extrapolate the varve thickness curve (Fig. 3) to yield a prediction of the weather of coming years. This is not to be recommended, however, considering the degree of error involved in the varve counts. Assuming an error of ±3 % then the cumulative error back to 1552 A.D. may be as much as ±13 years, almost half the period length. However, meteorological observations indicate that a period of sunnier summers reached its climax at the end of the 1960's and the beginning of 1970's, and that we are at present in a period of cloudier summers in northern Sweden (Fig. 5).

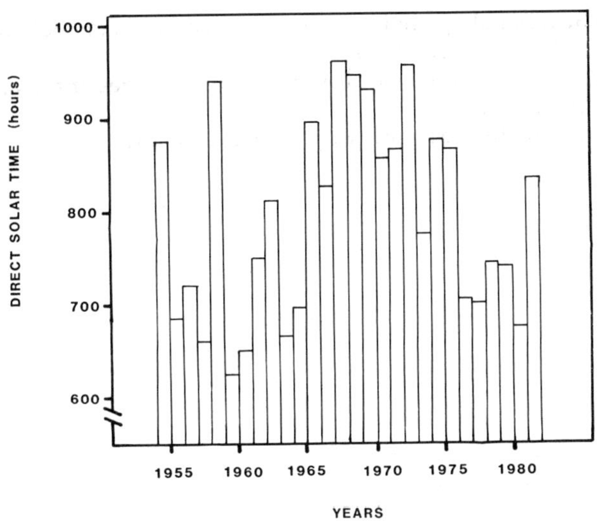

Fig. 5. Number of hours with sunshine during the period June - August 1955-1982 recorded at the Sundsvall meteorological station, which is the one lying closest to Lake Judesjön (SMHI Year-books).

ACKNOWLEDGEMENTS

This research was supported by the Swedish Natural Research Council. Special thanks are due to Göran Broström, Mats Nilsson, Philip Tallantire and Christer Wilhelmsson.

REFERENCES

Peck, R.M. 1973. Pollen budget studies in a small Yorkshire catchment. In: Quaternary Plant Ecology (eds. H.J.B. Birks & R.G. West), pp. 43-60. Blackwell Scientific Publications.

Renberg, I. 1981a. Formation, structure and visual appearance of iron-rich varved lake sediments. Verh. Internat. Verein. Limnol., 21, 94-101.

Renberg, I. 1981b. Improved methods for sampling, photographing and varve-counting of varved lake sediments. Boreas, 10, 255-258.

Renberg, I. 1983. Varved lake sediments - geochronological records of the Holocene. Geol. Fören. Stockh. Förh., 104, 275-279.

Segerström, U., Renberg, I. & Wallin, J.-E. 1984. Annual sediment accumulation and land use history; investigations of varved lake sediments. Verh. Internat. Verein. Limnol., 22, (in print).

SMHI. 1955-1982. Monthly and yearly summary of weather and water supply in Sweden. Yearbook 37-64.

Strahler, A.N. 1973. Introduction to physical geography. John Wiley & Sons, Inc.

Wetzel, R.G. 1975. Limnology. W.V. Saunders Co.

SUNSPOT CYCLES AND GLOBAL OSCILLATIONS

D.J. Schove

St David's College, West Wickham, Kent, UK.

Sunspot numbers can be constructed with fair reliability back to 1700, and for meteorological purposes relative values have been extended back to AD 1500 (Schove 1983a, 10, Table 1). Meteorological data in many parts of the world seldom go back before 1841, and attempts to find the admittedly weak and complex relationships with weather are often unsuccessful. In Sweden good records for temperature extend back into the mid-eighteenth century, and Svenonius and Olaussen (1979, cf. Schove 1983a, 233ff) have found correlations; we now need to know the seasonal pressure and wind relationships but for this we must wait until the pre-1860 barometric data for Lund and Upsala have been published (summer pressure can be estimated from the tables in Schove 1954). Sunspot maxima and minima are known approximately back to AD 300 and intermittently back to before 400 BC. Proxy time-series such as are being developed in Sweden by Bartholin and Karlén (this volume) reflecting summer temperature back to before AD 500 are thus of special interest. Varve series from the present day reflecting summer rains are being developed by Cato, who is replicating his results so that the errors characteristic of varve series have been eliminated. Renberg and Segerström (this volume) present a long lacustrine varve series where the estimated error is only 1%.

The relationships of sunspots and weather depend on the magnitude of the sunspot cycles: the sign of the relationship is reversed for strong and weak cycles in the case of the medieval Nile floods (Schove 1983a, 291; 1977, 51, fig. 4.8; Schove 1983a, 326 = 1964, 470). For magnitude of solar activity we have two independent methods - estimates based on auroral (and sunspot) observations and calculations based on tree-ring

N.-A. Mörner and W. Karlén (eds.), Climatic Changes on a Yearly to Millennial Basis, 257–259.

radiocarbon. Both are given in Fig.1. For proxy series since 1500 simple array analysis using quartiles is suggested (as given in Schove 1983b, 395).

High and low sunspot numbers lead to two- and three-year cycles respectively. Initially stratospheric, the varying proportions enable us to cross-date varves and tree-rings (Schove 1983, 352; 1978, 228). Tree-rings in Asia Minor show similar fluctuations between the two types of cycle as the Bristlecone pines in the United States (Schove, this volume).

Further cycles can be found in sunspot and auroral data. A short auroral cycle of 1.4 years has been found by Silverman (1983), who points out that this cycle is stronger when aurorae are weaker. If persistent solar activity through Silverman's 22-year periods occurred at specified solar longitudes this could be a beat effect between the lunar (29½ day) and a solar rotation cycle of average length just under 28 days. If this theory is correct the 1.4 year cycle should change length in the course of the solar cycle. If, however, this cycle does not depend on the invisibility of weak aurorae at New Moon, we have a new cycle for meteorological investigation.

Longer cycles and wave-trains include the 80-year cycle in solar phase (and solar wind?) and the 200-year cycle in both phase and amplitude (Schove 1983a, 194-195; 1955, 143). In amplitude there are several cycles in the 60-90 year range. Chinese scholars (Ding et al. 1982) have confirmed the cycle of about 60 years in the Far Eastern Sunspot record (their list includes several 'new' observations as in AD 434, 478, 479, 482, 489, 490 and 496). They also find a long wave-train of about 250 years, but this does not appear to be a persistent cycle affecting tree-ring radiocarbon. The meteorological effects of these cycles are not known and previous suggestions (Schove 1983a, 318; 1964, 451) are not significant.

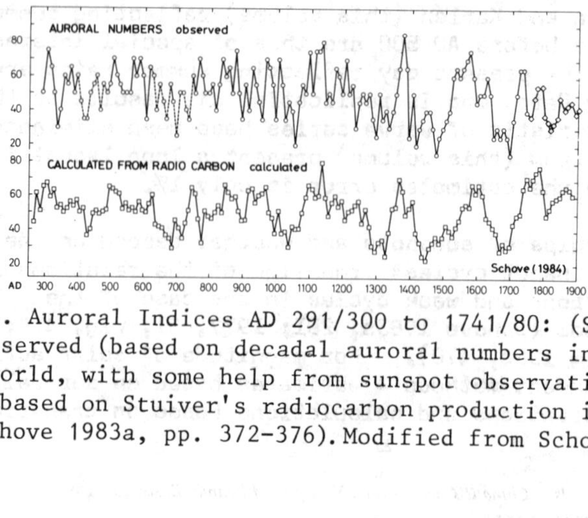

Fig 1. Auroral Indices AD 291/300 to 1741/80: (Sunspots 1750/1979) a) Observed (based on decadal auroral numbers in different parts of the world, with some help from sunspot observations). b) Calculated (based on Stuiver's radiocarbon production index as explained in Schove 1983a, pp. 372-376). Modified from Schove 1983b.

References

Ding, Y., Lou, B. and Feng, Y. 1982. The Ancient Solar Activity Maxima Epochs of Various Periods. Acta Astronomica Sinica, 23 (3), 287-298.

Schove, D.J. 1954. Summer Temperatures and Tree-rings in North Scandinavia, AD 1461-1950. Geografiska Annaler, Stockholm, 36, 40-80.

Schove, D.J. 1955. The Sunspot Cycle, 649 BC to AD 1986. Journ. Geophys. Res., 60, 127-146 = Schove 1983a pp.194-195.

Schove, D.J. 1964. Solar Cycles and Equatorial Climates. Geol. Rundschau, 54, 448-477. p.470 = Schove 1983a p.326, p.451 = Schove 1983a p.318.

Schove, D.J. 1977. African Droughts and the Spectrum of Time. In: Drought in Africa (ed. D. Dalby, R.J. Harrison-Church and F. Bezzaz), Afr. Envir. Spec. Rep. 6, 38-53.

Schove, D.J. 1978. Tree-rings and Varve Scales combined c.13500 BC to AD 1977. Pal. Pal. Pal., Vol.25, 209-233. p.228 = Schove 1983a p.352.

Schove, D.J. 1983a 'Sunspot Cycles' (ed. D.J. Schove) (Benchmark book published by Hutchinson Ross and distributed by Van Nostrand Reinhold).

Schove, D.J. 1983b. Sunspot, Auroral, Radiocarbon and Climatic Fluctuations since 7000 BC. Annales Geophysicae, Paris, 4-5, 391-396.

Schove, D.J. 1983c. Global Oscillations and the Absolute Dating of Varves, Ice-cores and Tree-rings. In: Abstracts of the Second Nordic Symposium on Climatic Changes. (Ed. K. Frydenthal, W. Karlen, N-A. Mörner and E. Rasmussen), 1983, pp.116-122. Danish Meteorological Institute, Copenhagen.

Silverman, S.M. and Shapiro, R. 1983. Power Spectral Analysis of Auroral Occurrence Frequency. J. Geophys. Res.

Svenonius, B. and Olausson, E. 1979. Ring Widths of Trees, Solar Activity and Weather Conditions in Sweden in the Period 1750-1975. Geologisk Foreningens i Stockholm Fordhandlinger, 100, 97-118.

DENDROCHRONOLOGY IN SWEDEN

Thomas Seip Bartholin

University of Lund, Sweden

The work within the Lund laboratory is based on chronologies for which the physical range is known through exact datings (Fig. 1). We expect that the range of the chronologies is actually much wider, but this theory has not yet been properly tested.

An international interest in the Lapland project is a reflection of the availability of unusually excellent source material from the northernmost treeline of pine. See the papers of Aniol and Karlén. Around the lake Torneträsk living pines can reach an age of over 600 years, and trees fallen during the Viking Age are still lying on the ground, easily taken for sampling.

A total of 277 bore cores were collected from 142 living trees, about two cores per tree, and an evaluation of these recent samples indicated that the region is dendrochronologically very homogeneous. A full 83% of the 273 subfossil samples were dateable, and the entire material, recent and subfossil, contained 131,348 rings. The finest 265 samples, totalling 64,028 rings, were finally selected for the construction of the standard curve for the area (Fig. 2), each single year covered by an average of 42 samples. Those samples not chosen for the standard curve were for the most part rejected because of missing rings.

It is notable that three times in the course of less than 50 years minimum values occur which are the lowest during the 1546 years investigated; in 1601, 1642 and 1645.

There are large numbers of stems several thousands of years old lying in the peat bogs and small lakes around Torneträsk, above the present tree line. This source material was explored in the summer of 1983 in the hope of being able to extend the existing standard chronology even further back in time.

N.-A. Mörner and W. Karlén (eds.), Climatic Changes on a Yearly to Millennial Basis, 261–262.

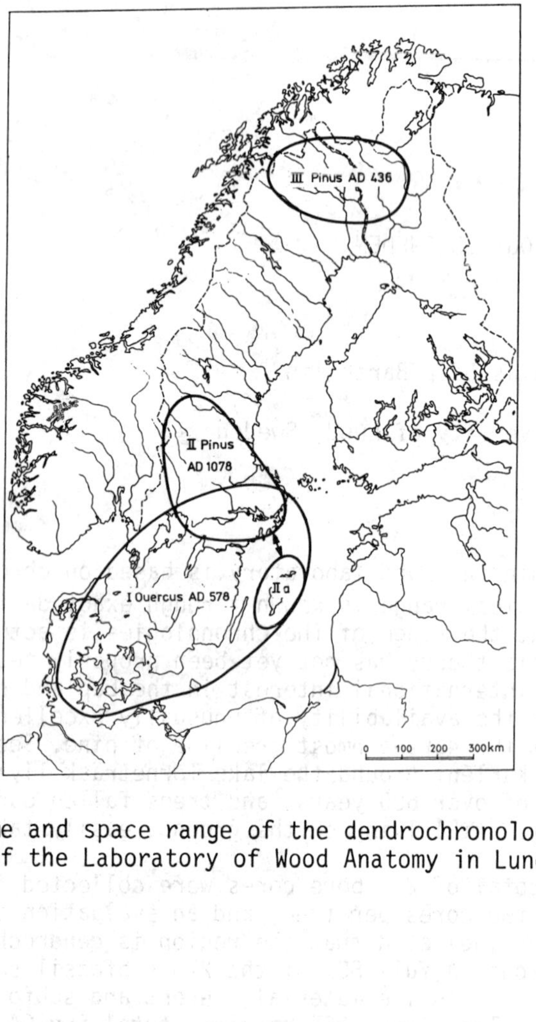

Figure 1. Time and space range of the dendrochronological standard curves of the Laboratory of Wood Anatomy in Lund.

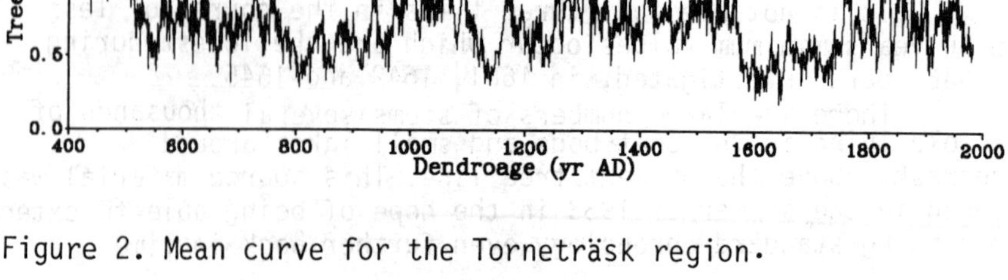

Figure 2. Mean curve for the Torneträsk region.

DENDROCHRONOLOGY, MASS BALANCE AND GLACIER FRONT FLUCTUATIONS IN NORTHERN SWEDEN

Wibjörn Karlén

University of Stockholm, Sweden

ABSTRACT

A chronology of glacier front fluctuations for northern Scandinavia derived from lichenometric and C14-dates on moraines is compared with information about mass balance inferred from dendrochronology. The glacier chronology is verified. Glacier advances not known from moraine stratigraphy in Sweden probably occurred around AD 1140 and AD 1240.

INTRODUCTION

The present knowledge of Holocene climatic variations in northern Sweden is largely inferred from glacier moraine stratigraphy. A recently published dendrochronology for the same area adds information about climate and facilitates a check of the stratigraphy. This is possible because the mass balance of one glacier in the area is known and tree ring width can be linked to mass balance and front variations.

DENDROCHRONOLOGY

The dendrochronology of northern Sweden depicts the annual tree ring width from AD 436 to the present (Bartholin & Karlén, 1983). Details about the dendrochronology are given elsewhere in this volume (Bartholin this volume, Aniol & Eckstein this volume). For this study it is important that the number of missing rings seems small (Bartholin this volume) and that the ring width appears to be largely determined by the July temperature (Aniol & Eckstein

N.-A. Mörner and W. Karlén (eds.), Climatic Changes on a Yearly to Millennial Basis, 263–271.

this volume). The temperature of other months, precipitation and the growth during previous years seem to have only a minor influence on pine tree growth in northern Sweden. A reconstruction of the July temperature for the period AD 1680-1979 is presented in this volume (Aniol & Eckstein this volume, Fig. 8). Periods favoring glacier expansion and contraction are, in this study, inferred from the dendrochronology, because the dendroclimatological study is not yet completed for most of the 1500 years for which the tree ring width is determined.

THE MASS BALANCE OF STORGLACIÄREN

The program at Tarfala research station includes a mass balance study of Storglaciären. The program was initiated in 1945 and is still in progress (Schytt, 1947-1981). This continental glacier is located within the Kebnekaise massif. The area of the glacier is about 3 km^2, which makes it slightly larger than the typical glacier of the area. After a study of the mass balance of this glacier during the period 1947-1971, Schytt concluded that variations in the mass balance were largely caused by variations in the summer temperature (Schytt, 1973, p. 78). A lowering of the Tarfala summer temperature (June, July and August) from the average +5.9°C to +4.7°C would lead to an ablation equal to the winter precipitation. An increase in winter precipitation of 40 % would be needed to balance the average 1947-1971 ablation.

This conclusion is supported by continued mass balance studies. During the 1946-1981 period the accumulation surpased the 1947-1971 average by 40 % on only three occasions. In the same period the summer temperature was below the critical +4.7°C during seven years. Further evidence indicating that fluctuations in the mass balance are largely determined by variations in summer temperature, is found in meteorological records. During a documented glacier advance in the beginning of this century the temperature at the Tarfala research station frequently was below the critical +4.7°C. (The temperature is calculated from observations made in Kiruna; the correlation coefficient is +0.9°C). Observed low precipitation in Kiruna during this period further supports the theory that changes in temperature, and not changes in winter precipitation, caused the positive mass balance and thus the advance of Storglaciären in the beginning of this century. The same sensitivity to temperature is assumed for other glaciers in the area. Glaciers near the west coast of Norway are mainly sensitive to changes in winter precipitation (Haakensen & Wold, 1981, p. 42).

Several consecutive years of positive mass balance will result in an advance of a glacier front. The advance may lag behind the climatic event. Observations from Norway (Liestöl, 1967) indicate a lag of four years for an advance in the beginning of this century. The extent of an advance will depend on factors such as the size of the glacier at the beginning of the climatic event, the duration of the event and the excess in the net mass balance. These factors, and complicating factors such as response time (Nye, 1965) makes it difficult to predict the extent of an advance from climatological data.

Since variations in summer temperature cause both changes in mass balance (and thereby glacier front fluctuations) and changes in the width of pine tree rings, a correlation between these factors is expected. However, while the glacier ablation is largely dependent on temperature conditions during the whole summer, tree ring width is considered a function of the July conditions only. Regression analysis suggests only a weak correlation (+0.5°C). The correlation supports the relationship observed between tree ring width and documented glacier behavior during the present century; narrow tree rings are associated with positive mass balance and glacier advances, while wide rings are associated with negative mass balance and glacier retreats.

GLACIER MASS BALANCE AND FRONT VARIATIONS INFERRED FROM DENDROCHRONOLOGY

Although the Storglaciären mass balance series is the longest in the world, the length is too short to permit the establishment of a good statistical relationship between tree ring width and ablation. The low correlation mentioned above, together with the effect of variations in winter precipitation, permit only an approximate division between years of positive and negative mass balance. During the period in which observations were made there were eight years with a positive balance. Disregarding one year when the balance was undoubtably caused by exceptional winter accumulation, the ring width of positive years averaged 0.6 mm. The ring width of years with a negative balance averaged 0.7 mm. Assuming no long term variations in precipitation, it seems likely that during periods with ring widths averaging less than about 0.6 mm over a period of several years, there was a positive mass balance. The dendrochronology now available suggests periods of such balance around AD 550, 800, 860, 910, 1140, 1240 and during 1570-1750 (Fig. 1). In addition, the mass balance probably was positive during several short intervals, particularly between 1770 and 1915. Extended periods of positive mass balance probably caused glacier expansions. Even short periods of positive mass balance may lead to advances and moraine formation. The moraines will be preserved only if the advance occurs during a general retreat, and this retreat is not followed by a large subsequent advance.

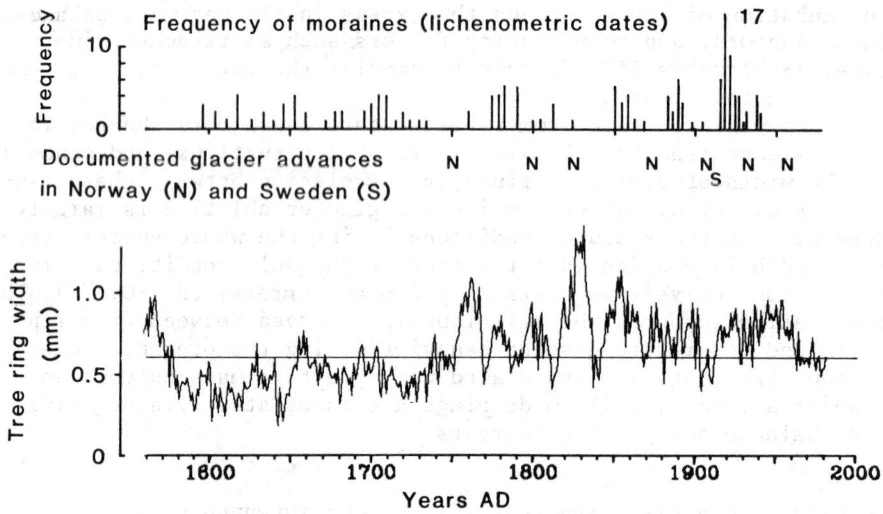

Fig. 1. Frequency of moraines in northern Sweden and the dendrochronology from northern Sweden 1560-1979. Assuming no long term change in precipitation, tree ring widths thinner than 0.6 mm are believed to indicate summer temperatures low enough to cause positive mass balance for Storglaciären and other glaciers in northern Sweden. Several consecutive years of positive mass balance will result in glacier advances. Lichenometry dates indicate a lag of about 10 years between periods of negative mass balance and glacier retreats.

GLACIER ADVANCES IMPLIED FROM MORAINE AND PROGLACIAL LACUSTRINE SEDIMENT STUDIES

A few glacier moraines in northern Sweden are C14-dated and about 200 moraines fronting over 50 glaciers are dated by the lichenometry technique. Several glacier advances are known from historical records and a large number of C14-dates associated with moraines have been obtained in Norway. The use of the lichenometric as well as the radiometric technique in moraine studies involves a number of debatable assumptions and problems. The interpretation of the results have been discussed (Giffey & Matthews, 1978; Matthews, 1980; Matthews, this volume). It is suggested that some C14-dates obtained on well-developed soils may be seriously in error.

A lichen growth curve covering the last 300 years is based on observations of lichen growth at waste heaps around mines within the mountains of northern Sweden. The short distance between the mines and the moraines is believed to permit relatively good dating. Glaciers seem to have retreated from moraines AD 1590-1620, around 1650, 1680, 1700-1720, 1780, 1800-1810, 1850-1860, 1880-1890 and 1916-1920 (Karlén, 1976, p. 31). These dates are largely supported by the dendrochronology (Fig. 1). Clusters of lichenometric dates on moraines occur about 10 years after periods of presumed positive mass balance. A lag of 10 years seems likely considering glacier response and also the fact that lichenometry yields minimum dates (Karlén, 1975).

The quality of the dates in the Swedish glacier chronology is fundamental to the discussion concerning fluctuations in the Swedish climate. Dendrochronology has reduced the importance of the glacier moraine chronology, because dendrochronology depicts climatic variations in a more detailed way and the chronology is precise. The good agreement between the chronologies implies that climatic information based on lichenometric dates of moraines may be better than generally assumed.

The lichen growth curve before AD 1600 is based on only a few C14-dates. Lichenometric dates on moraines formed before this date are thus less precise than those from the last few hundred years. Lichenometric dates on moraines and C14-dates associated with moraines in Norway and Sweden indicate glacier advances 1300-1000 BP (AD 650-950), (Karlén, 1976, p. 6, 1979, 1982, p. 30). The C14-dates are obtained on buried soils and sheared off tree-stumps (Giffey & Worsley, 1978). Dates on overridden trees are probably good, while other dates may be in error as mentioned above. Since the dates are not stratigraphically tied to each other, the similarity in age between dates on soils and treestumps may be a coincidence. However, glacier advances probably occurred around AD 800-900; the dendrochronology demonstrates that the climate at that time was approximately as cold for as long a time as during the 1600's and early 1700's (Fig. 2).

For the period AD 950-1590 there is only one good indication of a glacier advance in Sweden. This consists of a C14-date on a buried soil in front of Ritajekna in Sarek National Park (Karlén, 1981a), yielding a maximum date for one glacier advance of 640±90 BP (AD 1290). The site was overridden by an advance that formed a moraine farther downslope, which is dated by lichenometry to approximately AD 1600. Advances in Norway seem, if C14-dates on moraines can be trusted, to have been more distinct during this period. Seven C14-dates associated with moraines have yielded ages around 600 BP and 430 BP (AD 1330 and 1490). The dates are believed to depict a period of advances. A distinct cluster of dates seems unlikely if samples from buried soils frequently yield dates that are several hundred years off. Several Norwegian moraines are also dated to this period by lichenometry. In addition, a Norwegian document from AD 1340 mentions a glacier advance at that time (Eide, 1955).

The dendrochronology indicates that summers were cool but hardly cool enough to cause distinct glacier advances at this time. A winter precipitation greater than that at present, in combination with relatively cool summers, could cause glacier

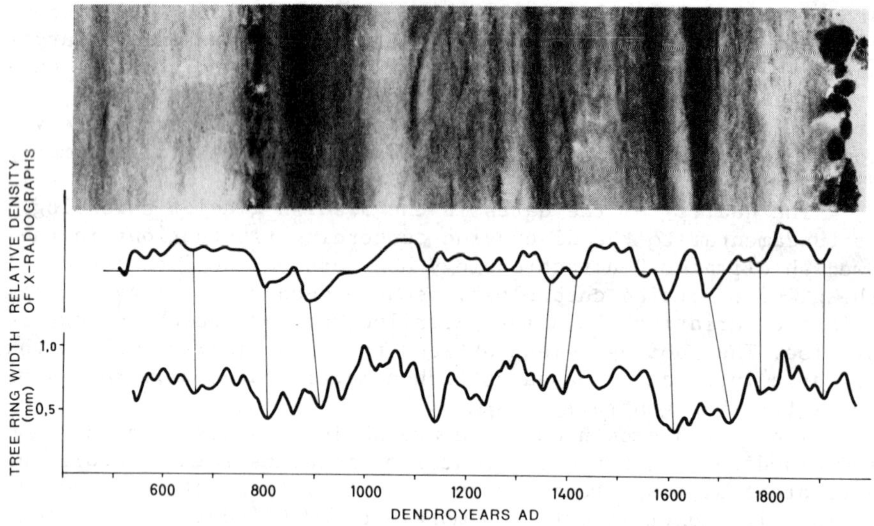

Fig. 2. Tree ring width variations are well correlated with fluctuations in the organic content of proglacial lacustrine sediments. At the top, a copy (positive) of a X-radiograph depicting the sediments of the uppermost 10 cm of a core from Vuolep Allakasjaure, a proglacial lake in northern Sweden. The upper curve shows the relative density of the X-radiograph. The light band (relatively rich in organic debris) above AD 1000-1100 is radiocarbon dated to about 900 BP. The lower curve shows variations in tree ring width in northern Sweden (31 year running mean). Periods of cool summers assumed to have caused positive mass balance for Storglaciären occurred around AD 550, 800, 860, 910, 1140, 1240, and during 1570-1750- Cool summers also occurred at other times and were particularly common between 1770 and 1915.

advances. Evidences from England indicate extremely wet conditions between AD 1300 and 1500 (Lamb, 1982, p. 187). Great amounts of precipitation, in combination with intense westerly winds during the 1300's and late 1400's (Lamb, 1982, p. 51) may have caused such large winter accumulation, particularly on the west side of the Scancinavian mountains. As mentioned above, mass balance studies on several Norwegian glaciers indicate that the balance of maritime glaciers is strongly affected by fluctuations in winter accumulation (Haakensen & Wold, 1981, p. 42).

So far there are no indications of glacier advances in Scandinavia around AD 1140 and 1240, two periods when the summer temperature seems to have been cold enough to cause positive balance. The warm period between AD 950 and 1140 is likely to have caused a considerable retreat. Probably glaciers were very small after this period of nearly 200 years, when summers were even warmer than the ones of the 1930's. The glaciers started the advances from withdrawn positions far behind the big moraine systems. Moraines formed during these periods therefore were overridden by later advances. Several glacier advances are dated at about 800 BP (AD 1150) in the Alps (Schneebeli & Röthlisberger, 1976). Baffin Island (Miller, 1973), Brooks Range (Calkin, 1982), New Zealand (Burrows & Gellatly, 1982) and South America (Mercer, 1982).

Long-term glacier fluctuations affect the composition of proglacial lacustrine sediments (Karlén, 1981b). The dating of such sediments is problematic and the low rate of sedimentation does not permit recognition of events of short duration. However, a somewhat questionable support for high glacier activity around AD 800, 1300 and the last few hundreds of years can be found in this type of material (Fig. 1), (Karlén, 1982).

TREE LIMIT VARIATIONS AND DENDROCHRONOLOGY

Holocene climatic fluctuations in northern Sweden are implied from variations in the distribution of pine trees (Karlén, 1976, 1982). The period covered by the dendrochronology to some extent overlaps periods from which C14-dates on fossil pine suggest a pine tree limit at a higher elevation. High pine tree limit is believed to be associated with warm summers. As expected, several dates on high pine tree limit coincide with the period of wide tree rings AD 950-1140. A few dates on high pine tree limit from around 1500 BP (AD 480) are contemporaneous with the beginning of the chronology. The beginning of the chronology is built on the tree rings of those pines which are found near or above the present pine tree limit. No conclusions should be drawn from this because the ring width measurements from about the first 100 years of the dendrochronology are made from a small number of samples and are therefore not comparable with the dendrochronology of the following 1400 years.

REGIONAL SIGNIFICANCE

The dendrochronology and other types of records have yielded information about summer temperature during the past 1500 years. Since the present summer temperature conditions are fairly homogeneous in northern Sweden, it is likely that the results are representative for this area also during preceeding periods. The present day conditions do not suggest a detailed similarity between northern and southern Swedish conditions. However, even if the regional significance of the short-term fluctuations are limited, there is a general similarity between dates on glacier advances in northern Sweden and several other areas.

CONCLUSIONS

The similarity between glacier chronologies inferred from dendrochronology and moraine stratigraphy studies suggests good precision of the lichenometric technique at least when applied on moraines from the last few hundred years. Cold climatic conditions inferred from glacier advances dated by the C14-technique to 1100 BP is confirmed. The similarity in age between C14-dated advances and narrow tree rings may be coincidental, but this is not likely.

Glacier advances in Norway known from historical documents and the C14-dates associated with some Norwegian moraines may be related to an increase in precipitation, since neither dendrochronology nor Swedish moraine stratigraphy indicate particularly cold climate. This conclusion is based on the observations that Swedish glaciers seem more sensitive to changes in temperature than precipitation, while Norwegian, west coast glaciers, seem most sensitive to changes in precipitation.

ACKNOWLEDGEMENTS

This study was made possible thanks to grants from the Swedish Natural Science Research Council and "Bendixsons Stipendiefond". Support from "Svensk Jordbruksbokföring" is also acknowledged.

REFERENCES

Bartholin, T.S., Karlén, W., 1983. Dendrokronologi i Lappland. Dendrokronologiska sällskapet, meddelande 5, 3-16.

Burrows, C.J. & Gellatly, A.F., 1982. Holocene glacier activity in New Zealand. In: Holocene Glaciers (ed. W. Karlén), Striae 18, 41-47.

Calkin, P.C. & Ellis, J.M., 1982. Holocene glacial chronology of the Brooks Range, Northern Alaska. In: Holocene Glaciers (ed. W. Karlén), Striae 18, 3-8.

Eide, T.O., 1955. Breden og Bygda. Norveg 5, 1-42.

Griffey, N.J. & Matthews, J.A., 1978. Major Neoglacial glacier expansion episodes in southern Norway: Evidences from moraine ridge stratigraphy with C14-dates on buried palaeosols and moss layers. Geogr. Ann. 60(A):1-2, 73-90.

Griffey, N.J. & Worsley, P., 1978. The pattern of Neoglacial glacier variations in the Okstindan region of northern Norway during the last three millenia. Boreas 7, 1-17.

Haakensen, N. & Wold, B., 1981. Glasiologiske undersøkelser i Norge 1979. Norges Vassdrags- og Elektrisitetsvesen. Rapport nr 3:81, 1-80.

Karlén, W., 1975. Lichenometrisk datering i norra Skandinavien - metodens tillförlitlighet och regionala tillämpning. Depart-

ment of Physical Geography, University of Stockholm, Report 22, 67.
Karlén, W., 1976. Lacustrine sediments and treelimit variations as indicators of Holocene climatic fluctuations in Lappland, Northern Sweden. Geogr. Ann. 58A:1-2, 1-34.
Karlén, W., 1979. Glacier variations in the Svartisen area, northern Norway. Geogr. Ann. 61A:1-2, 11-28.
Karlén, W., 1981a. A comment on John A. Matthew's article regarding C14-dates of glacial variations. Geogr. Ann. 63A:1-2, 19-22.
Karlén, W., 1981b. Lacustrine sediment studies. Geogr. Ann. 63A: 3-4, 273-281.
Karlén, W., 1982. Holocene glacier fluctuations in Scandinavia. In: Holocene Glaciers (ed. W. Karlén), Striae 18, 26-34.
Lamb, H.H., 1982. Climate, history and the modern world. Methuen, London and New York, 387.
Liestøl, O., 1967. Storbreenglacier in Jotumheimen, Norway. Norsk Polarinstitutt, Skrifter 141, 1-63.
Matthews, J.A., 1980. Some problems and implications of C14-dates from a podzol buried beneath an end moraine at Haugabreen, southern Norway. Geogr. Ann. 62A:3-4, 85-208.
Mercer, J.H., 1982. Holocene glacier variations in southern South America. In: Holocene Glaciers (ed. W. Karlén), Striae 18, 35-40.
Miller, G.H., 1973. Late Quaternary glacial and climatic history of Northern Cumberland peninsula, Baffin Island, N.W.T., Canada. Quaternary Res. 3, 561-583.
Nye, J.F., 1965. The frequency response of glaciers. J. Glaciol. 5:4, 567-607.
Schytt, V., 1947. Glaciologiska arbeten Kebnekajse. Ymer 1947, 18-42.
Schytt, V., 1959. The glaciers of the Kebnekaise-Massif. Geogr. Ann. 41:4, 213-227.
Schytt, V., 1973. Glaciologiska metoder i klimatforskningens tjänst. Svensk Naturvetenskap 1973, 77-88.
Schytt, V., 1981. The Net Mass Balance of Storglaciären. Kebnekaise, Sweden, related to the height of the equilibriumline and to the height of the 500 mb surface. Geogr. Ann. 63A, 219-223.
Schneebeli, W., & Röthlisberger, 1976. 8000 Jahre Walliser Gletschergeschichte. Die Alpen 52, 1-152.

DENDROCLIMATOLOGICAL STUDIES AT THE NORTHERN TIMBERLINE

Roland W. Aniol and Dieter Eckstein

University of Hamburg, Institute of Wood Biology

ABSTRACT

Analysing the growth pattern of pine trees near the northern timberline in Swedish Lapland a close relationship between measured temperatures in July was established. On the basis of this relationship the July temperature is reconstructed back to A.D. 1680.

INTRODUCTION

In many fields of research there is a growing interest in climatic fluctuations that have occurred throughout the post-glacial period. As instrumental observations extend back only two or three centuries at maximum, knowledge of the climate of the pre-instrumental time is still rather limited. Therefore it is necessary to extend the climatic records using other data available, so-called proxy data like tree-ring series.

The science of extracting climatological information stored in tree rings is dendroclimatology. During the last two decades it was developed and furthered by H.C. FRITTS and his group in Tucson/Arizona (FRITTS 1976). The present state of knowledge in this field is summarized in HUGHES et al. (1982).

This paper gives a brief introduction into the basic concept of dendroclimatology and the subsequent case study will illustrate the kind of information that can be obtained from tree rings.

N.-A. Mörner and W. Karlén (eds.), Climatic Changes on a Yearly to Millennial Basis, 273–279.

DENDROCLIMATOLOGICAL CONCEPT

Trees are long-lived organisms integrating and recording various external growth controlling influences during their lifetime. For dendroclimatological purposes a tree can be considered as a black box, in which the annual input of such external influences is converted into a visible and measurable output in form of the width of the annual growth layers (Fig. 1). The underlying metabolic processes need not be considered in detail.

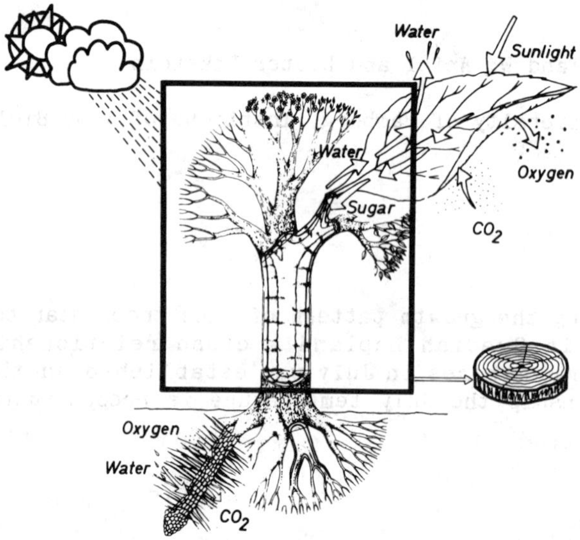

Fig. 1 Black box concept of dendroclimatology.

FRITTS (1976) showed how the climate of a given year t effects the ring width in a tree, but not only for that same year t, but through effects on buds, sugar, hormones, and the growth of needles and roots also in subsequent years so that the response reaches out up to several years. As each year t influences year t+1, t itself is influenced by the preceding year t-1.

The complex relationships make the extraction of climatic information from tree rings rather complicated. In order to obtain an estimate for the climate in year t the ring widths of year t, t+1 to t+k, and sometimes also of year t-1 have to be taken into account.

For a reconstruction of the climate the series of ring widths of many trees (Fig. 2) must be averaged to damp individual variations of growth and thus to gain a series showing the common response of a group of trees to climatic influences.

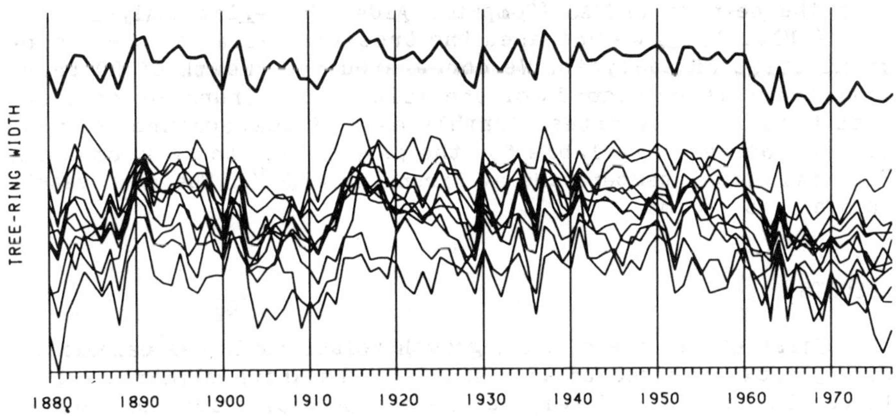

Fig. 2 Ten cross-dated tree-ring curves of pines from Northern Sweden and the resulting mean curve (thick line).

The strategy of dendroclimatology is to find a statistical relationship between biological output such as ring-widths and a set of climatic variables such as temperature and precipitation. This can successfully be done for a period, for which the values of the climatic variables have been measured. A part of this period is used to build up a model for the climate-growth relationship and is called the calibration period. If a good agreement between measured and estimated climatic values can be achieved one can use the model to compute estimates for the climatic variables in a second period for which measurements of the climatic variables are still available. If in this verification period the approximation fits equally well one can use the model to estimate values for the climatic variables in a third period for which no climatic measurements are available. This period is the reconstruction period.

MATERIAL

In the case study presented here we used the tree rings of pine (Pinus sylvestris L.) from four sites in the Torneträsk area in northernmost Sweden at the northern timberline around 400 m a.s.l. (BARTHOLIN, this volume). Since the area is still almost untouched by man it is especially suitable for such analysis. Mountain birch (Betula pubescens Ehrh.) is the main tree species, whereas pines are infrequent and show a relic type of distribution (SONESSON and LUNDBERG 1974). T.S. BARTHOLIN / University of Lund, and W. KARLEN / University of Stockholm did the sampling in the field and provided us with the tree-ring measurements (BARTHOLIN and KARLEN 1983) and the necessary climatic information.

With the help of CATRAS (Computer Aided Tree-Ring Analysis System) (ANIOL 1983) we prepared the tree-ring data for the subsequent steps of analysis. We correlated the growth of 60 trees with the weather records of the station in Kiruna 50 km south east from the tree sites. Monthly data of temperature and precipitation were available for the period 1901 to 1979. oceanic. The mean annual temperature at Kiruna is -1.6 °C. The mean yearly rainfall is 486 mm.

RESULTS

First of all the climate-growth relationship was calculated. Fig. 3 shows how the trees respond to the known climatic conditions. The effects of temperature and precipitation for each of the 14 months from previous July to current August on the variation in ring width are illustrated. The calculation is based on 43 periods of 14 months each from 1938 to 1980. A value above zero indicates a direct relationship and a value below zero indicates an inverse relationship. Three additional variables are

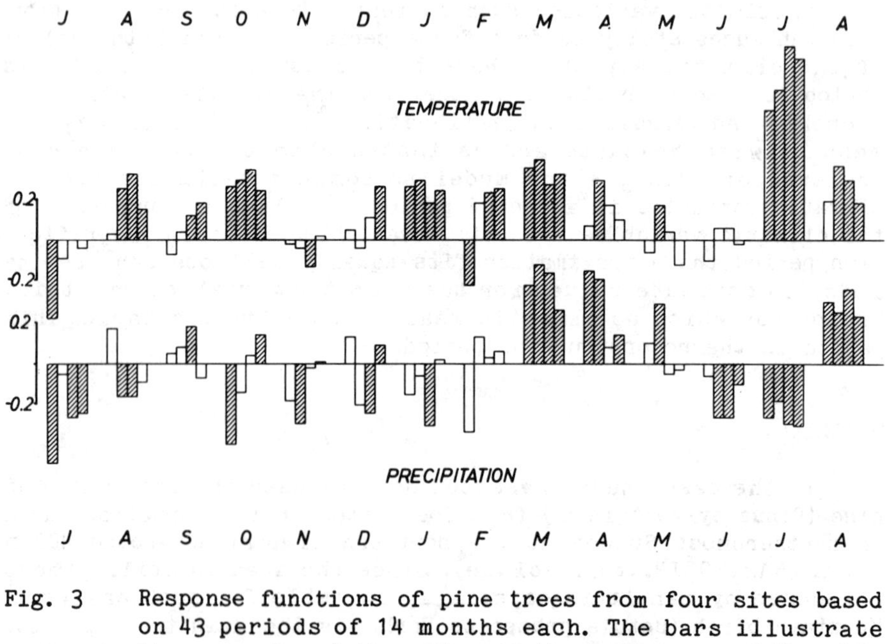

Fig. 3 Response functions of pine trees from four sites based on 43 periods of 14 months each. The bars illustrate the amount and direction of the influence of the climatic factors on growth separately for each of the four sites. If the influence at one site is significant at the 0.95 level the corresponding vertical bar is hatched.

taken into account, namely the ring widths of the years t-1, t-2, and t-3.

These response diagrams show that the temperature is of dominant importance for the growth at all four sites. If one considers temperature in detail it turns out that July temperature has by far the greatest influence. Consequently we have tried to reconstruct the July temperatures. The result is shown in Fig. 4. In addition to the reconstructed annual values of July temperature the low frequency fluctuations of July temperature have been calculated and plotted. Below average values are coloured black.

For verification we compared the reconstructed values both with the actually measured data of the station of Kiruna back to 1901 and of Karesuando near the Finnish border 120 km east from the tree sites. From the last mentioned station climatic data are available from 1979 to 1880. During this period there is a good similarity between measured and reconstructed temperatures with a correlation coefficient of 0.58. On the basis of this climate-growth relationship we extended the reconstruction of the July temperature as far back as 1680.

Beyond 1880 it is still not verified whether the striking deviations from average temperature around 1830 and 1750 have really happened or not. If these deviations can be verified by glaciological, historical or other studies, we will continue to estimate the July temperature back to A.D. 436. This is at present the total length of the pine chronology of Northern Sweden worked out by BARTHOLIN and KARLEN (1983).

CONCLUSIONS

To conclude, palaeoclimatic studies need proxy climatic data. In comparison with other proxy climatic data, tree rings have the advantage that they can be assigned accurately to a particular year and tree-ring series can be developed to cover several centuries. Therefore, the radial growth of trees could be considered as a coded year-book recording climatic data of the past.

In the case study presented here trees from the northern timberline have been analysed. The climatic signal in tree-ring series of trees from a timberline is higher than that in series from trees which had optimal growing conditions (ECKSTEIN and ANIOL 1981). Dendroclimatological studies can also be carried out at less marginal sites but require larger amounts of tree-ring data and more complex methods.

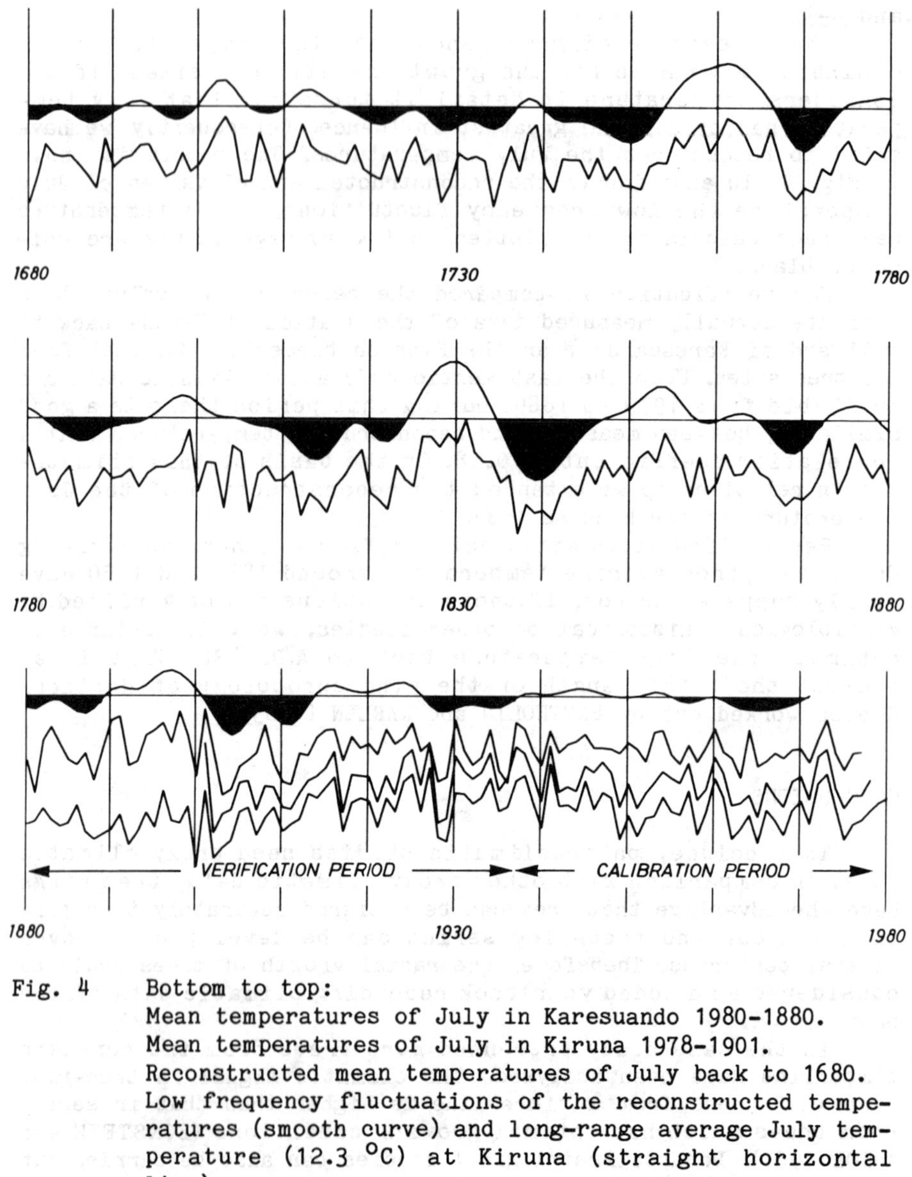

Fig. 4 Bottom to top:
Mean temperatures of July in Karesuando 1980-1880.
Mean temperatures of July in Kiruna 1978-1901.
Reconstructed mean temperatures of July back to 1680.
Low frequency fluctuations of the reconstructed temperatures (smooth curve) and long-range average July temperature (12.3 °C) at Kiruna (straight horizontal line).

ACKNOWLEDGEMENT

This study was partly supported by the Deutsche Forschungsgemeinschaft, Bonn, FRG.

REFERENCES

ANIOL, R.W., 1983: Tree-ring analysis using CATRAS. Dendrochronologia 1, 45-53.

BARTHOLIN, T.S. and KARLEN, W., 1983: Dendrokronologie i Lapland. Dendrokronol. Sällsk. Medd. 5, 3-16.

ECKSTEIN, D. and ANIOL, R.W., 1981: Dendroclimatological reconstruction of the summer temperatures for an Alpine region. Mitt. Forstl. Bundesvers. Anst. Wien 142, 391-398.

FRITTS, H.C., 1976: Tree rings and climate. London, New York, San Francisco, pp. 567.

HUGHES, M.K., KELLY, P.M., PILCHER, J.R. and LAMARCHE, V.C., 1982: Climate from tree rings. Cambridge, London, New York, pp. 223.

SONESSON, M and LUNDBERG, B., 1974: Late Quaternary forest development of the Torneträsk area, North Sweden. Structure of modern forest ecosystems. Oikos 25, 121-133.

LIMITATIONS OF ^{14}C DATES FROM BURIED SOILS IN RECONSTRUCTING GLACIER VARIATIONS AND HOLOCENE CLIMATE

John A. Matthews

University College, Cardiff, U.K.

ABSTRACT

The accuracy with which the date of burial of a palaeosol can be estimated is assessed using a very detailed radiocarbon-dating study of one buried soil horizon. Twenty six depth controlled dates have been obtained from physico-chemically fractionated samples of the organic surface (FH) horizon of a humo-ferric podzol buried by the southern Norwegian glacier Haugabreen at its 'Little Ice Age' maximum limit. The ^{14}C dates show an order of magnitude increase in age with depth from 485 ± 60 to 4020 ± 70 ^{14}C years B.P. over 14.5cm. Although this age range suggests potential errors of the order of thousands of years, the age/depth gradient appears to provide a basis for quite accurate estimates of the date of soil burial and of the date of moraine formation.

These results, combined with supporting relative-age evidence from a large sample of southern Norwegian glaciers, suggest that most glaciers in southern Norway reached their Neoglacial glacier maxima in the 'Little Ice Age'. This lack of synchroneity with other glaciated areas of Europe, including northern Scandinavia, may be due to regional differences in the pattern of Holocene climatic change.

1. INTRODUCTION

Glacier variations are an important source of information about Holocene climatic change, particularly in Scandinavia. The limitations of the techniques employed in the reconstruction and dating of glacier variations must therefore be critically evaluated by those interested in the implications for climatic change. In this

N.-A. Mörner and W. Karlén (eds.), Climatic Changes on a Yearly to Millennial Basis, 281–290.

paper one of the few methods for the absolute dating of Holocene glacier variations, namely the ^{14}C dating of soils buried beneath end moraines, is assessed in the light of an intensive dating study of one buried soil horizon. Emphasis is given to the potential of the method for obtaining accurate estimates of moraine age and to the implications for the chronology of Holocene glacier and climatic variations in Scandinavia.

2. ^{14}C DATING AT HAUGABREEN

The dated soil horizon is the surface organic (FH) horizon of an intact humo-ferric podzol buried beneath the outermost Neoglacial end moraine of Haugabreen (61°40'N, 6°44'E), an outlet of the Myklebust ice cap, north of Stardalen, southern Norway (Fig.1). The altitude of the site is about 660m above sea level, immediately above the local birch tree line, where the moraine ridge is about 2m high. Independent historical and lichenometric evidence suggests that the soil was buried when Haugabreen reached its 'Little Ice Age' limit in about A.D. 1750 (Matthews, 1980, 1982). The soil is developed in a pre-Neoglacial till, which is assumed to have been deposited about 9000 ^{14}C years B.P. (Andersen, 1980). Beneath the FH horizon the palaeosol consists of a 5cm thick bleached eluvial (Ea) horizon overlying illuvial B horizons of combined thickness 35cm. The stratigraphy of the site is shown in the inset to Fig.1. Full details have been given elsewhere (Matthews, 1980).

2.1 Methods

The pretreatment involved a two-stage chemical fractionation after removal of all visible rootlets from the sample by hand. Stage 1 was an acid hydrolysis (cf. Scharpenseel, 1977; Goh & Molloy, 1978), whereby acid-soluble 'fulvic acids' were separated by boiling in 2% HCl for ten minutes. Stage 2 consisted of two successive alkali extractions (cf. Kononova, 1961; Scharpenseel, 1971), each involving boiling the acid-insoluble residue in 0.5% NaOH for ten minutes; the alkali soluble/acid insoluble 'humic acids' being precipitated using concentrated HCl. The 'fine residual' fraction from the alkali extraction (<250μm) was also collected.

2.2 Results

All ^{14}C assays were carried out in a conventional gas proportional counter, were corrected for isotopic fractionation and were calculated using the 5568 yr half life. The first set of eight dates, from where the horizon was 5cm thick and the samples were 2.5cm thick, have been fully discussed in Matthews (1980). The second set of 18 dates relate to where the buried horizon was 14.5cm thick and dating involved samples of thickness 0.5 or 1.0cm. These latter

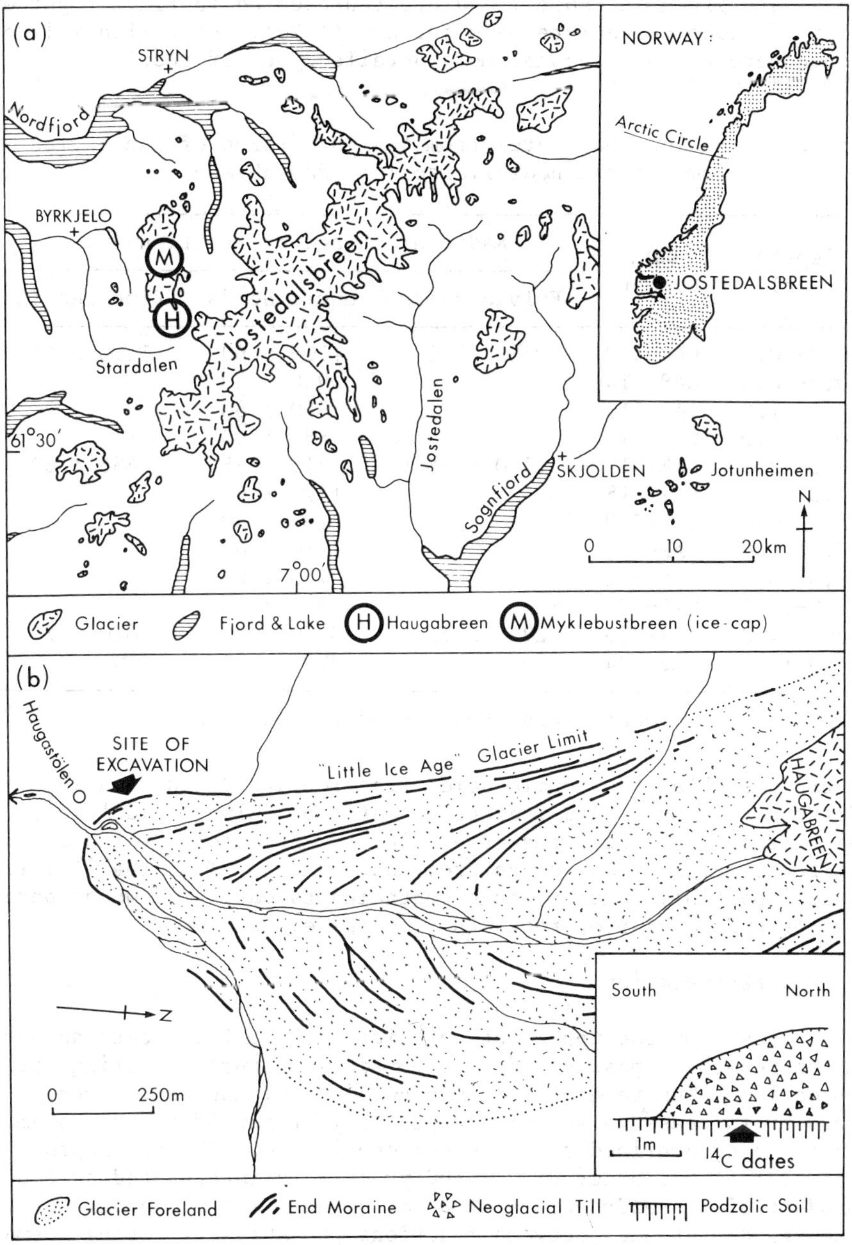

Figure 1. (a) Location of Haugabreen. (b) Site of the buried soil beneath the outermost Neoglacial end moraine in front of the glacier, and (inset) a simplified stratigraphic section through the moraine at the sample point for dating.

dates are given in Table 1 and are represented in Fig.2 together with 95% confidence intervals for age (± 2σ). Attention will be given here to this second, more detailed, set of results.

Table 1. Radiocarbon dates from the FH horizon of the buried humo-ferric podzolic soil at Haugabreen.

Depth* (cm)	Lab. Number	Radiocarbon age of fraction (± 1σ)		
		Fulvic acids	Humic acids	Fine residue
0.0- 0.5	CAR-513	525 ± 70	1105 ± 60	1110 ± 65
0.0- 0.5	CAR-514		980 ± 60	
0.5- 1.0	CAR-515		940 ± 60	
0.5- 1.0	CAR-516		885 ± 60	
3.0- 2.0	CAR-517	540 ± 55	745 ± 55	485 ± 60
3.0- 4.0	CAR-518		1275 ± 60	
5.0- 6.0	CAR-519		2135 ± 60	
7.0- 8.0	CAR-520		2110 ± 60	
9.0-10.0	CAR-521		2750 ± 65	
11.0-12.0	CAR-522		3335 ± 65	
13.0-14.0	CAR-523	2925 ± 75	3655 ± 70	4020 ± 70
14.0 14.5	CAR-524		3995 ± 70	

*Depth below the top of the horizon

Where possible, the 'humic acid' fraction was dated because the first set of dates had indicated that the 'fulvic acids' were significantly younger than the other two fractions (which did not differ significantly in age from each other). Thus the 'fulvic acids' possibly contain young contaminants carried down by percolating groundwater or rootlets from above.

2.3 Interpretation

The dates from the uppermost centimetre (Fig.2) indicate an age/depth reversal, possibly due to physical disruption during moraine formation. The remaining 'humic acid' dates suggest, however, a regular increase in age with depth, any effect of older contaminants being confined to the upper centimetre. The age/depth gradient of the order of 250 ^{14}C yr cm^{-1} is interpreted as the product of a steady accumulation of organic matter through time, the age differences between fractions probably reflecting normal decomposition, humification and translocatory processes prior to burial (cf. Jenkinson & Rayner, 1977; Stout et al., 1981).

The basal dates indicate that the horizon began to accumulate prior to 4000 ^{14}C years B.P., while the dates at a higher level

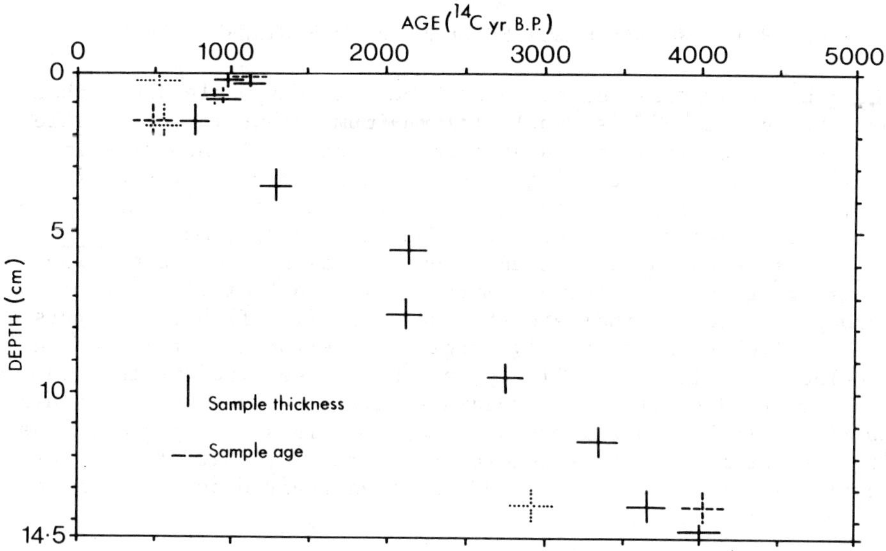

Figure 2. Radiocarbon age variation with depth in the buried soil horizon at Haugabreen. The three organic fractions dated are: 'humic acids' (——), 'fulvic acids' (.....) and 'fine residue' (— — —). See text for definition of fractions.

support the proposition that it ceased to form when it was buried by the 'Little Ice Age' advance of Haugabreen. The strong linear relationship exhibited by the 'humic acids' between 1.0-14.5cm depth ($r^2 = 0.98$, $p = <0.001$, $n = 8$) suggests considerable potential for the accurate dating of moraine formation and soil burial, despite age variations within the horizon of the order of 4000 years. For example, regression analysis of age on depth predicted the age of the soil at zero depth to be 253 ± 365 Calendar years B.P. (with 95% statistical confidence), an estimate in remarkably good agreement with the independent evidence for a 'Little Ice Age' glacier maximum in A.D. 1750, about 230 years ago.

In principal, therefore, the ^{14}C dating of at least certain kinds of buried soil horizon would appear capable of yielding a close approximation to the date of moraine formation. Realisation of the full potential of the technique would, however, seem to require a far more rigorous approach to soil sampling and analysis than has been attempted previously. In particular: (1) thin slices of soil near the former surface should be dated rather than bulk samples or even whole horizons; (2) eroded or otherwise disturbed buried soils should be avoided; (3) establishment of the scale and form of any age/depth relationship necessitates dates from more than two depths; and (4) dating of more than one soil organic fraction may add useful further information.

3. SUPPORTING RELATIVE-AGE EVIDENCE FROM SOUTHERN NORWAY

The ^{14}C dates presented above show beyond reasonable doubt that Haugabreen reached its Neoglacial maximum extent in the 'Little Ice Age'. Two lines of independent evidence - lichen size and an index of boulder weathering - suggest that this conclusion is applicable to southern Norwegian glaciers in general (details in Matthews & Shakesby, in press). The largest specimens of Rhizocarpon geographicum agg. found growing proximal to the moraine ridge crest on the outermost moraines in front of 91 southern Norwegian glaciers indicate that only at five of these glaciers are the lichens sufficiently large to question a 'Little Ice Age' Neoglacial maximum. Similarly, at only three glaciers in a sample of 34 where Schmidt hammer readings have been taken are R-values sufficiently low to suggest outermost moraines that pre-date the 'Little Ice Age'. It is concluded, therefore, that of this large sample of glaciers, at least 90% reached their Neoglacial maxima in the 'Little Ice Age'.

These data from southern Norway are in marked contrast to that from northern Scandinavia where, according to the work of Wibjørn Karlén (1973, 1975, 1976, 1979, 1982) and the results of the Reading University Okstindan Research Project led by Peter Worsley (Alexander & Worsley, 1973; Worsley, 1974; Griffey, 1976; Worsley & Alexander, 1976a, 1976b; Griffey & Worsley, 1978; Bertie, 1979), most glaciers reached their Neoglacial maxima before the 'Little Ice Age'.

4. DISCUSSION

The implications of this for studies of Holocene glacier and climatic variations are twofold: first, for the accuracy with which moraine age has been and can be estimated; and second, for the probability that closely adjacent regions are not entirely synchronous in their pattern of glacier variations, and hence climatic changes. Both implications are closely related because high spatial and temporal resolution in climatic reconstructions require an accurate chronology.

In Scandinavia, the chronology of Holocene glacier variations rests heavily on ^{14}C age determinations derived from buried soils. This has led to considerable controversy where a large number of glacier advances have been recognised based on a small number of ^{14}C dates (cf. Karlen, 1981; Matthews, 1982). The new dating results from the Haugabreen palaeosol clearly demonstrates potential errors of the order of thousands of years where palaeosols are thick and well developed, although they also demonstrate that estimates within a few hundred years are possible using sophisticated techniques under favourable conditions. This is not to say that all previous soil datings are subject to errors of this magnitude. Thin soils that have developed over a shorter period of time cannot be subject to such large errors from age/depth gradients that require thousands of years to develop. Examples are provided by the date of 245 ± 65 ^{14}C yr B.P. at Årjep Ruotesjekna, northern Sweden (Karlén and Denton, 1976), and a similar young date obtained at Steikvassbreen, northern Norway (Griffey and Worsley, 1978), both of which appear to relate to very thin embryonic soils. In contrast, some dates reported from thick soils are more likely to be subject to larger errors, such as those of 3170 ± 120 ^{14}C yr B.P. from Steikvassbreen (Griffey, 1976), 1200 ± 260 and 1490 ± 120 ^{14}C yr B.P. from Leirbreen, Jotunheimen (Griffey and Matthews, 1978), and 2970 ± 95 ^{14}C yr B.P. from Mikkajekna, northern Sweden (Karlén, 1981).

Other glacier variation chronologies may well be subject to similar errors (cf. Grove, 1979; Röthlisberger et al., 1980), although much of the detailed Alpine chronology is based on multiple thin soils in lateral moraines built up by intermittent superimposition or accretion, which could not have developed age/depth gradients to the same degree as soils developed over many thousands of years. In very thick soils produced by higher rates of soil development in lower latitudes, such as the andosols from Mexico reported by Heine (this volume), the problem would appear to be relatively easily avoided, because any age/depth gradient would be relatively gentle. It would be most unwise, however, to assume that the steep increase in age with depth reported here from the Haugabreen Podzol is unique to that site, particularly in view of the preliminary dates obtained from an arctic-alpine Brown soil beneath the outermost Neoglacial moraine at Vestre Memurubreen, Jotunheimen, which have revealed strong age/depth gradients in a very different kind of soil (Matthews, 1981).

Taken together with the supporting evidence, ^{14}C dating results from Haugabreen, supported by the lichenometric data, strongly suggest large differences between Neoglacial glacier variations in southern compared to northern Scandinavia. In the south the 'Little Ice Age' event appears to have been larger, relative to any other Neoglacial glacier variations. This regional difference in the relative scale of the 'Little Ice Age' glacier expansion episode may have been caused by a slight difference in palaeoclimatic change between the two regions. A difference in mean summer temperature change of less than 1.0°C between the two areas would seem to be a sufficient cause. Even this small value may be too large in view of the strong temperature dependence of glacier fluctuations in Scandinavia and estimates that cold phases of the 'Little Ice Age' were associated with summer temperatures no more than 2.0°C colder than those of today (Karlén, 1973; Karlén and Denton 1976; Matthews, 1976, 1977).

In retrospect, it is not altogether surprising that glacier and climatic variations are not identical in scale or timing throughout the Scandinavian peninsula. Inter-regional differences would indeed be expected to increase as shorter time scales are considered. Although it is only by the development and application of very accurate absolute dating techniques that such patterns will be fully resolved, in seeking glacier and climatic variation chronologies it would appear that a combination of approaches could be profitably employed. An example has been provided by this paper in which intensive absolute dating at one site has been combined with extensive relative-age dating at a large number of sites. Given suitably precise and accurate chronologies, regional palaeo-climatic patterns detectable on the intermediate time-scales of the Holocene, Neoglaciation and the 'Little Ice Age' may provide considerable insights into the larger problems associated with longer-term climatic change.

5. ACKNOWLEDGEMENTS

The contents of this paper are based on field work associated with the Jotunheimen Research Expeditions. Radiocarbon dating was carried out by the Radiocarbon Dating Laboratory, Department of Plant Science, University College, Cardiff, under the direction of Dr P. Quentin Dresser. In addition I thank Dr James R. Petch for comments on the manuscript and Dr Wibjørn Karlén for cordial discussions in Stockholm. The Royal Society provided a travel grant to enable my participation in the symposium.

6. REFERENCES

Alexander, M.J. & Worsley, P. 1973, Stratigraphy of a Neoglacial end moraine in Norway. Boreas, 2, 117-142.

Andersen, B.G. 1980, The deglaciation of Norway after 10,000 B.P.. Boreas, 9, 211-216.

Bertie, D. 1979, An investigation of Neoglacial stratigraphy and climatic change in the Arctic Circle area of Scandinavia. Ph.D. Thesis, University of Reading.

Goh, K.M. & Molloy, B.P.J. 1978, Radiocarbon dating of palaeosols using soil organic matter components. Journal of Soil Science, 29, 567-573.

Griffey, N.J. 1976, Stratigraphical evidence for an early Neo-glacial maximum of Steikvassbreen, Okstindan, north Norway. Norsk Geologisk Tidsskrift, 56, 187-194.

Griffey, N.J. & Matthews, J.A. 1978, Major Neoglacial glacier expansion episodes in southern Norway: evidences from moraine ridge stratigraphy with ^{14}C dates on buried palaeosols and moss layers. Geografiska Annaler 60(A), 73-90.

Griffey, N.J. & Worsley, P. 1978, The pattern of Neoglacial glacier variations in the Okstindan region of northern Norway during the last three millenia. Boreas, 7, 1-17.
Grove, J.M. 1979, The glacial history of the Holocene. Progress in Physical Geography, 3, 1-54.
Jenkinson, D.S. & Rayner, J.H. 1977, The turnover of soil organic matter in some of the Rothamsted classical experiments. Soil Science, 123, 298-305.
Karlén, W. 1973, Holocene glacier and climatic variations, Kebnekaise Mountains, Swedish Lappland. Geografiska Annaler, 55(A), 29-63.
Karlén, W. 1975, Lichenometrisk datering i Norra Scandinavien - metodens tillforlitlighet och regionala tillamping. Stockholm University, Naturgeografiska Institutionen, Forskningsrapport, 22, 1-70.
Karlén, W. 1976, Lacustrine sediments and tree-limit variations as indicators of Holocene climatic fluctuations in Lappland, northern Sweden. Geografiska Annaler, 58(A), 1-34.
Karlén, W. 1979, Glacier variations in the Svartisen, northern Norway. Geografiska Annaler, 61(A), 11-38.
Karlén, W. 1981, A comment on John A. Matthews's article regarding ^{14}C dates of glacial variations. Geografiska Annaler, 63(A), 19-21.
Karlén, W. 1982, Holocene glacier fluctuations in Scandinavia. Striae, 18, 26-34.
Karlén, W. & Denton, G.H. 1976, Holocene glacier variations in Sarek National Park, northern Sweden. Boreas, 5, 25-56.
Kononova, M.M. 1961, Soil organic matter: its nature, its role in soil formation and in soil fertility. Pergamon.
Matthews, J.A. 1976, 'Little Ice Age' palaeotemperatures from high altitude tree growth in S. Norway. Nature, 264,-245.
Matthews, J.A. 1977, Glacier and climatic fluctuations inferred from tree-growth variations over the last 250 years, central southern Norway. Boreas, 6, 1-24.
Matthews, J.A. 1980, Some problems and implications of ^{14}C dates from a podzol buried beneath an end moraine at Haugabreen, southern Norway. Geografiska Annaler, 62(A), 185-208.
Matthews, J.A. 1981, Natural ^{14}C age/depth gradient in a buried soil. Naturwissenschaften, 68, 472-474.
Matthews, J.A. 1982, Soil dating and glacier variations: a reply to Wibjørn Karlén. Geografiska Annaler, 64(A), 15-20.
Matthews, J.A. & Shakesby, R.A. in press, The status of the 'Little Ice Age' in southern Norway: relative-age dating of Neoglacial moraines with Schmidt hammer and lichenometry. Boreas, 13.
Röthlisberger, F., Haas, P., Holzhauser, H., Keller, W., Bircher, W. & Renner, F. 1980, Holocene climatic fluctuations - radiocarbon dating of fossil soils (fAh) and woods from moraines and glaciers in the Alps. Geographica Helvetica, 35, 21-52.

Scharpenseel, H.W. 1971, Radiocarbon dating of soils - problems, troubles, hopes. In: Palaeopedology: origin, nature and dating of palaeosols. (ed. D.H. Yaalon), International Society of Soil Science & Israel Universities Press.

Scharpenseel, H.W. 1977, The search for biologically inert and lithogenic carbon in recent soil organic matter. In: Soil organic matter studies II. (ed. for the International Atomic Energy Agency), IAEA Vienna.

Stout, J.D., Goh, K.M. & Rafter, T.A. 1981, Chemistry and turnover of naturally occurring resistant organic compounds in soil. In: Soil biochemistry, Volume 5. (eds. E.A. Paul & J.N. Ladd), Dekker.

Worsley, P. 1974, On the significance of the age of a buried tree stump by Engabreen, Svartisen. Norsk Polarinstitutt Årbok, 1972, 111-117.

Worsley, P. & Alexander, M.J. 1976a, Neoglacial palaeoenvironmental change at Engabrevatn, Svartisen, Holandsfjord, north Norway. Norsk Geologisk Undersøkelse, 321, 37-66.

Worsley, P. & Alexander, M.J. 1976b, Glacier and environmental changes - Neoglacial data from the outermost moraine ridges at Engabreen, northern Norway. Geografiska Annaler, 58(A), 55-69.

CLIMATE AND GLACIATION IN KONG KARLS LAND, EASTERN SVALBARD

Björn Holmgren, Abisko Research Station, Lappland, Sweden
Ingrid Olsson, Uppsala University
Erik Skye, Uppsala University
Göran Alm, Stockholm University

INTRODUCTION

The present paper presents some observations of lichens and mosses, their micro-climatic environment and relation to permanent ice fields in a selected area of Kongsöya, the largest of 3 main islands of Kong Karls Land in eastern Svalbard (Fig. 1), during the Swedish Arctic Expedition in 1980 (Ymer-80).

The ice fields on Kongsöya, which exist at elevations close to sea level, occupy at present only a few percent of the total land area and attain a maximum thickness of a few tens of meters. From this point of view they may seem rather insignificant. As will be made clear in the following the ice fields were more extensive not too long ago. The initial phase of glaciation and its corresponding climate is of considerable general interest.

PHYSIOGRAPHIC SETTING AND CLIMATE

Plateau mountains built in several steps consisting of basalt strata of Cretaceous age intruded between easily eroded sedimentary rocks of latest Triassic and early Cretaceous age (Nathorst, 1910), rise to 235 m a.s.l. in eastern Kongsöya and to 320 m in the western part. In our study area the surface consists of cobbles, mainly of basalt, and solid and frost-shattered rocks. Raised beaches are prominent features of the landscapes up to about 100 m a.s.l. Permafrost occurs everywhere below a depth of about 0.5 m, according to our subsurface temperature measurements at the end of summer at 3 places representing winter snow depths varying between 0 and 1 m. On the average Kong Karls Land is situated within the pack ice limit also at the end of summer according to sea ice lim-

N.-A. Mörner and W. Karlén (eds.), Climatic Changes on a Yearly to Millennial Basis, 291–302.

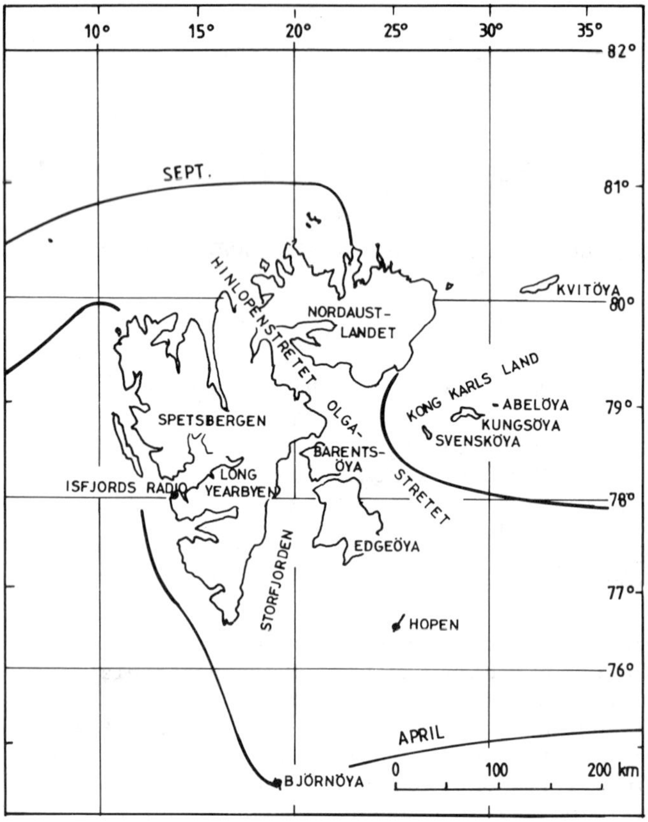

Fig. 1. Map of Svalbard with sea ice limits in April and September during 1946-1963. Redrawn after Lunde (1965).

its summarized after Lunde (1965) in Fig. 1.

The summer climate on Kongsöya may be described as Arctic maritime, similar to that existing over pack ice or near the pack ice edge: air temperature close to freezing, humidity near saturation, and winds light to moderate. High frequency of stratus and other low clouds mostly accompanied by wetting drops or a fine drizzle are other characteristic ingredients of the pack ice climate. The effect of solar radiation on the surface micro-climate is quite considerable (Fig. 2). In the summer of 1980 the average daily temperature of a snow-free surface was above 0 °C for more than 3 months and surface temperatures below freezing were infrequent. Another factor of importance e.g. for the growth of lichens is the high air humidity which combined with the slight drizzle made rock surfaces notably damp on most days. The micro-climate may accordingly be expected in some respects to be quite favourable

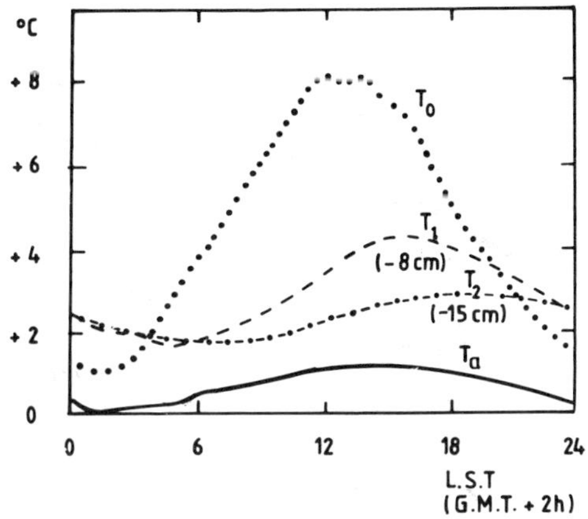

Fig. 2. Diurnal variation of air (T_a), surface (T_0) and subsurface (T_1, T_2) temperatures at a raised beach of basalt cobbles during 16-31 July, 1980. Average hourly values.

to vegetation in spite of maximum monthly temperatures of only about +1 °C in July and August.

PERENNIAL AND PERMANENT ICE

From snow courses we found that snow accumulation is dominated by snow drifting in relation to micro-, meso- and macro-scale topographical features. Ignoring snow and ice that may remain between the higher beach ridges during cold summers there are 2 main types of ice fields:

1. Ice banks on steep slopes sometimes limited upwards by escarpments. The ice banks may appear at any altitude between the mountain summits and sea level. We studied mass and energy budgets of 2 ice banks, Rektangelisen and Triangelisen (Fig. 3).
2. Ice cupolas on wide plateaus without sheltering topography. The genesis of ice cupolas is not associated with any increase of accumulation due to lee-effects. Ice cupolas may be considered as embryos for glaciers of ice cap type. No ice cupola in Kong Karls Land is found on plateaus below 100 m a.s.l. It is of more than curiosity interest to note that the higher plateaus above about 200 m, which are all relatively small, also lack permanent ice. This may be due to low accumulation by wind erosion and/or to relatively high ablation above the temperature inversion capping the marine boundary layer (see below). We studied one ice cupola, Rundisen, with its highest point at 103 m a.s.l. at the end of the 1980 summer.

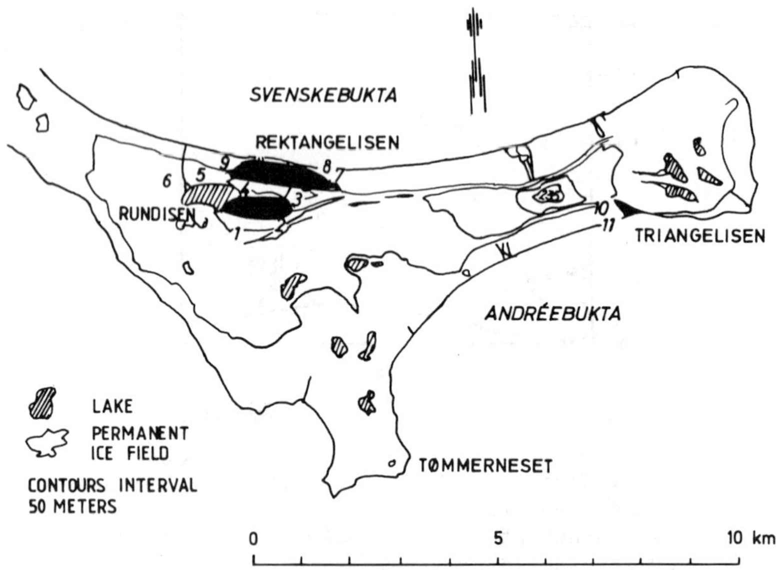

Fig. 3. Eastern Kongsöya. The numbers indicate sampling places for lichens and mosses used for ^{14}C dating (Olsson et al. this volume, Table 1).

All permanent ice fields on Kongsöya are build up by superimposed ice, i.e. by percolation of meltwater through the snow and refreezing on cold impervious ice below the snow pack. The average annual temperature a few m below the ice surface is around -10 °C. The maximum thickness of Rektangelisen is at present about 25 m, and that of Rundisen about 10 m.

MASS AND ENERGY BALANCES

Winter, summer and net balances are given in Table 1. Except for Triangelisen, which enjoys an exceptionally favourable position for accumulation by snow drifting, the ice fields on Kongsöya receive much less winter accumulation than glaciers in Spitsbergen (Liestöl, 1979) and in Nordaustlandet (Schytt, 1964). The winter balances of Rundisen and Rektangelisen are comparable with the amounts of winter precipitation measured at the weather stations on Hopen and Björnöya. Thus, the existence of the ice fields on small Arctic islands is not primarily an effect of high accumulation which is in agreement with previous findings for the Meighen Island Ice Cap (Paterson, 1969; Taylor-Alt, 1975) and the Storöya Ice Cap (Jonsson, 1982).

Table 1. Mass balances, Kongsöya 1979/80.

	Elevation	Area	b_w	n	b_ε	n	b_n	Acc^+	n
	m	km^2			cm w.eq.				
Rundisen	90-105	0.7	16	197	-73	19	-57	11	196
Rektangelisen	5- 90	1	25	290	-87	16	-62	18	233
Triangelisen	0- 60	0.1	56	188	-80	5	-24		

n = Number of measuring points
Acc^+ = Accumulation measured on April 23, 1982.

All ice fields on Kongsöya lost their seasonal snow and were moreover subject to a considerable loss of old ice in 1979/80. From barometric measurements made by Hamberg in 1898 and from barometric and theodolite measurements made by us in 1980, it appears that the crest of Rundisen was lowered from 133 to 103 m corresponding to an average annual loss of 0.33 m (w.eq.).

Measurements of the net energy gain at the ice surface of Rundisen shows a correlation of 0.74 with air temperature based on 103 mean daily values. Using 3-day averages the correlation increases to 0.81. These results are highly significant and agree with many other investigations in demonstrating the importance of air temperature in the ablation process (see e.g. Holmgren, 1971; Braithwaite, 1981).

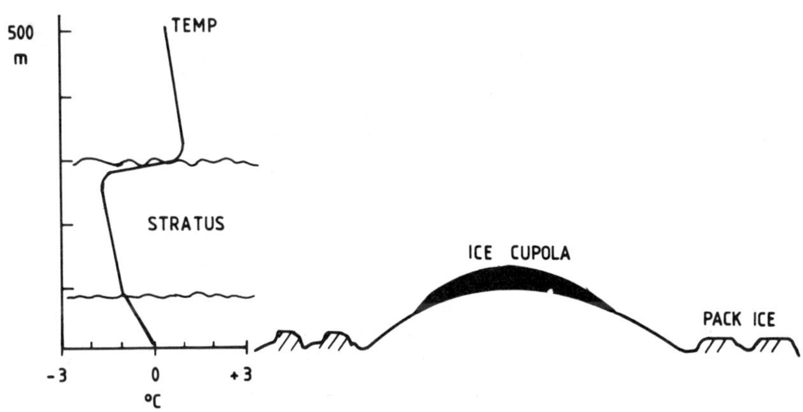

Fig. 4. Schematic of the relations between glaciation, temperature and cloudiness on small islands surrounded by melting pack ice in summer. The temperature profile is modified after Sverdrup (1933) and own measurements on Kongsöya in 1980.

The existence of the ice fields, especially the ice cupolas, at elevations close to sea level may be understood in terms of the properties of the marine boundary layer over melting pack ice in summer (Fig. 4).

THE LICHEN HABITATS

In relation to the ridges and troughs of the raised beaches one finds a conspicuous zonation of the lichen species growing on the surface rocks. The obvious reason for this is the snow depth variations which cause great variations of the micro-climate including the length of the snow-free period. In a transect perpendicular to the raised beaches there is in principle a repeated pattern of the lichen assemblages. Some species obviously prefer the crests while other prefer troughs or slightly sheltered places where the snow is not too deep. In some troughs the lichens appear partly damaged as indicated by missing pieces of thallus and a worn appearance generally. In deeper troughs there may be no sign of lichens at all or there may be rounded scars and discolorations on rocks demonstrating former occurrence of lichens. It seems evident that the living conditions in troughs are often marginal, probably because of long-lasting snow in cold summers.

For the lichens on crests of the raised beaches and on rock outcrops the exposure to winds and extreme temperatures in winter does not seem to cause any major survival problems; we often noted mature lichen covers in those positions. However, we also found that the crests of the small-scale topography around the major ice fields, especially Rundisen, were free of lichens.

By observing trim-lines, defined by the occurrence and non-occurrence of mature lichens on the crests of the small-scale topography in order to avoid the contribution of short-lived ice patches of insignificant size, we reconstructed the map of Fig. 5. This map should indicate a former maximum of continuous ice. Rundisen and Rektangelisen were joined and there were several separate arms of thin ice stretching into the surrounding depressions.

LIVING AND DEAD

Because of the thinning and recession of the ice fields during the last few years including 1980, we were able to observe how ice-buried lichens and mosses melted out from below Rundisen and the two ice banks under investigation. This vegetation, occurring in a border zone outside of but close to the ice edge had apparently been deep-frozen for a long time. It had not been deformed by any ice movement and could at superficial inspection have been confused with a healthy mature lichen cover of the type growing on the ridges that had not been covered by ice. It soon became apparent from lichens exposed in previous summer a little further away

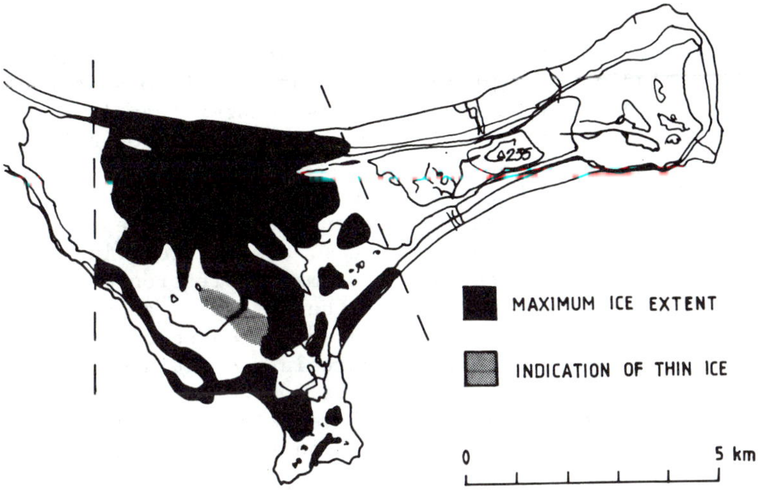

Fig. 5. Maximum ice extent in the delineated area as indicated by lichen trim-lines.

from the ice edge, that the emerging lichens must have been killed. By the distance from the ice edge we estimated that the dead lichens usually start to become fragmented some years or perhaps less after exposure to the atmosphere. In the more exposed micro-positions at the surface the degradation advances quickly, whereas portions of thallus remain longer when sheltered by pebbles. One may speculate that snow drifting, after partial chemical decomposition by fungi and bacteria, is an active agent in the physical removal of the killed lichens. Upon ice burial, the lichens may be expected to be completely enclosed by superimposed ice, occasionally with high bubble content, but generally without interconnecting veins, according to our observations of ice stratigraphy.

We collected 9 samples of cushion-type fruticose lichens mainly *Stereocaulon*, and 2 samples of moss, *Rhacomitrium* ssp, for ^{14}C dating. Of the lichens 7 samples had recently melted out of the ice while 2 were growing on raised beaches outside the former ice maximum (Fig. 3). Both moss samples had probably been exposed 2 weeks before collection. Each lichen sample consisted of about 10 cushions that had grown within an area of about 100 m^2. Excepting the lowest mm at the base, the whole cushions were generally used for dating. Three samples were parted in fractions for special analysis (Olsson et al., this volume).

DATING OF LICHENS AND MOSSES

The results of ^{14}C are given in Olsson et al., (this volume, Table 1) including interpretation of the data from a statistical viewpoint. It appears that sample 4 is younger than 1, 2, 3, 7, 8 and 9. This seems reasonable also from a glaciological viewpoint since the thickness of Rundisen is small and sample 4 was collected on a ridge. It is thus indicated that this ridge was covered by ice about 400 ^{14}C years later than the other sites. The samples 1, 2 and 3 relating to Rundisen and 7, 8 and 9 relating to Rektangelisen (Fig. 3) do not show any systematic difference although there is statistical evidence that these samples as a group represent a period instead of a single event. Because of the statistical spread, and because of the uncertainty concerning the ^{14}C age of different fractions of the lichens, we conclude that we are not justified to assume an age difference between samples 1, 2 and 3 on the one hand and 7, 8 and 9 on the other. Although there are several uncertainties in the interpretation, especially in this case, of ^{14}C age in calendar years, it seems that the dates would represent ice expansions during the period 1200-1650 AD (Olsson et al., this volume).

The ^{14}C activity of the dead mosses collected in front of Triangelisen shows a pronounced atomic bomb effect. Mosses may be expected to have a life time of about 3-5 years. Concurrently with the sharp rise of ^{14}C the nuclear tests brought about an increase of various radioactive isotopes like ^{137}Cs in the beginning of the 1960s.

An analysis of all lichen and moss samples for ^{137}Cs (which will be reported in detail elsewhere) showed, when combined with the ^{14}C results, that the mosses were covered by ice during an expansion of Triangelisen in 1964 or 1965. At this time the edge was thus at the same position as in the summer of 1980.

MASS BALANCE AND SIZE VARIATIONS OF ICE BANKS AND CUPOLAS

We discuss this schematically by assuming simplified topography and by ignoring ice movements. The net mass balance per budget year may be written

$$b_n = b_w + b_s; \text{ and } B_n = B_w + B_s$$

where small characters denote the local specific net, winter and summer balances, and capitals denote the corresponding balances integrated over the whole ice field.

1. For ice cupolas we approximate b_s and b_w

$$b_s = b_{so} + \alpha z, \text{ and } b_w = b_{wo} + \beta z$$

where index o denotes the balances at some reference level, α and β are constants and z is the vertical coordinate (Fig. 6).

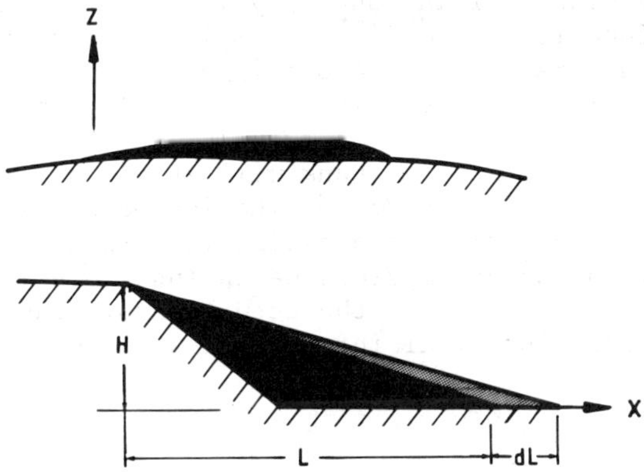

Fig. 6. Schematic ice cupola and ice bank.

We may then write a differential equation for b_n as

$$(1)\ b_n = \rho \frac{dz}{dt} = b_{wo} + b_{so} + (\alpha + \beta) z$$

where ρ = ice density and t = time.
2. For ice banks we assume a 2-dimensional x, z cross section according to the figure which also defines L and H. According to our observations the ice surface profiles are close to linear. We write

$$b_s = b_{so} + \alpha z, \text{ and } B_w = b_{wo} L + b_{wo} q$$

The winter balance is divided in 2 components: one corresponding to snow accumulation without snow drifting, and the other corresponding to accumulation by drifting. The factor q will vary with climate, mainly wind-speed and direction, and the local topography. From the figure we obtain

$$B_s = \int_o^L (b_{so} + \alpha z) dx = L(b_{so} + \frac{\alpha H}{2})$$

and

$$(2)\ B_n = \frac{\rho H}{2} \cdot \frac{dL}{dt} = L(b_{wo} + b_{so} + \frac{\alpha H}{2}) + q b_{wo}$$

In Eqs (1) and (2) z and L may be solved by prescribing relavant boundary conditions. The solutions may be studied by employing measured values of H, b_{so}, b_{wo}, q, α and β, or by assuming various values in order to study the effects on the size and thickness of the ice fields. For instance, by letting b_{wo} and b_{so} vary by time as sine functions of different amplitude and frequency further general information may be obtained.

Among the results we note that an ice cupola, once initiated on a wide plateau, will continue to grow and become a flowing glacier in a steady climate. An ice bank will attain a stable equilibrium as long as $(b_{so}+b_{wo}+\alpha H/2) < 0$. The ice bank will approach the equilibrium exponentially with a time lag expressed in the given parameters. In the conclusions we include some results obtained using the above scheme on the actual mass balance measurements and the glacier-historical information available on Kongsöya.

CONCLUSIONS

The existence of ice cupolas is mainly a consequence of the temperature decrease with elevation, whereas ice banks are signs of increased accumulation due to lee effects in connection with snow drifting. The occurrence of ice fields at elevations close to sea level is primarily related to the low air temperature within the atmospheric boundary layer above the melting pack ice in summer. Therefore, there exists an unusually direct coupling between the sea ice and the glaciers on small, low Arctic islands. During at least a few centuries before about 1200 AD (this date may have an uncertainty of a few hundred years) there was no ice field in the area where Rundisen is now situated. In fact there is indirect evidence that there was no precursor to Rundisen during the Holocene. Rundisen may to a certain extent be regarded as the response to the relative land emergence which is of the order of 100 m during the Holocene (Salvigsen, 1981).

Because of the great time lags involved Rundisen never turned into a real flowing glacier (although this may be a matter of definition) but remained as an ice field of insignificant flow. It is also of interest to note that the expansion of Rektangelisen to its maximum size should have taken place over several centuries, and that this expansion partly was a response to increased winter accumulation in comparison with the present climate according to our calculations using the scheme of the previous section and the glaciological data.

Since the initiation of Rundisen and expansion of Rektangelisen in late medieval time their extent appears continuously to have been as large or larger than in 1980. If the climate trend of the last 50 years continues, Rundisen will desappear in a few decades. Rektangelisen will diminish from its present length of about 500 m perpendicular to the slope to about 100 m. Triangelisen, now about 300 m long, is in an approximate equilibrium with the climate of the last 50 years.

In the near future it might therefore be possible to collect vegetation, which may successively melt out close to the edges of Rundisen and Rektangelisen and which may be used for more precise dating of the initial phases of ice expansion on Kongsöya.

ACKNOWLEDGEMENTS

Financial support was received from the Swedish Natural Science Research Council. The field work was part of the Swedish Arctic Expidition Ymer-80 under the scientific leadership of Professor Valter Schytt. Valuable advice was given by Professor David W. Goodall and Dr Vagn Alstrup.

REFERENCES

Andrews, J.T., Davis, P.T. and Wright, C. 1976: *Little Ice Age Permanent Snow-cover in the Eastern Canadian Arctic: Extent Mapped from Landsat-1 Satellite Imagery*.Geogr. Ann. 58A, pp 71-81.

Braithwaite, R.J. 1981: *On Glacier Energy Balance, Ablation, and Air Temperature*. J. of Glac. Vol 27, No 97, pp 381-391.

Holmgren, B.E. 1971: *Climate and Energy Exchange on a Sub-polar Ice Cap in Summer. Part F: On the Energy Exchange of the Snow Surface in Summer*. Meteorologiska institutionen, Uppsala universitet, Meddelande No 112.

Jonsson, S. 1982: *On the Present Glaciation of Storöya, Svalbard*. Geogr. Ann. 64A 1-2, pp 53-79.

Lieström, O. 1979: *Glaciological work in 1978*. Norsk Polarinstitutt. Årbok 1978. Oslo 1979.

Lunde, T. 1965: *Ice Conditions at Svalbard 1946-1963*. Norsk Polarinstitutt. Årbok 1963, pp 61-80.

Nathorst, A.G. 1910: *Beiträge Zur Geologie Der Bären Insel, Spitsbergens Und Des König-Karl-Landes*. Bull. Geol. Instn Univ. Uppsala 10, pp 261-416.

Olsson, I.U., Holmgren, B. and Skye, E. 1984: *Questions arising when using lichen for ^{14}C measurements in climatic studies*. (This volume).

Paterson, W.S.B. 1969: *The Meighen Ice Cap, Arctic Canada: Accumulation, Ablation and Flow*. J. of Glac. Vol 8, No 54, pp 341-352.

Salvigsen, O. 1981: *Radiocarbon Dated Raised Beaches in Kong Karls Land, Svalbard and Their Consequence for the Glacial History of the Barent Sea Area*. Geogr. Ann. 68A (3-4), pp 283-291.

Schytt, V. 1964: *Scientific Results of the Swedish Glaciological Expedition to Nordaustlandet, Spitsbergen, 1957 and 1958*. Parts 1 and 2. Geogr. Ann. 46. pp 243-281.

Sverdrup, H.U. 1933: *Meteorology Part 1, Discussion, The Norwegian North Polar Expedition with the "Maud" 1919-1925*. Scientific Results, Vol II. Geofysisk Institutt, Bergen.

Taylor-Alt, B. 1975: *The Energy Balance Climate of Meighen Ice Cap, N.W.T.* Vols I and II. Polar Continental Shelf Project, Ottawa, Ph. D. thesis in Meteorology at McGill University, Montreal.

QUESTIONS ARISING WHEN USING LICHEN FOR ^{14}C MEASUREMENTS IN CLIMATIC STUDIES

Ingrid U. Olsson, Björn Holmgren, and Erik Skye

Uppsala University, Sweden

Some lichen species survive long enough to be regarded as longlived samples. Since the life-time may vary from a few decades to hundreds of years it must somehow be estimated with the help of biological or other information in order to allow a reliable determination of the real weighted mean age with its uncertainty. This is particularly important for the last few hundred years with their significant variations in the ^{14}C activity of the atmoshpere. There is a probability of a reservoir age of the lichen because of the symbiosis between the fungus and the alga composing the lichen; the production of acids strong enough to weather rocks may be suspected also to dissolve any carbonate available. Fungi obtain their nutrients from the algae, and the water involved may contain carbon with less activity than that in the atmospheric CO_2. When the lichen deteriorates the material may be re-used to produce new parts of the particular organism with carbon which appears older than the contemporaneous air. It is known that bird droppings may enhance the lichen growth in some cases. They could be a source of low-activity carbon on sea-shores since birds feeding on fish should exhibit an apparent age.

Some experiments are reported to yield a basis for an evaluation of the reliability of age determinations on lichen.

THE SAMPLES

The samples consisted of cushion-type lichen, mainly *Stereocaulon*, collected close to Rundisen and Rektangelisen on Kongsøya, Svalbard (Holmgren *et al.*, 1984, this volume, fig. 3). Two of the samples, nos. 5 and 6, were living. As a comparison, two similar samples, *Cladina stellaris*, were collected in Sweden growing on

N.-A. Mörner and W. Karlén (eds.), Climatic Changes on a Yearly to Millennial Basis, 303–308.

Table 1. ^{14}C ages or excess activity of lichens, mainly *Stereocaulon*, and mosses, *Rhacomitrium* spp., from Kongsøya and lichens from Uppland.

	Sample	Fraction	Dating no.	^{14}C age B.P.	$\delta^{13}C$, PDB ‰
Kongsøya	1		U-4459	770±65	-25.8
	2		U-6084	770±70	-26.1
	3		U-4460	675±60	-25.4
	4		U-6085	465±90	-24.6
	4	top	U-6099	350±80	-27.5
	4	base	U-6098	440±75	-27.2
	4	<1mm	U-6101	340±80	-23.8
	4	1-2 mm	U-6100	260±90	-26.8
	7		U-4462	945±60	-23.8
	7	top	U-6104	550±80	-27.75
	7	<1mm	U-6103	600±130	-25.7
	8A		U-6086	500±110	-27.3
	8B		U-6102	715±70	-28.0
	9		U-6087	975±55	-24.6
Excess over standard %					
	5		U-4461	9.1±0.7	-24.9
	5	top	U-6148	13.2±1.1	-27.5
	6A		U-4458	15.2±1.1	-26.2
	6B		U-4463	15.1±1.4	-24.7
Uppland	Ryggmossen		U-6105	30.6±0.8	-25.2
Uppland	Smara		U-6106	30.7±1.0	-25.3
Kongsøya,	Triangelisen	10	U-4464	29.7±2.6	-24.9
		11	U-4465	43.5±1.7	-25.6

a hummock in the bog Ryggmossen and on an outcrop together with mosses respectively. Furthermore, two moss samples were collected on Kongsøya close to Triangelisen, further discussed by Holmgren *et al.* (this volume). Three samples were dated as different fractions and two were dated twice, the pretreatment and subsequent steps being performed using different but similar parts of the original sample. If the lichen has a biological age of about 100

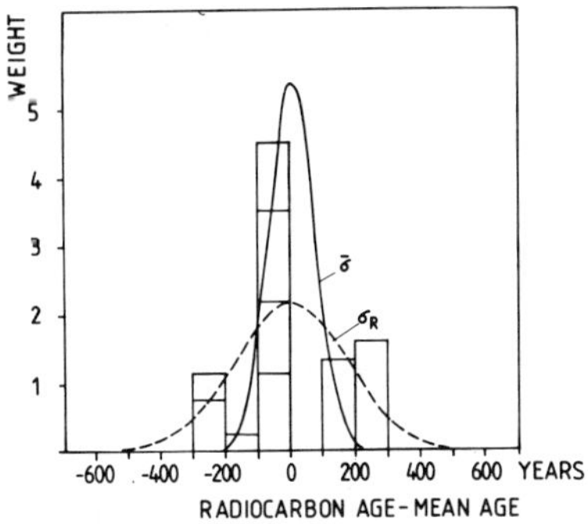

Figure 1. The distribution of the nine radiocarbon ages of samples 1,2,3,7,8, and 9. σ_R is the actual standard deviation and σ that calculated from quoted errors.

years or more it might be possible to detect different ^{14}C activities in different parts, since the growth of fruticose lichen takes place on the upper parts as a consequence of the assimilation. It is also accepted that old matter may be dispersed from the basal parts, but weathering may cause the outer parts to be lost. A ^{14}C age is a measure of a mean biological age plus the time which has elapsed since the death of the lichen. The growth, the weight gain, probably increases with the age in the beginning to reach a maximum and perhaps decrease later on.

Each laboratory number designates not only a new measurement but also a new pretreatment and a new combustion. Each sample was pretreated with dilute hot HCl and dilute warm NaOH as described by Olsson (1968). The organic content of the original dry samples varied, but was usually about one quarter of the sample. The insoluble fraction was used. The soluble fraction was hardly detectable.

THE ^{14}C RESULTS

The results are given in Table 1. The $\delta^{13}C$ values are of different accuracies since the first determinations were made in one laboratory and the last in another with a different mass spectrometer.

From Table 1 it seems justified to assume that sample 4 in reality is younger than the others. This is corroborated by the location (Holmgren *et al.*, this volume). The mean value of all measurements of samples 1,2,3,7,8, and 9 is 775±25 ^{14}C years. The spread is, however, such that this mean value should be regarded as a mean of 775 ^{14}C years with a spread, σ, of ±175, dating a period instead of a single event. The σ calculated from the single

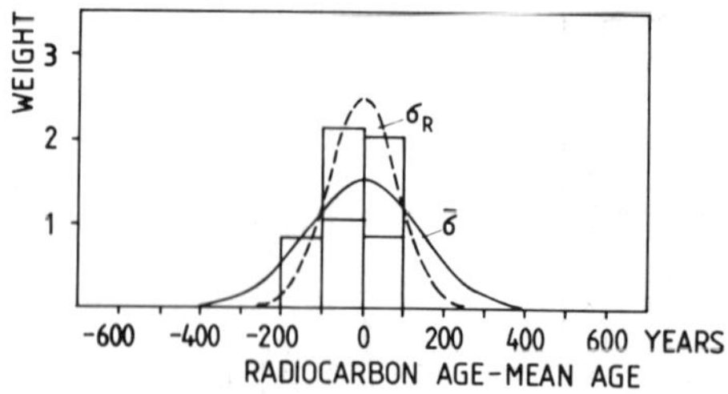

Figure 2. The distribution of the radiocarbon ages of the different fractions of sample 4. σ_R is the actual standard deviation and $\bar{\sigma}$ that calculated from quoted errors.

measurements is ±70 ^{14}C years (Fig. 1). The mean value of the five independent measurements of sample 4 is 375±40 ^{14}C years. The spread is given in Fig. 2, indicating a σ of ±85 years. The σ calculated from the single measurements is ±140 years. The number of measurements is, however, only five, thus too small to draw further conclusions. Details of the calculation of the actual standard deviation and that calculated from the quoted errors are given by Stuiver (1982). The two living samples from Svalbard show such a low activity that they obviously started to grow well before the atom-bomb era - thus before *c*. 1950. Since the mean ^{14}C activity of the atmosphere has had an excess of about 0.4 times the standard activity during this time, and these two samples had an excess of about 10 to 15%, it seems realistic to assume a biological age in the order of 100 years. An accurate calculation includes estimates of the mean excess for the actual growing periods and data on the mass increase during the life of the lichen. The growth must increase in the beginning but is certainly different for different species. Moreover some old parts are naturally rejected. In an Arctic area losses may also be due to snow drifting. The two living samples of another species from Sweden, collected in 1983, have an activity indicating that they derive from the last five to 10 years. The two measurements of sample 6 agree very well, whereas the two measurements of sample 5 differ with 4.1±1.3% of the standard activity. No real difference between the aliquots of sample 6 can be expected. The two parts of sample 5 differ insofar as the top of the lichen was used for the second dating, giving higher activity. The difference could imply that the top is younger than the basal parts. This is not contradicted by the results for sample 7 with a difference in ^{14}C age of 395±100 years, thus about 4 σ. The shape of the calibration curve is, however, such that the difference is between 200 and 400 calendar years. Sample 4 neither supports nor contradicts the assumption, since the ^{14}C ages have a normal distribution of the different fractions.

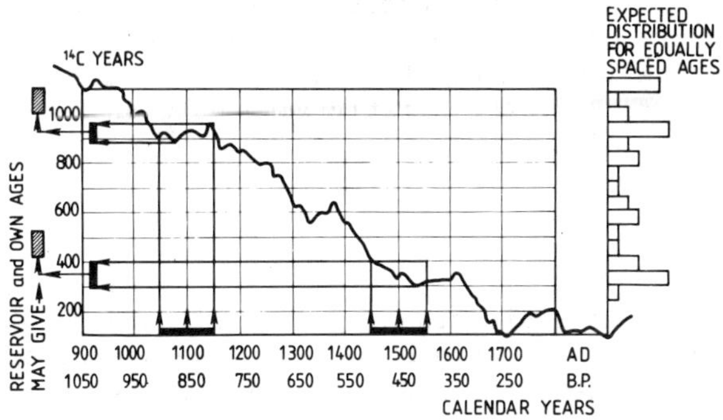

Figure 3. The calibration curve (Stuiver, 1982) with the normally expected ^{14}C age for two age ranges in calendar years and the expected ^{14}C age for the lichens on Svalbard. The last correction is due to the possibly lowered activity on Svalbard and the longevity of the species investigated (mainly *Stereocaulon*). During the last 1000 years some clustering of dates is to be expected with maxima around 1100, 900, and 300 ^{14}C years ago.

CALIBRATION OF THE ^{14}C RESULTS

It is difficult to interpret radiocarbon dates of recent samples like these because of the variations of the ^{14}C level of the atmosphere. Stuiver (1982) has submitted high precision data in the form of points in diagrams giving radiocarbon age and dendroyear AD. When using these values it must be remembered that they are not directly applicable to the present dates since a lichen sample yields a ^{14}C age as a weighted mean of the ^{14}C ages of the different parts of the total sample with a biological age of perhaps 200 years. Then smoothed curves should be used (Olsson, 1974; Olsson & Florin, 1980; de Jong, 1981).

Since measurements of the very recent ^{14}C activity at Svalbard and islands in the Arctic indicate a lower activity than over Scandinavia (Olsson, 1983) any calibration will be still more complicated (Fig. 3). It seems reasonable to assume that a short-lived sample growing 900 to 800 B.P. in Scandinavia would have a ^{14}C age of about 930 years (B.P.). Because of the possibly depressed ^{14}C activity and the longevity of the present lichens the data given by the laboratory would approach 1000 or maybe 1100 ^{14}C years B.P. The upper limit for the age span presented here is thus lowered to about 750 years when converted to calendar years. A normal sample growing 500 to 400 years ago would have a ^{14}C age about

360 years B.P. but the ^{14}C depression in the Arctic is more likely to cause the date of a lichen sample to be about 500 ^{14}C years B.P., so that the lower limit for the age presented here is about 300 calendar years B.P. A realistic span between the extreme dates seems to be about 450 years.

CONCLUSION

When discussing radiocarbon ages of lichens we must consider not only the secular variations but also the source of carbon, the biological age of the lichens and global and regional variations. The points of importance then are the species used, the environment and what parts of the cushion are used. It must also be remembered that there might be a bias in the laboratory making the measurements and that making the calibration curve, and that sigma should be estimated realistically, i.e. not only the statistics for the number of disintegrations counted but also the other uncertainties encountered in the laboratory should be included.

ACKNOWLEDGEMENTS

Financial support for these datings was received from the Swedish Natural Science Research Council. For assistance in the radiocarbon laboratory thanks are due to Tomas Kronberg, Maud Söderman, and Birgitta Hansson.

REFERENCES

Holmgren, B., Olsson, I.U., Skye, E., & Alm, G., 1984. Climate and glaciation in Kong Karls Land, Eastern Svalbard. This volume.

de Jong, A.F.M., 1981. Natural ^{14}C variations. Thesis, Groningen.

Olsson, I.U., 1968. Modern aspects of radiocarbon datings. Earth-Science Reviews, 4, 203-218.

Olsson, I.U., 1974. Some problems in connection with the evaluation of C^{14} dates. Geol. Fören. Stockh. Förh., 96, 311-320.

Olsson, I.U., 1983. Radiocarbon dating in the Arctic region. Radiocarbon, 25, 393-394.

Olsson, I.U. & Florin, M.-B., 1980. Radiocarbon dating of dy and peat in the Getsjö area, Kolmården, Sweden, to determine the rational limit of *Picea*. Boreas, 9, 289-305.

Stuiver, M., 1982. A high-precision calibration of the AD radiocarbon time scale. Radiocarbon, 24, 1-26.

SOME STUDIES OF THE LITTLE ICE AGE OF RECENT CENTURIES AND ITS GREAT STORMS

Hubert H. Lamb

University of East Anglia, Norwich, England.

ABSTRACT

The long series of monthly mean temperatures in central England since 1659, produced by the patient researches over many years of the late Professor Gordon Manley, establishes the reality of the colder climate of recent centuries. Estimates of the probable error margins of the values in this series are given. And so the series gives us our most reliable estimate of the magnitude of the temperature depression in England and neighbouring countries. In northern Scotland, southern Norway and Iceland there are indications of a significantly greater depression of the prevailing temperatures. This seems to have been associated with an important deviation of the ocean currents and surface temperatures in north-eastern parts of the Atlantic. The enhanced thermal gradient between latitudes about 50° and 60-65°N in this part of the world is thought to have provided a basis for the development of some greater wind storms in these latitudes than have occurred in most of the last 100 years, though there are signs that in about the last decade or two storminess has been increasing again. Daily synoptic weather maps can be reconstructed which appear to establish the strength of some of the great storms of the Little Ice Age in the North Sea region. The manner of construction of the isobar analysis which permits the wind strength estimates is explained, and some indication of the probable error margins is given.

INTRODUCTION

We may use the reconstructed record of prevailing temperatures in

N.-A. Mörner and W. Karlén (eds.), Climatic Changes on a Yearly to Millennial Basis, 309–329.

central England in successive half centuries over the last thousand years (Fig. 1 in Climate and History ...... in this volume) to define the Little Ice Age period. Recent study suggests that the individual monthly values of mean temperature given by Manley (1974) for each year since 1659 have a standard error of about 0.1°C back to around 1730, increasing to about 0.2°C in the 1720s and perhaps 0.5° in the earliest part of the record before 1680. This means that 10-year averages of the monthly values probably have a standard error of about 0.1°C in the 1690s and no more than 0.2°C in the 1660s. The 50-year means should be within about 0.1°C at that stage of history.

Temperature records, and records of various phenomena - e.g. glaciers, snow data, wine harvests - that can serve as proxy temperature records from elsewhere in Europe provide general confirmation, especially of low temperatures around 1700 and in the first half of the nineteenth century (e.g. Labrijn, 1945; Matthews, 1977; Schweizerische Met. Anstalt, 1957).

The derivation of prevailing central England temperatures over eight centuries before the earliest thermometer records, from the statistics of descriptive accounts of months of various well-defined character, has been described elsewhere (Lamb, 1965, 1977 pp. 31-35). As there is rather little difference in the quantity of manuscript source material reporting the seasons in England between about AD 1100 and 1550, the error margins of the temperature values so derived are nearly constant over this span of time. Hence, there is no doubt about the coldness of the period 1550 to 1700 in this part of the world, nor about the severity of the cooling between the thirteenth and fifteenth centuries.

Iceland and Greenland seem to have had the peak of their medieval warmth a century or more earlier, beginning cooling sharply by AD 1200. Iceland may have had some temporary recovery of warmth in the fifteenth century rather than in the early sixteenth century as in Europe. How far the magnitude of the changes shown by the central England temperature record should be considered representative will be discussed in the next section.

CHANGES IN THE OCEAN

The long continuous record of sea surface temperature measurements at the Faeroe Islands, going back to 1867, shows that within that time the warmest and coldest 5-year periods differed from each other by 1°C, i.e. twice the range of the air temperatures prevailing in England or central Europe. This must be largely due to variations in the ocean, particularly the volume of polar water in the East Greenland Current and the branch of it that heads towards, and sometimes passes, eastern Iceland.

The position of the boundary in the ocean surface between

this polar current and the warm water of Gulf Stream origin has varied hugely over long periods of time. The reconstruction (Lamb 1979) of sea temperatures, from fisheries and ice records, that established the seventeenth century position means that over the 30 years between 1675 and 1705 average sea surface temperatures between the Faeroe Islands and southeast Iceland were probably about 5°C below the twentieth century normal. This surely implies that the 30-year average air temperatures in northern Scotland and southwest Norway at that time must have been about 2 to 2.5°C below the modern averages - in agreement with the contemporary reports of permanent snow on the tops of the highest mountains in central and eastern parts of northern Scotland and glacier observations on Hardangervidda in southwest Norway (O. Liestøl, personal communication, 13 September 1977, see also Lamb, 1979). In eastern Iceland the prevailing air temperatures in those years should probably have been 3°C below today's. It was probably the severest phase of the Little Ice Age at least in these parts of Europe.

The enhanced thermal gradient between the northernmost Atlantic, about Iceland and the Faeroe Islands, and the regions near latitude 50°N might provide an energy base for the development of occasional storms in European longitudes of greater intensity than in warmer periods.

EVIDENCE OF ENHANCED STORMINESS IN THE NORTH SEA AND SURROUNDING REGIONS

There is evidence of an era of enhanced storminess about the North Sea and the coasts of northwest Europe roughly coinciding with the deterioration of climate (falling temperature levels and increased wetness of the peat bogs that have been studied) that ended the medieval warm period and also coinciding with the coldest phases of the climate between 1550 and various parts of the eighteenth and early nineteenth centuries. The evidence is of at least three kinds, enumerated below.

(i) Great sea floods devastating the low-lying coastlands around the North Sea, often with great loss of life and establishing some new water bodies that survived until the twentieth century (Zuider Zee in the Netherlands, Jadebusen in northwest Germany, the Norfolk Broads in England). A great amount of land, with many villages, was also lost from the west coast of Denmark (Jutland and the, at that time Danish, province of Slesvig/Schleswig). Corresponding erosion of the coast of England destroyed two great ports and took all the land about them (Ravenspur or Ravensburgh, which stood east of Hull on land beyond the present Spurn Point, and Dunwich on the coast of Suffolk, which also extended farther east than it does today).

(ii) Storms of blowing sand. These established a whole new range of coastal dunes in Belgium, the Netherlands and northwest Germany from AD 1150 or 1200 onwards after some centuries during which the older dunes had become stabilized by vegetation. The phenomenon extended also inland in these countries and deposited layers of drifted sand. Similar troubles with drifting sand were reported in the seventeenth century in East Anglia (Norfolk). There were many local disasters through sand drift and advancing dunes that changed the coasts in various places in northwest Europe, sometimes closing former harbours.

(iii) Between 1690 and 1703 at least four storms were reported that impressed the people then living as of exceptional severity even for those times. (In two of these cases the precise date has not yet been identified.)

(a) Reputedly in the autumn of 1694 the Culbin estate, on the south coast of Moray Firth, in northeast Scotland, with sixteen good farms covering a total area of 30 km^2 was buried by blown sand, according to one account, in a single night (Edlin, 1976). The land remained a wilderness of shifting sand dunes until it was successfully planted with forest in this century.

(b) On 12 September 1695 (possibly Old Style - Julian calendar) a violent storm at night did great damage in London, overturning a ship in the Thames, dashed 70 ships (colliers setting out empty back to Newcastle) on the beaches about Yarmouth with the loss of most of the crews, drove a fleet of Dutch ships back to The Downs (on England's south coast) and stranded many ships on both sides of the Channel (Lindgren and Neumann, 1981). This seems to be the storm that Daniel Defoe remembered[1] thirty years afterwards as an unparalleled disaster at sea, with another fleet of coal ships coming south from Newcastle also being wrecked by the NE gale as they approached the coast of Norfolk. Altogether 200 ships and more than a thousand lives were lost off Norfolk that night.

(c) In the autumn of 1697 a great storm overwhelmed with sand a settlement at Udal, on North Uist, in the Outer Hebrides, that seems to have known human occupation continuously for four thousand years (Morrison, 1967-8).

(d) The great storm that crossed southern England on the night of 6-7th December 1703 (New Style - Gregorian calendar) causing enormous damage to buildings and trees, in the cities and in the country, and an exceptional tide in the Thames, has been generally assessed as the severest storm of wind ever reported in England. It was described in great detail by Defoe (1704).

Any fuller account of the events producing the coastal changes (Lamb, 1981) which provide two of these three categories of evidence must notice that the risk to the coasts varies not only with the incidence of storm surges but also with (a) long-term changes of sea level, consisting of both eustatic and isostatic components, and (b) long-term variations of the tidal force (associated with lunar orbit evolutions) and resulting variations in the range of the tides (Pettersson, 1914; summarized briefly in Lamb (1972(b), pp. 220-222) and also (c) any long-term changes in the tides produced by alterations of the bottom topography of a shallow sea such as the North Sea (Nielsen, 1938: I am indebted to Dr E. Rasmussen for drawing my attention to the latter, some effects of which are evidenced in his own 1958 paper). Mörner (1980, and at this symposium) has interestingly drawn attention to the fact that the detailed histories of sea level in the northwest European/northeastern North Atlantic region (e.g. northwest England, southern shores of the North Sea, Kattegat and north Norway) worked out by numerous contributors seem to present a regional eustatic history, which may differ somewhat from the global average and which agrees in notable detail with the region's temperature-climate history. Such a regional eustatic history may, according to Mörner (loc. cit.), indicate changes in the shape of the geoid, a subject to which Maksimov (e.g. 1971) drew attention many years ago.

SOME NOTABLE STORMS ANALYSED: SAMPLE CASES FROM THE LITTLE ICE AGE PERIOD

It is to be hoped that some day it may be possible to submit the great storms we have named above to meteorological analysis.

In the meantime, other storms within the cold climate period have been analysed.

(a) The Stormy Summer of 1588

An analysis has been published (Douglas et al. 1978) of the famous Spanish Armada storms in 1588, using the winds and weather reported, albeit without instrument measurements, by the ships. The first analysis was tested by its ability to explain the weather observed on the then Danish island Hven, in Øresund, daily by the astronomer Tycho Brahe. As this was satisfactory on 72% of the days, despite the fact that the analysis had not been designed to cover the Danish area, the whole series of maps was then re-analysed with the Danish observations incorporated and with the greatest care for logical consistency. The result is a series of synoptic

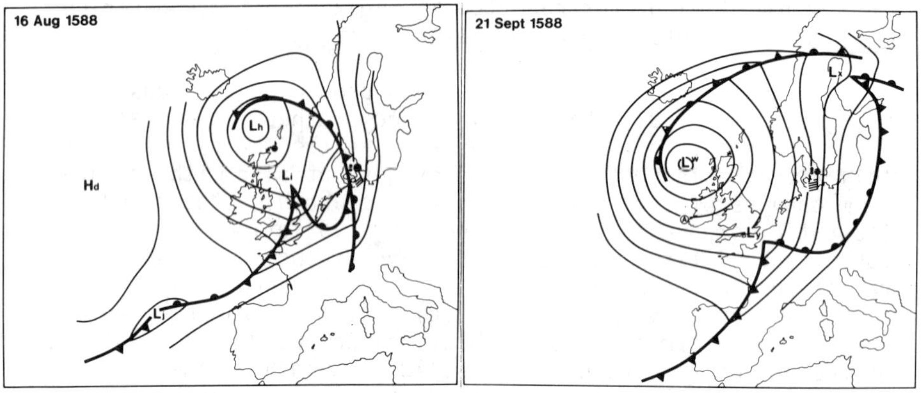

Fig. 1: Synoptic meteorological analysis for 16 August and 21 September 1588 (both dates adjusted to the modern Gregorian calendar). The latter was the date of the greatest storm which wrecked most of the ships of the Spanish Armada on the west coasts of the British Isles. The positions of the ships which reported the weather, near the Orkney Islands, west of the Hebrides and southwest Ireland are marked and, Tycho Brahe's observations in Denmark are plotted.

weather maps (two samples showing storms are illustrated in Fig. 1, respectively the severest storms in August and September 1588), with suggested isobars, for 60 individual days between 12 May and 26 October 1588. The positions of depression centres passing across the central part of the mapped area are clear enough to make it possible to measure their speed of travel. Hence, by statistical relationships due to Palmen (1928) and Chromow (1942), indications can be gained of probable jet stream and surface gradient wind speeds.

The observation coverage of these maps for storms affecting the British Isles, North Sea and northern Europe is best between late July and end of September 1588. During those weeks in that single summer the study indicates that at least six times the winds at jet stream level were close to or somewhat exceeded the probable upper limit of the modern (1949 to 1970) frequency distribution for that time of the year. Around 15 September the southwesterly jet stream over the eastern North Atlantic is indicated as having a probable speed of 130 knots and on 8-9 August 150-160 knots is indicated as the most probable speed. The strongest surface pressure gradient winds in the postfrontal cold air in the storms of 18 and 21 September on the coasts of the British Isles should probably be estimated at 80 to 90 knots.

Judged by the details reported by the Spanish ships on the west coast of Ireland on 21 September 1588, and the wreckage of most of them, the (westerly) storm on that date (Fig. 1) was one of the severest (Douglas et al. 1978).

(b) The Great North Sea Storms on 21-22 March 1791 and in December 1792

To analyse the wind and weather situations occurring from the 1780s and 90s onwards, methods very close to those applied to present-day synoptic meteorology can be used. At the date of this storm, observations were available from fifty places in Europe, ranging from northern Iceland to Rome and from Barcelona to St. Petersburg and northern Sweden. In addition, occasional reports from ships at sea, or in port, at useful positions, including the Faeroe Islands, strengthen the analysis. (I am indebted to my friend Dr K. Frydendahl and the archives of the Danish Meteorological Institute for the most valuable ships' data.) A gazeteer of the observing stations, their positions and their practices, in those times has been built up in the Climatic Research Unit, Norwich, by Mr. J.A. Kington who produced a daily weather map series from 1 January 1781 to 31 December 1786; and this has been added to by the present studies. Data and information about the observations have been graciously given by collaborators too numerous to name individually here, but my thanks are due for a particularly big contribution which came from Dr. E. Hovmøller working in Swedish archives.

The procedure starts by accepting the barometric pressures reported by a strategically distributed network of stations for which the details known and the quality of the record suggest confidence: these pressures are then corrected to standard gravity and sea level. On this basis a map of the mean pressures prevailing over Europe and the eastern Atlantic in that month can be drawn. The average correction required to make the mean barometric pressures reported at other places fit the map can then be deduced. These corrections are then applied routinely for every day of the month, and the additional stations are entered on the maps. Finally, for every station on the map, including those provisionally accepted as the basis of the monthly average map, the frequency distribution and range of apparent errors of their fit on the daily maps is studied. The standard errors calculated supply an index of reliability which is a valuable addition to the station gazeteer.

The results of this reliability test may be somewhat flattering to stations at the edge of the mappable area, which tended to be relied on because of their position, so long as the quality of their records inspired confidence; but no such criticism can apply in areas where the observation network was quite dense, even near the edge of the map. From this study it appears that in the 1970s the Swedish stations, Trondheim, Gordon Castle in northern Scotland and the stations used in France were among the best, mostly with standard errors of about one millibar or less.

A separate test of the reliability of the analysis of the synoptic maps with regard to pressure gradients and the storm winds over the North Sea and surrounding regions was performed by first analysing the maps for the whole month of March 1791 with only four stations (Trondheim, Stockholm, St. Petersburg and Copenhagen) in northern Europe north of 55°N apart from the British Isles stations. These places were as far apart as the places on opposite sides of the North Sea:

e.g.	Stockholm - Trondheim	600 km
	Copenhagen - Trondheim	830 km
	Stockholm - St. Petersburg	720 km

compared with

Edinburgh - Hamburg	600 km
Norfolk - Hamburg	600 km
Northeast Scotland - Trondheim	1000 km

Later, the observations at six more places in Sweden and Finland were entered. Unfortunately, the barometer used at Gothenburg was not one of the best, its record being marred by failure to respond fully to the quickest changes of pressure. Eliminating the minority of days, 8 out of 31 in that stormy month, when the Gothenburg barometer readings were obviously badly wrong because of this defect, the standard error of the pressures indicated by the first analysis of the maps at that place was 1.24 mb: part of this may still be due to the poor behaviour of the Gothenburg barometer. Overall, the test suggested standard errors of 1 to 1.5 mb at points in the middle of the biggest gaps in the network over the northern half of Europe. The corresponding errors of the pressure gradients over the North Sea or in gaps of that size within the network would be about 1 mb/200 km, equivalent to about 10 knots (5m/sec) in the gradient winds indicated.

But the maps deteriorate rapidly outside the area of the observations network: bigger errors occurred at no great distance beyond the edge of the network; thus, at Härnösand, 250 km outside the edge of the network used for the test analysis, the standard error was 3.7 mb. A similar test on the May 1795 maps revealed standard errors of the daily maps amounting to 3.8 to 4.4 mb at points 500 to 600 km east of the limit of the observation network used over central-eastern Europe.

In March 1791 the first warning of disturbance to the anticyclonic situation with light winds which had been dominating northern Europe since the 16-17th came with the passage of a cold front southwards over Iceland on the night of the 18-19th, followed by three days of hard N'ly wind with bitter frost and blowing snow there. Temperatures as high as 14 to 16°C in the warm SW'ly airstream over the continent as far north as Brussels and Hamburg on the afternoon of the 20th, and 8 to 10° in most of Britain, were replaced by N'ly winds with wintry showers and afternoon temperatures in England mostly between 3 and 6°C on the 21st. A strong N'ly gale reached the British Isles on the evening of the 20th after falls of pressure of 20 to 30 mbs in 24 hours, followed by equally rapid rises. At Hamburg the pressure fell 37 mbs in 24 hours between the afternoons of the 20th and 21st; and around 5h on the morning of the 22nd, after a night of storm, the tide in the River Elbe reached an exceptional level (Rohde and Petersen 1977, and contemporary accounts in the old Hamburg newspaper Hamburgische Adresse-Comptoir Nachrichten). This storm maintained gale strength for three days over some parts of the North Sea, probably the longest duration of any of the severe storms of that century, owing to the slow movement of the cyclonic system from the 20th onwards in the increasingly meridional situation.

On this occasion, and during storms in the month of December 1792, geostrophic winds reaching 130 to 150 knots at some points over the North Sea and neighbouring coasts were indicated by the surface pressure map. Since in all these cases we are dealing with "straight" - i.e. great circle - windstreams in the regions where the pressure gradients were strongest, the gradient winds were presumably also of about this strength. Strongest gusts at the surface over the sea and exposed areas near to the coasts may also have been over 100 knots, perhaps much over 100 knots[2].

No gale matched this in the twentieth century over the same region with the exception of the recent storms of 23-25 November 1981 and 18 January 1983, when the strongest gradient winds again reached 130 knots or rather more.

The meteorological situation that developed on 10-11 December 1792 (Figs. 2, 3) produced the highest sea flood of the eighteenth century around Hamburg, exceeding the highest tide level on 22 March 1791 (and earlier flood levels in 1751 and 1756) by about 10 cms. The strongest gradient winds are estimated as westerly 150 $\pm$ 30 knots over the Faeroes-Shetland region on the morning and early afternoon of the 10th and westerly to northwesterly 140 $\pm$ 20 knots over the southernmost North Sea and the German coast in the late evening of the 10th till about midnight. (A 21h map on the 10th was analysed.)

Tide levels in the Elbe caused alarm in two earlier storms on the 5th and 7-8th of the month, nearer the time of neap tides. The strongest gradient winds at the critical time were again westerly to northwesterly, 100 to 140 knots.

It was a remarkable month for the frequency of gales and storm-force winds in many parts of Europe. Although the exact equivalents of the scale of wind force then in use are not known, and the scale may not have been so professionally used as the later Beaufort scale has been, stormy winds were reported by observers in northern Holland on 10 days of the month, in Paris on 8 days, at Marseille 7 days, in Moscow 9 and in Hamburg 6. The pressure gradients suggest that Beaufort force 9 or more occurred on at least 12 days of the month over the sea areas off north Holland, Friesland and northern Germany and also over northernmost parts of the North Sea.

There was a great thermal contrast between the warm and cold airmasses involved in this month. Temperatures up to 11 to 13°C occurred in the warmest airmasses passing over England and western parts of Europe south of latitude 53°N and up to +6°C over southern Sweden. Yet, despite the lack of periods of still, clear weather in the cold airmasses an afternoon temperature of -9°C was registered in a N'ly wind at Stockholm on the 12th. And a day or two later the same airmass produced afternoon temperatures of -13°C at Trondheim and -20°C at Härnösand (62°N).

(c) The Storm of 8-9 May 1795

This storm is chiefly known for the destruction wrought in the forests of central Sweden. The month opened with a remarkably warm southerly to southwesterly airstream over much of Europe: 26°C was measured in Stockholm on 1 May. A fresh northerly outbreak from the Arctic appeared over Iceland on the 5th, with prolonged snowfall followed by hard frost. On the 7th-8th a strong N'ly outbreak developed

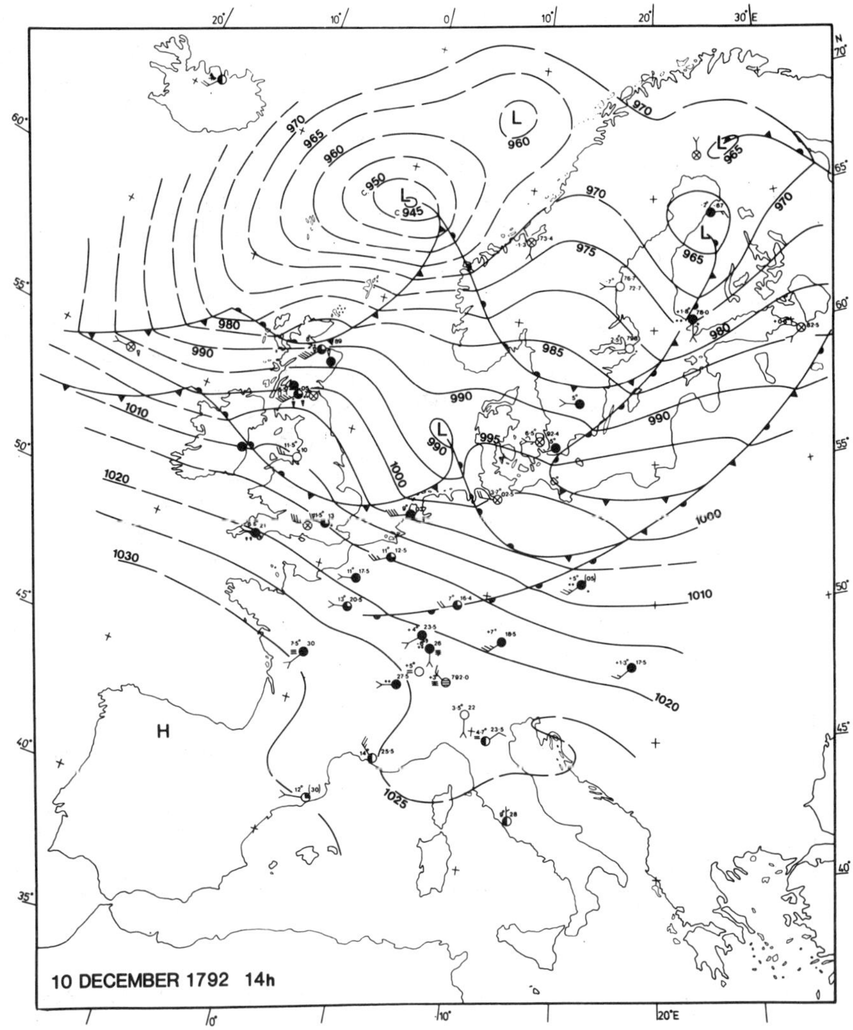

Fig. 2: Synoptic weather map and analysis for 14h on 10 December 1792.

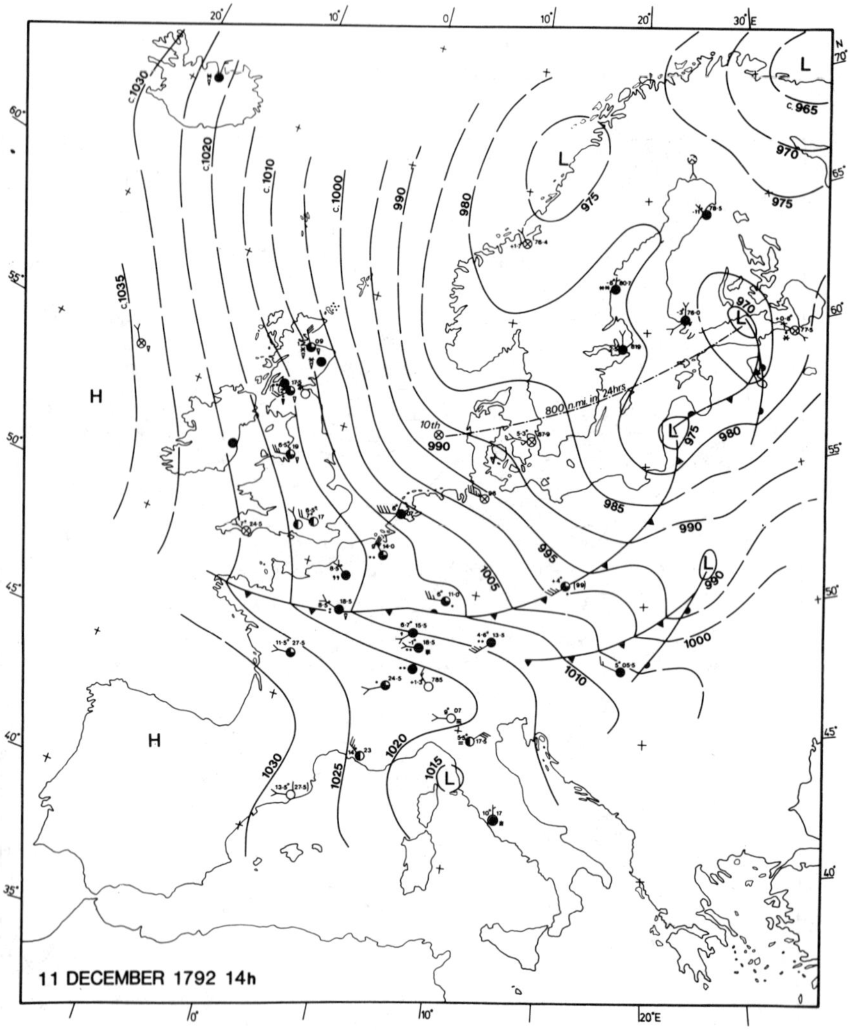

Fig. 3: Synoptic weather map and analysis for 14h on 11 December 1792.

over the whole Norwegian Sea and North Sea, where it rejuventated a small depression that proceeded to cause strong winds and heavy snowfall over central Sweden for several days. The damage to the forests was undoubtedly partly due to the weight of snow. Geostrophic winds of up to 100 knots are indicated over central, southwestern and southern Sweden on the 8th, but the gradient winds may have been no more than 60-80 knots. The pressure gradients are particularly well authenticated by the fairly dense network of observations over the area. However, the severest gusts and squalls of wind at the surface near the fronts over Sweden may well have ranged up to about 90 knots.

SUMMARY

Owing to the near-impossibility of maintaining constant exposure of anemometer sites - perhaps the only examples of long homogeneous series of wind-strength measurements are from off-shore lighthouses - probably the only way of obtaining satisfactory comparisons over long stretches of time of the windiness prevailing is by measurements of the barometric pressure fields (Lamb and Weiss 1979). That is the basis of the studies here reported.

In this work so far, 31 severe windstorms over the North Sea and neighbouring regions of northern, central and western Europe have been analysed. Identification of the severest cases for study in centuries before the present one has been from the greatest reported disasters by coastal floodings, shipwrecks and damage to coasts and inland. This method of selection, particularly because the reports of coastal flooding are generally the most serious and likely to be the most complete, will probably not bring to light all the severest storms but will certainly provide a worthwhile sample list of outstanding storms. Within the present century the list is likely to be more nearly complete, and effort has been made to analyse all the very severe cases, preferably including examples of as many as possible of the different types of meteorological situation that may produce the extremes. This should provide the best possible basis for comparisons with earlier times.

The severe storms analysed so far were distributed as follows:

Sixteenth century	2 cases	1900-49	4 cases
Eighteenth century	7 cases	1950-83	15 cases
Nineteenth century	3 cases		

The results of the analyses so far performed may be tabulated as follows:

Table 1. Number of cases in different categories

Period	Total Numbers of Strong Storms Analysed	Main Direction of Storm Winds at Surface					Strongest Gradient Wind			
		SW	W	NW	N	SE	70-90kt	90-110kt	110 130kt	Over 130kt
1588-1800	9	1	2	4	2	0	4	1	1	2
1800-1899	3*	0	0	1	1	0	2	0	1	0
1900-1950	4	0	0	1	1	2	3	1	0	0
1950-1983	15	2	3	6	2	1	9	4	1	0
Total	31	3	5	12	6	3				

* In one case the location of the cyclonic centre precluded estimation of the surface wind direction.

Table 2. Frequency of different maximum gradient wind strengths as a proportion (%) of the severe storm cases analysed.

Period	Strongest Gradient Wind 70-90kt	90-110kt	110-130kt	Over 130kt	Extreme
1588-1800	44	11	11	22	c. 150kt (twice)
1800-1899	67	0	33	0	c. 120kt
1900-1950	75	25	0	0	c. 100kt
1950-1983	60	27	7	0	100-130kt

All the storms which lasted 3 days or more were either NW to N or SE'ly storms.

The month of December 1792 was characterized by remarkably frequent storms, several of them of brief duration because of the rapidity with which even the occluded cyclones were moving. This points to a very strong jet stream. Using the statistical relationships found by Palmen and Chromow (loc. cit.), already referred to in connection with the 1588 storms, the speeds of movement of ten depressions between 5 and 19 December 1792 were found to indicate jet streams moving at over 130 knots in 5 cases and over 150 knots in 3 cases, the two strongest cases suggesting figures of 175 ± 25 knots and (on 10-11 December 1792) 260 ± 30 knots. This last figure probably exceeds somewhat the strongest jet stream observed in the present century over this part of the world, though not equalling the extremes of 300 knots reported over the South Indian Ocean and over Japan. It is to be noted that even the August and September cases in 1588 indicated jet streams of 115 to 130 knots over the northeast Atlantic-northern Europe region, also exceeding the probable twentieth century maximum for those months.

Gradient winds near the surface of up to 120 knots in a warm sector on 10-11 December 1792 were indicated by the isobars, but as in most other great North Sea storms the strongest gradient winds were in the cold air at the rear of the depression. So the probability of speeds of up to 150 knots or thereabouts on that date is confirmed by this method.

The predominance of winds from the quarter between west and north in the severe storms analysed may be a peculiarity of the North Sea region, though the association between outbreaks of air from the Arctic and great storms in these latitudes is certainly a more general experience. Inland and on other coasts in England the strongest winds generally come from between S and W. This is a widely known feature in tornadoes also. But in the North Sea and Norwegian Sea on the whole the roughest weather occurs with NW'ly and N'ly winds with their long fetch over open sea. And these are the directions that tend to produce the worst sea floods about the coasts of the southern North Sea. In various parts of the world mountainous coasts tend to channel the wind and, through convergence, produce the highest speeds quite near the coasts, in directions parallel to the coast or the flank of the mountains. This is the explanation of the secondary maximum of storm winds in the North Sea coming from the SE, in nearly all cases strongest near the coast of Norway. The occurrences of great storms from that quarter seem to be a feature of runs of years when blocking anticyclones over Scandinavia or north Russia are frequent. They also tend to be associated with very cold air from the continental interior channeled between the mountain mass and the warmer airmasses approaching with warm-type fronts from the Atlantic and over the British Isles.

The association of great storms in this region with outbreaks of Arctic air from the Norwegian Sea is illustrated by the monthly mean sea-level pressure maps for March 1791, December 1792, and May 1795 seen in Fig. 4. These three months provided seven of the great storms analysed. Indications of meridionality are also seen in the frequencies of the different weather pattern types (Lamb 1972(a)).

CONCLUSIONS

These studies have used observational material never previously submitted to scientific analysis. The meteorological analysis provides a number of indications that some of the storms in the region studied during the Little Ice Age period were more severe than twentieth century experience, although an increasing tendency since 1950 has been reported. It is also shown that outbreaks of Arctic air from the north in the Norwegian Sea and an enhanced thermal gradient between latitudes 50 and 60 to 65°N in the east Atlantic-British Isles-central European sector were involved.

ACKNOWLEDGEMENTS AND DATA SOURCES

The research on the severest North Sea storms of the past was funded by Shell Exploration Ltd. from 1975 to 1982, and the author acknowledges his indebtedness.

He also expresses his gratitude for generous gifts of the basic weather observation data from the late Professor Gordon Manley who supplied the records for Cambuslang (near Glasgow),

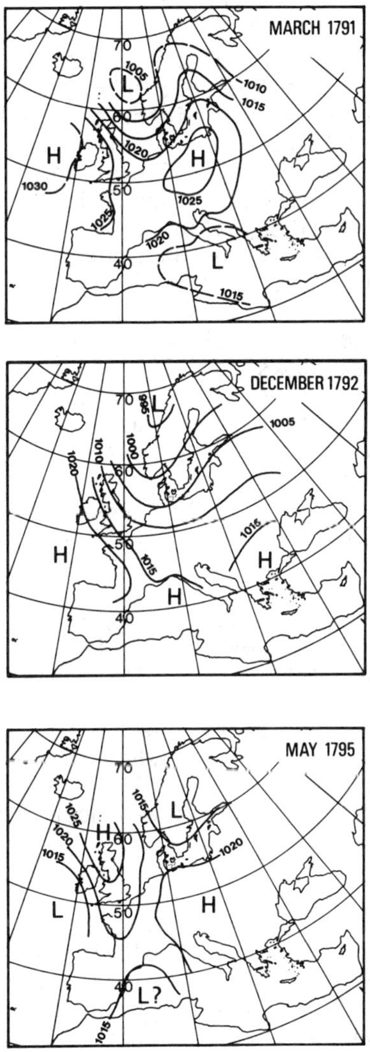

Fig 4: Maps of monthly average barometric pressure at mean sea level for
(a) March 1791
(b) December 1792
(c) May 1795

Liverpool, Lyndon (Rutland), Stroud (near Gloucester), Modbury (near Plymouth), and London; to Dr. Harald Nissen and the University Library, Trondheim for the Trondheim record; to Dr. K. Frydendahl and the Danish Meteorological Institute for the observations from Copenhagen and from some very usefully placed Danish ships at sea; to Dr. E. Hovmøller and the Swedish Meteorological Institute for observations from ten places in Sweden and Finland; to Dr. P.K. Rohan, as Director of the Irish Meteorological Service, for the observations in Dublin and occasionally elsewhere in Ireland; to Mr. J.A. Kington of the Climatic Research Unit for use of the observations which he had obtained from Paris. Poitiers and Montdidier (in northern France) and from Milan; to the Director of the Schweizerische Meteorologische Anstalt, Zürich for the observations from four places in Switzerland and from Mülhausen in Alsace; to Professor Manuel Puigcerver of the University of Barcelona for the observations from that city; and to Dr. G. Zanella of the University of Parma, northern Italy for the observations made there.

Most of the other observations mapped came from the collection in the yearbooks for 1791 and 1792 (the last year) of the old Societas Meteorologica Palatina, Mannheim. Gilbert White's weather observation record at Selborne, Hampshire was kindly made available for copying by the Trustees at Selborne. And a valuable record for Gordon Castle in northeast Scotland, running from 1781 to 1827, was obtained on microfilm from the Meteorological Office. The writer's thanks also go to the descendant relatives and Trustees of persons who kept weather diaries at Kemnay in Aberdeenshire and at three places in Norfolk (Parson Woodforde at Weston Longville, Mary Hardy at Letheringsett and the Blofeld family at Hoveton). The agreement between the weather reported in these last three diaries, kept at places 20 to 30 kms apart lends confidence to the use of such records when it is plain that the diarist was a meticulous person. This agreement was 100% except on days when there was an obvious meteorological explanation of the differences - as, for instance, greater shower activity and windiness near the North Sea.

[1] See Defoe's Tour through the whole island of Great Britain, in three volumes (1724-1726). Defoe actually dates the storm "about thirty years ago in 1692, I think that was the year..." but the details tally with other accounts of this storm on 12 September 1695.

[2] Note that the strongest gusts of the surface winds at exposed sites sometimes exceed the gradient wind speed by a substantial margin. In the storm of 16-17 February 1962, which also produced a disastrous sea flood in the Hamburg region, 154 knots

was measured in Shetland when the gradient wind was only 80 to 90 knots; and on 9 April 1933 on the mountainous island of Jan Mayen (71°N 8°W) 163 knots was measured by a standard anemometer.

REFERENCES

Chromow, S.P. 1942. Einfuhrung in die synoptische Meteorologie. Vienna (Springer, 2nd edn.) 532 pp.

Defoe, D. 1704. The Storm. London.

Douglas, K.S., Lamb, H.H., and Loader, C. 1978. A meteorological study of July to October 1588: the Spanish Armada storms. Climatic Research Unit Research Publication No. 6 (CRU RP6), Norwich (University of East Anglia).

Edlin, H.L. 1976. The Culbin Sands. In: Environment and Man, vol. 4 (Reclamation). Ch. 1. pp. 1-31. (Editors J. Lenihan and W.W. Fletcher) London (Blackie).

Labrijn, A. 1945. Het klimaat van Nederland gedurende de laatste twee en een halve eeuw. Mededelinger en verhandelingen, 49, No. 102. De Bilt (Koninklijk Nederlands Meteorologisch Institut).

Lamb, H.H. 1965. The early medieval warm epoch and its sequel. Palaeogeography, Palaeoclimatology, Palaeoecology, 1, 13-37. Amsterdam (Elsevier).

Lamb, H.H. 1972(a). British Isles weather types and a register of the daily sequence of weather patterns 1861-1971. Geophysical Memoirs No. 116. London (Her Majesty's Stationery Office for Meteorological Office).

Lamb, H.H. 1972(b). Climate: present, past and future, vol. 1 (Fundamentals and Climate Now). London (Methuen) 613 pp.

Lamb, H.H. 1977. Climate: present, past and future, vol. 2 (Climate History and the Future). London (Methuen) 835 pp.

Lamb, H.H. 1979. Climatic variation and changes in the wind and ocean circulation: the Little Ice Age in the northeast Atlantic. Quaternary Research, 11, 1-20.

Lamb, H.H. 1981. Climatic fluctuations in historical times and their connexion with transgressions of the sea, storm floods and other coastal changes. In: Transgressies en occupatie-geschiedenis inden Kustgebieden van Nederland en België.

pp. 251-290, (edited by A. Verhulst and M.K.E. Gottschalk). Ghent, Belgium (Rijksuniversiteit).

Lamb, H.H. and Weiss, I. 1979. On recent changes of the wind and wave regime of the North Sea and the outlook. Fachliche Mitteilungen, Nr. 194. Traben-Trarbach (Amt für Wehrgeophysik - Geophys. BDBw -FM 194, ISSN 0343-6025). 108 pp.

Lindgren, S. and Neumann, J. 1981. The cold and wet year 1695: a contemporary German account. Climatic Change, 3, (2), 173-187.

Maksimov, I.V. 1971. Causes of the rise of sea-level in the present century. Okeanologiya, II, 530-541. Moscow (Akad. Nauk. Okean. Kom.) In Russian.

Manley, G. 1974. Central England temperatures: monthly means 1659 to 1973. Quart. J. Roy. Meteorol. Soc. 100, 389-405.

Matthews, J.A. 1977. Glacier and climatic fluctuations inferred from tree-growth variations over the last 250 years, central southern Norway. Boreas, 6, 13-24.

Mörner, N.-A. 1980. The northwest European "sea-level laboratory and regional Holocene eustasy. Palaeogeogr., Palaeoclim., Palaeoecol., 29, 281-300.

Morrison, A. 1967-8. Harris estate papers. Transactions of the Gaelic Society of Inverness, 45, 57-8.

Nielsen, E. Steeman. 1938. De danske farvandes hydrografi i Litorinatiden. In: Meddelelser fra Dansk Geologisk Forening, vol. 9, Hefte 3, pp. 337-350. Copenhagen.

Palmen, E. 1928. Zur Frage der Fortpflanzungsgeschwindigkeit der Zyklonen. Meteorologische Zeitschrift, 45, 96-9.

Pettersson, O. 1914(a). Climatic variations in historical and prehistoric time. Svenska Hydrografisk-Biologiska Kommissioneens Skrifter, Häft V.

Pettersson, O. 1914(b). Etude sur les mouvements internes dans la mer et dans l'air. Svenska Hydrografisk-Biologiska Kommissionens Skrifter, Häft VI.

Rasmussen, E. 1958. Past and present distribution of Tapes (Venerupis) Pullastra (Montagu) in Danish waters. Oikos, 9 (1), 77-93.

Rohde, H. and Petersen, M. 1977. Sturmflut. Neumünster (Karl Wachholz Verlag). 148 pp.

Schweizerische Meteorologische Anstalt 1957. Der Temperaturverlauf in der Schweiz seit dem Beginn der Meteorologischen Beobachtungen. Annalen der Schweiz. Met. Anstalt, Nr. 11. Zürich.

THE POTENTIAL OF DOCUMENTARY DATA FOR THE RECONSTRUCTION OF PAST CLIMATES. EARLY 16TH TO 19TH CENTURY. SWITZERLAND AS A CASE STUDY.

Christian Pfister

Historical Institute and Geographical Institute,
University of Berne

The methodology

The paper describes the methods adopted to reconstruct past weather and climate from historical sources (Pfister 1984). The creation of the national weather service in Switzerland (1864) serves as the dividing line between the periods of "historical" and "modern" data.
The collected body of historical evidence was first classified into natural and human data, in accordance with its origin and the place, where it is found. The human data can be further classified into documentary and physical sources, depending on its form and based on its content they can be separated into explicit weather data and proxy-data. The former compromise quantitative or verbal descriptions of weather observations and instrumental measurements; the latter describe such phenomena as snowfall, snowcover, the freezing of lakes, and the blossoming and ripening of plants, which are governed by meteorological factors. The relationship of human data to specific weather elements, as in the case of natural proxy-data, is determined by calibrations.

The human evidence which documents the study has originated principally from chronicles, from personal papers (diaries, calendars) containing intermittent meteorological entries, from non-instrumental weather diaries, and from weather-logs that contain both instrumental measurements and visual observations. In the reconstruction of the climate from 1525 to 1863, more than 118 manuscripts and 150 printed sources with weather

N.-A. Mörner and W. Karlén (eds.), Climatic Changes on a Yearly to Millennial Basis, 331–337.

evidence were found in the Swiss archives. For the period prior to 1750, all the available meteorological evidence was utilized; for the subsequent period, the evidence was used more selectively. More than 30'000 records were derived from the entire body of evidence.

Gaps in the records are virtually nonexistent for the summer months; the missing data are concentrated in the months of October and November.

The historian's standardized methodology for the evaluation and verification of sources was applied to the entire body of evidence. As is commonly known, the majority of chroniclers compiled their accounts in part from older accounts that often contained errors in dates, names, etc. Accordingly, the secondary source material found in chronicles was either excluded or clearly designated as such.

The thermometric evidence dates from 1755 and is based on the Basel series. The Basel time series was also used to calibrate the proxy-data. The measurement of rainfall dates from 1708; brief, scattered, and incomplete series are available up to 1778. The Geneva precipitation series begins in that year, and continues without interruption to the present time.

Those notations in the descriptive sources that permit estimates of temperature, such as the freezing of lakes and the duration of snowcover, have been extracted and set apart. The same procedure was followed for the reports of spring, summer and autumn phenophases, and for the data that allow estimates of precipitation, such as flooding and low water marks. The number of days containing references to rain and snow were computed from the weather diaries.

Sporadic reports on snowcover are available for anomalous winters from the 16th century. From the end of the 17th century, regular descriptions of snowcover can be found in a number of weather diaries.
The sequence of severe winters can be documented from reports of the freezing of Swiss lakes. It has been established that lakes in the alpine borderland freeze in a specific rank order, according to their surface, depth, and individual characteristics.

Evidence that bears on floods and low water levels can be found in the written sources and also as marks on buildings.
A variety of proxy-data were utilized as a substitute for thermometrical measurements during the months of vegetative growth, notably observations of the blossoming, ripening, and

growth, notably observations of the blossoming, ripening, and harvesting of crops, the indications of grapevine yields, and the reports on the quality (i.e. sugar content) of wines.

On balance, the temperature values of the winter months could be documented on the basis of snow and ice features, while the spring-summer temperatures could be documented on the basis of plant indicators. Estimates of temperature for October and November were difficult to calculate, however, because reports are not available for many years, either on vegetation or snowcover. Estimates of precipitation are based on the number of days with rain and snow noted in weather diaries, and also on the evidence of floods and low water, in the absence of more precise measurements.

The collective body of data has been analyzed by computer programs. The descriptive evidence (without the "serial" data) was first standardized by the use of a numerical code, which contains more than 400 positions. The evidence was then sorted in accordance with the chronology and by type of data, and subsequently retranslated into a readable documentation (Fig. 1).
The synchronous comparison of all types of evidence that are available for a designated time interval permits mutual control of the data, in particular of descriptive and proxy-data. This procedure also makes it possible to refine the quality and the time resolution of the interpretation of the data.

Temperature estimates based on proxy-data cannot be made with sufficient precision for periods of less than two or three months. Although descriptive evidence has a high time resolution and yields a detailed account of weather patterns, it cannot be quantified. But the two types of data supplement each other, so that the temperature and precipitation characteristics of individual months could be determined from the evidence. The methodological approach utilizes a weighted index. The value of 0 was adopted for "normal" weather conditions and for the months for which the evidence is missing; the values of +3 and -3 were adopted for "extreme months", and the values of +1, -1, +2 and -2 for the less marked gradations.

In a final procedure, the decennial means of the indices for the instrumental period were compared with the measured decennial means in regression equations. Temperatures and precipitation levels could be estimated for the preindustrial period from these equations.

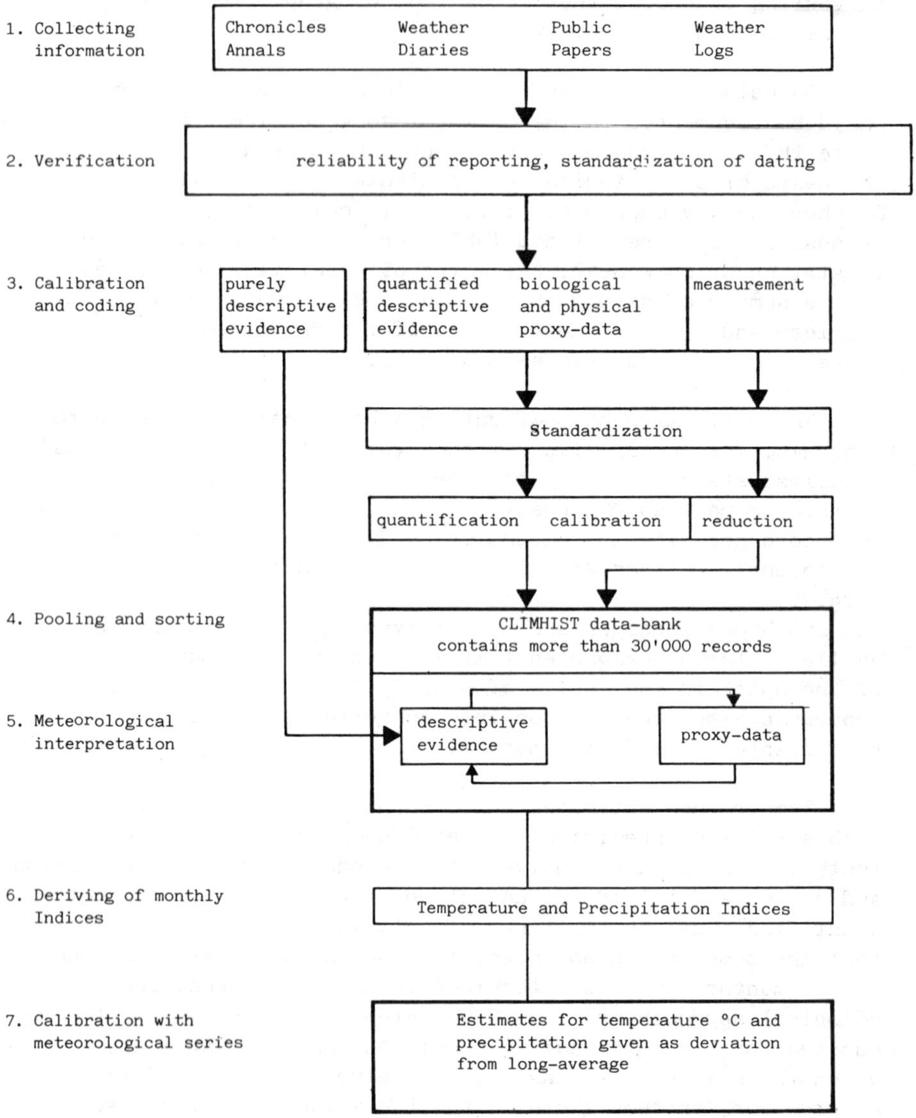

Fig. 1 Methodology for deriving meteorological data from documentary evidence expressed as departures from means 1901-1960. Temperatures: Basel, Precipitation: average of 6 stations, Standard error within the period of measurement: Temperature: 0.12 to 0.25°, Precipitation: 4,2 to 9,5%, The standard errors prior to the period of measurement depend from the quality of the evidence.

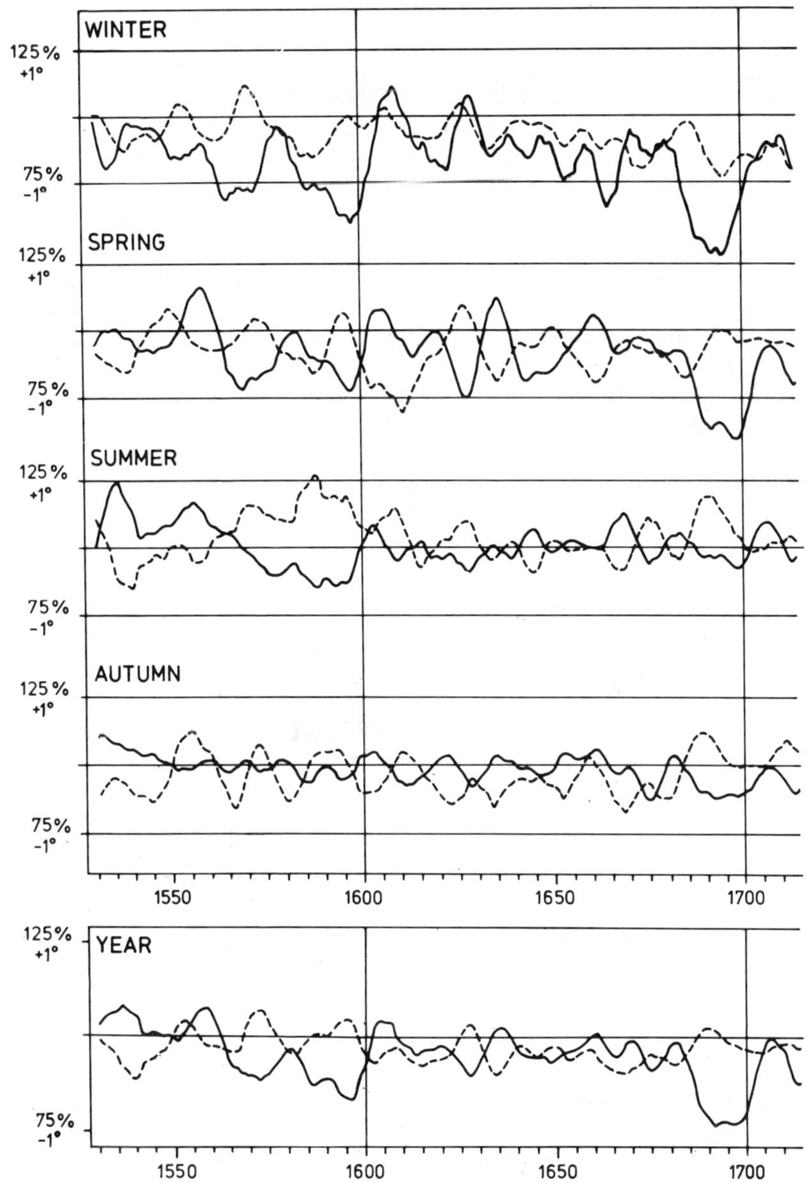

Fig. 2. Estimates of temperature (full line) and precipitation (dashed line) in the Swiss Plateau 1525 to 1900 expressed as departures from the 1901-1960 means. Temperature refers to Basel and precipitation to the average of six stations.

The Structure of the "Little Ice Age"

Although the impression of diversity predominates when the temperature and precipitation patterns of the four seasons are compared (Fig. 2), one common feature of the "Little Ice Age" climate stands out in contrast to the 20th century: winters were colder and drier (o.7° and 5% on average). The same holds for springs that were 0.3° colder and 5% drier. The long term warming of spring began in 1885 and has ended after 1950. The striking increase of temperature and wetness in winter from 1895 however, has continued in the last two decades. Compared to the average of the last 450 years, winters between 1965 and 1979 were 1,3° warmer and 25% wetter. The climatic trends in summer coincide roughly with glacier fluctuations in the alps (ZUMBUEHL, 1980).

Fig. 3 Temperature, precipitation and glacier fluctuation 1530–1630

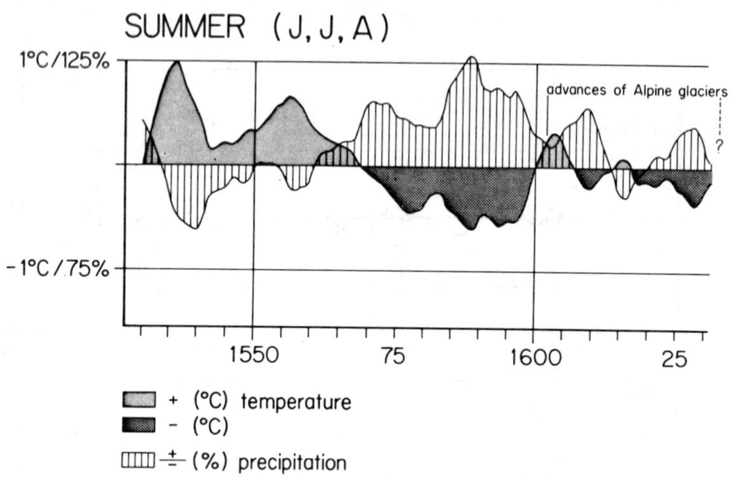

Between 1530 and 1564 summers were 0,4 to 0,6° warmer, from 1565 to 1601 they were 0,3° cooler and more than 10% wetter than in the period 1901–1960 (Fig. 2). The summers of the 17th century do not differ much from those of the present time. Those in the 18th century were about 5% wetter, those from 1800 to 1860 were about 0,3° cooler and 5% drier. The far reaching glacier advances around 1600 (Pfister, Zumbühl, in preparation) seem to coincide more with a series of very wet summers (Fig. 3), while those in the early 19th century are more connected to low temperatures (Fig. 4).Temperatures in autmun fluctuated around the level of the present century up to 1686. From 1686 to 1920

Fig. 4 Temperature, precipitation and glacier fluctuation 1775-1875

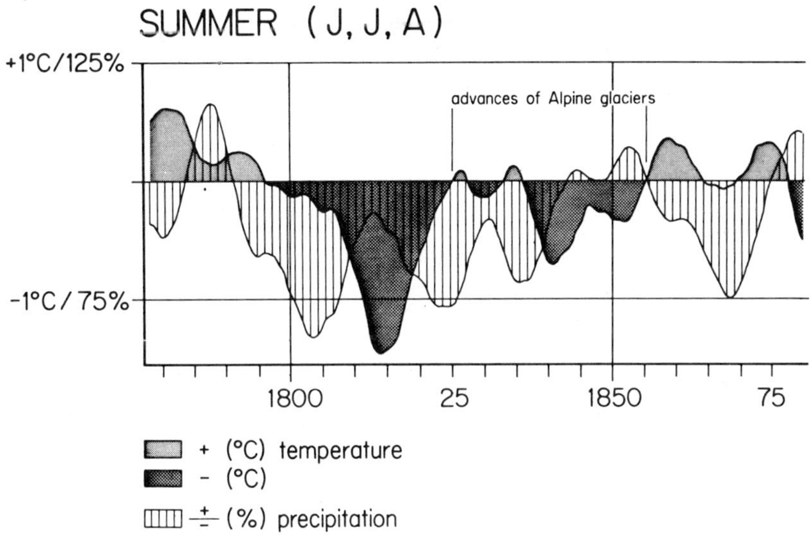

they were 0,7° to 0,8° cooler. From the yearly curves the climatic natures of the five centuries may be summarized: 16th century: cooling, 17th century: dry, 18th century: cool, 19th century: cold and dry, 20th century: warming.

References:

Pfister, Ch., 1984 Das Klima der Schweiz von 1525-1863 und seine Bedeutung in der Geschichte von Bevölkerung und Landwirtschaft, 2 Vol., Haupt.

Zumbühl, H., 1980 Die Schwankungen der Grindelwaldgletscher in den historischen Bild- und Schriftquellen, 111 p., Birkhäuser.

A CONTRIBUTION TO THE RECONSTRUCTION OF WEATHER IN CENTRAL EUROPE AT THE END OF THE 16TH CENTURY /with regard to meteorological observations of Tycho de Brahe on Hven/

Jan Munzar

Geographical Institute ČSAV, Brno, Czechoslovakia

General investigation concerning the sources of meteorological data from the territory of present-day Czechoslovakia for the years 1582 - 1597 /i.e. for the years when Tycho de Brahe made his observation on the isle of Hven/ has been carried out. Most valuable seem to be the regular observations of weather, contained in the diaries of Karel Starší of Žerotín for the years 1588 - 1589 and 1591 /in Latin and Czech/. We present both the examples of daily observations of weather and the monthly summaries of weather character. The reliability of the records has been verified, the terminology used is discussed.

Among the most important sources for the study of weather and climate in Europe at the end of the 16th century belong undoubtedly the meteorological observations made by Tycho de Brahe on the island of Hven in the years 1582 - 1597. These served as the basis for a reconstruction of the weather and of synoptical situations in Western Europe from May to October 1588, in connection with the decisive naval battle fought between the Spanish and the English fleets /Douglas et al 1978/.

From the same time originate the observations made by David Fabricius in north-western Germany in the years 1588 - 1613 and by Leonhard III Treutwein in Bavaria in the years 1587 - 1593 /Lenke 1968/. Less known are the daily records of weather made in Basle by the Czech physician Matyáš Borbonius of Borbenheim in the years 1596 - 1597 /Pejml and Munzar 1968/.

N.-A. Mörner and W. Karlén (eds.), Climatic Changes on a Yearly to Millennial Basis, 339–342.

As a supplement to Brahe's observations from the period 1582 - 1597, the author has made a preliminary investigation of the meteorological information available from the latter period for the area covered by present day Czechoslovakia. /It is a well-known fact that shortly after completing his observations at Uranienborg on the island Hven, Tycho de Brahe came to Prague. It is also in Bohemia that his most fruitful life was ended in 1600. It is nearly certain that Brahe continued his meteorological observations in Bohemia, too, but no results of this have come down to us./

The most valuable observations preserved from this epoch are obviously the meteorological records of Karel Starší /Senior/ of Žerotín /1564 - 1636/, the well-known representative of the Moravian nobility. Meteorological observations recorded by him daily comprise the period from January 1, 1588 to April 15, 1589 /of Gregorian calendar/. They are continued by irregularly made records from the year 1591, which are being deciphered at present. Data from Žerotín's diaries are in Latin /1588, 1591/, and from 1589 are in Czech. This is probably the first long-term daily observations of weather carried out on the territory of Bohemia and Moravia.

Unfortunately, Žerotín's observations were not made at one place, since the author travelled. In the first period mentioned above, which covers 471 days, the observations mostly originate from Náměšt nad Oslavou in Moravia /187 days, but also are from Prague /63 days/, various places in Germany in the summer of 1588 /88 days/, and from Austria /10 days/. The author of this paper hopes to be able to successively publish these records in extenso; this brief report is just an opportunity to present excerpts.

The passage dealing with daily records made at Náměšt nad Oslavou /Lat. = 49°10'N, Long. = 16°10'E, H = 390 m/:

January 1, 1588: "Tenuem pluviam nocturnam coelum mane serenissimum insecutum est asperioris venti flatibus nubes coactas discutientibus." /After a slight night rain the sky was quite clear in the morning when the blowing of a hard wind was dispersing the accumulated clouds./

At the end of the individual monthly periods Žerotín tries to evaluate the course of the weather, which is interesting for us. January 1588 was assessed as follows: "Totius huius mensis constitutio exceptis paucis aliquot diebus fuit valde amoena et serena, non alia fere, ac verno aut autumnali tempore solet. Audio in hac provincia a multis annis talem tempestatem tamque moderatam non fuisse." /The impression gained from this month, with the exception of a few days, was that it was most pleasant and clear, thus being similar to that of usual spring or autumn periods. I hear this kind of weather, mild like this, has not been witnessed in this country for quite a few years./

The vocabulary used in the weather records is copious. Žerotín mostly pays attention to the presence of clouds and the character of weather. The latter is often evaluated using such words as "valde mitis" /very nice/, "tristis ac inamoenus" /sad and unpleasant/ etc. When observing precipitation he notes, besides rainfall, rain with snow, snow, hailstorms, snowstorms and hoarfrost. He also notes thaws, icy ground and fogs. His frosts are labeled as intense, very intense, average, stronger than average one, piercing, sharp and piercing and very heavy.

Karel of Žerotín evaluated the wind only from the point of view of its force, its direction being noticed only in exceptional cases. He uses such terms as a strong, harsh, sharp, dry, cold wind etc. The wind direction is recorded only on March 2, 1588 /northerly wind/ and on December 11 and 12, 1588 /southerly wind/. In his evaluation of the month of April 1588 the north and south wind directions only disclose a prevailing meridional circulation.

As far as the time verification /the calendar/ is concerned, the use of the Gregorian one is confirmed by the notice about an eclipse of the moon on the night of March 13, 1588, or by the identification of Easter Sunday with the 17th of April 1588.

As far as verification of the facts is concerned, an example can be given. In the history of Náchod, a town in eastern Bohemia, there is mention of a cloud-burst and a flood in the region of near-by town, Police, in May 1588; the inundation washed away two bridges. Žerotín, who was by coincidence in eastern Bohemia at that time, has left the following note: May 23, 1588: "Aer toto fere die serenissimus, quamvis sub vesperam ab ingenti calore nubes coactae in impetuosissimum et densissimum imbrem resolutae sint." /The weather was clear for nearly the whole day, though there were some clouds in the evening caused by the extreme heat, which broke in a very violent and dense rainfall./

The author of this observation arrived at Hradec Králové at sunset "in ipsa decidentis imbris furia" /amidst a persistent fury of the falling rain/. The correspondence is very good, the observation testifying probably to a local heat thunderstorm.

These passages have perhaps given the reader a sufficient idea of the valuable material that has been brought to light. At present

the verification and generalization of these observations are being continued. Attempts are being made to widen the list with materials coming down from the whole pre-instrument period, i.e. approximately from the 15th century, the aim being to successively evaluate climatic fluctuations in historic times in the territory of the present Czechoslovakia.
The materials presented here could serve to reconstruct the weather and climate for all of Europe, too.

References

Diarium Caroli Liberi Baronis a Zierotín, ex autographo fol. Cerroni II, Nr. 114. State Archives Brno.

Douglas, K.S., Lamb, H.H., Loader, C., 1978: A meteorological study of July to October 1588: the Spanish Armada storms. Res. Publ. /CRU RP 6/. Norwich: University of East Anglia, Climatic Research Unit, 87 pp.

Dreyer, I.L.E., Raener, I., 1927: Tychonis Brahe Dani Opera Omnia, Vol. 9. Hauniae, p. 1 - 146.

Lenke, W., 1968: Das Klima Ende des 16. und Anfang des 17. Jahrhunderts nach Beobachtungen von Tycho de Brahe auf Hven, Leonhard III Treutwein in Fürstenfeld und David Fabricius in Ostfriesland. Berichte des Deutschen Wetterdienstes Bd. 15, Nr. 110, Offenbach a.M., 49 pp.

Pejml, K., Munzar, J., 1968: Das Wetter in Basel in den Jahren 1596 - 1597 nach dem Tagebuch des Mathias Borbonius von Borbenheim. Vierteljahrsschrift der Naturforschenden Gesellschaft in Zürich, Jg. 113, No. 4, p. 407 - - 416.

THE IMPACT OF CLIMATE ON GRASS GROWTH AND HAY YIELD IN ICELAND: A.D. 1601 TO 1780

A.E.J. Ogilvie

Climatic Research Unit,
University of East Anglia, U.K.

ABSTRACT

The people of Iceland have always relied on livestock, especially sheep, for food and other essential products. Animal husbandry was of particular importance in earlier times, when other resources, such as fish, were less readily available than they are today. The annual hay crop, needed to provide extra winter fodder, was vital to this type of farming. In this paper, relationships between grass growth and hay yield, and the climate variables of temperature and precipitation are examined for the period 1601 to 1780 by means of simple statistical tests. The results show a strong link between climate and grass growth and hay yield. Considerable regional variability is noted, however, with the northern part of Iceland clearly more sensitive to climatic fluctuations than the southern part.

INTRODUCTION

Several of the most recent studies of the impact of climate on society have stressed the difficulties of determining the exact nature of the influence of climate on human activities (de Vries, 1980; Anderson, 1981; Ingram et al., 1981; Ogilvie, 1981). The major obstacle is how to assess and isolate possible climatic impact from an infinite variety of other possible factors. Before undertaking a climate-impact study it is therefore necessary to consider how to provide the most accurate picture of what really occurred. This requires a carefully considered approach, and a rigorous methodology. Central issues are: location, the amount and quality of the data available, and the timescales of impact.

N.-A. Mörner and W. Karlén (eds.), Climatic Changes on a Yearly to Millennial Basis, 343–352.

Iceland in the seventeenth and eighteenth centuries is particularly appropriate for a study of past climate impact. There is a wealth of reliable data available, both on climate, and on socio-economic matters. Furthermore, Iceland is a prime example of a society living in conditions marginal for agriculture and habitation; conditions which are generally thought to be those in which climate-society interactions should be easiest to delineate (Parry, 1981). Here, one specific area of climate impact is considered: the impact of climate on grass growth and hay yield.

DATA

Prior to the nineteenth century, most data on climate and climate impact in Iceland are of a qualitative nature; documentary accounts of weather events and social phenomena. Seventeenth- and eighteenth-century sources include annals, private and official correspondence, travel books, investigations undertaken by the Danish government, diaries, land registers and censuses. The major sources used for this study are the later Icelandic annals (Annálar 1400-1800) and unpublished reports written by eighteenth-century Icelandic officials (Bréf til Stiftamtmanns) available in the National Archives, Reykjavík.

The problems associated with the use of documentary sources, and the importance of analysing all such sources with regard to reliability have been considered elsewhere (Vilmundarson, 1972; Bell and Ogilvie, 1978; Ogilvie, 1981). However, it is appropriate to point out some of the advantages of such sources. Unlike other types of indirect data, such as glacial- and pollen-cores and tree-rings, documentary evidence can provide very precise and detailed information on climatic and other events. Furthermore, no elaborate statistical techniques are required in order to extract this information. It should also be noted that, because the sources are abundant, it is possible to ensure accuracy by cross-checking accounts.

The first step in analysis of the data used here was to file information from different sources for every year from 1601 to 1780. After comparing the accounts, the information was then categorized and summarized. For example, seasons were described as "good" (mild), "severe" (cold) etc., and the grass growth and harvests as "good", "poor" and "average". The harvest descriptors are related to yield, but the relationship is a complex one partly determined by the success and/or ease of harvesting. This point is discussed further below. Examples of the type and availability of data are shown in Table 1 below where twenty years of summarized data from the north of Iceland are given. Separate data files were drawn up for the south, east and west.

YEAR	WINTER	SPRING	SUMMER	AUTUMN	SEA ICE				GRASS-GROWTH	HARVEST
					W	SP	SM	A		
1721	Stormy then good	Good	Variable, wet	Variable then severe					Good	Wet but reasonable
1722	Severe	Severe	Good	Good then severe					Quite good	Good
1723	Average then good	Very good	Very good	Good to variable					Good	Good
1724	Severe	Good	Good, some rain	Good, calm					Good	Good
1725	Some snow and storms	Severe April to June	Good						Poor	Better than expected
1726	Very good	Very good	Dry and good	Variable		X			Poor	Poor
1727	Very good	Mainly good	Dry then very wet	Variable to bad	X				Poor	Poor
1728	Severe	Severe	Dry, good then wet	Variable to good		X	X		Poor	Poor
1729	Quite good	Variable then severe	Good	Good then cold	X	X	X		Poor	Mainly poor
1730	Severe	Mainly severe	Cold and wet	Wet then good					Poor	Ruined
1731	Good		Good						Very good	Good
1732	Good but stormy		Cold, snow, rain	Good		X	X			Spoiled by rain
1733	Very good	Good	Good	Variable			X		Average - good	Spoiled by rain
1734	Sleet, frost, snow	Severe	Good then wet	Good and calm					Poor	Spoiled by rain
1735	Harsh, snows, storms	Hard, storms	Variable, wet	Variable then good					Poor	Average to very bad
1736	Good		Good	Good then stormy					Quite good	Good
1737	Quite severe	Severe	Cold	Good, later storms					Poor	Reasonable
1738	Mainly severe	Severe	Cold	Good					Unusually poor	Very poor
1739	Severe	Good	Quite good	Very good					Good	Poor
1740	Good	Good	Frost in July						Poor	

Table 1. Seasons and Harvests in Northern Iceland 1721-1740.
"W" = winter, "SP" = spring, "SM" = summer, "A" = autumn.
"X" denotes presence of sea ice off the coast.

THE IMPACT OF CLIMATE ON GRASS GROWTH AND HAY YIELD

In the seventeenth and eighteenth centuries, Iceland's economy was based on animal husbandry with sheep and cattle the most important animals. The annual hay crop, needed to feed the livestock over the winters when there was little or no vegetation available, was vitally important. A failure of the hay crop could lead to losses of livestock and to further consequences for the human population, and an analysis of the contemporary relationships between crop yield and climate is an essential first step in establishing less direct climate-society links.

The present-day relationship between climate and grass growth and hay yield, which has a firm physiological basis related to the shortness of the growing season, has been well established (see, for example, Bergthórsson, 1976; Fridriksson, 1969, 1973). It is still of interest, however, to test this relationship in the past when different agronomic and farm management strategies prevailed, and when climatic conditions were different. Distinct advantages in doing this are that the data series may be longer (in this case 180 years) and relationships are not complicated by modern technological innovations such as the use of nitrogen fertilizers.

Although the seventeenth- and eighteenth-century data analysed for this study are, as stated above, mostly of a qualitative nature with information sometimes incomplete, it is still possible to use simple statistical techniques to test the reality of climate impact in the past. For example, relationships between temperature, precipitation, grass growth and the hay harvest can be tested by means of contingency tables. Thus, in order to test, for example, whether grass growth is significantly related to spring temperature, the data can be grouped into cold, mild and average temperature classes for the weather, and poor, average and good for the grass growth. A similar procedure can be used to examine other relationships and a number of variables were chosen for analysis: winter, spring and summer temperature and grass growth; winter, spring and summer temperature and harvests; summer rainfall and grass growth; summer rainfall and harvests.

An example of such an analysis is shown below (Table 2). The table gives the observed number of occurrences in each joint category and the expected number of occurrences (in brackets) obtained by multiplying the observed column and row frequencies together. The expected numbers so calculated are those which would occur if there were <u>no</u> relationship between the row and column variables (in this case grass growth and spring temperature). Since the table is 3 x 3, the number of degrees of freedom for a chi-squared (hereafter chi^2) test is 4 ((3-1) x (3-1)). For 4 degrees of freedom a chi^2 value in excess of 9.49 is expected to occur by chance only 1 time in 20 (i.e. a value above 9.49 is "statistically significant at the 0.05, or 5 per cent level"). A chi^2 value above 13.3 would occur by chance only 1 time in <u>100</u> and is significant at the 1 per cent level.

Table 2. Spring Temperature and Grass Growth. All Regions 1601-1780. chi^2 = 100.6.

Grass growth \ Spring temperature	Cold	Average	Mild	Totals
Poor	95(59.3)	30(53.4)	17(29.3)	142(30.9%)
Average	76(93.9)	116(84.6)	33(46.5)	225(48.9%)
Good	21(38.8)	27(35.0)	45(19.2)	93(20.2%)
Totals	192(41.7%)	173(37.6%)	95(20.7%)	460(100%)

In the table presented here a significant difference between observed and expected frequencies, based on the value of the test statistic

$$chi^2 = \sum_{i=1}^{9} \frac{(O_i - E_i)^2}{E_i}$$

(where the sum, i=1 to 9, is over the number of boxes in the contingency table), indicates a relationship between variables. The value of the test statistic is tested for significance using a chi^2-distribution. In this example the value is 100.6, highly significant and indicating a strong relationship. From the table it can be seen that the observed incidence of poor grass growth following cold springs is much larger than would be expected if cold springs had no influence.

Although the data collected here for the period 1601 to 1780 were sufficient to allow statistical tests to be performed, data were not always available for every year. The period 1601 to 1680, for example, is not as well documented as 1681 to 1780. Also, certain regions (in particular the east and, to a lesser extent, the south of Iceland) have large gaps in information. Because of this some years either had no data, or had data pertaining to only one of the variables under consideration. Obviously, if there were data for both variables, they were included under the relevant headings. Where information for one of the variables was lacking, the missing value was assumed to be "average". Because of the nature of the sources used, and because <u>some</u> information is available, the assumption of "average" when a specific description is absent is reasonable. Note, however, that this is a "conservative" assumption, in that it would tend to lower the value of the chi^2 test statistic. Furthermore, a number of cases were examined with and without this assumption, and the qualitative conclusions were unchanged.

The table shown here considers data from all regions together. Of course, climatic conditions differ noticeably from region to region, and there may well be regional differences in the precise relationships between climate and hay yield or grass growth (see, for example, Gunnarsson, 1983). However, these relationships are sufficiently strong that the link remains clear even when data from different regions are combined. In order to obtain further insight into the differences between regions (and an indication of the regionality of vulnerability to climatic stress) tests were also made using data from the north, west and south of Iceland separately. Insufficient data from the eastern part of Iceland prevented any analysis of this region alone.

Because there were insufficient data to analyse the period 1601 to 1680 separately, these were combined with the period 1681 to 1780. When comparing results for 1681 to 1780 with 1601 to 1780, we expect the longer period to give slightly higher values of chi^2, simply because the longer period has more data overall. However, if the longer period shows a lower chi^2 value than the shorter one, it can be inferred that the relationship in 1601 to 1680 was weaker than in 1681 to 1780, even though we cannot say anything quantitative about the strength of the relationship in the earlier period.

The results of these contingency table analyses are summarized in Tables 3 to 6 below. A comparison of the results for 1601 to 1780 and 1681 to 1780 shows that they are, on the whole, similar. This suggests that the relationships between climate and grass growth and harvests (i.e. final yield) must have remained much the same over the whole period 1601 to 1780. An exception is the effect of winter temperature on grass growth and harvests, particularly in the south. Here, the relationship between winter temperature and grass growth is not statistically significant and is noticeably weaker in the longer period, 1601 to 1780. Such a conclusion is, however, not warranted. The fact that there are fewer data from the earlier period may make the relationships simply appear weaker: the assumption that conditions were average when no information was available was made more often prior to 1680 and this, in itself, may account for the changed significance levels. Winter temperature is also significantly related to harvests in all regions during 1681 to 1780 (at the 5 per cent level) but the links are not statistically significant over the longer period 1601 to 1780. However, in view of the above argument, and the fact that the relationship remains strong when all regions are grouped together, it cannot be concluded that there was any significant change in the winter temperature-harvest link.

When considered overall, these statistical tests show strong crop-climate relationships for almost all of the variable combinations considered. Winter temperature is significantly related to grass growth in the north and west, but not in the south. The final harvest is, however, related to winter

Table 3. Seasonal Temperature and Grass Growth: Comparative chi^2 results.

Time	1601 - 1780				1681 - 1780			
Region	All	North	West	South	All	North	West	South
Winter	43.0**	19.1**	11.1*	3.3	36.0**	12.4*	11.3*	7.3
Spring	100.6**	43.3**	32.2**	13.9**	101.7**	16.6**	15.9**	9.0
Summer	84.0**	44.9**	23.9**	36.1**	70.7**	22.5**	16.1**	20.2**

Table 4. Seasonal Temperature and Harvests: Comparative chi^2 results.

Time	1601 - 1780				1681 - 1780			
Region	All	North	West	South	All	North	West	South
Winter	33.6**	8.1	8.1	5.8	27.6**	12.7*	11.2*	11.5*
Spring	48.7**	37.5**	27.7**	7.4	53.3**	15.3**	16.5**	5.1
Summer	48.0**	28.2**	19.2**	29.5**	49.7**	21.3**	18.8**	16.3**

Table 5. Summer Rainfall and Grass Growth: Comparative chi^2 results.

Time	1601 - 1780				1681 - 1780			
Region	All	North	West	South	All	North	West	South
Summer	47.8**	23.9**	14.7**	12.3*	38.8**	14.6**	22.1**	15.2**

Table 6. Summer Rainfall and Harvests: Comparative chi^2 results.

Time	1601 - 1780				1681 - 1780			
Region	All	North	West	South	All	North	West	South
Summer	108.9**	47.9**	47.9**	16.5**	83.3**	34.5**	42.1**	14.7**

** = significant at the 1% level
 * = significant at the 5% level

temperature in all regions including the south. When data from the whole of Iceland are combined, then both the winter temperature/grass growth and winter temperature/harvests are highly significant. Spring and summer temperatures are both strongly related to grass growth and harvest (and both are more important than winter conditions). However, there is some regional variation. The relationships with summer temperature and spring temperature are similar in the north and west; but in the south, summer temperature appears to be more important than spring temperature. In fact, the spring temperature relationship fails to reach a statistically significant level for harvests and grass growth in the south. In summary, temperatures in nearly all seasons and districts are significantly related to harvest and grass growth, with lower temperatures being detrimental.

Rainfall was considered only for the summer. This was because of lack of adequate rainfall data for other seasons and because it was felt that during winter and spring the effects of temperature would dominate rainfall effects. The results of statistical tests are shown in Tables 5 and 6. Summer rainfall was found to be statistically significantly related to both grass and harvest quality in all regions. For grass growth, the relationship is strongest in the south. For harvests, the relationship is stronger in the north and west.

Table 7. Summer Rainfall and Grass Growth. All Regions 1681-1780. $chi^2 = 38.8$.

Grass growth \ Summer rainfall	Wet	Average	Dry	Totals
Poor	40(26.5)	29(52.0)	27(17.5)	96(33.1%)
Average	30(36.1)	79(70.9)	22(23.9)	131(45.2%)
Good	10(17.4)	49(34.1)	4(11.5)	63(21.7%)
Totals	80(27.6%)	157(54.1%)	53(18.3%)	290(100%)

Both high or low rainfall tended to produce poor growth of grass (see Table 7). However, the relationship of rainfall to harvests appeared to be rather different; high rainfall was associated with a poor harvest but little rainfall with a good harvest (Table 8). This relationship may be a result of the way contemporary observers perceived the grass growth and harvest; wet weather during the harvest often meant that it would be spoiled and a dry harvest may frequently have been referred to as "good", simply because it was dry and the hay easily harvested even though the actual yield may in fact only have been average. However, it also seems likely that warm temperatures

Table 8. Summer Rainfall and Harvest. All Regions 1681-1780. $Chi^2 = 83.8$.

Harvest \ Summer rainfall	Wet	Average	Dry	Totals
Poor	68(34.9)	29(52.7)	10(19.4)	107(38.8%)
Average	21(32.6)	58(49.3)	21(18.1)	100(36.2%)
Good	1(22.5)	49(34.0)	19(12.5)	69(25.0%)
Totals	90(32.6%)	136(49.3%)	50(18.1%)	276(100%)

and dry weather during the harvest period (July-August) may have improved a crop of grass that, for a variety of reasons, had not grown well during the spring and early summer. There can be no doubt, however, of the highly statistically significant relationship between wet weather during the harvest period and a poor crop.

Despite the significant results of grass-climate relationships achieved here, it is not possible to determine the direct influence of rainfall or temperature by this means of analysis because these variations are not independent. Part of the rainfall relationship, for example, may arise because high rainfall is often associated with low temperatures. However, as noted above, rainfall almost certainly had a direct influence when it occurred during the harvesting period.

In all of these analyses, almost invariably the most highly statistically significant results are those from the north. This is a strong verification of the north's more agriculturally marginal climate, particularly with regard to temperature and related variables, like the length of the growing season. Almost all of the chi^2 results obtained here were statistically significant at better than the 1 per cent level. The only results which failed to reach the 5 per cent significance level in either of the two periods considered were from the south: winter temperature/grass growth and spring temperature/harvest. This suggests that the south is noticeably less sensitive to variations in climate, a conclusion borne out by modern studies of present day Icelandic climate. It is interesting, however, that summer temperatures are more important for grass growth in the south than in other regions. This must be largely because, in other regions, the grass crop has already been preconditioned by the climate of the earlier seasons so that the residual influence that summer climate can have is reduced. Finally, it should be noted that the strong relationships observed between winter conditions and grass growth and harvest quality are not necessarily direct (although a cold winter may well have meant lower soil temperatures during the early stages of grass growth). Because cold winters were often associated with the presence of sea ice which tended to persist through to the spring, winter and spring temperatures are not independent. A cold winter was frequently a precursor of a cold spring.

Taken as a whole, the chi^2 results presented here clearly show a strong relationship between crop growth and climate. Although the tests used here are fairly crude, there can be no doubt that severe seasons did mean that both grass growth and the harvest were likely to be affected. A wet harvest period almost invariably led to crop failure. Because of the central position the harvest had in the Icelandic economy it is only to be expected that climate had a considerable impact on society as a whole.

REFERENCES

Anderson, J.L., 1981. History and climate: some economic models. In: Climate and History (ed. T.M.L. Wigley, M.J. Ingram & G. Farmer), pp. 337-365. Cambridge University Press.

Annálar 1400-1800., 1922- . Hinu Íslenzka Bókmentafélagi, Reykjavík.

Bell, W.T. and Ogilvie, A.E.J., 1978. Weather compilations as a source of data for the reconstruction of European climate during the medieval period. Climatic Change, 1, 331-348.

Bergthórsson, P., 1976. Tödufall á Íslandi frá aldamótum. Freyr, 72, 250-263.

Fridriksson, S., 1969. The effects of sea ice on flora, fauna, and agriculture. Jökull, 19, 146-157.

Fridriksson, S., 1973. Crop production in Iceland. Int. J. Biometeor., 17, (4), 359-362.

Gunnarsson, G., 1983. Grasspretta, nýting og heyfengur 1630-1900 samkvæmt sögulegum heimildum. Freyr, 79, 250-255.

Ingram, M.J., Farmer, G. & Wigley, T.M.L., 1981. Past climates and their impact on Man: a review. In: Climate and History (ed. T.M.L. Wigley, M.J. Ingram & G. Farmer), pp. 3-50. Cambridge University Press.

Ogilvie, A.E.J., 1981. Climate and society in Iceland from the medieval period to the late eighteenth century. (Unpublished Ph.D. thesis, University of East Anglia, Norwich, U.K.)

Parry, M.L., 1981. Evaluating the impact of climatic change. In: Consequences of Climatic Change (ed. C. Delano Smith and M. Parry), pp. 3-16. Department of Geography, University of Nottingham, U.K.

Vilmundarson, Th., 1972. Evaluation of historical sources on sea ice near Iceland. In: Proceedings of an International Conference (ed. T. Karlsson), pp. 159-169. National Research Council, Reykjavík.

De Vries, J. 1980. Measuring the impact of climate on history: the search for appropriate methodologies. J. of Interdisciplinary History, 10, 599-630.

CLIMATIC CHANGES DURING RECENT CENTURIES AS INDICATED BY DENDROCHRONOLOGICAL STUDIES, MOUNT BAKER, WASHINGTON, U.S.A.

Olavi Heikkinen

University of Helsinki, Finland

Mount Baker (3284 m) is one of the ice-capped volcanoes in the Cascade Range. Many of the 13 glaciers on the mountain have oscillated two to three kilometres over the past century (e.g. Long, 1953). The mountain has a maritime climate. The precipitation at Mount Baker Lodge (1265 m), approximately 2800 mm a year, falls mainly as snow and sleet. The maximum snow cover of 4.5 to 6.1 m is reached in late March or early April (U.S. Weather Bureau, no date).

The heavy snow cover of the subalpine zone does not fully melt before summer. Abundant snowfall is probably one of the main reasons why the timberline in the Cascades is a mosaic of tree clumps and subalpine meadows (Franklin & Dyrness, 1973). On Mount Baker the timberline is at an altitude of 1400 to 1750 m.

During the last hundred years many subalpine meadows have been invaded by trees in the Pacific Northwest. This dynamic change in the forest-tundra ecotone has been reported from British Columbia (Brink, 1959), Washington (e.g. Franklin et al., 1971), and Oregon (Vale, 1981). The recent forest invasion, glacier fluctuations and tree-growth variations are probably all associated with climatic changes.

1. PURPOSE

The main purpose of this paper is to detect and date climatic changes on Mount Baker since 1500 A.D. This was done by (1) timing fluctuations of Coleman Glacier, (2) interpreting secular variations in tree-growth rate, and (3) determining the age distribution of recently established trees. The study area is depicted in Fig. 1.

N.-A. Mörner and W. Karlén (eds.), Climatic Changes on a Yearly to Millennial Basis, 353–361.

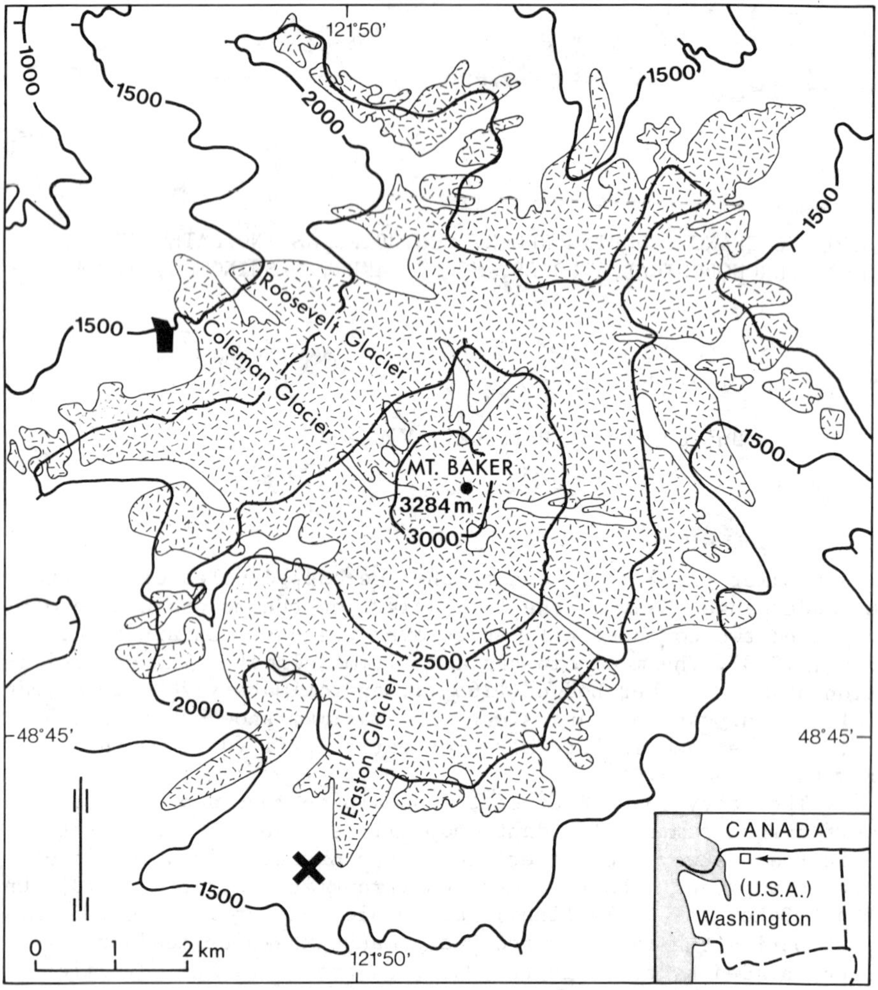

Fig. 1. Location map of Mount Baker. The cross refers to forest groves where cores for tree-ring chronology (Fig. 3) and response-function analysis (Fig. 4) were collected. The blackened area near Coleman Glacier indicates the ridge where recent forest invasion (Fig. 5) was studied. Shaded areas are covered with perennial ice. Contours are expressed in metres.

2. FLUCTUATIONS OF COLEMAN GLACIER

To examine fluctuations of Coleman Glacier eight terminal moraines younger than 500 years were dated. The minimum moraine ages are based on the number of tree rings in the oldest trees on each moraine (Table 1). Four moraines were dated by defining the years when the glacier damaged trees on the distal side of these moraines; the damage is indicated by narrow, unsymmetrical tree rings (cf. Lawrence, 1950). One moraine was dated from a photograph, and one by early glacier observations (Easton, no date) and from my own estimations. The ages of the two oldest moraines were obtained by adding about 30 years to the numbers of tree rings counted in the oldest trees on the moraines. The additional years are an attempt to compensate for the ecesis time and tree rings missed in coring; see Heikkinen (1984).

Table 1. Dates and dating technique for post-1500 moraines fronting Coleman Glacier. The ages in parentheses are derived from the number of tree rings counted in the oldest tree discovered on each moraine. See Fig. 2.

Dates (A.D.)		Object or method of dating
1978-79	(-)	ice-damaged tree
c. 1922	(1950)	observations and estimations
c. 1908-12	(1931)	photograph
c. 1886-87	(1892)	ice-damaged tree
c. 1855-56	(1863)	ice-damaged tree
c. 1823	(1838)	ice-damaged tree
c. 1740	(1771)	oldest tree on moraine, "corrected" age
early 1500	(1565)	oldest tree on moraine, "corrected" age

To make the chronology of the Coleman Glacier variations less discontinuous some known positions of the glacier have been included in the diagram in Fig. 2. The year 1949 was a turning point in the history of Coleman Glacier, when a long-lasting withdrawal was replaced by a fairly rapid advance (Bengtson, 1956; Harrison, 1970). The location of the glacier terminus in summer 1982, the zero point of the distances in Fig. 2, was observed by the author.

3. RELATIONSHIPS BETWEEN TREE GROWTH AND CLIMATE

A set of cores was extracted from ten mountain hemlocks (Tsuga mertensiana) in the subalpine zone (1600 m) near Easton Glacier (Fig. 1). The trees sampled were at the upper part of the timberline since they can be expected to respond sensitively to climatic variations (e.g. Fritts, 1976).

The ring widths measured were standardized to indices by fitting a negative exponential curve or a straight line to each ring-width series in order to remove non-climatic variations from the series (Fritts, 1976; Graybill, 1982). The standardized indices of all the cores were then averaged year by year to obtain the site chronology (Fig. 3). The average indices for the 1500s and 1600s are based on a few cores only; later centuries are replicated markedly better.

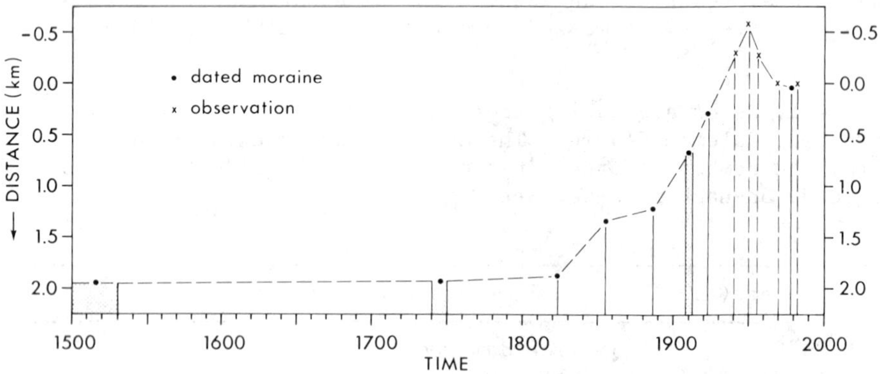

Fig. 2. Chronology of fluctuations of Coleman Glacier based on tree-ring counts and observations. The distances are related to the location of the glacier terminus in summer 1982.

One way to study tree growth-climate relationships is by response-function analysis based on the eigenvector technique, in which eigenvectors, derived from certain climate variables, act as statistical predictors of ring width (Fritts, 1976). The eigenvector analysis is applied here to the period 1931-80 of the site chronology (Fig. 3A). The eigenvectors are associated with mean monthly temperatures and total precipitations of 14 months from June of the preceding year to July during the year of growth. The climatic data are from the climatic region "Cascade Mountains West", within which the study area is situated.

The first 20 eigenvectors (their amplitudes) were included in the stepwise multiple regression analysis whereas the eight least important eigenvectors were discarded as a noise. Tree-ring indices for the five previous years were also included to make it possible to estimate the tree growth dependent on prior growth (cf. Fritts, 1976).

The stepwise regression analysis was finished when no additional predictors exceeded the F ratio = 1. This situation was reached when amplitudes of 12 eigenvectors and the indices for prior growth at lags of 1, 2 and 5 years were entered. The final regression equation, the response function, for ring widths is presented graphically in Fig. 4. The weights or response-

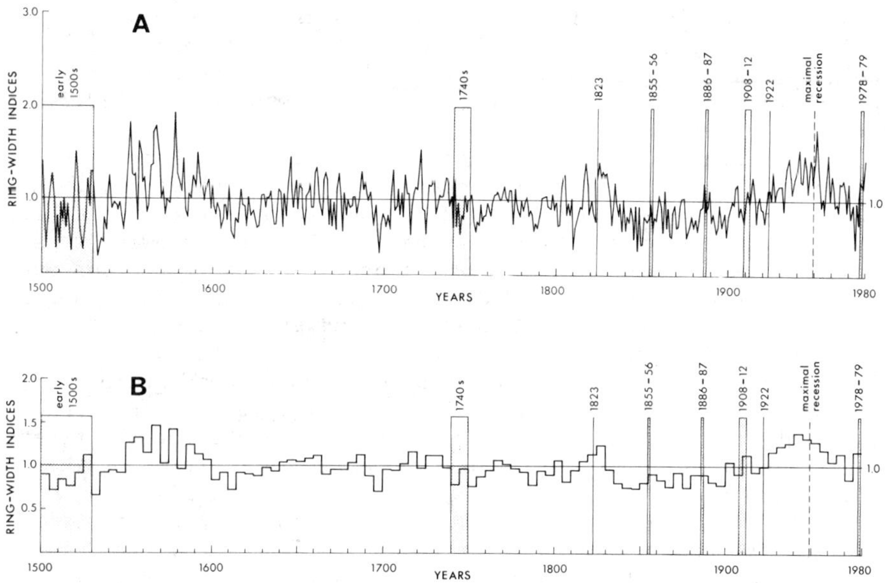

Fig. 3. Site (mean) chronology for mountain hemlock. Ring-width indices are presented as (A) annual values and (B) pentad averages. The bars with ages refer to the Coleman Glacier fluctuations described in Fig. 2.

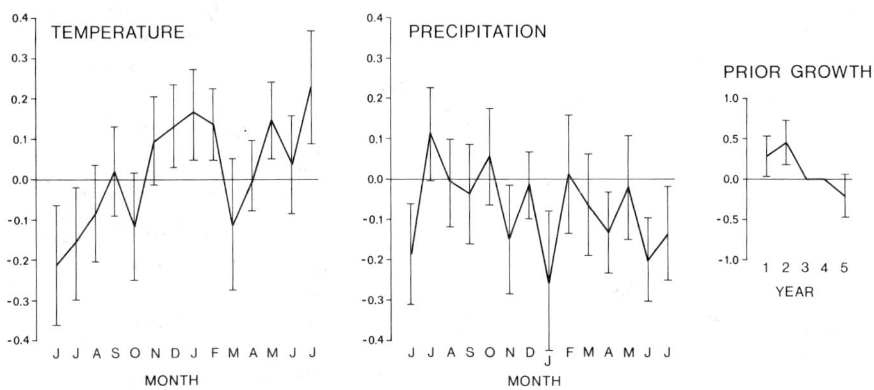

Fig. 4. Response function for ring-width indices of mountain hemlock against regional climatic data ("Cascade Mountain West") for the period 1931-80.

function coefficients for each month are associated with the original climate variables. The vertical lines through each weight show 95 % confidence limits. Climate accounts for the growth variance of 47.3 % whereas prior growth explains 20.7 %, a total explained variance of 68.0 %.

The left graph in Fig. 4 shows that tree growth tends to be enhanced by above-average temperatures. One can conclude, for instance, that warm weather from May on promotes the melting of the heavy snow cover, which in turn lengthens the growing season and increases photosynthesis, whose products add to the current growth (see Garfinkel & Brubaker, 1980). The significant inverse correlation between March temperature and growth could reflect the danger of frost-drought damage at that time of the year. The inverse relationships to the prior summer temperatures may be due primarily to the statistical technique used here, since it is obvious that the pronounced positive weight for the first prior growth consists mainly of the regression portion of the climate in the previous summer (Fritts, 1976, p. 382).

The second plot in Fig. 4 shows that the response function has mostly negative weights for precipitation. Thus, ring widths on maritime Mount Baker have a predominantly inverse relationship to precipitation. For instance, average precipitation during the current spring and summer is excessive for optimal tree growth.

Because tree rings are generally wide when the weather of the current year is warmer and drier than average, and narrow when it is cooler and wetter than average, sections of wide rings should coincide with periods of glacial recessions and sections of narrow rings with glacial advances. It is important, however, to keep in mind that the chronology of Coleman Glacier reacts to climatic events with a lag of one to a few years as can be inferred from Harrison's (1970) observations.

4. RECENT TREE INVASION

One hundred recently established conifers on a subalpine ridge (1400 to 1600 m) near Coleman Glacier (Fig. 1) were dated by tree-ring counts. The trees were distributed equally in ten sample plots with a radius of 5 m. Invasive tree species are as follows: Pacific silver fir (*Abies amabilis*), subalpine fir (*Abies lasiocarpa*), and mountain hemlock. Figure 5 gives the age distribution of the dated trees without reference to tree species. The oldest seedlings had taken root by 1886 and the youngest ones were established in 1960.

The lack of old trees or their remnants suggests that the ridge was treeless before the present forest cover. It is difficult to assess for how long because, for instance, the deglaciation of the area was not dated. Whatever the case, forest occupation took place after the "Little Ice Age" in this mille-

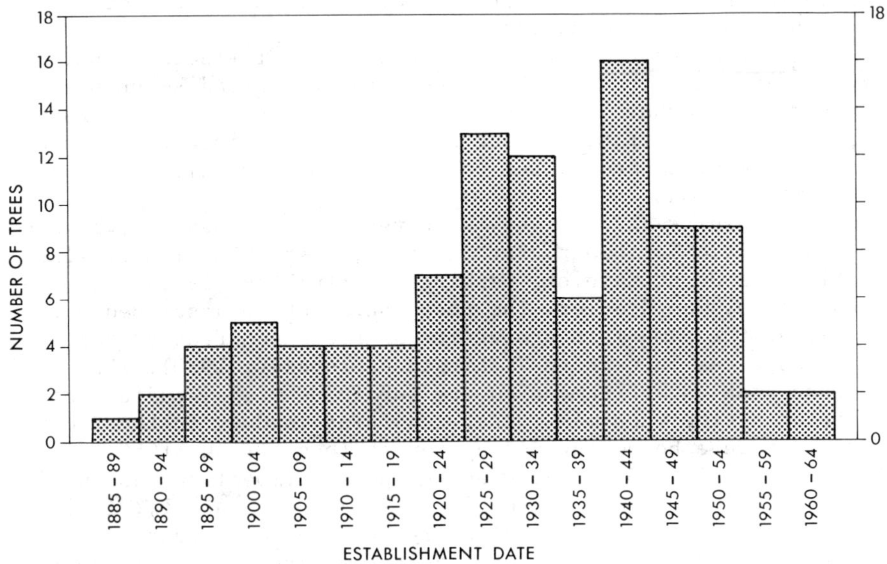

Fig. 5. Age distribution by pentad averages for 100 studied trees established on a subalpine ridge of Mount Baker during recent forest invasion.

nium, principally during the decades when Coleman Glacier was retreating (Fig. 2) and the rate of tree growth was at its peak for centuries (Fig. 3). Tree establishment seems to have been triggered and maintained by relatively warm and, apparently dry climates.

5. CONCLUSIONS

The three, overwhelmingly most distant moraines studied beyond Coleman Glacier date between the early 1500s and early 1800s (Fig. 2). In spite of the discontinuous character of the moraine chronology this might indicate that average temperatures were low during those three centuries. Rapid shrinkage of the glacier between the accumulation of the 1886-87 moraine and the mid-1900s was presumably due to the relatively warm, partly dry period after the "Little Ice Age" which ended about a century ago. The expanding trend of the glacier in the latter part of the present century, except for the few last years, indicates that the falling temperature of the past decades has most recently been followed by a rising one. The annual tree-growth rate (Fig. 3) and tree establishment (Fig. 5), both of which are enhanced by a warm, dry climate and reduced by cool, wet weather, are in moderate agreement with the climatic inferences drawn from the glacier fluctuations.

Climatic interpretations based on glacial fluctuations, tree growth, and tree establishment lead to some discrepancies, however, i.e., they "produce" somewhat divergent climatic patterns. The dissimilarities are understandable since these three natural phenomena are rather complex and have no simple mutual or climatic links.

The results agree widely with results obtained elsewhere in the Pacific Northwest. For instance, post-1800 glacier fluctuations on Mount Rainier, Washington, (Burbank, 1981) resemble those of Coleman Glacier. The response function described here has about the same main properties as those constructed for numerous sites in western Washington (Brubaker, 1980). The intense tree establishment on the subalpine ridge studied took place at a time when many other subalpine meadows in the Pacific Northwest were being invaded by forest (e.g. Brink, 1959; Franklin et al., 1971) and when the polar timberline was pushing ahead in northern Europe (Hustich, 1958). Consequently, it is no suprise that present results are broadly consistent with temperature changes in the past century in the Northern Hemisphere (e.g. Hansen et al., 1982).

REFERENCES

Franklin, J.F. & Dyrness, C.T., 1973. Natural vegetation of Oregon and Washington. USDA Forest Service General Technical Report, PNW-8, 417 pp.

Franklin, J.F., Moir, W.H., Douglas, G.W., & Wiberg, C., 1971. Invasion of subalpine meadows by trees in the Cascade Range, Washington and Oregon. Arctic and Alpine Research, 3, 215-224.

Fritts, H.C., 1976. Tree rings and climate, 567 pp. Academic Press: London.

Garfinkel, H.L. & Brubaker, L.B., 1980. Modern climate - tree-growth relationships and climatic reconstruction in sub-Arctic Alaska. Nature, 286, 872-874.

Bengtson, K.B., 1956. Activity of the Coleman Glacier, Mt. Baker, Washington, U.S.A., 1949-1955. Journal of Glaciology, 2, 708-713.

Brink, V.C., 1959. A directional change in the subalpine forest-heath ecotone in Caribaldi Park, British Columbia. Ecology, 40, 10-16.

Brubaker, L.B., 1980. Spatial patterns of tree growth anomalies in the Pacific Northwest. Ecology, 61, 798-807.

Burbank, D.W., 1981. A chronology of late Holocene glacier fluctuations on Mount Rainier, Washington. Arctic and Alpine Research, 13, 369-386.

Easton, C.F., no date. Mt. Baker, its trails and legends. Whatcom Museum of History & Art. Bellingham, Washington.

Graybill, D.A., 1982. Chronology development and analysis. In: Climate from tree rings (ed. M.K. Hughes, P.M. Kelly, J.R. Pilcher and V.C. LaMarche Jr.), pp. 21-30. Cambridge University Press: Cambridge.

Hansen, J., Johnson, D., Lacis, A., Lebedeff, S., Lee, P., Rind, D., & Russell, G., 1981. Climate impact of increasing atmospheric carbon dioxide. Science, 213, 957-966.

Harrison, A.E., 1970. Short notes - fluctuations of Coleman Glacier, Mt. Baker, Washington, U.S.A. Journal of Glaciology, 9, 393-396.

Heikkinen, O., 1984. Dendrochronological evidence of variations of Coleman Glacier, Mount Baker, Washington, U.S.A. Arctic and Alpine Research, 16, (in press).

Hustich, I., 1958. On the recent expansion of the scotch pine in Northern Europe. Fennia, 82, 1-25.

Lawrence, D.B., 1950. Estimating dates of recent glacier advances and recession rates by studying tree growth layers. American Geophysical Union Transactions, 31, 243-248.

Long, W.A., 1953. Recession of Easton and Deming Glaciers. The Scientific Monthly, 64, 241- 247.

U.S. Weather Bureau, no date. Mount Baker station, Washington. Climatological summary: means and extremes for period 1927-1951, and narrative climatological summary by E.L. Phillips, 2 pp.

Vale, T.R., 1981. Tree invasion of montane meadows in Oregon. The American Midland Naturalist, 105, 61-69.

THE COLD WINTER 1808-09: A RUSSIAN FORCE CROSSES THE ICE-BOUND GULF OF BOTHNIA AND TAKES (THE FIRST?) AIR-TEMPERATURE MEASUREMENTS IN HISTORY FOR PURPOSES OF A MILITARY OPERATION

S. Lindgrén & J. Neumann

Inst. of History, Univ. of Helsinki, Helsinki
Dept. of Atmospheric Sciences, Hebrew University, Jerusalem; now visiting with the Dept. of Meteorology, University of Helsinki, Helsinki

In March 1809 a Russian force crossed the ice-bound Kvarken area of the Gulf of Bothnia from the Finnish city of Vaasa to the Swedish city of Umeå. That winter was one of the coldest winters of the instrumental record in Northern Europe (Leningrad, Stockholm, Uppsala). The ice-crossing to Sweden formed the end phase of the 1808-09 war in Finland of Russia against Sweden. The Russian Generalmajor von Berg, who commanded a column of the force, left behind a "Journal" in manuscript, kind of a diary, for the period 16 March-1 April 1809 (N.S. dates), including the days of the ice march. For almost all of these days he records notes on the weather, and, surprisingly enough, air temperatures. It is almost certain that these air-temperature measurements, taken for the purposes of a military operation, are the first of their kind in history, not counting meteorological observations by naval forces of some nations.

There is evidence that another Russian force, which advanced along a land route to Sweden about the same period, also took air-temperature observations. In his conversation of 5 Feb. 1809 of Alexander I with the French Ambassador in St. Petersburg, the Tsar mentions that the low temperature of -36° (C) at Tornio (at the head of the Gulf of Bothnia), hampers operations.

It is plausible that the temperature observations were made in order to allay the fears of the Russian generals in the field that the ice cover of the sea may fracture under their units or that they may get stranded and taken prisoners in Sweden if the ice breaks up.

In addition to the unpublished "Journal" of Generalmajor von Berg, this paper makes use of four sets of unpublished meteorological observations at places close to the military operations, including observations at Umeå which was the end point

N.-A. Mörner and W. Karlén (eds.), Climatic Changes on a Yearly to Millennial Basis, 363–364.

of the Kvarken crossing. All of the above mentioned observations are contemporary with the Russian military measurements.

ON THE AVAILABILITY AND HOMOGENEITY OF INSTRUMENTALLY OBSERVED CLIMATOLOGICAL RECORDS

Raino Heino

Finnish Meteorological Institute, Finland
WMO Rapporteur on Climatological Observing Network Design

1. INTRODUCTION

International co-operation is an essential feature of meteorology. Vast amounts of synoptic data are daily exchanged between the Meteorological Services of the world. Strict rules have been developed for these purposes and information on observing stations is updated regularly.

In addition to the international network of synoptic stations, most countries have a network of climatological stations. The information generated by them is primarily for domestic use only. Consequently, observing practices are quite heterogeneous in spite of several recommendations made by the World Meteorological Organization (WMO).

Greater demands are being made on climatological data on the regional or global scales, however, and more uniform climatological observing practices and better information on climatological stations are becoming increasingly important for the implementation of the World Climate Programme.

2. AVAILABILITY OF CLIMATOLOGICAL DATA

Station networks for climatological purposes can be classified as follows:

N.-A. Mörner and W. Karlén (eds.), Climatic Changes on a Yearly to Millennial Basis, 365–370.

a) Synoptic stations

The network of synoptic stations has been established primarily for purposes of weather forecasting and the synoptic data is available in most countries via meteorological telecommunications an hour or so after the observations are made. There are about 10,000 synoptic stations at present. The observing practices of the synoptic network are strictly defined internationally, and up-to-date catalogues on the locations and observation programmes are produced regularly by the WMO (No. 217). Because of frequent observations of most climatic elements, the synoptic network simultaneously forms the basis for climatological services and research.

b) CLIMAT stations

Part of the synoptic network is used for climatological purposes on a real-time basis. Monthly means of temperature and atmospheric pressure, monthly precipitation totals and other data (known as CLIMAT messages) are sent via meteorological telecommunications to all countries at the beginning of each month. There are about 1500 CLIMAT stations at present, but there are plans to increase this number. The data is also published some months later in the Monthly Climatic Data for the World (NOAA-WMO).

c) Climatological stations

In addition to the synoptic stations every country usually has a large number of stations for domestic climatological purposes. Observations are not made as often as at the synoptic stations and the number of climatic elements observed is also limited. The number of stations making at least one observation per day of temperature and precipitation is estimated to be around 40,000. In addition, there are approximately 100,000 stations which only record precipitation observations. Because the data from climatological and precipitation stations is used only domestically the present numbers and observation programmes of these stations are not very well known.

d) Reference climatological stations

In order to study climatic changes, long and reliable

observation records are required for some stations around the world. According to the present WMO definition (No. 544) a reference climatological station is "a station, the data of which are intended for the purpose of determining climatic trends. This requires long periods of homogeneous records, where man-made environmental changes have been and/or are expected to remain at a minimum. Ideally, the records should be of sufficient length to make possible the identification of secular changes of climate". Each country is further recommended to establish and maintain at least one reference climatological station.

Today, the status of reference climatological stations, in particular, is far from satisfactory. Most long-term climatic series are from urban areas and do not fulfil the criterion of minimum man-made changes to the environment. On the other hand, the records from most rural stations are still quite recent. In any case, information on potential reference climatological stations is needed so that the best possible network may be selected for the studies of past and future climatic changes. Strict criteria for the selection of reference climatological stations of the world are currently under consideration by the WMO in order to guarantee the continuity and uniformity of these stations in the future, too.

In order to investigate the present situation and to fulfil the data requirements of the World Climate Programme the WMO Secretariat has recently initiated an inventory of the principal and ordinary climatological stations of the world. This survey will also help national meteorological services to keep their own inventories up-to-date in a systematic way and to develop more uniform observing practices. More than 100 countries have so far sent information on their climatological stations, including the main features of the station histories and observation programmes, too. The information received has been transformed into a computerized data-base.

The format of the WMO station data base is similar to the one used in the WMO Catalogue of Meteorological Data for Research (No. 174), which includes a list of meteorological stations with observational series over 80 years or more. This list dates back to 1961 and the new information can be used as a basis for an updated version of the catalogue.

In addition to the lists of stations with long observation series, other data on station networks can be prepared; these could be of use in the planning of scientific and operational activities on regional or global scales. Information to be published will consist of different maps, graphs and tables selected according to

- areas (e.g. different continents or regions, latitude-longitude grids, etc.)
- elements observed (temperature, wind, radiation, precipitation, etc.)
- observation periods (e.g. 1931-1960, more than 100 years, etc.).

The information included in the WMO inventory will thus form a basis for various regional or global activities. On the national level, however, more exact information is necessary. The difficulties in constructing station inventories and applying climatological data include the changing of observation sites, instruments and methods. All these may cause serious inconsistencies in the observation records.

3. HOMOGENEITY OF CLIMATOLOGICAL DATA

Meteorological observations have primarily been made for weather forecasting purposes, where the reliability and uniformity of the actual station network is extremely important. Consequently, changes in locations, instruments, etc. have not been regarded as sufficiently important.

Climatological observations, on the other hand, were originally created to determine patterns of climatic classifications and different kinds of shorter-term statistics and not neceassarily to indicate climatic changes. This being the case, climatological data are nowadays being used in studies for which they were not originally designed.

In practice it is very difficult to obtain homogeneous climatic data series. Various kinds of factors, such as changes in

- observation sites and their environment
- instruments and their placement
- observing methods and observers
- times of observations
- etc.

will introduce inconsistencies into the data. In the case of long-period temperature change the situation is even more complex as shown by MITCHELL (1953).

The inhomogeneity may be in the form of

- trend
- impulsive change of central tendency
- some kind of oscillation

or their combinations. There are many statistical methods that can be used for testing the homogeneity of climatological observation series. Several such tests have also been presented in the meteorological literature (e.g. WMO,1966, ABBOTT).

The purpose of the homogeneity tests is to indicate whether any bias is included in the data series and to study its form and significance. The extent of the changes being sought is often lower than the level of noise and the true signal of the data. Most homogeneity tests thus aim to remove the true signal, e.g. with the help of running means or by comparison with nearby stations.

Another way to find possible points of inhomogeneity is a careful study of the history of each observing station. Any changes in the location of the station or in the instruments and observing methods used, etc. must be checked as a possible source of inhomogeneity.

When the change has been diagnosed it is necessary to estimate its significance. Most well-known statistical tests (e.g. t and F tests or variance analysis) require the normality of the distribution. If the distribution is not normal, several distribution-free tests (e.g. Wilcoxon and Kolmogorov-Smirnov tests) can be used.

If inconsistencies are found and corrections made, the original data should not be destroyed and replaced with the homogenized data, because future studies may require the original data.

REFERENCES

Abbott, P.F. Quality control of long periods of climatological data (to be published by WMO).

Mitchell, J.M. Jr. 1953. On the causes of instrumentally observed secular temperature trends. Journal of Meteorology, 10, 244-261.

WMO, 1966. Climatic change, Technical Note No. 79, WMO-No. 195.

CLIMATE VARIATION IN NORTHERN EUROPE DURING THE PAST CENTURY
Evidence from a Danish record

E.W. Peterson[+] and S.E. Larsen

Risø National Laboratory. Denmark
+Present affiliation· Oregon State University, Corvallis, U.S.A.

At a Danish station, meteorological parameters have been measured during the period 1872-1980. A description of the variation of 30-year running means of some of these variables is presented. The correlation between the individual variables is presented, and their variations are compared with observations elsewhere.

I. INTRODUCTION

The island of Fanø lies on the west coast of Jutland, just outside the port of Esbjerg. Since November 1872 a continuous record of weather observations has been kept. Observations have been made three times a day, at 0800, 1400, and 2100 hours, and include measurements or observations of temperature, relative humidity, wind force and direction, visibility, precipitation, pressure, maximum and minimum temperature, snow cover, snow depth past weather, and cloudiness (Danish Meteorological Institute, and Brown et al. (1983)). Since there are essentially no breaks in the record for most variables, visibility being a notable exception, it has been possible to analyze the time series of the weather variables for annual and long term variations. This paper is concerned with some elements of the climatological-scale variability of four independently measured variables, namely temperature, cloudiness, pressure, and precipitation.

A fairly complete analysis of the annual and climatological variation in the various measures of weather observed at Fanø can be found in Peterson (1983). This article presents one of the more striking results of the analysis: the rather clear cor-

N.-A. Mörner and W. Karlén (eds.), Climatic Changes on a Yearly to Millennial Basis, 371–379.

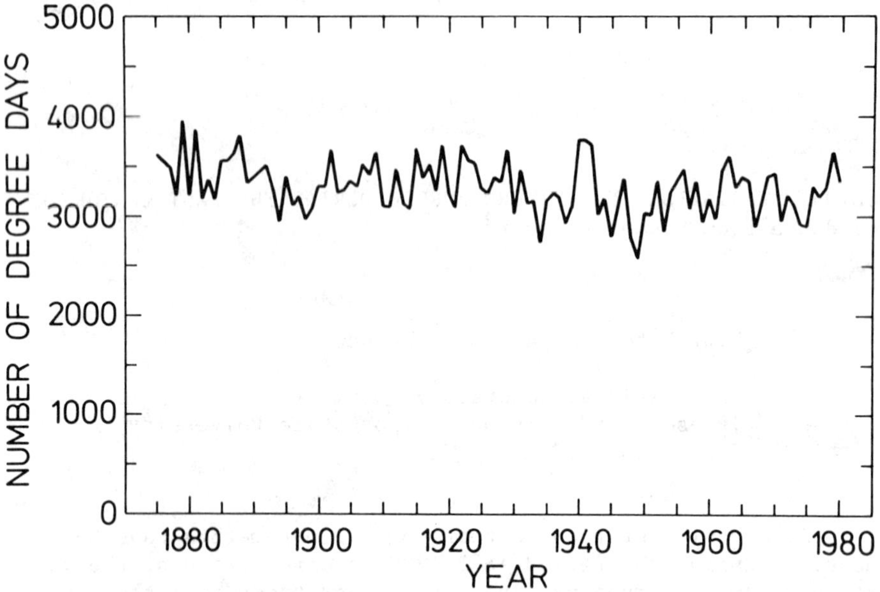

Fig. 1a. Annual number of heating degree days at Fanø, as a function of year.

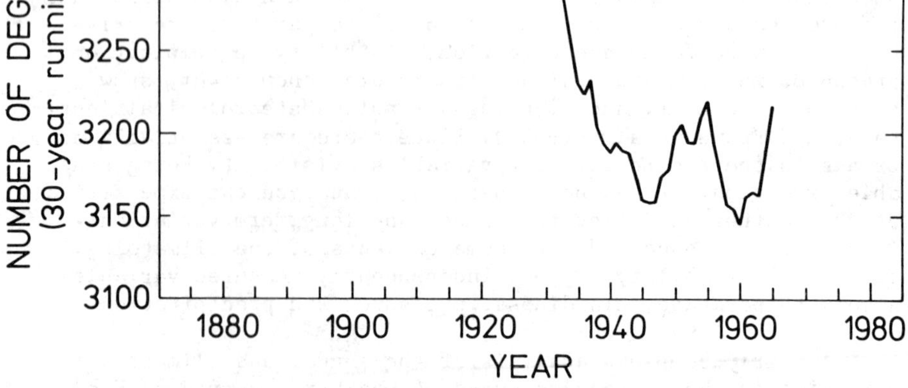

Fig. 1b. 30-year running mean of the annual number of heating degree days at Fanø.

relation between the climatological averages of the independently measured variables, temperature, cloudiness, pressure, and precipitation, over at least the first half of the 20th century.

II. DATA ANALYSIS

Annual averages of the various observations were computed, and, where it was more reasonable, such as for the temperature, the year was defined as beginning in October so as not to break up the winter season. 30-year running means were then computed, resulting in continuous time-series of each variable, extending from 1890 to 1965, of the 30-year averages. (Obviously the definition of the year is of no real consequence when it comes to 30-year averages. However it does mean that there are slight differences in the temperature and precipitation curves which are presented in Figs. 2 and 3, which were developed from data based on the October to September year, and those in Fig. 5, which are based on the calendar year.) 30-year running means were chosen for aggregating the data because it has become conventional to define climate in terms of 30-year means. Also, the interannual variations in the annual averages of the various weather elements are so large as to completely mask any climatological trends, and thus they must be removed by long-term averaging in order to discern any trends in the climate. Because of the huge interannual variations in weather, compared to trends in climate, the statistical "significance" of climate trends over periods as short as this record, one century, is quite small. One can only develop confidence that the trends are real by comparing trends in independent observations from the same station or by comparing records from other stations in the same climatological area.

Figures 1a and 1b show how application of a 30-year running mean on the series of total number of heating degree days for a year clearly brings forward a slow trend in the data, which is otherwise masked by the large interannual variation. The number of heating degree days each day is defined as the number of degrees the mean temperature is below 17°C and is zero otherwise. In Figs. 1a and 1b this signal is chosen to illustrate as well that the climate variations over the period considered might have social and economical implications; here particularly for the energy needs for house heating.

Figures 2 show the annual variation of the 30-year average temperature. Figure 2a shows the 30-year running mean of the annual average temperature, while Figs. 2b and 2c show the 30-year running means of the summer average and the winter average temperatures. (Summer is defined here as May-September, and winter as November-March. October and April were disregarded since they are transitional months with neither clear summer nor winter

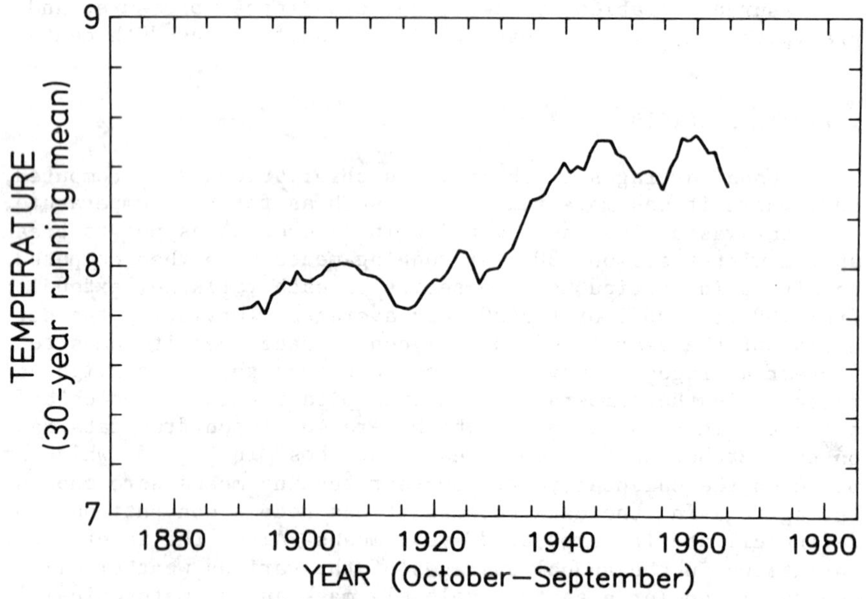

Fig. 2a. 30-year running mean of the mean annual temperature (°C) at Fanø as a function of year.

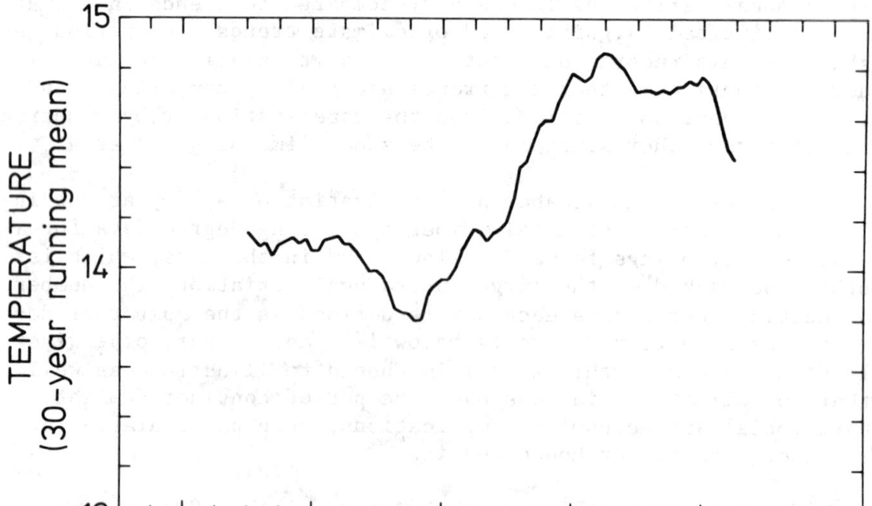

Fig. 2b. 30-year running mean of the summer temperature (°C) at Fanø as a function of year.

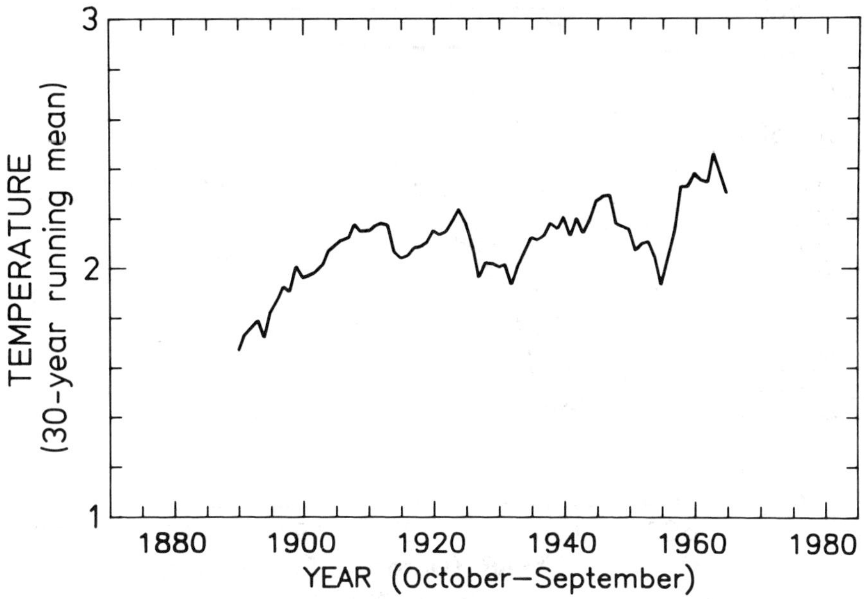

Fig. 2c. 30-year running mean of the winter temperature (°C) at Fanø as a function of year.

characteristics. In order to avoid the bias which would be introduced through use of the temperatures recorded at the three observation times in determining the seasonal or annual averages, the daily mean temperature, defined as the average between the daily maximum and minimum temperatures, was used in determining seasonal and annual average temperatures.)

The trend toward increasing climatological mean temperature during the first half of the 20th century is readily apparent in all three curves, however it is most noticeable in the trend of the summer temperatures. These trends are consistent with the general trends observed elsewhere in the northern hemisphere (Lamb, 1969, p. 176).

Figure 3 shows a rather striking increase in the 30-year mean annual precipitation up to about 1930. Since many factors can affect the the accuracy of precipitation measurements, one was at the first tempted to be suspicious of such a large increase in precipitation over such a short period (30 years). However there is corroborating evidence from other sites supporting this trend (Lamb, 1969, p. 197).

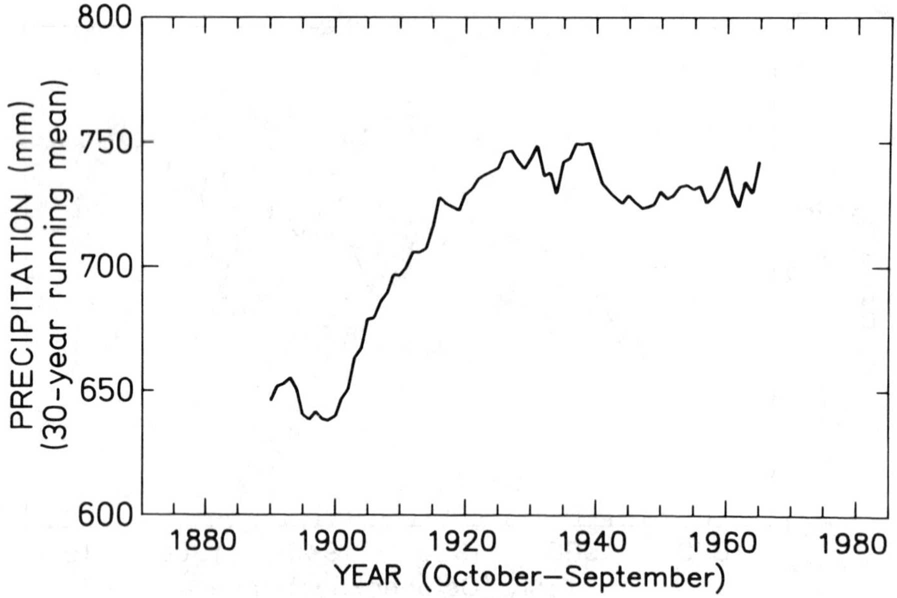

Fig. 3. 30-year running mean of annual precipitation (mm) at Fanø.

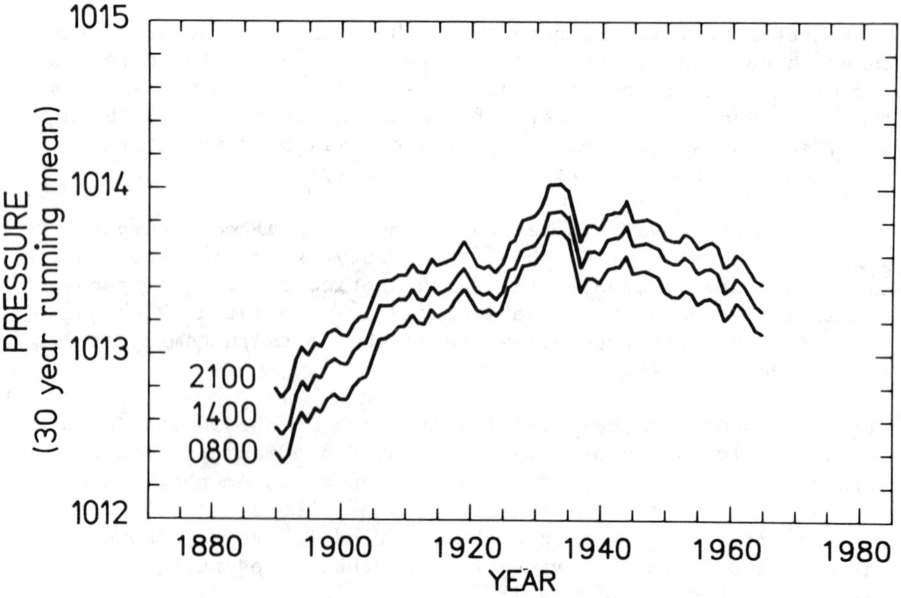

Fig. 4. 30-year running mean atmospheric pressure (mbar) at Fanø for the observation times. 0800, 1400, and 2100.

Figure 4 shows the 30-year running mean atmospheric pressure for the three observation times. Although it is beyond the scope of this article to discuss it, a diurnal tide in atmospheric pressure is easily seen. Of more importance here is the obvious increase in the climatological mean pressure up to about 1935. The data are in accordance with the movement of the storm track or polar front, as noted by Lamb (1969, p. 219). The polar front, along which the midlatitude cyclones move, is a region of low pressure compared to the higher pressure found both poleward and equatorward of it. Thus a movement of the mean position of the front away from the observing station would result in an increase in the average observed pressure at the station.

Figure 5 shows the 30-year running mean of the cloudiness for the three observation times. There is a maximum at about 1910 and a minimum at about 1930. This correlates well with the pressure trend, since higher pressure generally means less cloudiness.

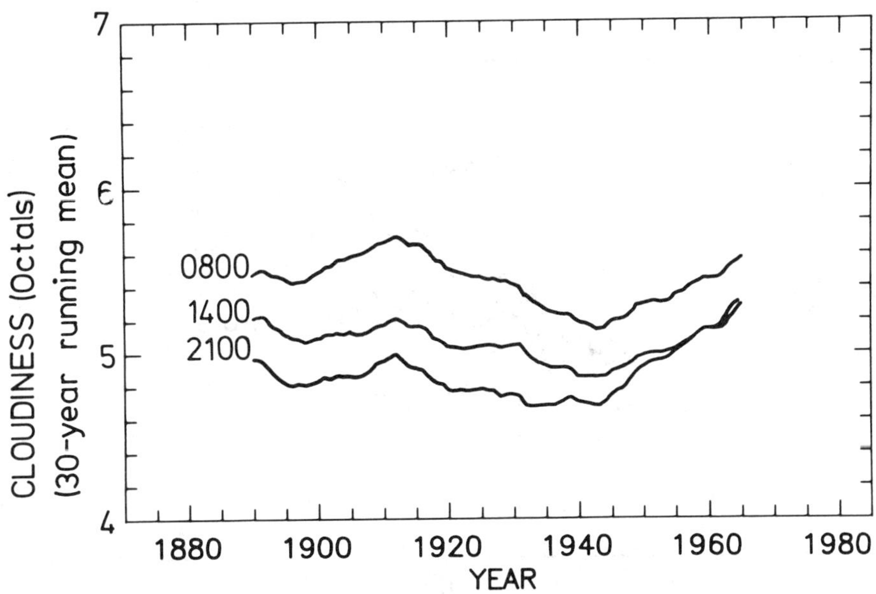

Figure 5. 30-year running means of cloudiness (octals) observed at 0800, 1400, and 2100, at Fanø.

III. DISCUSSION

The means and standard deviations of the yearly averages of temperature, precipitation, pressure, and cloudiness were computed and used to normalize the 30-year averages of their respective variables. Figure 6 shows the trends in the normalized climatological averages of these independently measured variables. (Cloudiness was replaced by its negative, cloudlessness, since it would be expected to be negatively correlated with temperature and pressure). The increased precipitation in the face of higher pressure and fewer clouds is apparently contradictory. However, in these latitudes, precipitation is limited by temperature: the moisture content of the air greatly increases with small increase in temperature when the supply of water vapor is unlimited, as it is in a marine environment such as that in which Fanø is situated. Thus it is physically realistic to have increased precipitation along with less storminess and less average cloudiness if the temperature has increased. In Peterson (1983) it is shown that, climatologically, there has been an increase in the number of days of shower activity and a decrease in the number of cyclone passages at Fanø during the period in which the temperature, pressure, precipitation, and cloudlessness were increasing. The rather good correlation between the records of these independently measured weather variables tends

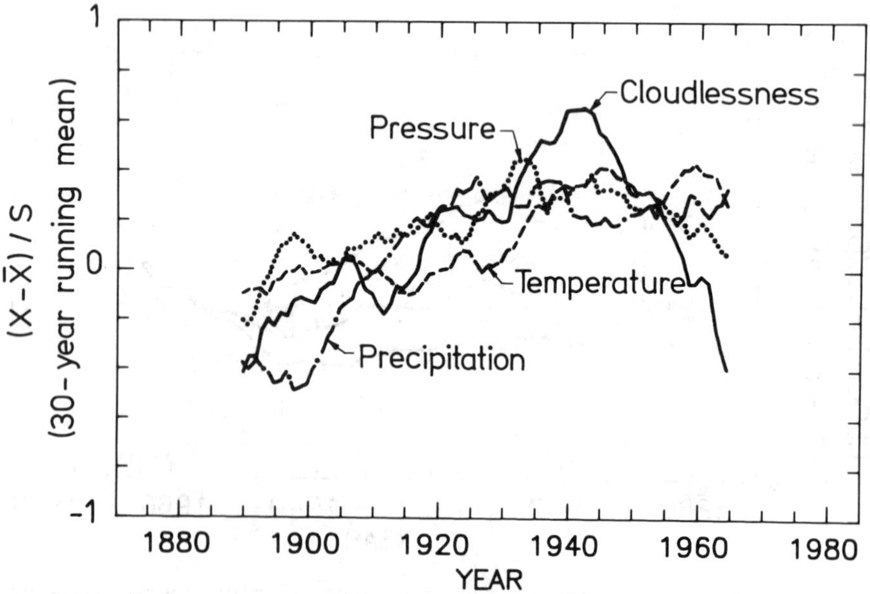

Fig. 6. Normalized 30-year running means of pressure, temperature, precipitation, and cloudlessness from Fanø. Variables were normalized by their respective means and standard deviations. X is the mean of the variable over the whole record (1875-1980) and S is the standard deviation of the yearly average of the variable.

to support the validity of each measure, each of which could be subject to some suspicion when standing alone because of the myriad problems involved with the process of making consistently valid observations over a long period of time. Since many factors affect these weather elements, one cannot expect a particularly strong correlation between them. However the correlation between the climatological means of these elements, which was discovered in the study reported here, does appear to make physical sense, and it is the first such correlation to be presented in the literature to the authors' knowledge.

References

1. Brown, N., Olesen, H.R., Prahm, L., and Reiter, A., 1983. Kvalitetskontrol af Fanø Klimadata. MST-Luft-A75, Miljøstyrelsens Luftforureningslaboratorium, Forsøgsanlæg Risø.
2. Lamb, H.H., 1969. Climatic Fluctuations. WORLD SURVEY OF CLIMATOLOGY, Vol. 2, (H. Flohn, Ed.), Elsevier, Amsterdam, 266 pp.
3. Peterson, E.W., 1983. A study of the weather record from Fanø (1872-1980), including an analysis of climate variation. Risø Report R-483, 70 pp.

CERTAIN RELATIONSHIPS BETWEEN LAKE LEVEL VARIATIONS AND SOME CLIMATIC FACTORS IN FINLAND

Reijo Keränen

University of Oulu, Dept. of Geography, Finland

ABSTRACT

The material in this paper consists of a study of variations in high-water level and in five climatic factors for six Finnish lakes at two Meteorological Stations during about a hundred years. Correlations and autocorrelations with a lag of 36 years are used as primary methods.

The periodicities in water level and climate partly seem to follow the periods of solar activity (11 and 22-23 years). Moreover, there are obvious similarities also in the periodicity of lunar circulation (18 and 31 years).

INTRODUCTION

The purpose of this study is to make a general comparison between lake level and climatic variations in Finland. The material consists of a study of the variations in the high-water level (HW) for the lakes Oulujärvi (1866-1979), Kallavesi (1864-), Saimaa (1847-), Päijänne (1871-) and Vanajavesi (1875-) (see Fig. 1). The catchment areas and the percentage of lakes are as follows: L. Oulujärvi (F=19890 sq. km, L=12.7 %), L. Kallavesi (F=16270, L=15.3), L. Saimaa (F=61265, L=19.9), L. Keitele (=6305, L=18.0), L. Päijänne (F=26480, L=19.5), L. Vanajavesi (F=8615, L=14.4). Lakes Oulujärvi and Vanajavesi have catchment basins of their own, whereas L. Kallavesi forms the head-water of L. Saimaa and L. Keitele that of L. Päijänne (see Fig. 1). Lake Oulujärvi and Vanajavesi discharge in the direction of the greatest land uplift (Kääriäinen, 1966), which results in a slow rise in water level. Other lakes flow in the direction of the least uplift, which in

N.-A. Mörner and W. Karlén (eds.), Climatic Changes on a Yearly to Millennial Basis, 381–389.

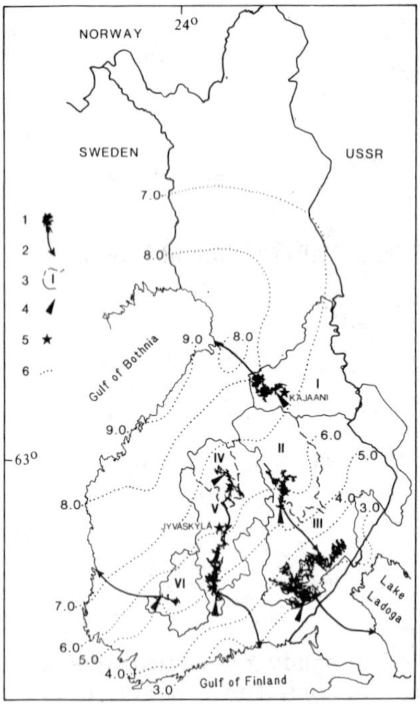

Figure 1. Location of lakes. I=L. Oulujärvi, II=L. Kallavesi, III=L. Saimaa, IV=L. Keitele, V=L. Päijänne, VI=L. Vanajavesi. 1=lake, 2=direction of outlet, 3=border of catchment area, 4=gauge, 5=meteorological station, 6=isobase of land uplift (according to Kääriäinen, 1966).

turn causes a slow fall in the water level. The transgression and the regression will best be seen in the low-water (LW) trends (see Keränen, 1982).

The climatic observations for L. Oulujärvi were obtained from the Meteorological Station at Kajaani and for the other lakes from Jyväskylä (see Fig. 1). Initially, 11 independent variables were used: annual precipitation, spring precipitation (I-IV), precipitation of previous autumn (VII-XII), annual mean temperature, means for diurnal minimum and maximum temperature, sums of diurnal minimum and maximum temperature for the whole year as well as for spring and previous autumn and, finally, the relative number of sunspots. By reference to the correlation matrix between these variables and the water level observations, the following were selected to be used as the base: 1. annual precipitation (P, mm), 2. precipitation of the previous autumn (Pa, mm), 3. spring precipitation (Ps, mm), 4. mean of diurnal minimum temperature (Tm, °C), 5. relative number of sunspots.

VARIATION OF WATER LEVEL AND CLIMATIC FACTORS

In order to elucidate the periodicity of the lake level, 5-yr moving means were calculated from the observed values (Fig. 2) and from these means, autocorrelations with a lag of 1-36 years were calculated (Fig. 4). The autocorrelation diagrams for lakes Kallavesi, Saimaa, Keitele, Päijänne and Vanajavesi are very similar. From two variation periods of different length, the first and most distinct is 23-26 years and the second, less distinct 30-32 years. In addition to these, at least L. Kallavesi shows a very slight indication of a period of 5-7 years. The autocorrelation for L. Oulujärvi gives evidence of a totally different variation: the lengths of the periods are 9-11 and 18-20 (22) years and then there is a weak period of over 30 years. Furthermore, the high-water trend for L. Oulujärvi shows a relatively sharp decline (Keränen, 1982) which can also be seen in Fig. 2.

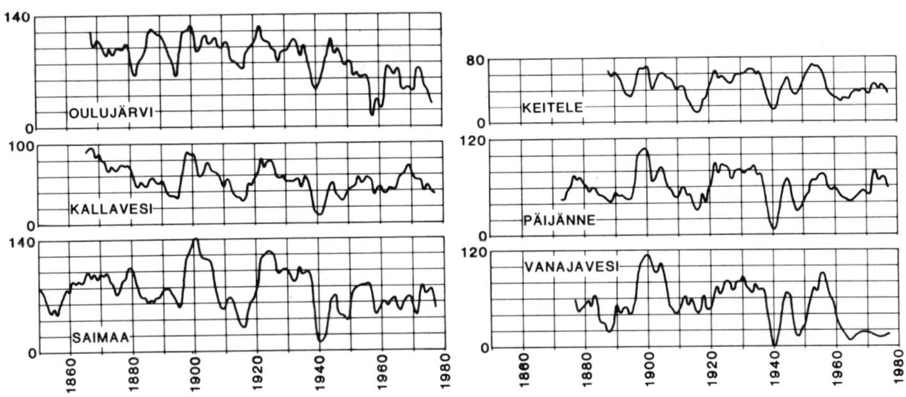

Figure 2. Variation of lake high-water level (HW, cm) in terms of 5-yr moving means.

The moving means for the climatic factors and the autocorrelations (Figs. 3 and 5) do not in every respect follow the same periodicities as shown in the water level variation. Analogously, the variation of 9-11 years for L. Oulujärvi seems to correspond only with the mean for diurnal minimum temperature (Tm in Fig. 5) and possibly with spring precipitations (Ps), while on the other hand annual precipitation (P), precipitation of the previous autumn (Pa) and spring precipitation (Ps) appear to correspond with the period of 18-22 years. In view of the climatic variables, a weak period of over 30 years occurs only in the precipitation rate of the previous autumn (Pa). According to the climatological observations from Jyväskylä, the usual lake level periodicity of 23-26 years seems to emerge most clearly in connection with the spring

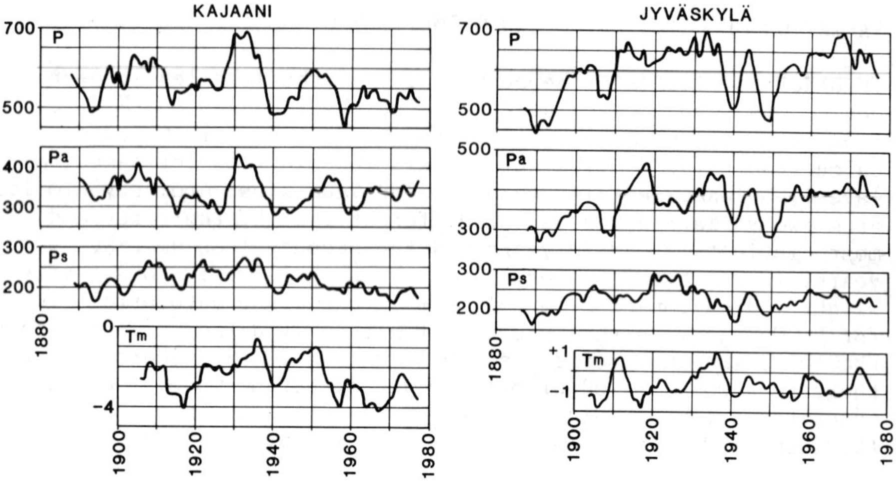

Figure 3. Variation of climatic factors at Kajaani and Jyväskylä in terms of 5-yr moving means. P=annual precipitation (mm), Pa= precipitation of previous autumn (VII-XII, mm), Ps= spring precipitation (I-VI, mm), Tm=mean of diurnal minimum temperature (°C).

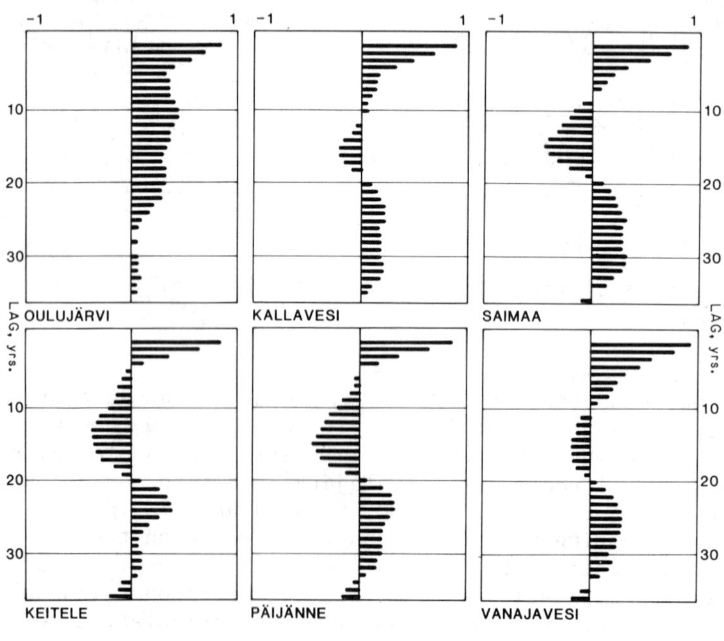

Figure 4. Autocorrelations for lake high-water levels (5-yr moving mean) with a lag of 1-36 years.

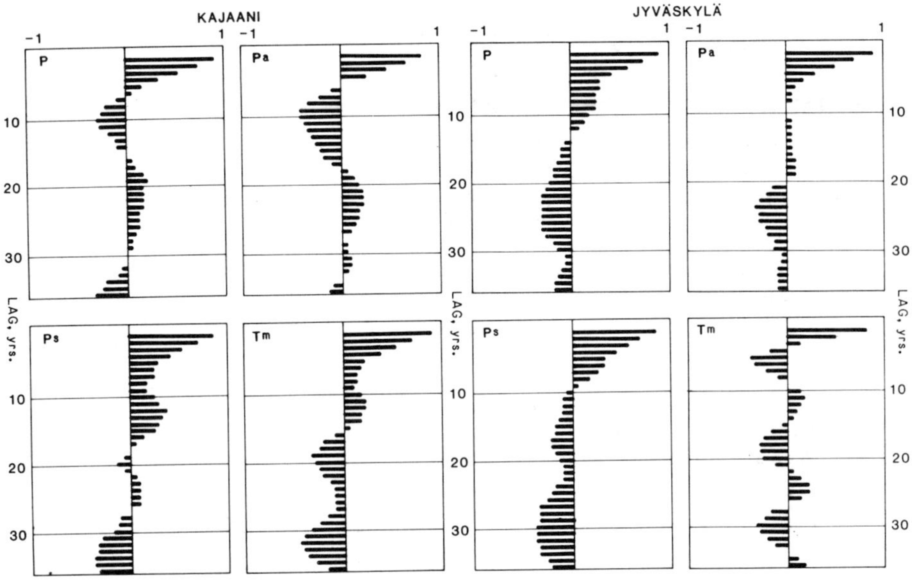

Figure 5. Autocorrelations for climatic factors (5-yr moving means) with a lag of 1-36 years. Explanations in Fig. 3.

precipitation and the diurnal minimum temperature rates. Instead, the periodicity of autumn and annual precipitation could account for water level periods of over 30 years. The means for diurnal minimum temperature observed at both Jyväskylä and Kajaani occur in particularly regular cycles of 11-12 and 24-25 years.

In order to clarify the covariation of climatic factors and lake levels, correlation coefficients between the variables were calculated (Table 1) that reveal the significance of the precipitation rate of the whole year and of the spring precipitation rate. The relatively weak correlations between the small L. Vanajavesi in the south (see Fig. 1) and all climatic factors are also worth noting, especially when considering the precipitation rate for the whole year. Correspondingly, the correlations of the northernmost lake, L. Oulujärvi, are relatively high.

When the periodicities of lake levels and four climatic factors (Table 2) are compared with the observations from elsewhere, we see a few partial similarities. Lamb (1972), for example, has noted lengths of periods from various sources, the commonest of them being 2-3, 5-6, 10-12, 18-24, 80-100 and about 170-200 years. The comparatively frequent cycle of approx. 30-33 years shown in Table 2 has nevertheless hardly occured. Simojoki (1955) has observed a cycle of over 30 years for L. Saimaa - a cycle noted also in this paper - and a corresponding phenomenon in the mean temperature of Helsinki. A distinct periodicity of approx. 30

	P	Pa	Ps	Tm
OULUJÄRVI	52	26	65	50
KALLAVESI	40	27	60	31
SAIMAA	35	39	40	35
KEITELE	30	16	44	37
PÄIJÄNNE	44	38	55	37
VANAJAVESI	25	30	34	21

Table 1. Correlation coefficients between high-water levels (HW) and climatic factors. Explanations in Fig. 3.

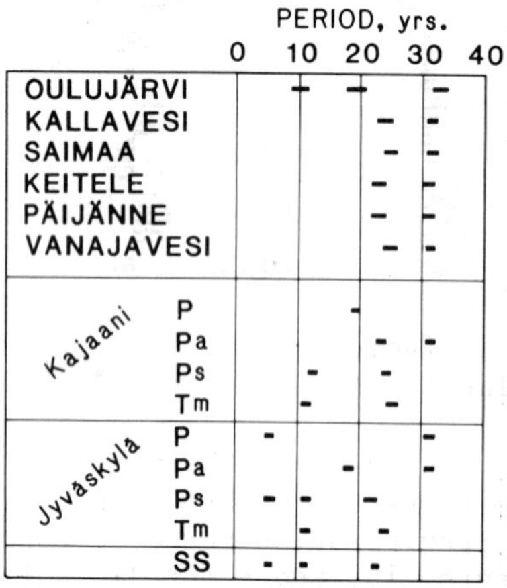

Table 2. Periodicities based on autocorrelations between lake high-water levels and climatic factors. Periods of sunspots (SS) according to Lamb (1972). Explanations in Fig. 3.

years has also been found in the variation of tree ring width for trees in Lapland as well as in the thickness of Estonian glacial clay varves (Sirén, 1961; Sirén & Hari, 1971). Fairbridge (1983) puts forward the idea that the lunar effect would be the cause of at least periods of 18.5 years (Lunar Nodal Cycle) and of 31 years (Perigee-Syzygy Cycle). The isolated variation and the covariation of the planets have also been given as causes for cycles of various length (e.g. Wood, 1972).

SOLAR ACTIVITY

Although the cyclicity of Finnish lake level variation does not seem to be completely similar to solar activity cycles (5.7, 11 and 22-23 years; Lamb, 1972), it is, however, interesting to take a look at the relationship between solar activity, hydrology and climatic factors (see Koutaniemi, 1983). The relative frequency of the sunspots is used as one of the reflectors of solar activity. The numbers of spots varies in the fairly even period of 11 years. Additionally there is a 22-year period (at least since 1835; Willett, 1949), the Hale-period, in connection with which occurs the variation in the polarity of the magnetic field of the spots. The Hale-period (also known as Double Sunspot Cycle) consists of a smaller and greater spot maximum and of a minimum between the two.

The correlations between the relative number of sunspots and lake levels with different lags (Table 3) show that the effect is not necessarily the strongest in a 'zero year'. The strongest correlation (Max in Table 3) is achieved in all other lakes 9 years before the 'zero year' except for L. Oulujärvi where the maximum dependency occurs 11 years after it. The coefficients in Table 3 are mostly negative; as solar activity increases, water level sinks and vice versa.

	LAG, yrs.					
	+2	+1	0	-1	-2	Max
OULUJÄRVI	-10	-22	-30	-29	-27	-32
KALLAVESI	07	-05	-15	-14	-13	-28
SAIMAA	00	-12	-25	-27	-23	-35
KEITELE	13	-10	-23	-24	-18	-29
PÄIJÄNNE	11	-06	-20	-27	-23	-39
VANAJAVESI	15	01	-18	-26	-27	-41

Table 3. Correlation coefficients between lake high-water levels and relative sunspot number with a lag of ± 2 years. Max=highest value of coefficient with a lag of ± 36 years (see text).

In order to clarify the influence mentioned above, the sunspot spectrum was divided into periods of 5-6 years so that either a spot minimum or maximum came in the middle of each period (see e.g. Simojoki, 1957). Mean high-water (MHW) (Table 4) during these maximum and minimum periods was calculated for different lakes. In all lakes, water level was lower during maximum periods than during minimum ones. The differences for L. Oulujärvi and L. Keitele are statistically significant (**, t-test) and for L. Saimaa and L. Päijänne almost significant (*). There is further a similar relationship between climatic factors and solar activity (Table 5). Precipitation rate is greater in periods of small activity compared with the peak periods. The mean for diurnal minimum temperature acts meteorologically in a corresponding way; it is colder (less cloudy) during greater activity than at other times. Only the differences within the annual precipitation at Jyväskylä are statistically almost significant (*).

	SSmax	SSmin	HALEmax	HALEmin
OULUJÄRVI	123.10	123.33	123.07	123.21
KALLAVESI	82.06	82.16	82.06	82.14
SAIMAA	76.01	76.21	76.03	76.09
KEITELE	99.57	99.72	99.50	99.63
PÄIJÄNNE	78.52	78.67	78.47	78.56
VANAJAVESI	79.84	79.95	79.86	79.83

Table 4. Lake mean high-water levels (MHW, m a.s.l.) during sunspot maxima and minima and during greater and smaller maxima of the Hale-period (see text).

		SSmax	SSmin	HALEmax	HALEmin
Kajaani	P	541	570	517	565
	Pa	330	350	326	334
	Ps	206	223	193	219
	Tm	-2.70	-2.42	-2.99	-2.48
Jyväskylä	P	576	622	567	584
	Pa	224	235	370	348
	Ps	359	376	214	233
	Tm	-0.85	-0.47	-0.70	-0.95

Table 5. Values for climatic factors during sunspot maxima and minima and during greater and smaller maxima of the Hale-period (see text). Explanations in Fig. 3.

It is to some extent possible to clarify the effect of the Hale period, too, by dividing the data similarly into periods of HALE max and HALE min in Tables 4 and 5, which stand for a greater and smaller sunspot maximum. In other lakes except in L. Vanajavesi the mean high-water (MHW) has been higher during smaller maximum periods than at other times. The observed differences are not statistically significant. In view of the climatic factors (Table 5), the data from Kajaani supports the idea of interdependence given above. According to the material from Jyväskylä, precipitation of the previous autumn and the mean for diurnal minimum temperature do not, however, show the same covariation.

CONCLUDING REMARKS

The short-term variation of high-water level (HW) for large Finnish lakes seems to follow several different periodicities, the cycle of about 25 years being the most common. In all lakes under examination there was slight indication of a periodicity of a little over 30 years. The cycles are somewhat more incoherent in climatic factors than in water level variation. According to the results it is possible to tie the water level variation, and also climatic variation, at least to the periodicity of the solar activity and the lunar circulation. However, the common cycle of 23-26 years of the water level is slightly too long in comparison with the Hale-period (22-23 years).

REFERENCES

Fairbridge, R.W., 1983. Planetary periodicities and Terrastial climate stress. In: Abstracts of the Second Nordic Symposium on Climatic Changes and Related Problems, (eds. Frydendahl et al), 29 p.

Keränen, R., 1982. Water level variations of Lake Oulujärvi, Finland. Nordia, 16:4, 227-242.

Koutaniemi, L., 1983. Complexity of solar variability, hydrology and climatic conditions as evidenced in the case of the Oulujoki and Kemijoki river basins, Northern Finland. Fennia, 161:2, 289-301.

Kääriäinen, E., 1966. The second levelling of Finland in 1935-1955. Veröff. Finn. Geod. Inst. 61. 313 pp.

Lamb, H.H., 1972. Climate: present, past and future. Vol. 1. Fundamentals and climate now. 613 pp. Methuen & Co.

Simojoki, H., 1955. Long rhythmic water level variation in Lake Saimaa compared with fluctuations in some climatological elements. Geophysica, 5:2, 107-112.

Simojoki, H., 1957. On the consideration of the double sunspot cycle in climatic investigations. Geophysica, 6:1, 25-29.

Sirén, G., 1961. Skoggränstallen som indikator för klimat fluktuationerna i norra Fennoskandien under historisk tid. Commun. Inst. For. Fenn., 54:2, 66 pp.

Sirén, G. & Hari, P., 1971. Coinciding periodicity in recent tree rings and glacial clay sediments. Rep. Kevo Subarctic Res. Stat., 8, 155-157.

Willett, H.C., 1949. Long-period fluctuations of the general circulation of the atmosphere. Jour. Met., 6, 34-50.

Wood, K.D., 1972. Sunspots and planets. Nature, 240:5367, 91-93.

LONG-TERM VARIATIONS IN SOME PHYSICAL PARAMETERS OF THE BALTIC SEA

Lasse Makkonen, Jouko Launiainen, Kimmo Kahma
and Pekka Alenius
Institute of Marine Research, Finland

ABSTRACT

Properties of time-series of water temperature, salinity, river runoff, sea level and ice conditions from the Baltic Sea area were studied. Air and water temperatures reveal a strong air-sea coupling and the water salinity has been found to have a slight increasing trend during the first half of this century. An estimate of the river runoff to the Baltic Sea shows a decreasing trend at the same time. The variance of sea level changes do not show any noticeable trend after 1920's, suggesting that there recently has been no significant change in the cyclonic activity in the Baltic Sea area. The data on ice conditions in Helsinki show a decrease of $1\frac{1}{2}$ months in the length of the ice covered season from 1860's to 1970's. The maximum annual ice extent of the Baltic Sea also reveals a decreasing trend in the 19th century. It correlates ($r = -0.89$) with the mean winter air temperature in Helsinki, which allows prognosis of the future mean maximum ice coverage on the basis of predicted mean air temperatures.

1. INTRODUCTION

There is an increasing interest in climatic changes at high latitudes, mainly because of the predicted high sensitivity of the climate at these latitudes to the increase of CO_2 in the atmosphere (e.g. Manabe and Wetherald, 1980). In addition, the interaction between sea ice conditions and thermal regime of the atmosphere increases the motivation for climatic studies at high latitudes. There are further reasons that make the Baltic Sea

N.-A. Mörner and W. Karlén (eds.), Climatic Changes on a Yearly to Millennial Basis, 391–399.

region an especially interesting area as far as the signal of the anthropogenic climatic change is concerned: Both diagnostic studies (Jäger and Kellogg, 1983) and first model approaches (Mitchell, 1983) to the geographical distribution of the sensitivity of mean air temperature to global warming suggest that this region belongs to an area of the highest sensitivity. The North-European region where the Baltic Sea is located offers one of the longest series of data on climatic changes in sub-arctic regions.

In this paper time-series on water temperature, salinity, river runoff, sea level variability and ice conditions are presented, and relations of these parameters to meteorological variables are examined. These oceanographic time-series have certain advantages over the meteorological time-series, e.g. air temperature, wind speed and precipitation. They have larger spatial and temporal coverage which smooths out local effects and, in addition, the oceanographic time series are generally less dependent on observation praxis and, therefore are more homogeneous. Oceanographic quantities such as ice extent are quite sensitive to climate, and therefore are good indicators of climatic changes (see Barry, 1983).

2. WATER TEMPERATURE, SALINITY AND RIVER RUNOFF

A strong air-sea coupling in the Baltic Sea is shown in Fig. 1, where 5-year floating averages of summer air and sea surface temperatures are presented. Fig. 1 suggests that summer surface water temperatures are presently $1^{\circ}C$ higher than during the second half of the 19th century and the beginning of the 20th.

An attempt to construct a long time-serie of the total river runoff to the Baltic Sea is shown in Fig. 2. The total runoff before 1920 was calculated by using the regression (Launiainen and Koljonen 1982) between the runoff of the river Vuoksi (data available from 1849) and the total runoff (data by Mikulski 1980 available from 1921 to 1970). There seem to be large variations in the total runoff on a time scale of 10-30 years and a decreasing trend in the long-term averages (30 y) during the 20th century until the 1950's. This trend as well as smaller-scale variations in the runoff seem to be associated with the reverse changes in water salinity observed by Matthäus (1979), at least in certain locations (Launiainen and Koljonen, 1981, 1982). It is interesting to note that according to Budyko (1982) an increase in the atmospheric CO_2-content may result in a considerable increase in the river runoff in this area. For the time being, however, time-series for verification of this prediction, is too short.

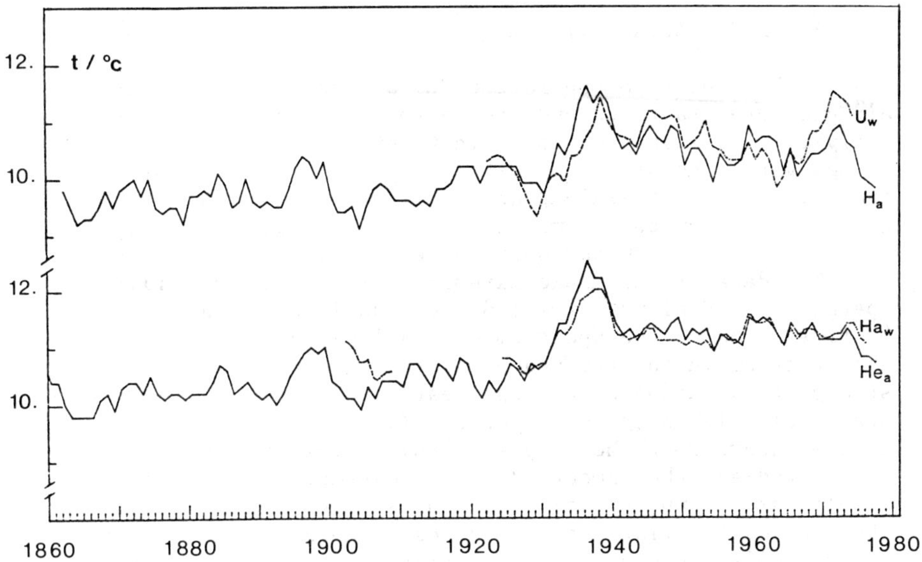

Figure 1. Long-term variation of summertime air and surface water temperature (5 y running averages). The temperatures are: U_w = water at Ulkokalla, Bay of Bothnia, H_a = air in Haparanda (June - Oct) and Ha_w = water at Harmaja, Gulf of Finland, He_a = air in Helsinki (June - Nov). (From Launiainen and Koljonen, 1981.)

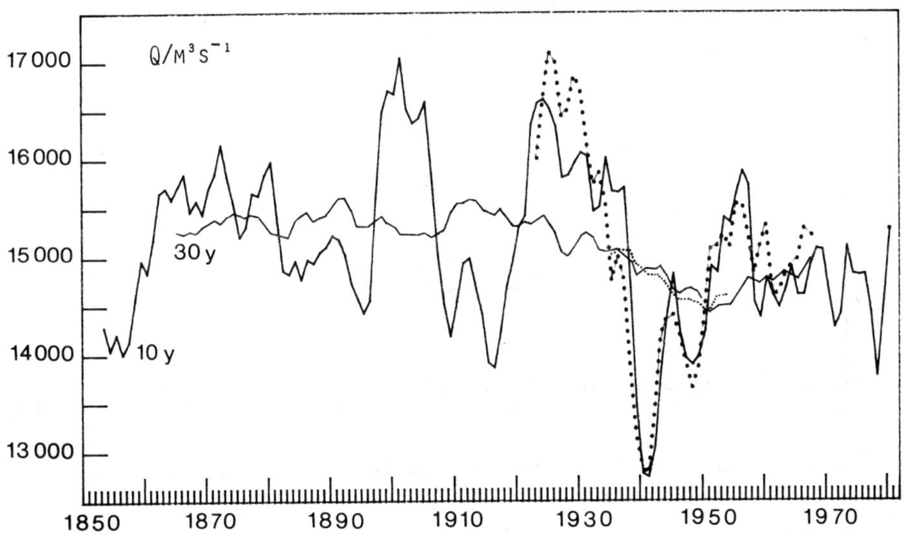

Figure 2. Total river runoff (Q) to the Baltic Sea. Ten (10 y) and thirty (30 y) year running averages. The dotted curve is calculated from the observed data 1921-1970 (Mikulski 1980) and the continuous line gives an estimate using the correlation between the discharge of the river Vuoksi and the total runoff.

3. SEA LEVEL VARIABILITY

Sea level changes in the Baltic Sea are mainly due to wind and air pressure gradients and are therefore related to the cyclonic activity; for instance, the height interval of the monthly extreme values varies considerably within a year. Sea level data along the whole coast of Finland are available from the beginning of the 1920's. From that time the height interval of monthly extreme values does not show any noticeable trend on a yearly time scale.

The data of the Hanko mareograph (which was the first one operating in Finland) from 1887 were analysed in more detail. We first studied the spectrum of sea level heights and found that the spectrum in the frequency interval from 1 to 8 days had the strongest variation within the year. The standard deviation of sea level heights in 8-day periods (which approximately corresponds to the square root the wave spectrum at this interval) was calculated and then smoothed by 30-year moving averages. The results are shown in Fig. 3 for the season October - January, when the sea level variations are largest. There is again no trend after 1920's. Before that the variations of the sea level seem to be larger. This suggests that there has been no significant long-term change in cyclonic activity in the wintertime after 1920's. The decreasing trend before 1920's coincides with the decrease of the maximum annual ice extent in the Baltic Sea (cf. Fig. 5).

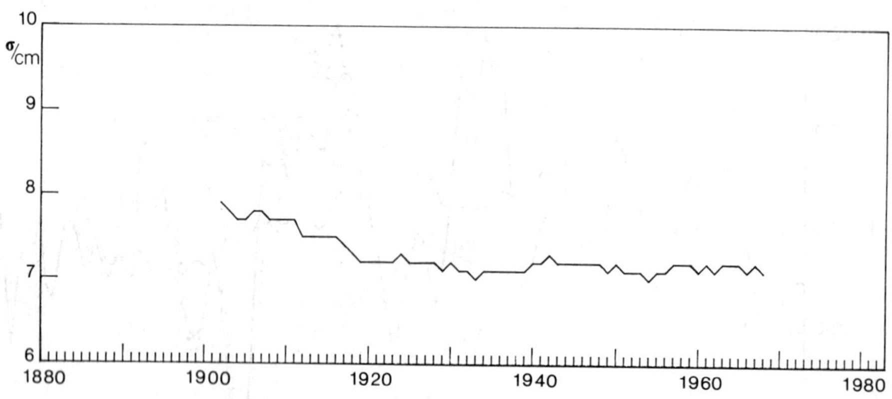

Figure 3. Standard deviation of sea level in Hanko, 8 days period. Averaged over October - January, 30-year running means.

4. ICE CONDITIONS

The data on ice conditions in Helsinki in Fig. 4 show a pronounced decrease in the length of the ice season: the duration of ice cover has shortened by 1½ months from the 1860's to the 1970's.

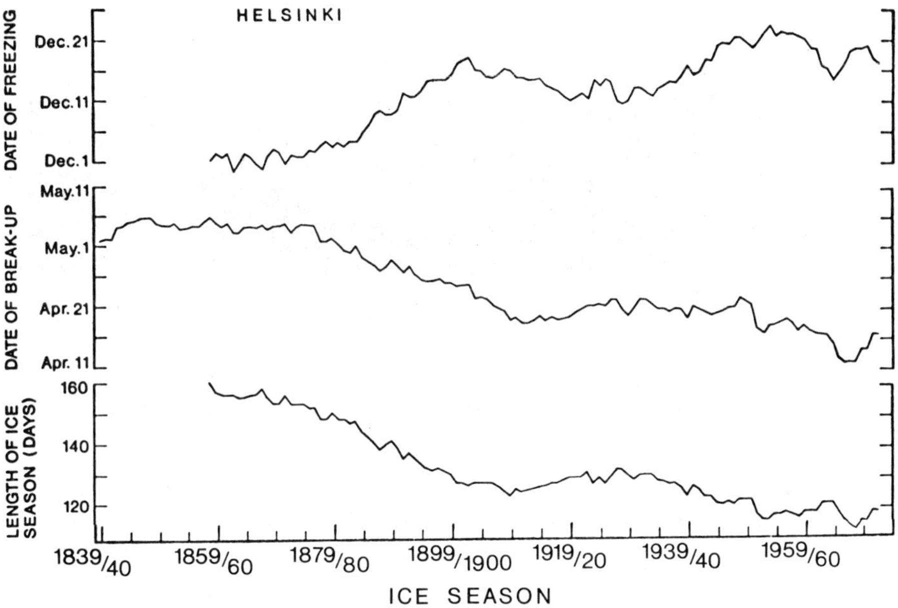

Figure 4. Time-series of the data of freezing and ice break-up and the length of ice season at Helsinki (20-year running averages). (From Leppäranta and Seinä, 1984.)

The maximum annual ice extent of the Baltic Sea in Fig. 5 also reveals a decreasing trend in the 19th century. It correlates ($r = -0.89$) with the mean winter air temperature in Helsinki (Fig. 6), which allows prognosis of future mean maximum ice covers on the basis of predicted mean air temperatures. Fig. 7 shows an example of such a prognosis where a 1.8^{o} C increase in the mean winter air temperature as was predicted by Budyko et al. (1981) is assumed. As can be seen, this anthropogenic warming of climate would dramatically alter the ice conditions in the Baltic Sea.

5. CONCLUSIONS

There has been an increase in surface water temperature and a decrease in the length of the ice covered season as well as in the annual maximum ice extent in the Baltic Sea during the last 100 years. Further, a slight increase in water salinity and a small decrease in river runoff has been observed.

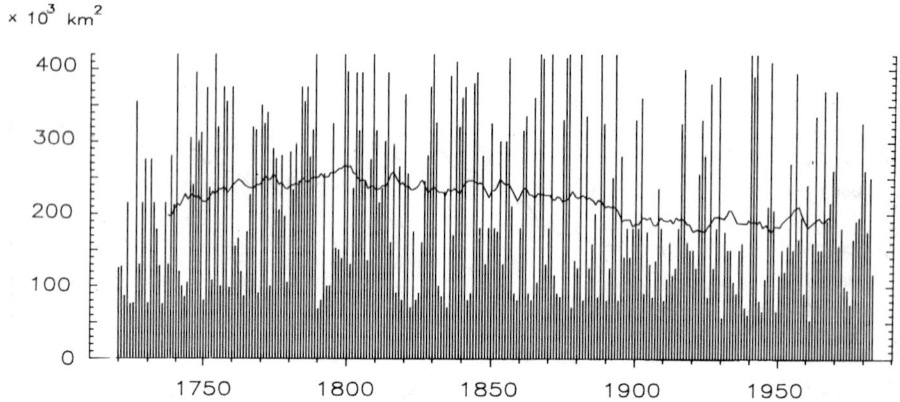

Figure 5. Maximum ice extent in the Baltic Sea 1720-1983. Yearly values and 33 year floating averages (cf. Alenius & Makkonen 1981).

The cyclonic activity, as indicated by sea level variability, was greater in the end of the 19th century and in the beginning of the 20th than it has been after that.

All the above reveals a close relationship between oceanographic and meteorological parameters and shows changes in oceanographip conditions in the Baltic Sea during the last century. As a result, the increase in the atmospheric CO_2 content with the predicted consequences in sea ice conditions, among others, indicates the need for further careful monitoring for finding possible signals of climatic changes.

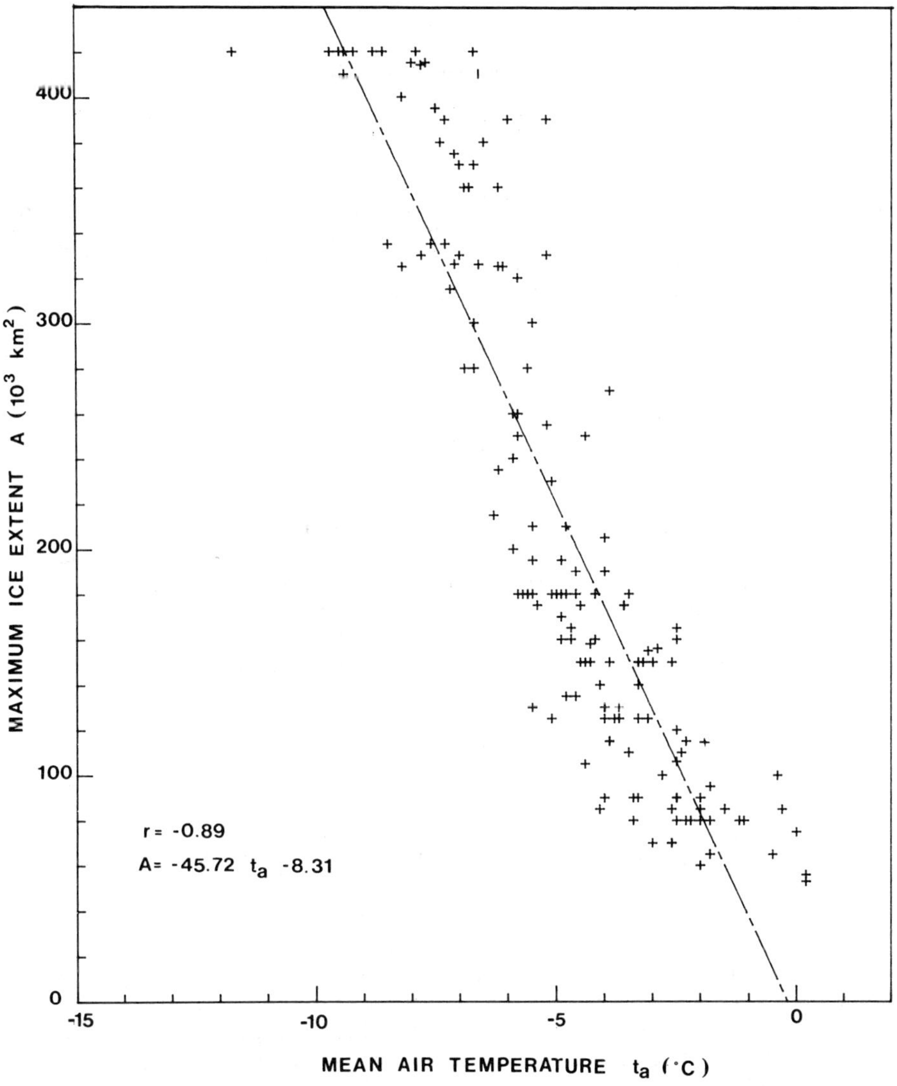

Figure 6. Maximum ice extent in the Baltic Sea versus mean winter air temperature (Dec - March) in Helsinki.

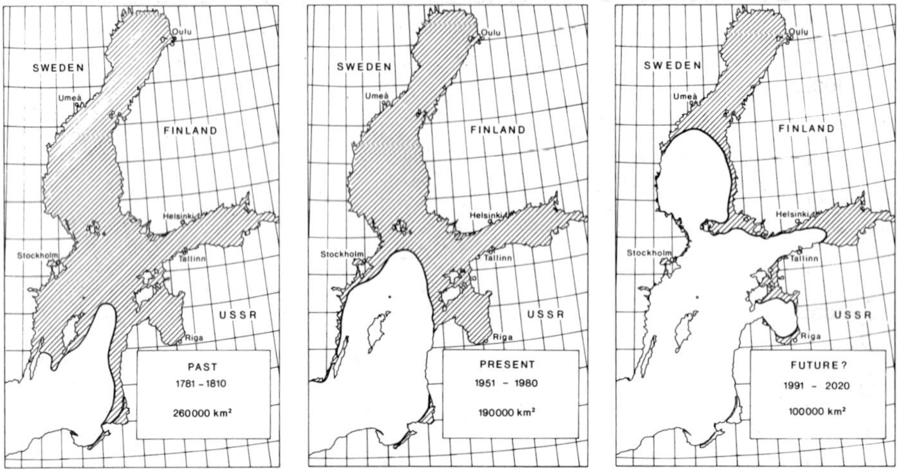

Figure 7. Typical ice situations in the Baltic Sea corresponding to the mean annual maximum ice cover of three 30-year periods. The mean maximum ice cover 100 000 km^2 for the period 1991-2020 is based on the increase in mean winter air temperature at the latitude of Helsinki by 1.8^o C during 1900-2005 adopted from the prognosis for the anthropogenic climatic change by Budyko et al. (1981).

REFERENCES

Alenius, P. & Makkonen, L., 1981. Variability of the Annual Maximum Ice Extent of the Baltic Sea. Arch. Met. Geoph. Biokl., Ser. B, 29, 393-398.

Barry, R.G., 1983. The nature and significance of cloud-cryosphere interactions in the marginal snow/ice zones. The Second Nordic Symposium on climatic changes and related problems.

Budyko, M.I., 1982. The earth's climate: past and future. International Geophysics Series vol 29, 307 pp. Academic-Press.

Budyko, M.I., Byutner, E.K., Vinnikov, K.Ya, Golitsyn, G.S., Drozdov, O.A. & Karol, I.L., 1981. Anthropogenic changes in the global climate. Meteorologiya i Gidrologiya, 8, 5-14.

Jäger, J. & Kellogg, W.W., 1983. Anomalies in temperature and rainfall during warm arctic seasons. Climatic Change, 5, 39-60.

Launiainen, J. & Koljonen, J., 1981. Variations of temperature and salinity at Finnish fixed hydrographic stations in the Baltic Sea. 69th ICES Statuory Meeting, 5-14 October 1981, Woods Hole, USA, C.M. 1981/C:25, 15 pp. (mimeogr.)

Launiainen, J. & Koljonen, J., 1982. Variation of salinity at Finnish fixed hydrographic stations in the Gulf of Finland

and the river runoff to the Baltic Sea. Meri 11, 39-50.
Leppäranta, M. & Seinä, A., 1984. Long-term variations in the sea ice conditions in the northern Baltic Sea. To appear.
Mikulski, Z., 1980. River inflow to the Baltic Sea. University of Warsaw, Faculty of Geography and Regional Studies, and Polish national committee of the IHP/UNESCO, 17 pp. (mimeogr.)
Manabe, S. & Wetherald, R.T., 1980. On the distribution of climate change resulting from an increase in CO_2 content of the atmosphere. J. Atmos. Sci., 37, 99-118.
Matthäus, W., 1979. Langzeitvariationen von Temperatur, Salzgehalt und Sauerstoffgehalt im Tiefenwasser der zentralen Ostsee. Beiträge zur Meereskunde, 42, 41-93.
Mitchell, J.F.B., 1983. The seasonal response of a general circulation model to changes in CO_2 and sea temperatures. Quart. J. R. Met. Soc., 109, 113-152.

STRATEGIES FOR ASSESSING IMPACTS OF CLIMATIC CHANGE IN MARGINAL AREAS

T.R. Carter and M.L. Parry

International Institute for Applied Systems Analysis, Austria

Greater rigour is required in designing research strategies to assess possible impacts of climatic change on society. We can identify relatively disadvantaged or marginal areas which may be particularly sensitive to climatic fluctuations. This vulnerability to climate is best described as a level of risk of significant impact, e.g. crop failure. Changes in climate are accompanied, therefore, by changes in the level of risk. These changes may be very rapid in marginal areas and are characterised by a very steep, quasi-exponential "risk-surface". With suitable models linking climate to crop growth potential, boundaries of agroclimatic risk can be mapped geographically. Temporal shifts of these boundaries identify areas of potential impact. The approach is illustrated in examples from North America and Northern Europe.

The overriding problem facing any study of climate and society is the awesome complexity of the interactions. On occasion we have side-stepped this issue and resorted to investigating merely the synchrony of climatic and social events, with little scrutiny of their connection other than that they occur at the same time and in the same place. The assumption has sometimes been that synchronous events are events which necessarily have a causal connection - an assumption which is clearly false. We need to increase the rigour of our research strategy and thus cope with the complexity of the interactions.

One means of (at least partly) achieving this is to employ a predictive approach to climate impacts in marginal areas, assuming that marginal areas are a particularly suitable laboratory because they are the first to be affected and the most severely affected by climatic anomalies: i.e. they exhibit a high

N.-A. Mörner and W. Karlén (eds.), Climatic Changes on a Yearly to Millennial Basis, 401–412.

degree of risk resulting from climatic change and variability. For this reason marginal areas have been selected by the World Climate Impact Program to be the focus of a study conference on CO_2-induced climate impacts (at Villach, Austria, September 1983) and a two-year research project at IIASA on the vulnerability of food production to climate change.

In this paper we review a number of different strategies for evaluating climate impacts in marginal areas. Our thesis is that changes in climate can usefully be analysed, firstly, as temporal changes in risk and, secondly, as spatial shifts in the probability of pay-off. We will illustrate this contention by reference to a number of case studies.

MARGINALITY AND CLIMATE

We can identify three types of marginality - spatial, economic and social. The first type relates to locations and areas at the edge of their ideal climatic region, where systems of marginal agriculture are frequently ill-adapted to their environmental resource base - for example, where warmth or moisture is frequently insufficient (or, conversely, frequently excessive) for an adequate return to particular types of farming. But whether yield-levels are adequate or barely adequate (i.e. are marginal in an economic sense) is culturally determined - it depends on farming expectations and perceived alternatives. Spatial marginality can thus be resolved into economic marginality. It is also possible to identify marginal groups which, as a result of their social rather than intrinsically economic disadvantage, may be equally vulnerable to unfavourable climatic anomalies or fluctuations. None of these marginal areas or groups is strongly buffered against change of the environment and may thus be particularly sensitive to variations of climate.

CLIMATE CHANGE AS CHANGE IN RISK AND PROBABILITY OF PAY-OFF

For a number of reasons the impact of climate variations on marginal areas can most effectively be measured as a change in the level of risk, i.e. in the probability of an adverse event such as the probability of crop "failure", or net loss, or shortfall from some critical level of output. First, marginal farmers, by definition, operate towards the limits of profitability, have a slender buffer against hardship and thus are more concerned with survival than with wealth. Secondly, the profit-maximising farmer (including those in non-marginal areas) knows well that net returns are not simply a function of average yield, but also of the balance he strikes between gambling on "good" years and insuring against "bad" ones (Edwards, 1978). Thirdly, the pay-off boundary for particular farming activities may depend on the frequency of "good" or "bad" weather; for example, a major constraint on profitable wheat production in Alberta is related to the probability of first autumn freeze (Robertson, 1973). At some locations the parameters of climate which frequently have a major influence on rates of plant growth (e.g. temperature, precipitation, solar radiation) decrease in a roughly linear fashion towards the margin of cultivation. For example, in areas where cereal cropping is limited largely by temperature (viz. at high latitudes and high elevations) accumulated warmth decreases approximately linearly with increasing elevation and increasing latitude. While this is, of course, a generalization, the point is that, assuming annual levels of warmth or moisture to be normally distributed from year to year, the probability of a minimum level of warmth or moisture required to avoid failure, loss or critical shortfall would increase, not linearly towards the margin of cultivation but in an S-shaped curve which is characteristic of the cumulative frequency of a normal distribution (Fig. 1). At the lower end of this curve

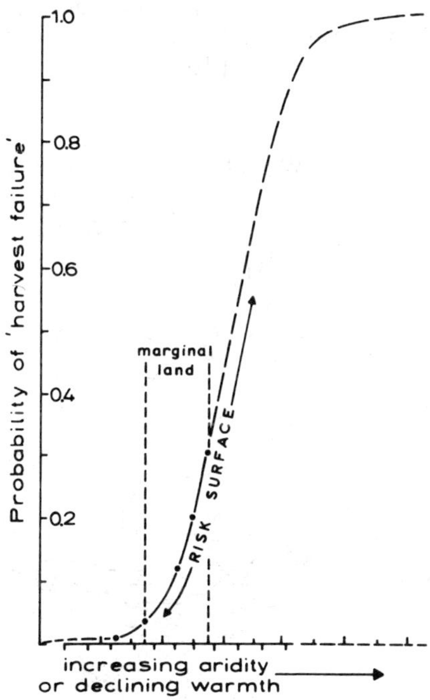

Figure 1. Risk surface due to probability of crop "failure", net loss or critical shortfall, with linear (and normally distributed) gradient of aridity or warmth. Probabilities of harvest "failure" are for oats (var. Blainslie) in S. Scotland (adapted from Parry, 1976).

there is a marked, indeed quasi-exponential, increase in the probability of failure; and it will be shown that it is precisely at this part of the curve that marginal land is frequently located. It seems, therefore, that marginal areas are frequently characterized by a very steep "risk surface". A consequence of this is that any changes in average warmth or aridity, or in their variability, would have a marked effect on the level of risk.

CLIMATE CHANGE AND THE SECULAR SHIFT OF PAY-OFF BOUNDARIES

Climatic variability can therefore be viewed as a risk surface upon which there occurs a varying probability of pay-off. Real "boundaries" of pay-off can be mapped empirically from real enterprise boundaries; alternatively theoretical boundaries can be selected on the basis of notional critical levels of risk tolerance (e.g. a frequency of 1 in 5 failure). In either case climatic change can be evaluated as a shift of these pay-off boundaries:

a) Canadian Great Plains. The shift of critical isopleths has been used to determine the effect on Canadian wheat and barley production of a 1°C downturn of temperature. Biophotothermal timescale equations have been employed to estimate if and when

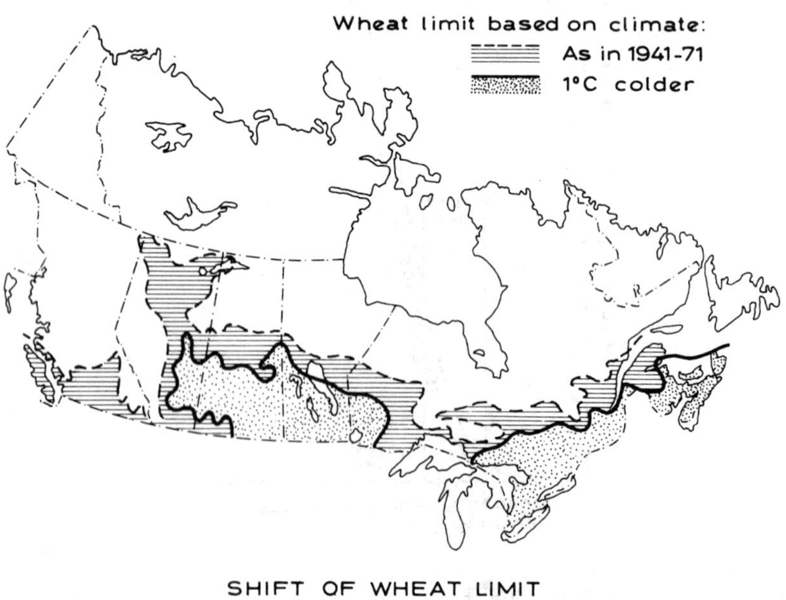

Figure 2. Effect of 1°C cooling on wheat limit in Canada (after Williams & Oakes, 1978).

these crops would normally reach various phenological stages at each of 1100 stations in Canada (Williams & Oakes, 1978). To compute the climatic resources for a cooler climatic regime, 1°C was subtracted from the temperature normals for every month. This made the assumed planting date later, extended the time required to mature as computed by the biophotothermal timescale equations and brought forward the date of first fall freeze. Figure 2 illustrates the shift of isopleths bounding the wheat-maturing zone: the area suited to wheat production (not constrained by soil or terrain) would be reduced by one-third. The area suited for barley would contract by only one-seventh because it extends further north and therefore is more limited by terrain than by temperature. These are, of course, average estimates; no account has been taken of changes in the degree of risk.

b) U.S. Corn Belt. A second variant of the isopleth-shift approach can be illustrated by reference to work on the U.S. Corn Belt (Fig. 3). Newman (1980) applied daily differences of ± 1°C to growing degree-days (GDD) for 18 stations in Indiana over a 10-year period in order to simulate the spatial shift of corn belt boundaries for a 1°C-warmer and drier climate, which is a plausible scenario for the future given continued increases in the CO_2 content

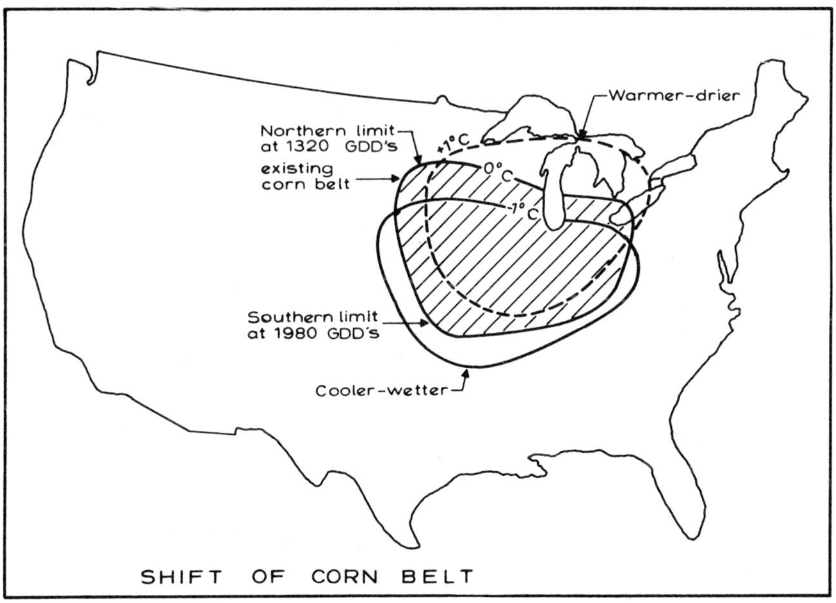

Figure 3. Simulated geographical shift in U.S. corn with temperature changes (based on frost-free growing season thermal units, GDDs) (after Newman, 1980).

of the atmosphere (Kellogg & Schware, 1981), and for a 1°C-cooler and wetter climate, which is a plausible simulation of conditions which probably occurred for some cool decades in the seventeenth century.

c) Northern Europe. We can also simulate shifts in the probability of harvest failure with changes in temperature alone. In northern Britain a 1°C increase in mean monthly temperature throughout the growing season would, ceteris paribus, lead to a 140-metre upward shift of the probability isopleths regarded as critical for successful farming (a failure frequency of 1 in 3.3) (Fig. 4). Across the British Isles as a whole there would be regional variations in the lapse rate of temperature with elevation. In total about 2 million ha (about one-third) of Britain's unimproved

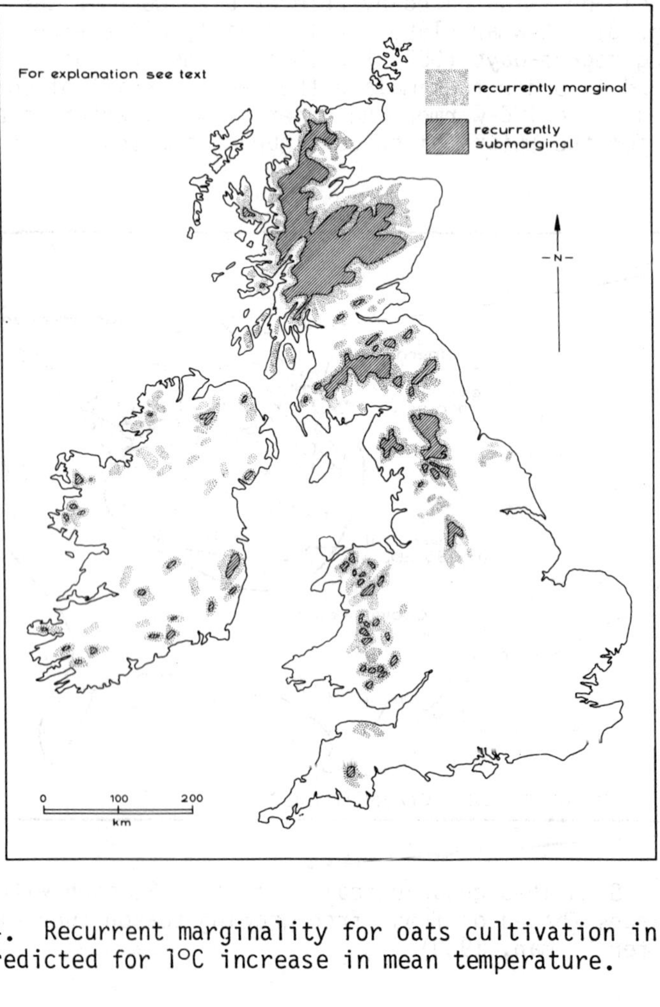

Figure 4. Recurrent marginality for oats cultivation in British Isles predicted for 1°C increase in mean temperature.

moorland, which are at present submarginal for cereals in terms of summer warmth, would become marginally viable for cereal farming.

CLIMATIC VARIABILITY AS THE INTER-ANNUAL SHIFT OF PAY-OFF BOUNDARIES

The foregoing analysis is seriously weakened by its focus on average conditions (of yield, pay-off, etc.) and by its failure to consider that, in reality, pay-off boundaries are shifting annually and that the boundaries between, for example, different farming regions reflect a response to the perception of these inter-annual variations. We can remedy this failure by mapping the pay-off boundary for each year and analysing its inter-annual variability (and any changes in its variability). For example, we can identify, for each year, the elevation at which cereals will ripen in northern Europe. For example, during the run of warm years 1788-1792, crops would have ripened above 550 m in southern Scotland. In other years (e.g. 1816 and 1817) crops would have failed here even at elevations of only 180 m (1816) and 260 m (1817). In fact, there was extensive famine, bankruptcy and abandonment of marginal farmland throughout the U.K. at this time (Parry, 1978).

Given adequate data on the farming system and on climatic variability in marginal areas it is possible to construct scenarios of the impact on marginal agriculture of the weather of individual years, or "runs" of years or of longer term climatic fluctuations. For each of these scenarios we can predict a pay-off boundary at a particular position on the gradient of the agricultural frontier. In the present example in S. Scotland, for "warm" years (>1700 day-degrees C) we can predict a pay-off boundary at about 400 m. For "cool" years (<1400 day-degrees C) it would fall below 300 m. To the cereal farmer above 300 m in "cool" years the result would be a net loss. Over both warm or cool phases we can say that the pay-off boundary will "average out" between these elevations and that, above it, cereal farming might cease.

YIELD SIMULATION MODELLING AND THE SHIFT OF PAY-OFF BOUNDARIES

In many situations, a fuller understanding of crop/climate relationships is necessary, both: a) for a better appreciation of climatic influences on contemporary crop production, and b) to assess the probable impacts of future climatic fluctuations of crop cultivation and yield. Indeed, before considering the effects of non-climatic factors on crop growth, it is important that the climatic potential of an area is evaluated, to provide a background (albeit fluctuating) resource base. Only then may other factors be superimposed, e.g. disease, soil type, management, etc., some detrimental to crop growth, others implemented to enhance development. These are assumed to be standard in the model outlined below.

The model. The cereal yield simulation model is an attempt to identify and model those seasonal climatic conditions which are important in the development and yield of a cereal crop (winter wheat in this study). It was developed initially by Dr. M. Hough (Meteorology Section, M.A.F.F., Harrogate, North Yorkshire) but has undergone extensive reworking and amendments to incorporate recent field and laboratory observations.

The model simulates a crop's growth as the sum of photosynthesis and respiration processes. The rate at which a plant's weight increases is limited by the rate at which it can assimilate carbon dioxide from the atmosphere for reduction to carbohydrate. The rapidity of this process (photosynthesis) depends largely upon the intensity of solar radiation, the leaf area available for interception and the temperature of the plant's environment, with an additional limiting factor of water stress.

Not all the carbohydrate produced during photosynthesis contributes directly to the growth of the plant. A proportion is used up by respiration, a temperature-dependent process involving the making of new cells and maintenance of existing plant structure. Therefore, subtracting the respiration from gross photosynthesis leaves total dry matter production. Grain yield can be estimated by further subtracting the dry matter weights attributable to roots, stem and leaves using indices derived from operational and experimental observations. However, this study is restricted to consideration of total dry matter weights, hereafter referred to as "yields".

Simulating potential cereal yields in Northern England.

a) Sowing date. The sowing date has an important effect on the timing of leaf development during the optimum growth period, and in most cases the earlier the crop is sown, the higher is the potential dry matter yield. For simplicity, each model run simulates crop development commencing at 4 arbitrary sowing dates representing the observed range (late September to mid-November).

b) Model operation. Simulations are executed on computer for weekly time increments and outputs include tables and graphs showing the weekly accumulation of dry matter throughout the growing season. So, at the site of any meteorological station providing suitable data, an indication of the climatic yield potential for winter wheat may be gained. This study uses daily data from a number of stations in northern England at a range of altitudes and for a period of up to 21 years.

While it might seem nonsensical to expect cereals to be cultivated above 500 m in the Pennine uplands because of unsuitable soils, leaching, poor drainage, inaccessibility, etc., the meteorological data from an upland station are perfectly applicable for climatic modelling. In fact, the use of data representing somewhat severe conditions is an effective test of the model's sensitivity and validity.

Comparison of typical simulations at a lowland site (6 m) and an upland site (556 m) reveal two noteworthy features. Firstly, the growing period is considerably longer at the upland station for equivalent sowing dates (11-12 weeks). Secondly, the yields predicted for the upland station are lower than those for the lowland station (about 1 T/ha).

The sensitivity of the model may be demonstrated by comparing upland and lowland yields over two contrasting seasons (Fig. 5). The first illustrates the predictions for a cooler than average season. The development of the upland crop is considerably retarded and the yield much reduced compared to the average. The latter effect is largely a result of late development and the

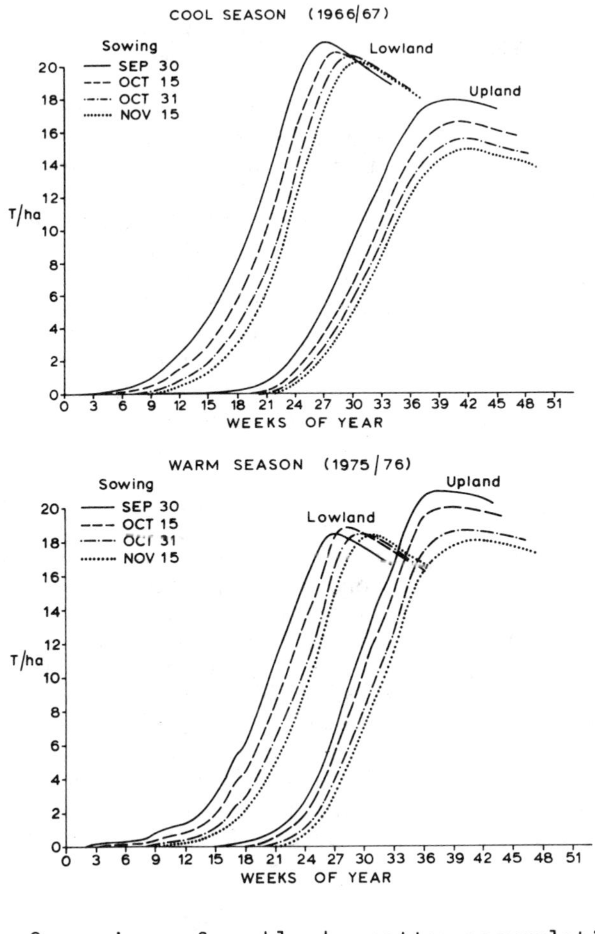

Figure 5. Comparison of weekly dry matter accumulation of winter wheat at lowland and upland stations for cooler than average and warmer than average conditions.

inability of the crop to utilise fully the benefits of higher solar radiation in the summer months. This point is exemplified for the drought year, 1976, where the modelled yields are greater in the uplands than the lowlands. Development is more rapid at both sites with increased radiation interception, but high summer temperatures in the lowlands have actually restricted development whilst in the cooler uplands, growth conditions are close to optimum.

In these examples, the model has simulated growth conditions for a fully irrigated crop (i.e. no water stress). In most years, however, there is a marked water deficit in lowland eastern England whereas crops in an upland location are usually able to transpire at their potential rate. Thus, yields are depressed to a greater extent in the lowlands although in practice the effect is commonly offset by irrigation.

Yield thresholds, length of growing season and harvest failure. Two preconditions are now introduced which must be satisfied by the modelled crop to prevent harvest failure.

The first premise, although it may not hold for smaller scale activities, is reasonable for a majority of commercial farming operations given that government incentives are frequently available for particular crops. It is assumed that there is a positive relationship between level of crop yield and financial return, and below a certain yield threshold the harvest may be considered to have failed (for whatever reason, e.g. insufficient profit margin, net financial loss, net loss of seed grain, etc.).

A second constraint is imposed by the length of the growing season. For a given sowing date, a cut-off date may be defined after which harvesting is considered either impossible or unprofitable. The very latest harvest date which is allowable is that one year after sowing, otherwise cropping in successive years would be progressively retarded. However, other criteria are likely to restrict harvesting to an earlier date including waterlogging, moisture content of the grain, autumn frost, etc.

When the two constraints are imposed on an annual dry matter growth curve, four possible conditions may be defined, one resulting in a successful harvest (i.e. sufficient yield and suitably early harvest date), the remaining three describing harvest failure.

Frequency and probability of harvest failure. This analysis may be replicated for all station-years providing an indication of the frequency of climatically-induced harvest failure at each location which, given adequately long-term records, can be converted to a probability of harvest failure at each station.

If the station probabilities are now mapped, isopleths of equal probability may be constructed, producing a risk surface of harvest failure.

At this stage the idea of a probability threshold may be introduced, i.e. the probability of harvest failure above which the risk of failure is too great for cultivation to be undertaken (pay-off boundary). This may be delimited on the risk surface and represents the probability threshold during those years for which the model was operated.

Climatic change and the shift of isopleths. Since the probability threshold is representative of mean climatic conditions in only one selected time period, longer term climatic fluctuations (in the order of decades) will only become apparent if another time period is analysed. Unfortunately, suitable detailed meteorological data from earlier periods are limited in coverage, duration and availability. However, it is quite plausible to speculate about future climates using, for example, the projections of CO_2/climate modellers as inputs to the cereal yield simulation model.

The effects of these climatic changes on modelled yields would be manifest in the shifting of isopleths of harvest failure. The areas delimited by the shifting isopleths alternate between states of unacceptable risk of harvest failure in one period to acceptable risk in another. Thus the impact of climatic change on agriculture is likely to be greatest in this area of recurrent marginality.

SUMMARY

The strategy outlined above may be neatly summarised as a flow diagram (Fig. 6). Although analysis of preliminary data has only recently begun, the general approach appears to be appropriate and

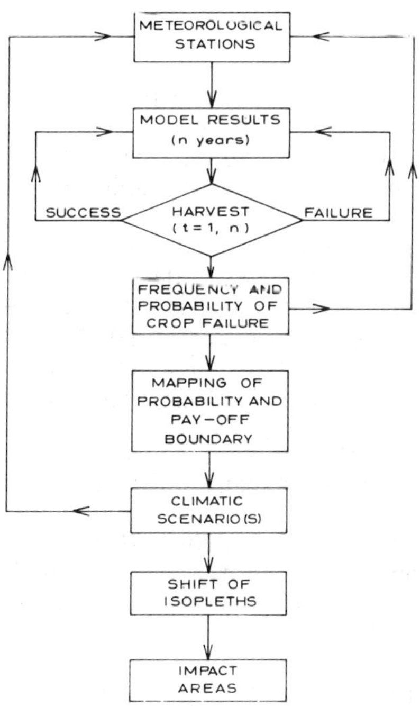

Figure 6. Steps in the identification of climate impact areas.

workable, offering a suitable framework for further studies of climatic assessment in marginal areas.

CONCLUSION

More than twenty years ago participants at the Aspen Conference on the climate of the eleventh and sixteenth centuries (Proceedings, 1963) recommended marginal areas as appropriate "laboratories" for the measurement of climatic impacts. We have made some progress since then, but the science of climate impact assessment is still embryonic and far from perfect. We have, however, the beginnings of a methodology and of an armoury of techniques. Over the next two years it will be the aim of a WCIP research project at IIASA to employ this methodology and these techniques and to develop them further in order to evaluate the impact on food production of changes in climate in the future.

ACKNOWLEDGEMENTS

Mr. Carter acknowledges financial support from the Natural Environment Research Council during tenure of a Research Studentship.

REFERENCES

Edwards, C., 1978. Gambling, insuring, and the production function. Agric. Econ. Res., 30, 25-28.
Kellogg, W.W. & Schware, R., 1981. Climate change and society. Consequences of increasing atmospheric carbon dioxide. Westview.
Newman, J.E., 1980. Climate change impacts on the growing season of the North American Corn Belt. Biometeorology, 7 (2), 128-142, Supplement to Int. J. Biometeor., 24.
Parry, M.L., 1976. The significance of the variability of summer warmth in upland Britain. Weather, 31, 212-217.
Parry, M.L., 1978. Climatic change, agriculture and settlement. Dawson.
Proceedings of the conference on the climate of the eleventh and sixteenth centuries, Aspen, Colorado, 1963. NCAR, Technical Notes, 63-1.
Robertson, G.W., 1973. Development of simplified agroclimate procedures for assessing temperature effects on crop development. Plant response to climatic factors. Proc. Uppsala Symp., 1970. UNESCO.
Williams, G.D.V. & Oakes, W.T., 1978. Climatic resources for maturing barley and wheat in Canada. In: Essays on meteorology and climatology: In honour of Richard W. Longley (eds. K.D. Hage and E.R. Reinelt), University of Alberta, Studies in Geography, Monograph 3.

IV. Models and Mechanisms

MECHANISM FOR LINKING SOLAR ACTIVITY TO WEATHER-SCALE EFFECTS, CLIMATIC CHANGES AND GLACIATIONS IN THE NORTHERN HEMISPHERE

Václav Bucha

Geophysical Institute of Czechosl. Acad. Sci.,
Czechoslovakia

INTRODUCTION

As the results of palaeomagnetic research into sedimentary rocks have shown, the geomagnetic pole was located several times and for long periods in the Pacific during the interval 18 - 12 thousand years ago. This pole moved from the Pacific to the eastern part of North America in the interval 12 to 10 thousand years ago when also the last period of glaciation terminated. In order to prove whether the geomagnetic pole plays any role in the longterm climatic changes, the mechanism causing short-term changes in climate and weather should be found.

Using processed sets of geomagnetic and meteorological data, we demonstrate that increased corpuscular radiation indicated by geomagnetic activity and characterized by enhanced penetration of electrons into the middle atmosphere and stratosphere, by tho generation of bremsstrahlung and the formation of ion clusters at altitudes of around 15 km, results in increased temperatures and atmospheric pressure in the stratosphere and troposphere. It will be shown that these effects also lead to an immediate increase in temperature and pressure at the 500 mb level in the auroral oval and cause the meridional type of atmospheric circulation to change to zonal. The results substantiated by statistical calculations indicate that these changes in flow gradually influence pronounced changes in temperature and pressure in the northern polar cap, increase the cyclogenesis and are responsible for sudden stratospheric warmings especially in polar regions under increased corpuscular radiation.

With a time lag of 13 - 20 days relative to the commencement of the geomagnetic storm, depending on the rate at which the me-

N.-A. Mörner and W. Karlén (eds.), Climatic Changes on a Yearly to Millennial Basis, 415–448.

teorological processes involved proceed, a pronounced increase in temperature occurs in Europe and North America due to the prevailing zonal flow across the Atlantic.

When the corpuscular (geomagnetic) activity displays a marked decrease, meridional flow increases in intensity again and with the same time lag relative to this decrease Arctic air penetrates into Europe and eastern North America, where this is then reflected in an abrupt and pronounced drop in temperature. As the distribution of temperature deviations from normal shows, pronounced negative temperature deviations occur practically on the same territories of Europe and North America which were subjected to glaciation during the last glacial period.

Under very low geomagnetic activity the processes in the auroral oval are weak, meridional flow being prevelant in the northern hemisphere; atmospheric blocking develops especially over the eastern part of the North Atlantic. Cooling of the northern hemisphere can be observed, which is the most expressive in Europe and in the eastern part of North America. At the same time, temperatures in the western United States (Pacific region) tend to be significantly higher than the long-term averages; drought periods in the Sahel zone in Africa occur in periods with low geomagnetic activity.

Under high geomagnetic activity zonal flow in the North Atlantic region intensifies considerably; there is a relative absence of atmospheric blocking in the Atlantic and Pacific regions, sudden stratospheric warmings in winter are relatively frequent, temperatures in Central Europe and in the eastern part of North America are higher than the long-term averages. In the Pacific region of North America cold winters occur and rainfall increases in the Sahel.

Let us try to prove the suggested link of geophysical and meteorological processes in connection with changes in the corpuscular radiation (geomagnetic activity).

1. EFFECT OF CORPUSCULAR RADIATION ON PROCESSES IN THE MIDDLE ATMOSPHERE AND TROPOSPHERE

At extratropical latitudes in every season there are basically two types of atmospheric circulation (Chromov, 1964): the meridional type with blocking pressure highs over the Atlantic, but also the Pacific, extending northwards, and pressure lows extending from the polar region (l.h.s. of Fig. 1), and the zonal type with a central cyclone over the polar region (r.h.s. of Fig. 1), with one type changing to the other many times in the course of the year. Two examples are given in Figs. 2,3. We know that the temperature difference, which we observe when the meridional flow changes into zonal, is quite appreciable in some regions; the increase in temperature due to the zonal flow from the Atlantic may amount to as much as 20°C in Europe. At the same time cold air comes to the western part of North America. The abrupt decrease of temperature when corpuscular radiation is low occurs due to a sudden invasion of arctic air and may be as much as 30°C in Europe and in the east of North America. Simultaneously, the warm air from the Pacific comes to the western part of North America where we observe increased temperatures as was proved mainly for winter periods.

Our results indicate that the reason for this peculiar behaviour of the atmospheric circulation are the processes which take place along the auroral oval at the time of enhanced corpuscular (geomagnetic) activity characterized by the penetration of energetic particles and bremsstrahlung as far as the lower stratosphere (Clauer, 1980), resulting in direct increases of temperature and pressure even in the troposphere, and causing the zonalization of atmospheric flow to increase (Bucha, 1983). Let us try to prove this statement.

Relatively unstable active regions, particularly variable filaments, occur at the solar disk centre prior to the enhancement of geomagnetic activity (Figs. 4,5) (Bucha, 1980 a,b).

The solar-energy flux for meteorological phenomena is

$$P_{EM} = \pi r_E^2 F(1 - A) = 8.9 \times 10^{16} \text{ W}$$

(if we assume the Earth to have an albedo A = 0.5, r_E is the radius of the Earth, F is the solar constant). The available corpuscular energy flux is less than one millionth that of the solar electromagnetic energy flux absorbed by the Earth (Dessler, 1974).

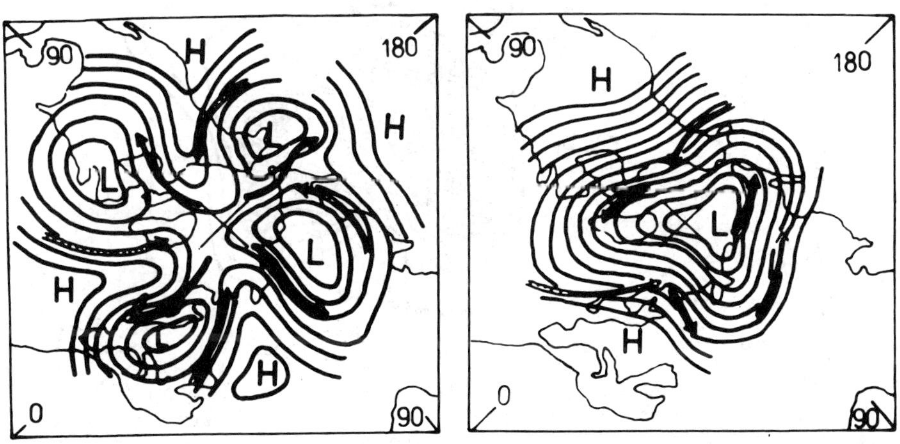

Figure 1. Example of meridional (l.h.s.) and zonal (r.h.s.) type of atmospheric circulation shown on a 500 mb level map (Chromov, 1968).

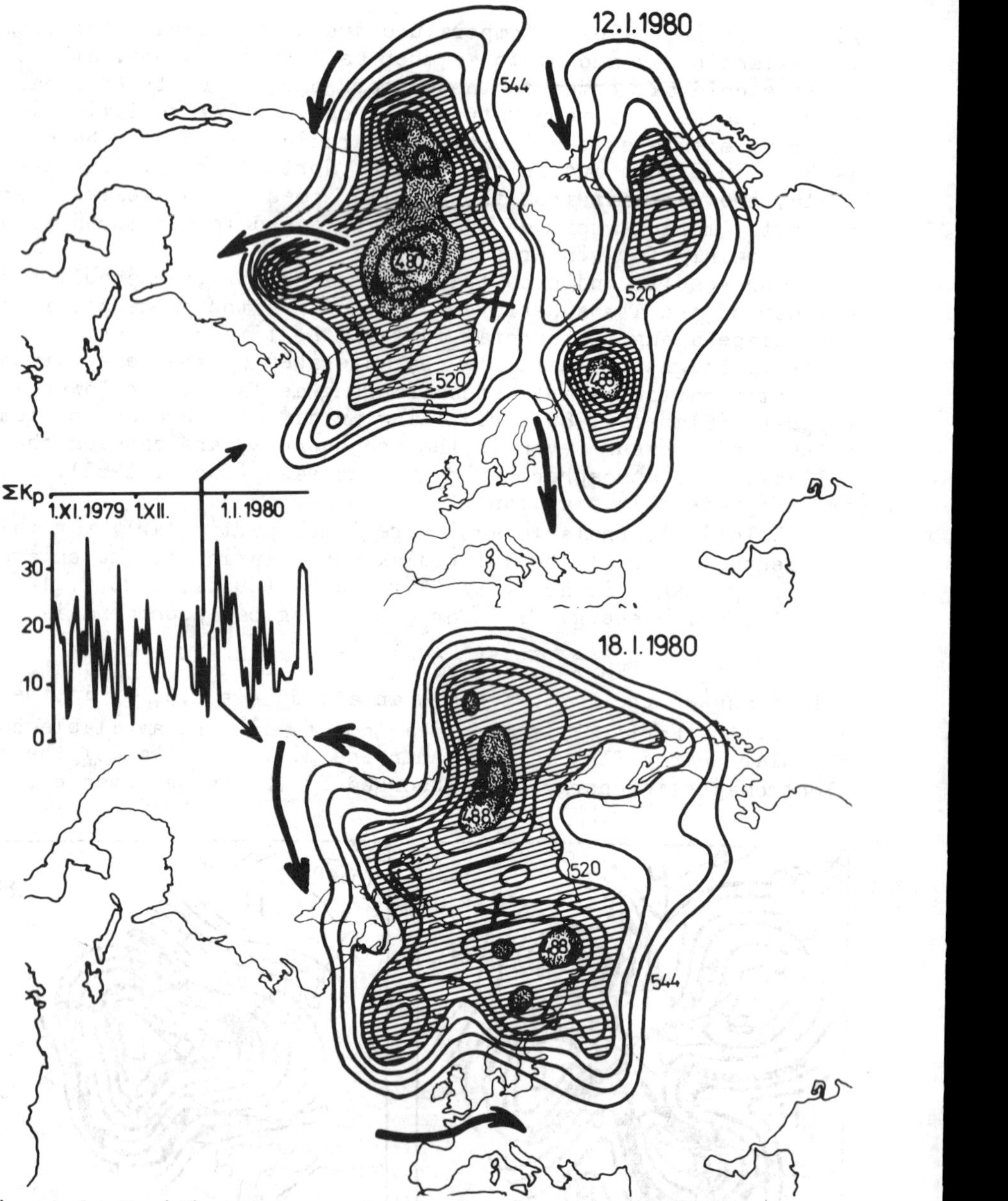

Figure 2. Meridional type of atmospheric circulation (12.1.1980 - top) as a result of decreased geomagnetic activity (on 22 to 25.12.1979) and zonal flow (18.1.1980 - bottom) which developed at a time of enhanced geomagnetic activity (on 29 to 31.12.1979). The arrows indicate cold, Arctic (top) and warm, westerly (bottom) flow in Europe and eastern part of North America (500 mb level).

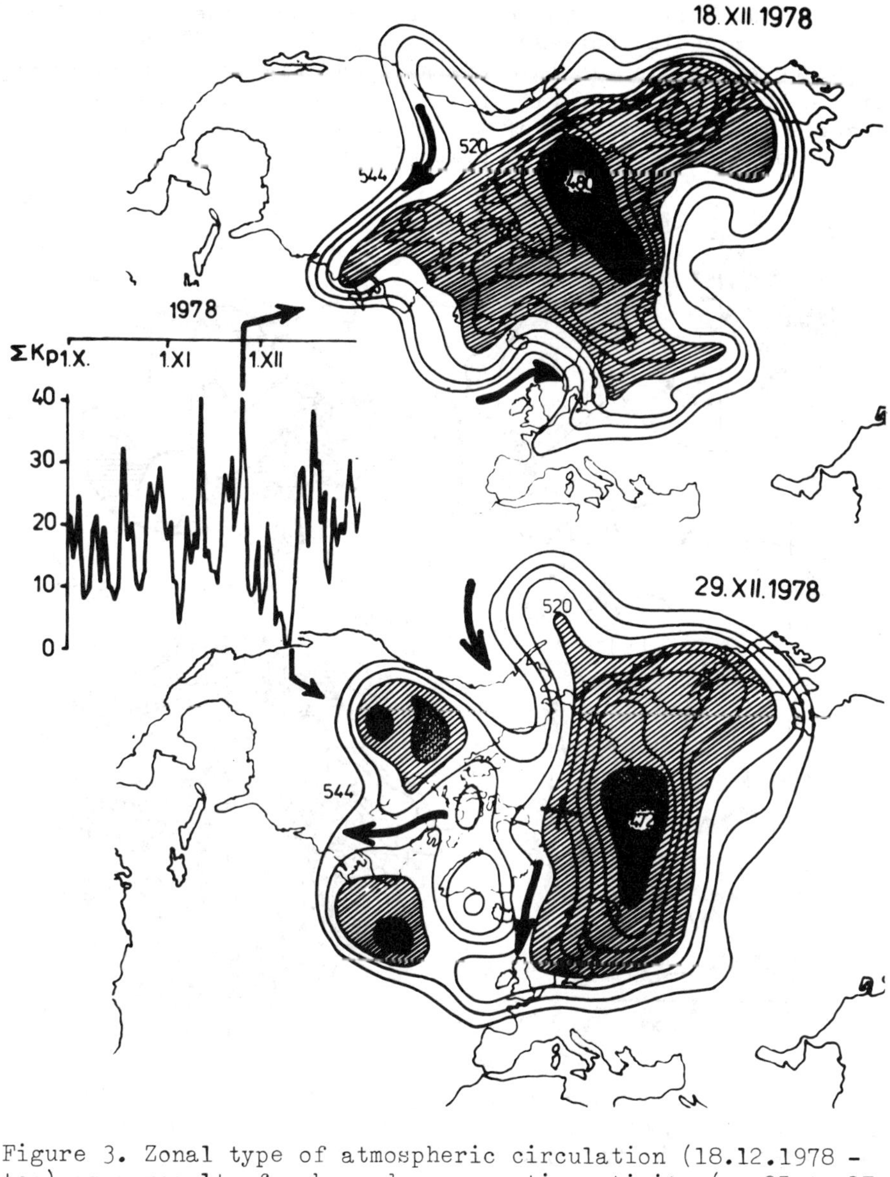

Figure 3. Zonal type of atmospheric circulation (18.12.1978 - top) as a result of enhanced geomagnetic activity (on 25. to 27. 11.1978) and meridional type (29.12.1978 - bottom) which developed at a time of very low geomagnetic activity (on 10. to 12.12. 1978). The arrows indicate warm, westerly (top) and cold, Arctic (bottom) flow in Europe and eastern part of North America (500 mb level).

For the winter hemisphere, however, P_{EM} might drop to 6×10^{15} W. During an intense magnetic storm, the corpuscular energy flux could increase to $P_{c(max)} = 10^{12} - 10^{14}$ W, which might be enough to use this energy as a trigger. Thus there appears to be enough power within the magnetosphere to cause such changes in the vorticity of the lower atmosphere if the power can be directed and coupled effectively.

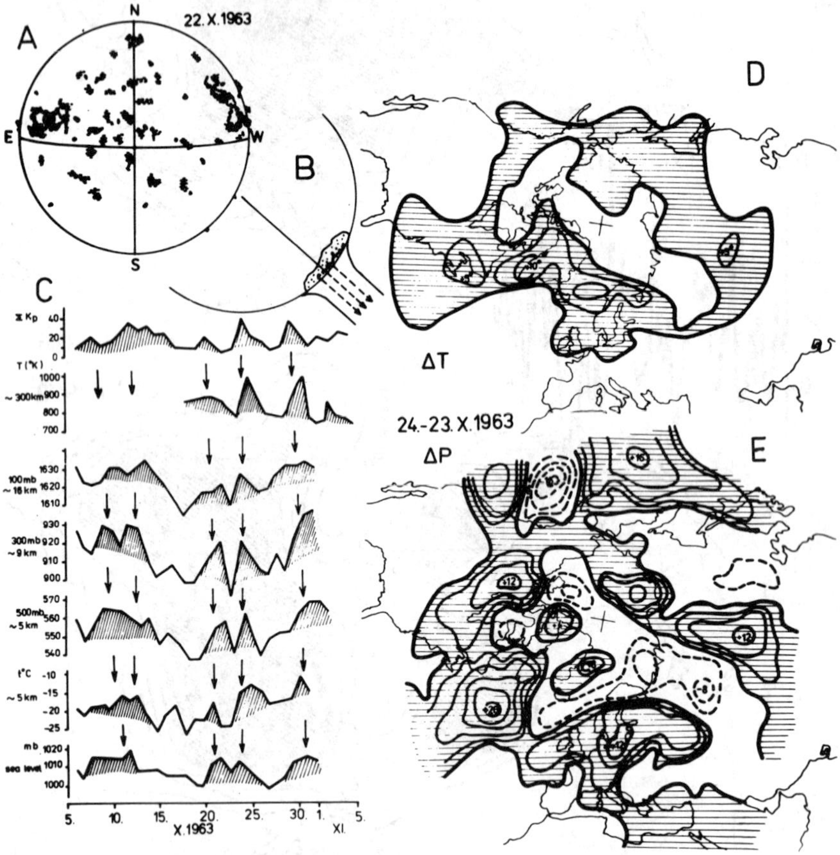

Figure 4. A) Occurrence of a vanishing filament above the centre of the solar disk on 22. October 1963, preceding an increase of geomagnetic activity, B) schematical representation of the corona and of a corpuscular stream directed at the Earth, C) relation between enhanced geomagnetic activity and increased temperature and pressure along a vertical section in the auroral oval (Fig. 2 in Bucha, 1983), D) temperature differences - warming (hatched) and E) pressure increases along the auroral oval.

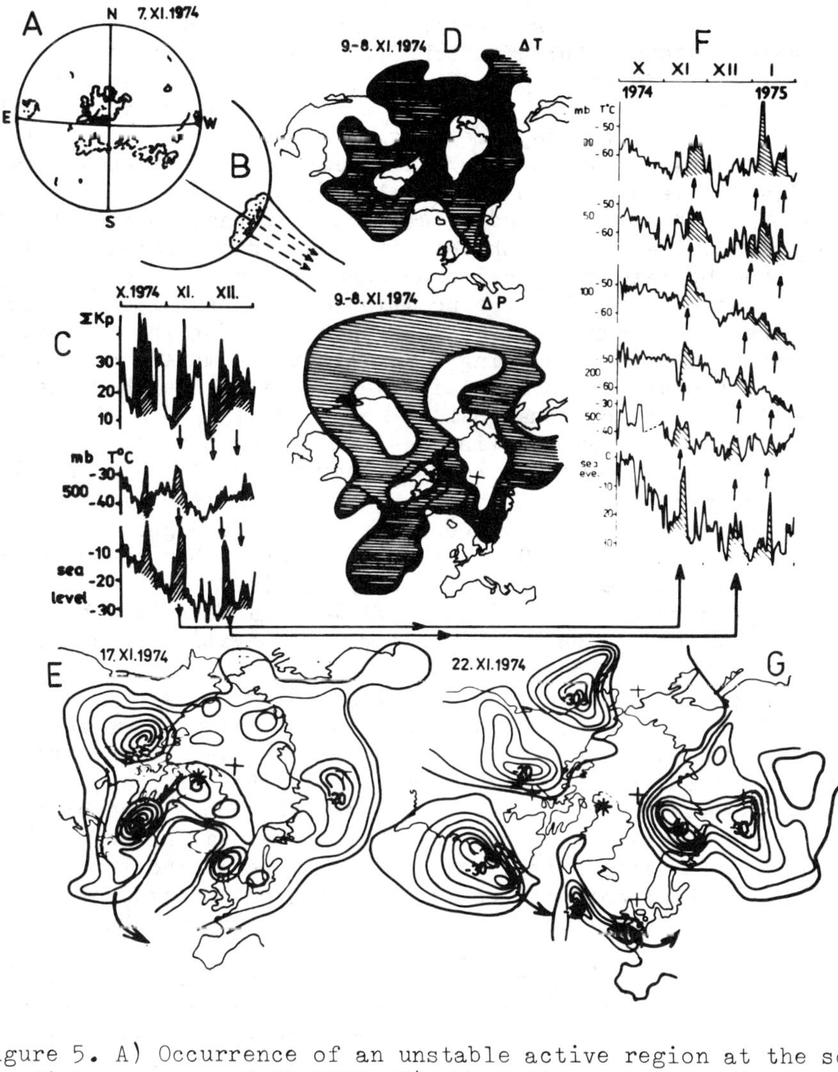

Figure 5. A) Occurrence of an unstable active region at the solar disk centre on 7.11.1974, B) schematic representation of the corona pointing towards the Earth, C) geomagnetic activity (ΣK_p) and temperature changes in the auroral oval (observatory 78°N, 60°E) at the 500-mb level and at the sea level, D) temperature differences - warming (hatched - top) and pressure increases (bottom) at 500-mb level, E) increased cyclogenesis and transfer of air into the polar region (17.11.1974), F) sudden stratospheric warming (Bucha, 1980a), G) flow to the south between Grenland and Canada, across the Atlantic (17.11.) and together with the warm air from the Carribean sea towards Europe (22.11.1974).

Together with the increased penetration of corpuscular particles into the Earth's magnetosphere, we may observe a marked increase in temperature and considerable changes in pressure not only at altitudes of 270 - 400 km above the Earth's surface as recorded by Explorer 17 (Jacchia et al, 1967), but also in the stratosphere and troposphere (Fig. 4D) (Bucha, 1976, 1980 ab), where temperature and pressure increases are observed at practically all levels including sea level, e.g. along parallel 50°N between meridians 30°W and 55°W (east of Canada in the Atlantic - - average values); here the effect being investigated was least perturbed by meteorological processes (Bucha, 1983).

If we believe that the position of the geomagnetic pole as a centre of the auroral oval is responsible for the long-term changes in climate including glaciations, we must prove the function of the auroral oval in influencing also the short-term climatic and weather changes.

Let us consider our findings concerning the temperature increase of 2 - 10°C and pressure increase of 10 - 20 millibars along the auroral oval in the northern hemisphere at the 500-mb level immediately after the enhancement of geomagnetic activity, from day to day (Fig. 6); on the other hand, when the geomagnetic activity decreases conspicuously, we immediately observe the tendency to revert to meridionalization characterized by the ray pattern of the regions with positive and negative pressure differences (Bucha, 1983).

In order to prove that meridional flow changes to zonal as a result of penetrating electrons and bremsstrahlung, leading to an increase in temperature and pressure even in the troposphere, we determined the average pressure values read at 12 points along the auroral oval (Fig. 7). Here, the consequences of pressure effects are most pronounced after the heating in the higher layers of the atmosphere, being directly reflected in the troposphere in the change of the meridional type of atmospheric circulation into zonal (Fig. 5). It is known (Bucha, 1983) that very pronounced meteorological changes, occurring particularly as a result of Rossby's waves, obscure with their much larger amplitudes the said effect which can, however, be immediately observed at nearly all points (a pressure increase of 4 - 8 mb). In an effort to prove that enhanced geomagnetic activity, indicating enhancement of corpuscular radiation, leads to an increase of atmospheric pressure along the auroral oval and to a change of meridional flow into zonal, we calculated the correlation coefficient between geomagnetic activity (Kp-indices) and the average values of atmospheric pressure first derived for 12 points at the 500 mb level, for the period of September to December 1977. The correlation coefficient equal to 0.643 proves that this correlation is relatively good (Fig. 7).

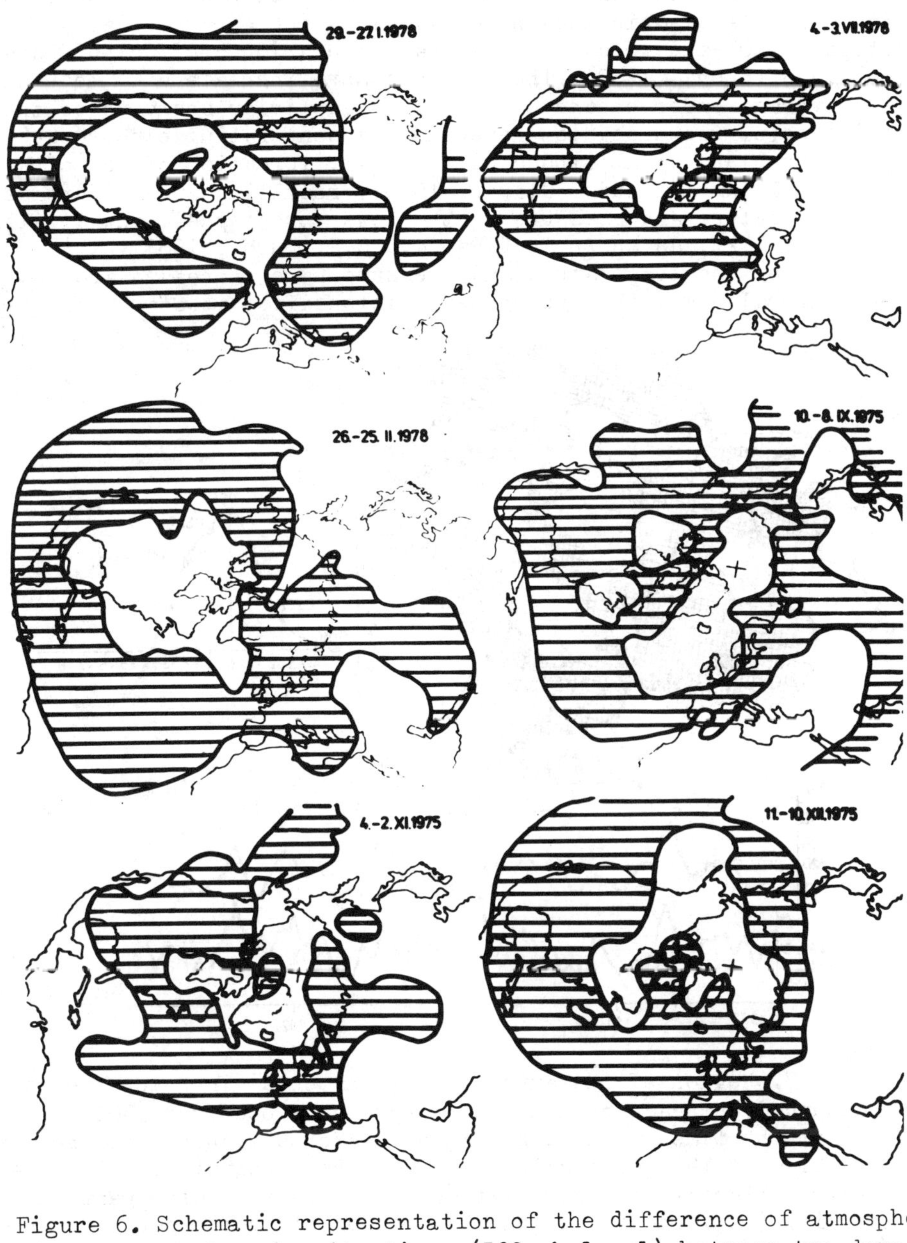

Figure 6. Schematic representation of the difference of atmospheric pressure for six situations (500-mb level) between two days prior to and at the beginning of a magnetic storm. One can observe that the region of increased pressure forms ovals in the form of annuli with the pressure increasing by 8-mb on the average.

We then determined the average values of pressure recorded at 36 points distributed uniformly along the auroral oval at steps of 10° and we came up with a correlation coefficient equal to 0.593 (Bucha, 1983). The said pressure increase of 4 mb on the average under enhanced geomagnetic activity seems to be the evidence of the mechanism generated by penetrating electrons, bremsstrahlung at altitudes below 20 km, formation of ion clusters at 15 km altitude (Clauer, 1980), resulting in an increase of temperature and pressure in the lower troposphere and at altitudes of 5 km, as shown in Figs. 4,5,6,7 (Bucha, 1983).

After the temperature and pressure have increased around the auroral oval, the cyclogenesis intensifies at sea level du-

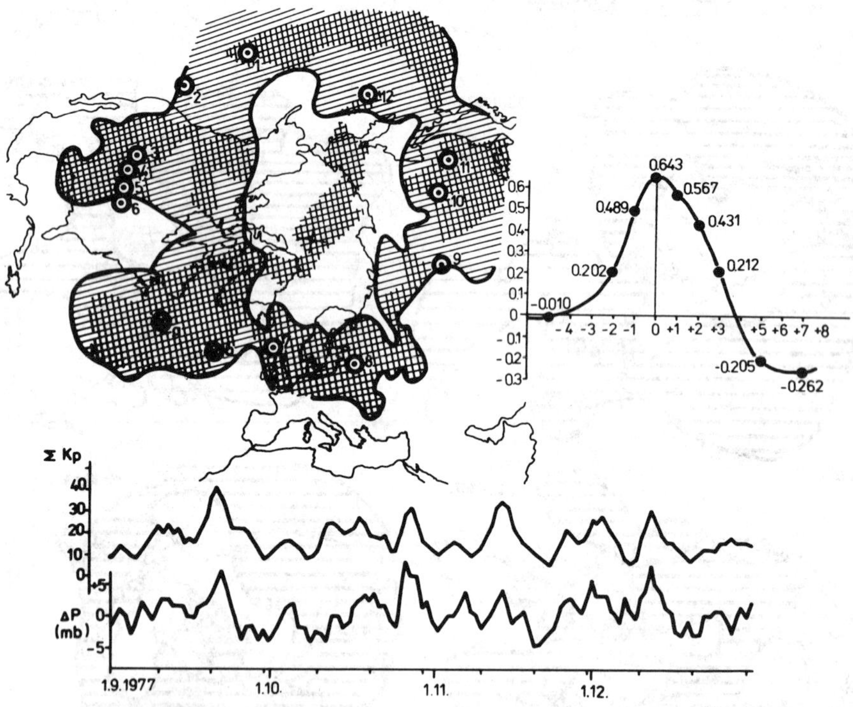

Figure 7. Region of the auroral oval (hatched) in which the most pronounced increase of atmospheric pressure occurred (500-mb level) at the time isolated geomagnetic storms occurred in the latter third of 1977. The values of pressure at points 1 - 12 were used to calculate the correlation coefficient between geomagnetic activity (ΣK_p) and deviations of pressure from the normal (r.h.s.). The graph of the changes of geomagnetic activity (ΣK_p - 3 daily moving averages) and pressure deviations (averages from points 1 - 12) given at the bottom of the figure indicates for zero time lag the correlation coefficient equal to 0.643.

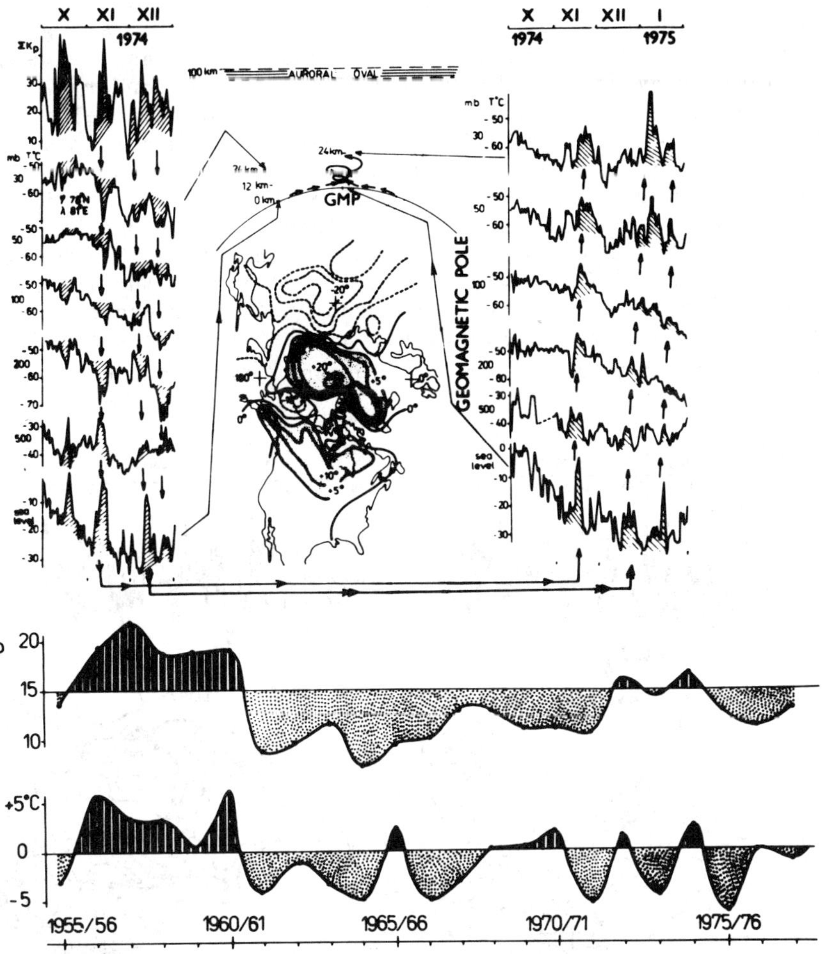

Figure 8. Top: Geomagnetic activity ΣK_p, temperature changes in the auroral oval (observatory 78°N, 60°E - l.h.s.), over the geomagnetic pole where the sudden stratospheric warming is observed (r.h.s.) and distribution of anomalous temperatures at sea level during the geomagnetic storm (Nov. 9,1974 - in the middle). Bottom: Comparison of geomagnetic activity (Ap-indices - averages for November, December, January, February and March) with the average temperature values above the North Pole at the 30-mb level for 1955 - 1978 (Labitzke, 1979). The correlation coefficent obtained (0.649) indicates that the enhanced geomagnetic (corpuscular) activity results in an zonalization of atmospheric circulation and leads to the intensification of sudden stratospheric warming - increases of temperature at the 30-mb level (see 1956 - 1960) as a result of the processes described in this paper. Low temperatures in the stratosphere correspond to low geomagnetic activity.

ring the following days and the pressure lows that are generated move to the north, towards the geomagnetic pole as the centre of the auroral oval; this leads to a pronounced secondary warming (as much as 30°C) along the inside of the auroral oval with a time lag of 2 - 4 days after the increase of geomagnetic activity, as a result of warm air penetrating from the Atlantic and Pacific to the north (correlation coefficient 0.65). A low is created in the region of the pole and this leads, on the one hand, to the occurrence of sudden stratospheric warmings (Fig. 8 - correlation coefficient 0.649) (Labitzke, 1979) and to the intensification of the flow from the above-mentioned polar cyclone due south, between Canada and Greenland, across the Atlantic and into Europe, on the other hand (Fig. 5). The warm air from the Caribbean,

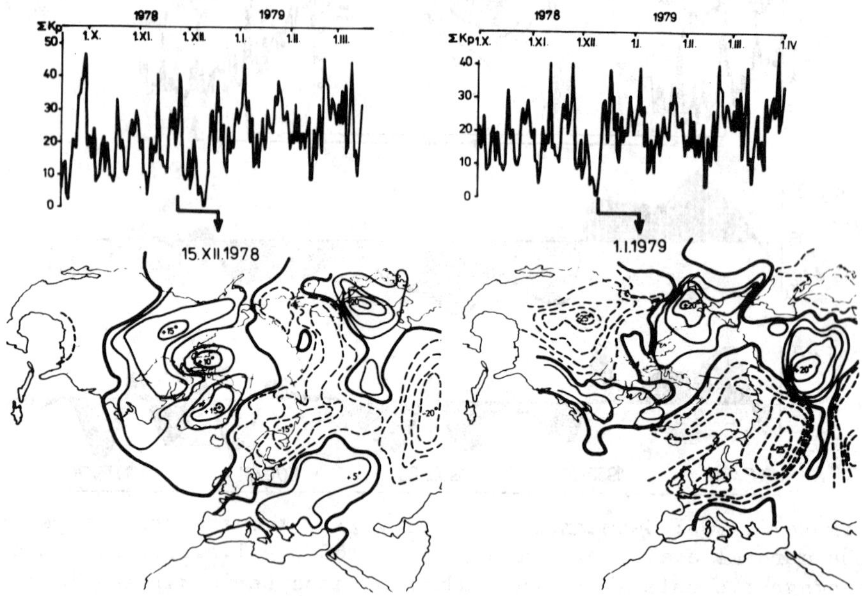

Figure 9. Association between several days of enhanced geomagnetic activity and pronounced positive temperature deviations from the normal, occurring simultaneously on the territory of Europe and North America with a time lag of 13 - 20 days after activity increase (l.h.s. - 15.12.1978): on the other hand, the several days of decreased geomagnetic activity result in the occurrence of exclusively negative temperature deviations from the normal in Europe and North America with a time lag of 13 - 20 days relative to the beginning of decreased activity (r.h.s.).

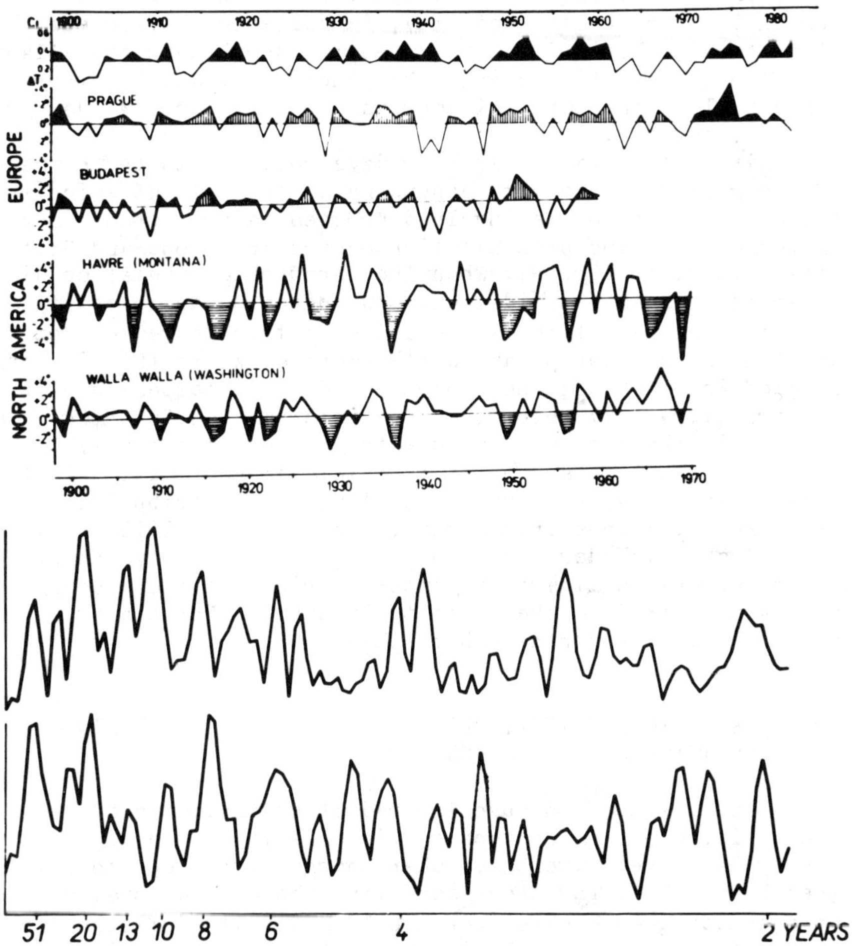

Figure 10. Top: Comparison of geomagnetic activity (C_i indices - averages for November, December, January, February) with the average temperature values-deviations from the normal in Europe (Prague, Budapest) and in the Pacific region of the U.S.A. (Havre, Walla - Walla). The positive correlation coefficient was obtained for C_i-indices and temperature in Prague (+ 0.632 for the interval 1949 - 76) and the negative coeficient for Havre (- 0.457 for the interval 1949 - 64). Bottom: Fourier analysis of geomagnetic data (C_i-indices) and temperature data in Prague (averages for two winter months December - January and period 1885 - 1975). The graph displays equal statistically significant spectral density periods of approximately 51, 20, 13, 10, 8, 6 and 2 years for both quantities.

which proceeds along the east coast of North America to the north during low geomagnetic activity, is deflected to the east by the flow from the Davis Straits proceeding south over the Atlantic once the geomagnetic activity has increased, and together they proceed along the parallels across the Atlantic towards Europe (Fig. 5, bottom).

With a time lag of 13 - 20 days relative to the commencement of the geomagnetic storm, depending on the rate at which the meteorological processes involved proceed, a pronounced increase in temperature and precipitation occurs in Europe and North America (l.h.s. of Fig. 9). When the corpuscular radiation displays a marked decrease, meridional flow increases in intensity again and with the same time lag relative to this decrease Arctic air penetrates into Europe and North America, where this is then reflected in an abrupt and pronounced drop in temperature (r.h.s. of Fig. 9). We may refer to these intervals as "short glaciation periods in the present" because they cover practically the same regions which were subjected to glaciation during the last glacial period (Fig. 21)(Bucha, 1983) as the distribution of temperature deviations from normal, e.g. on 1. 1. 1979 (r.h.s. of Fig. 9) shows. Under very low geomagnetic activity the processes in the auroral oval are very weak, meridional flow, carrying cold Artic air from the north to Europe and North America, being prevelant in the northern hemisphere.

2. GEOMAGNETIC ACTIVITY, GLOBAL CHANGES IN ATMOSPHERIC CIRCULATION AND IN TEMPERATURE

The effect of corpuscular radiation indicated by geomagnetic activity on weather changes is expected to be higher for the winter hemisphere when the solar-energy flux drops to 6×10^{15} W (Dessler, 1974). The processes along the auroral oval at the time of enhanced geomagnetic (corpuscular) activity influence the meridional type of atmospheric circulation to change into the zonal type and vice versa. Such a decisive change should be reflected in expressive long-term temperature changes.

As given in Fig. 10, enhanced temperatures occurred in Europe in winter when the zonal flow from the south Atlantic prevailed as a consequence of the increased corpuscular radiation. At the same time in the western part of North America decreased temperatures are observed (due to the cold flow from the Northern Pacific). Vice versa, below-normal temperatures are found in Europe due to invasions of Arctic air into Europe when, at the time of low geomagnetic activity, the deceleration of westerly momentum associated with the formation of atmospheric blocking in the Atlantic and Pacific regions occurs. Simultaneously, above-normal temperatures in the western part of North America are observed as a result of the warm flow from the south (l.h.s. of Fig. 1).

Another example of a relationship between corpuscular (geomagnetic) activity (curve a) and temperature changes on the nort-

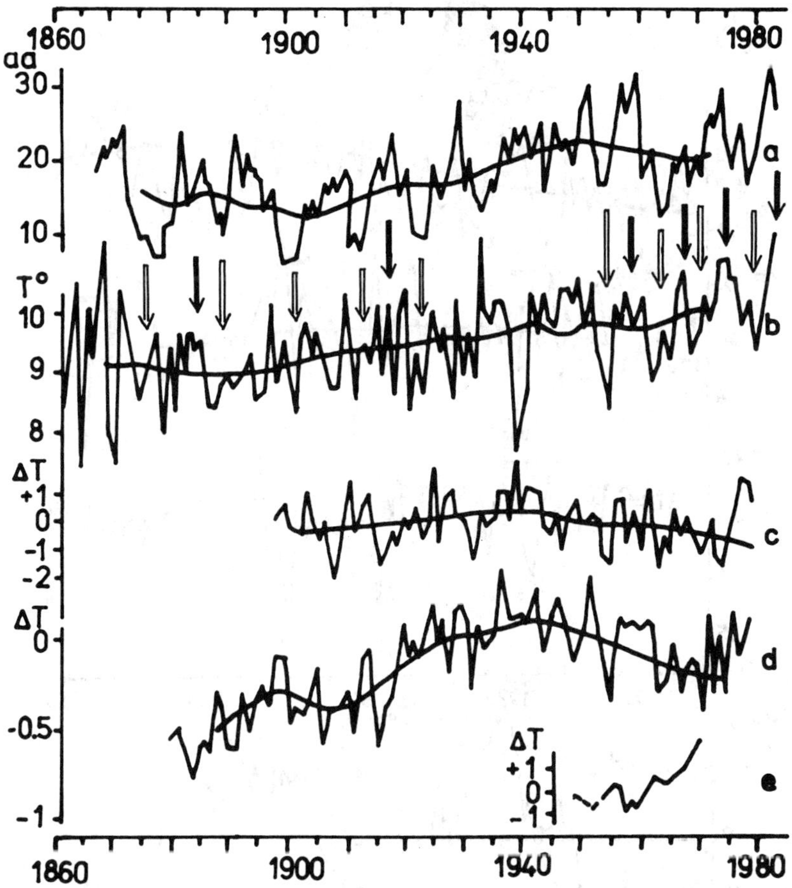

Figure 11. Annual changes in the geomagnetic activity (aa indices) and in the temperature deviations in Europe (Hydromet.Inst., Prague) and on the northern hemisphere (according to Jones et al, 1982 and Budyko, 1969), both showing that increased values of geomagnetic activity (curve a) correspond to increased temperatures in Europe (Prague) (curve b) and to decreased temperatures in Alaska (curve c) due to the intensified zonalization of atmospheric flow on the northern hemisphere. While there is a decrease of temperature between 1960 and 1972 in the northern hemisphere (curve d) we observe that the temperature in Antarctica is increasing between 1960-1970 (curve e - averages from 7 observatories).

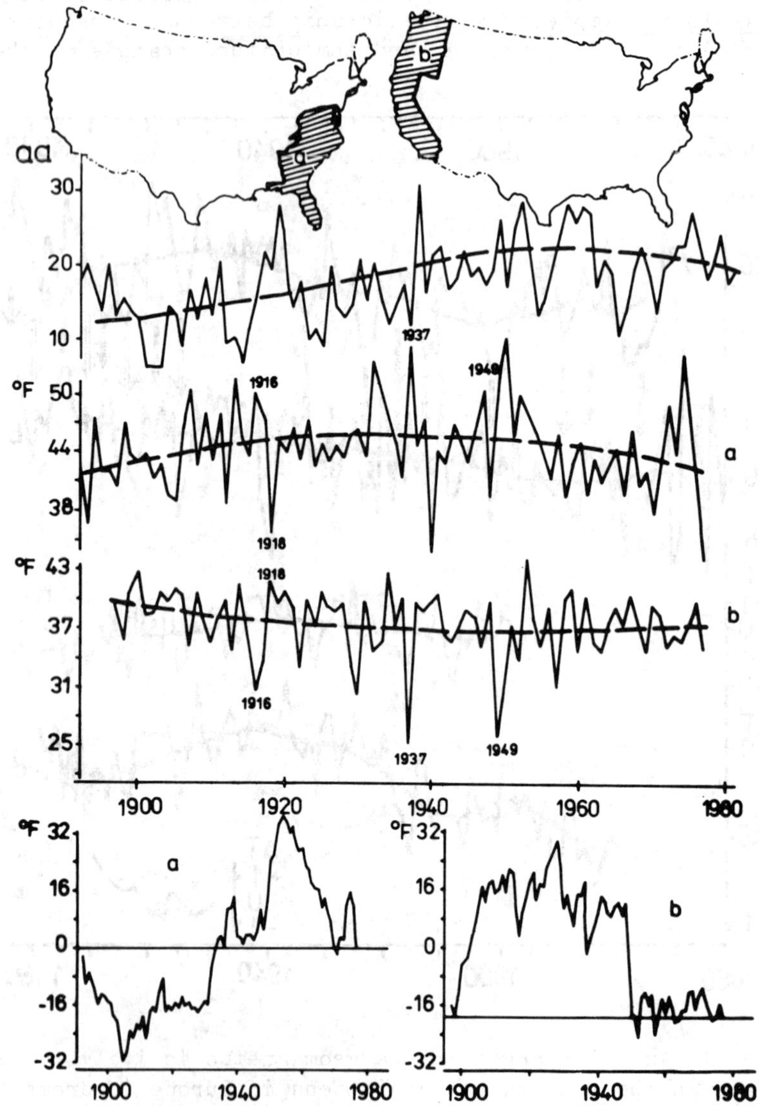

Figure 12. Annual values of geomagnetic activity (aa indices), average areally weighted January temperature (^{o}F) for South Atlantic (curve a) and Pacific (curve b) regions of the U.S.A. (top of Fig.) The smooth trend curve is a third-order polynomial least-squares fit. Cumulative January temperature departures from the long-term mean for South Atlantic (l.h.s. bottom) and Pacific (r.h.s. bottom) regions, a and b having opposite trend (Diaz & Quayle, 1978).

hern hemisphere (Kelly et al, 1982, Budyko, 1969) is given in Fig. 11. The geomagnetic activity was increasing since 1900 till 1960. Between 1961 - 1971 a sharp decrease was observed. Similar features are characteristic of temperature deviations in central Europe (Prague - curve b). Moreover, we can observe similar fluctuations also with shorter periods. The temperature in Alaska, however, has an opposite trend (curve c).

The increase of temperature values on the northern hemisphere can be explained by the contraction of the northern polar vortex at the time of enhanced geomagnetic activity when part of warm air from the equatorial area moves due to intensified cyclogenesis in the northern hemisphere to the north. Vice versa, the expansion or even the splitting of the northern polar vortex at the time of decreased corpuscular radiation (geomagnetic activity) can lead to the meridionalization and slowing down of meteorological processes, enabling the transport of warm air from the equator to the south. This conclusion is supported by the fact that the sea-surface temperature in the northern polar region decreased between 1958 - 1972 while enhancement of temperature was observed in Antarctica during this period (Voskresenskij, 1977) (Fig. 11). On the other hand, there was warming on the northern hemisphere between 1880 - 1939 and decrease of temperatures in Antarctica (Zacharov, 1976).

A strong feature evident when we compare the temperature profiles for the Pacific regions with those of the eastern United States (South Atlantic) is the negative correlation (Fig.12) (Diaz and Quayle, 1978). During cold western winters, temperatures in the east tend to be significantly warmer than the long--term averages. Again the comparison shows that with the relatively low geomagnetic activity (1899 - 1920) the temperatures in the Pacific region were above normal (due to the prevaling flow from the south) while those in the East (South Atlantic) were subnormal (invasions of Arctic air to the middle and eastern part of North America). Vice versa, after the geomagnetic activity had increased (1938 - 1975) the prevailing meridional flow changed to the zonal circulation across the Pacific, North America and the Atlantic; this could be, similarly as in Europe (see Fig. 10), the cause for the warming in the Atlantic region of the U.S.A. due to the prevailing zonal flow from the Caribbean Sea and simultaneously for the cooling in the Pacific region (cooler air coming from the Northern Pacific).

3. INFLUENCE OF GEOMAGNETIC ACTIVITY ON ATMOSPHERIC BLOCKING AND ON DROUGHT PERIODS IN THE SAHEL

Atmospheric blocking can be defined as a quasi-stable circulation arising from orographic instability with strong thermal driving (Charney and Strauss, 1980). There are two preferred regions for blocking, the eastern part of the Atlantic region and the western part of the Pacific region. The Rocky Mountains and the Tibetian Plateau are considered to play an important role in both onset and maintenance of the block. The deceleration of the westerly momentum associated with the formation of blocking on the northern hemisphere is alternated by the relatively long small amplitude waves in the westerlies (Fischer, 1983). The duration of a blocking episode as well as of westerlies can be long or short-lasting with one type changing to the other many times in the course of the year. The interannual appearance, however, can be considerably different. In Fig. 13 the comparison between average geomagnetic activity in winter and the percentage of days with blocked flow is given. We can conclude that the enhan-

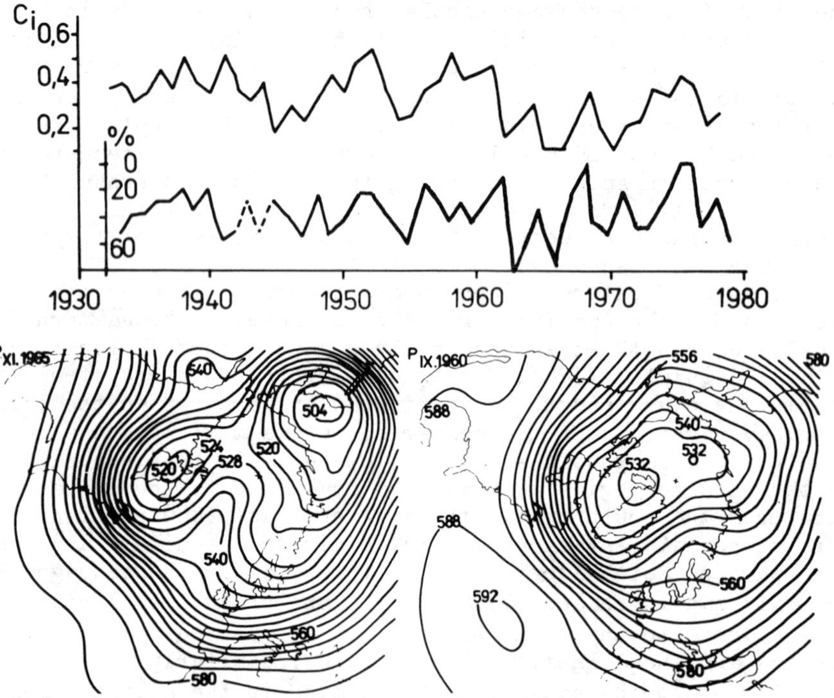

Figure 13. Comparison of geomagnetic activity (C_i-indices - averages for November, December, January, February) with the variation in blocking activity in the Atlantic area (Rex, 1950, Lejenäs, Okland, 1983); winter trends in the percentage of days dominated by Atlantic blocking are given. Preliminary results show that the number of days with blocked flow in the Atlantic region has its maximum at the time of low geomagnetic activity (1963 - 66) and its minimum when the geomagnetic activity was enhanced (e.g. 1974 - 76) (r.h.s., bottom). Bottom: Two cases (monthly averages) are shown - blocking in the Atlantic (Nov.1965) when geomagnetic A_p index equalled to 6 (l.h.s.) and increased zonal flow (Sept. 1960 - A_p equal to 20, r.h.s.).

ced corpuscular radiation has a strong influence on the temperature and pressure due to the processes in the auroral oval and causes the zonalization of atmospheric flow (long small amplitude waves) to prevail in both the Atlantic and Pacific regions. Vice versa, as given in Fig. 13 formation of atmospheric blocking corresponds to the low geomagnetic activity when the processes in the auroral oval are weak. A long-term decrease of geomagnetic activity (e.g. in November 1965 - monthly average when the geomagnetic Ap index was only 6) corresponds to the long-term blocking in the Atlantic with four low pressure areas extending from the polar region (l.h.s. of Fig. 13, bottom), while at the time of increased geomagnetic activity (e.g. Sept. 1960 when Ap index was 20) there was no blocking and the polar vortex was contracted (r.h.s. of Fig. 13, bottom).

Also the annual average temperature and pressure deviations display a conspicuous blocking meridional pattern for the lowest average annual geomagnetic activity, with a gradual increase of zonalization with enhancing geomagnetic activity, as we can see,

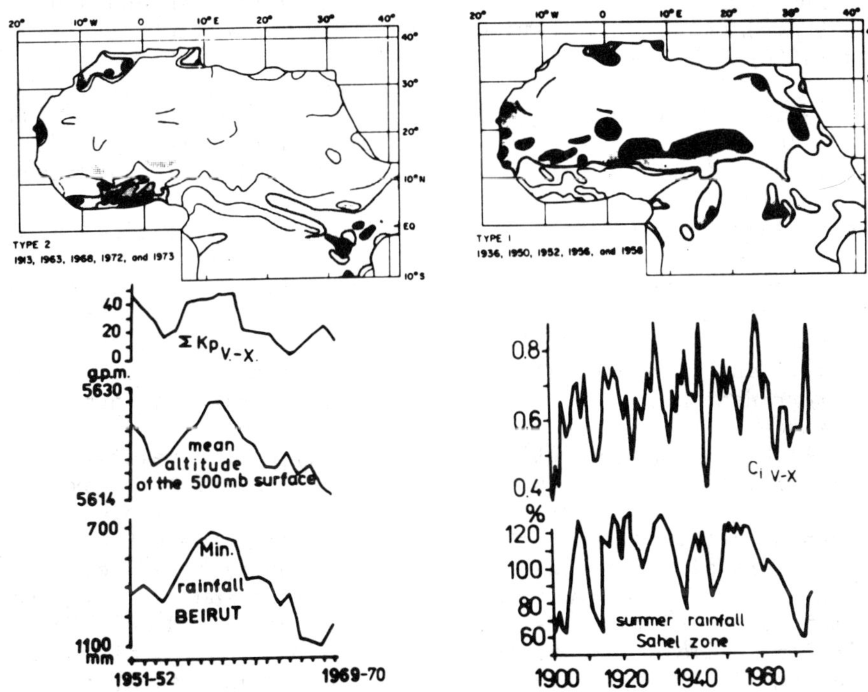

Figure 14. Comparison of geomagnetic activity (C_i-indices for May to October-r.h.s.) with seasonal rainfall in the Sahel zone (Nicholson,1979 a,b). At the time of low geomagnetic activity the processes in the auroral oval are weak and the rainfall decreases in the Sahel. Under enhanced geomagnetic activity monsoon rainfall in the Sahel increases.Negative correlation in Beirut(l.h.s.).

e.g. from the gradual changes of the distribution of deviations of annual temperature (sea level) and pressure (500-mb level) averages, which gradually change from the meridional type corresponding to low geomagnetic activity in 1965 to the zonal type corresponding to the high average annual geomagnetic activity in 1960 (Fig. 10 in Bucha, 1983).

We can also study the causes of severe droughts in the Sahel zone in Africa (Faure et al, 1983) and in north-west India. According to Winstanley (1973) "at present there is no uniformly accepted theory of the general circulation of the atmosphere to include the fluctuations on the different time scales; therefore, physically based climatological rainfall forecasts cannot be produced. Climate does change and it would be of great benefit to long term economic planning to know the future climatic changes".

Even though the solution of the problem is very difficult and will need a very thorough discussion some preliminary conclusions based on the correlations of geomagnetic activity and seasonal rainfall can be suggested. We can see in Fig. 14 that drought periods in the Sahel zone investigated by Nicholson (1979 a,b) occurred in periods when the geomagnetic activity was low (1900-1902, 1911-1914, 1922-1924, 1941-1944, 1962-1972). Let us try to find a possible explanation. As we have shown, in the period of increased geomagnetic activity zonal circulation prevails on the northern hemisphere without blocking. For the Sahel zone it means that relatively long small amplitude waves in the circumpolar westerlies tend to be restricted to mid-latitudes; that is, according to Winstanley (1973), the circumpolar vortex is contracted. There is a northward shift of the principal climatic zones taking place also in Africa. Thus, when the zonal circulation in the northern hemisphere is increasing in strength (r.h.s. of Fig. 14, top), monsoon rainfall in the Sahel (r.h.s.) and in northwest India increases and rainfall over the Mediterranean and the Middle East (l.h.s. of Fig. 14, bottom) decreases.

With a low geomagnetic activity, i.e. with a relatively weak zonal circulation there are relatively short large amplitude waves in the circumpolar westerlies; the circumpolar vortex is expanded (l.h.s. of Fig. 1, top). Weather systems in mid-latitudes are slower moving and persistent anticyclones develop, which "block" the eastward moving depressions and deflect them to the north and to the south. There is a southward shift of the principal climatic zones. Thus, when the zonal circulation is decreasing in strength, i.e. meridionalization increases, rainfall decreases in the Sahel (r.h.s. of Fig. 14, bottom) and in north-west India and increases over the Mediterranean and the Middle East (l.h.s. of Fig. 14, bottom). Within an expanded or contracted circumpolar vortex the regional variations of seasonal rainfall are determined by the preferred longitudinal positions of the troughs and ridges.

4. CHANGES IN GEOMAGNETIC ACTIVITY, TEMPERATURE AND AGRICULTURAL PRODUCTION IN EUROPE

As we showed in Figs. 1 - 7 (Bucha, 1983), the increase in temperature and pressure along the auroral oval significantly affects the atmospheric circulation in the troposphere immediately on enhancement of corpuscular radiation (geomagnetic activity) and is responsible for the change from the meridional type to the zonal type, for the intensified cyclogenesis and it gradually leads to the intensification of westerlies. With a time-lag of 13 - 20 days relative to the increase of geomagnetic activity, a pronounced increase in temperature occurs in Europe and in North America. Vice versa, when the corpuscular radiation displays a marked decrease, meridional flow increases again, blocking patterns are formed in the Atlantic and Pacific and with the same time-lag the drop in temperature can be observed in Europe and in the eastern part of North America due to invasions of Arctic air. A comparison of geomagnetic activity (ΣK_p -indices, upper curve in Fig. 15) with the daily values of temperature (departures from normal) drawn with a time-lag of 15 days (lower curve in Fig. 15) shows that under low geomagnetic activity the winter temperatures in Central Europe (Prague) are considerably below normal (Fig. 15, graphs to the left). On the other hand, when the geomagnetic activity is enhanced in average and decreases for 1 - 3 days only, then the whole winter period used to be well above normal (Fig. 15, graphs to the right). The effect of increased (decreased) geomagnetic activity on the increase (decrease) of temperature with a time-lag of 15 - 20 days can be observed in all four cases given in Fig. 15, and especially expressively for the winter period 1962 - 63.

As implied by the results of the spectral analysis, applied to a set of diurnal data for a 4-year period (1962 - 65) (Fig. 16) (Bucha, 1980 b), there is a certain relation between the variations of geomagnetic activity and variations of atmospheric pressure over the geomagnetic pole (at sea level as well as at 500 mb level). As regards geomagnetic activity (M_{Kp}), Fig. 16 displays statistically significant spectral density periods of 27, 13.5, 9 and 6.7 days. The most pronounced maximum was observed at 13.5 days. A similar pronounced maximum was found on the spectral curve of atmospheric pressure variations at sea level, as well as at an altitude of 5 km. The subsidiary values of the increased spectral density in the neighbourhood of this maximum at 13.5 days is evidence that in some cases the relation is looser. The same is true of the periods of 27 days, which occur within the interval of 25.5 to 30.5 days with the 5 km pressure data.

If we look at the graph representing the gross agricultural production in Czechoslovakia (Fig. 17), in years when an increased level of geomagnetic activity, i.e. strong zonal atmospheric circulation was recorded in the month of May (representing the main critical period of the growth of grain and other products in central Europe and Canada), the gross agricultural production will be seen to be higher on the average (e.g., in 1956 - 60, 1967 - 68, 1973 - 74). On the other hand, when the geomagnetic activity was low (decreased zonal flow), there was a pronounced

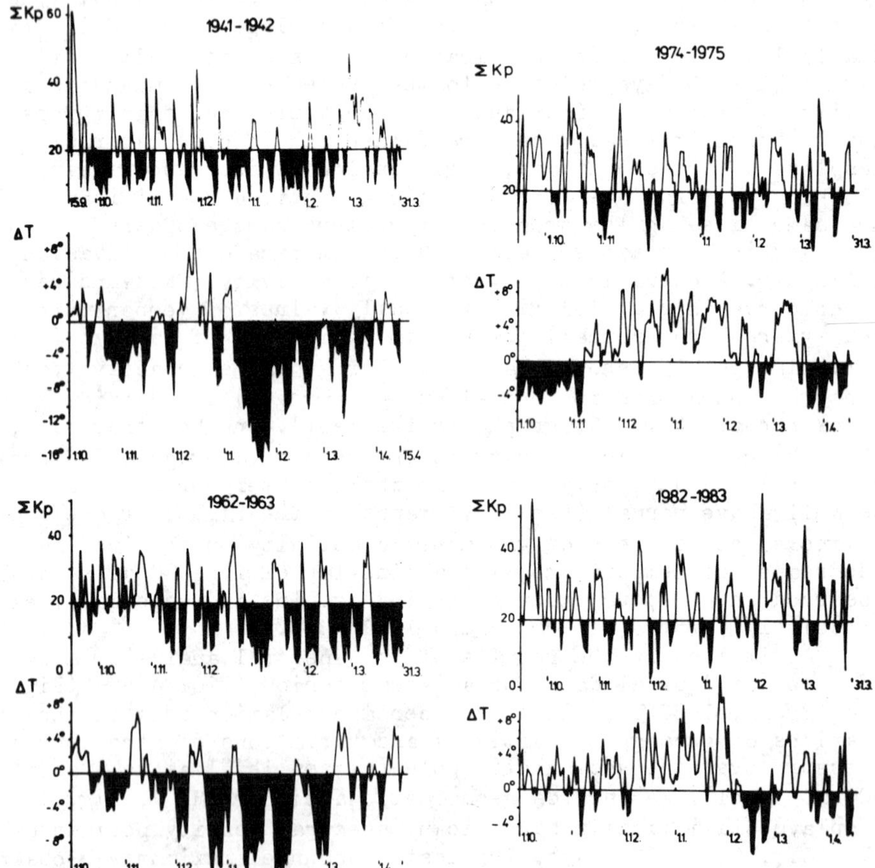

Figure 15. Changes of the diurnal values of geomagnetic activity (ΣK_p -indices) and of temperature (deviations from the normal) in Prague for the period September - April. The shift of 15 days between the geomagnetic activity and temperature curves represents on the graph the average time lag between the increase of geomagnetic activity and increase of temperature in Prague. Severe winters correspond to very low values of geomagnetic activity (graphs to the left). Mild winters occurred at the time of enhanced geomagnetic (corpuscular) activity (graphs to the right).

decrease in production (e.g., in 1954, 1962 - 65, 1970, 1972) due to subnormal temperatures especially in winter and spring. Similar results were found for Canada, USSR and FRG (Bucha, 1980 b). The correlation coefficient between geomagnetic activity and

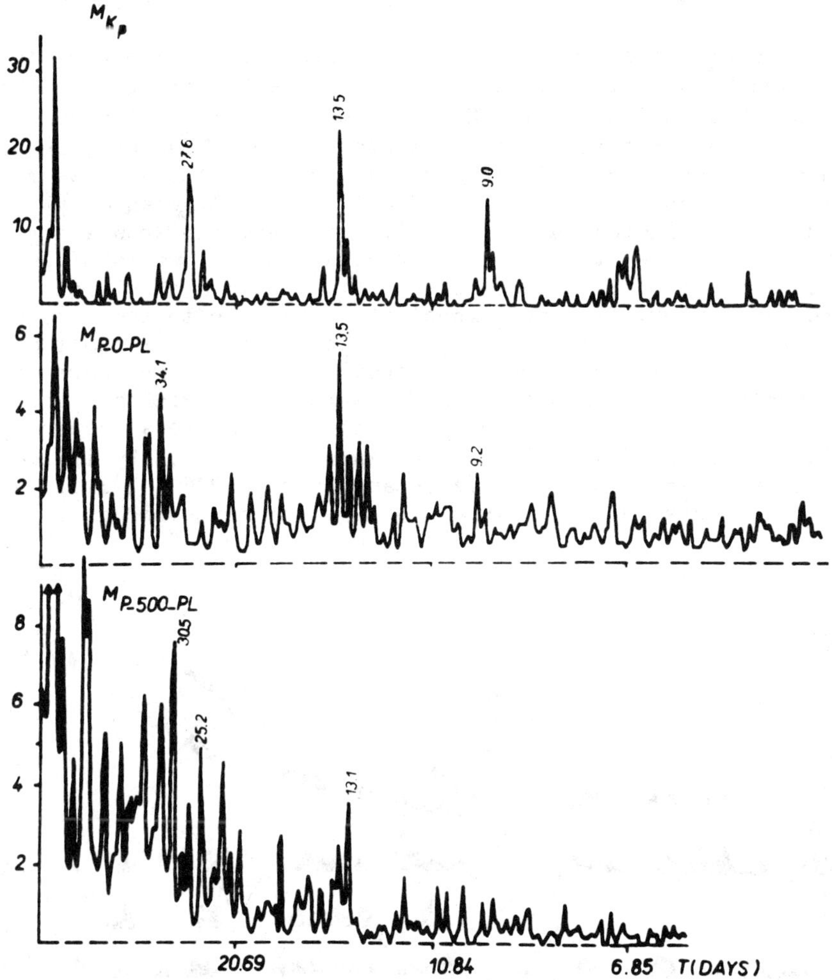

Figure 16. Spectral function (maximum entropy) of geomagnetic activity M_{Kp}, of the atmospheric pressure over the geomagnetic pole at sea-level (M_{P-0-PL}) and at the 500 mb level ($M_{P-500-PL}$) for the period 1962 - 65. The most pronounced coinciding maximum was found for a period of 13.5 days and is evidence of the relation between geomagnetic activity and atmospheric pressure over the geomagnetic pole (Bucha, 1980 b).

agricultural production for Czechoslovakia, (Fig. 17) was found 0.462 for the period 1949 - 1982.

5. CHANGES IN GEOMAGNETIC ACTIVITY, TEMPERATURE AND WIND VELOCITY, ITS CONNECTION WITH THE SEISMICITY IN CENTRAL EUROPE

Recently a catalogue of earthquakes was published by Schenková and Kárník (1981). If we consider the occurrence of all 518 earthquakes in Bohemia and Moravia according to individual months, it turns out that in the three winter months (November, December, January) there were 207 earthquakes, which is three times more compared to summer months (June, July, August), when only 79 earthquakes originated (Bucha, 1981). This seems to show a certain periodicity of earthquakes. If we consider the swarms of earthquakes as one event, the dependence still shows a doublefold occurrence in winter. We have demonstrated that due to the penetration of energetic electrons to the altitude below 20 km from the magnetosphere during enhanced geomagnetic activity, an increase of temperature and pressure along the auroral oval takes place (Figs. 6, 7). This effect influences and intensifies the atmospheric circulation and wind velocity as we can observe not only in the Atlantic but also in the Pacific ocean (Bucha, 1976 - 1981). This effect is more intensive in winter than in summer because the ra-

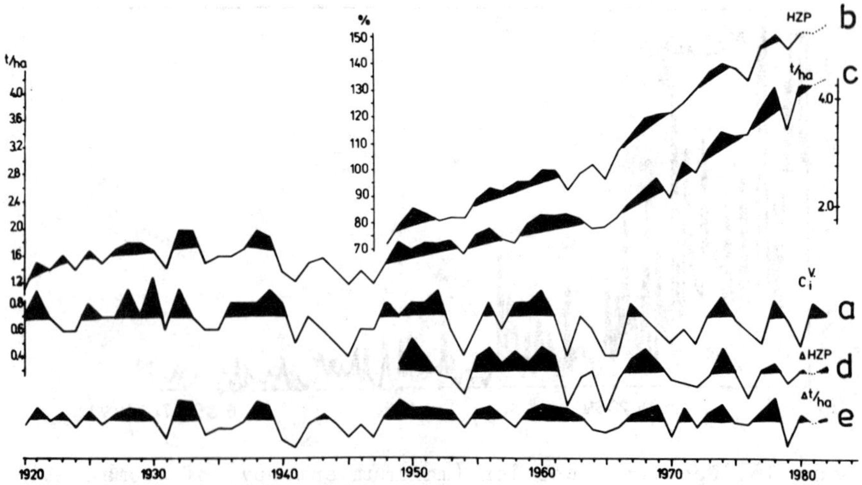

Figure 17. Dependence of the gross agricultural production in Czechoslovakia on geomagnetic activity. Curve a represents the geomagnetic activity in May, curve b the actual gross agricultural production, curve c the actual grain yields, curve d the deviation of the gross agricultural production from the average and curve e the deviation of the grain yields from the average.

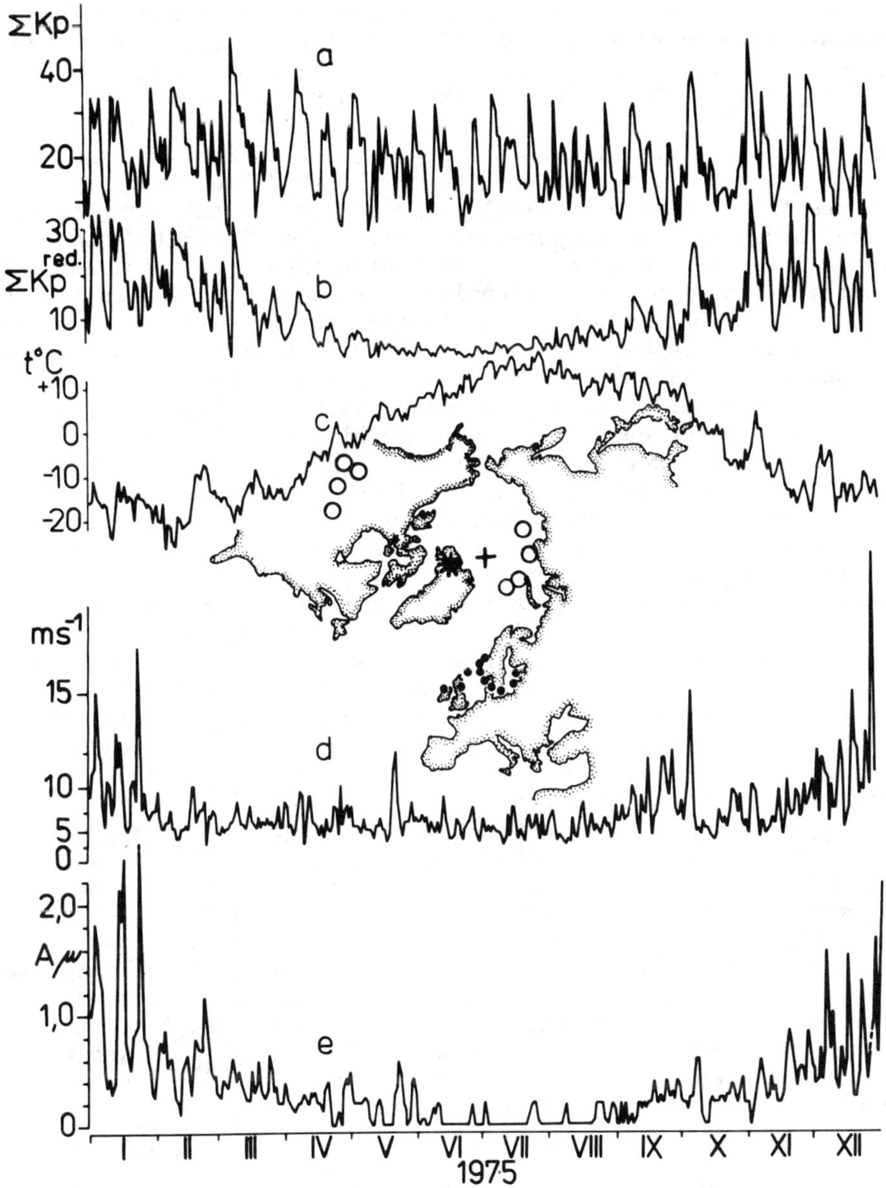

Figure 18. Geomagnetic activity ΣK_p 1975 (curve a), reduced corpuscular energy flux ΣK_p^{red} (curve b), averages of temperature for stations (o) along the auroral oval (curve c), averages of wind velocity for stations (•) (curve d), amplitudes of microseismic activity at Průhonice near Prague (curve e).

tio between the corpuscular and solar-electromagnetic energy flux is ten times higher in the northern hemisphere in winter than in summer. This is shown in Fig. 18 where the amplitudes of temperature fluctuations in the auroral oval caused by corpuscular radiation are 15 - 20°C, while in summer only 2 - 4°C and where the correlation coefficient between the geomagnetic activity (curve b) and temperature at 8 stations (curve c) is 0.65. We can see further that also the wind velocity (curve d) represented by mean values from 11 meteorological observatories on the coast of northern Europe and the microseismic activity in Central Europe (curve e) is much more expressive in winter than in summer. This wind velocity represents the intensity of seawaves (tides) on the rocky coast of Scotland, Norway, Denmark and the Baltic Sea. This tide seems to cause oscillations of certain parts of individual blocks of the earth's crust exposed to this effect and leads to the origin of microseisms, as can be seen from the comparison of the curve characterizing the average wind velocity (d) with the curve of amplitudes of microseismic activity (e) (Fig. 18). The correlation coefficient is 0.70 and the significance level 0.1 %.

This complex corpuscular, geomagnetic and atmospheric effect mainly occurs in the coastal areas of the Atlantic and Pacific oceans, i.e. not only in Europe but also in Japan and in the whole of southeast Asia, as turns out from the charging wind speed. The geomagnetic activity shows the so called half-year period (maxima at the time of equinoxes) (Fig. 19 top - curve 1). Then the effect of the enhanced corpuscular radiation (indicated by geomagnetic storms) on the increased wind velocity and on the intensity of seawaves (tides) on the rocky coast of Scotland, Norway, Denmark and the Baltic coast (curve 2) and on the origin of microseisms in Europe (curve 3) takes place with a lag of about 15 days, i.e. in April and October. In this way one can also explain the two secondary maxima on the curves of microseismic activity in Japan (curve 4) and in Europe (curve 5) and earthquakes in Bohemia and Moravia (curve 6 - Fig. 19).

The similarity of curves characterizing the amplitudes of wind velocity (Fig. 19 - curve 2) and microseismic activity (Fig. 19 - curves 3 - 5) with the curve expressing the frequency of earthquakes in Bohemia and Moravia (curve 6) seem to be an indicator of a trigger mechanism which could take part in the generation of earthquakes in Europe and East Asia under specific conditions as described in this paper.

6. LONG-TERM CHANGES IN THE GEOMAGNETIC FIELD AND CLIMATE, GLACIATIONS IN THE NORTHERN HEMISPHERE

Climatic changes include a relatively broad spectrum of periods starting with hundreds of millions of years up to the fluctuations with several years periods. Let us try to discuss the main of them and to contribute to the explanation of their causes.

It is known that during the Phanerozoic only three times the continents were repeatedly glaciated. It was between 800 - 600 million years ago (Precambrian - Cambrian), 320 - 250 million (Carboniferous - Permian) and during the past 10 million years. The most probable cause taken into account is the continental drift, the main characteristics of which can be explai-

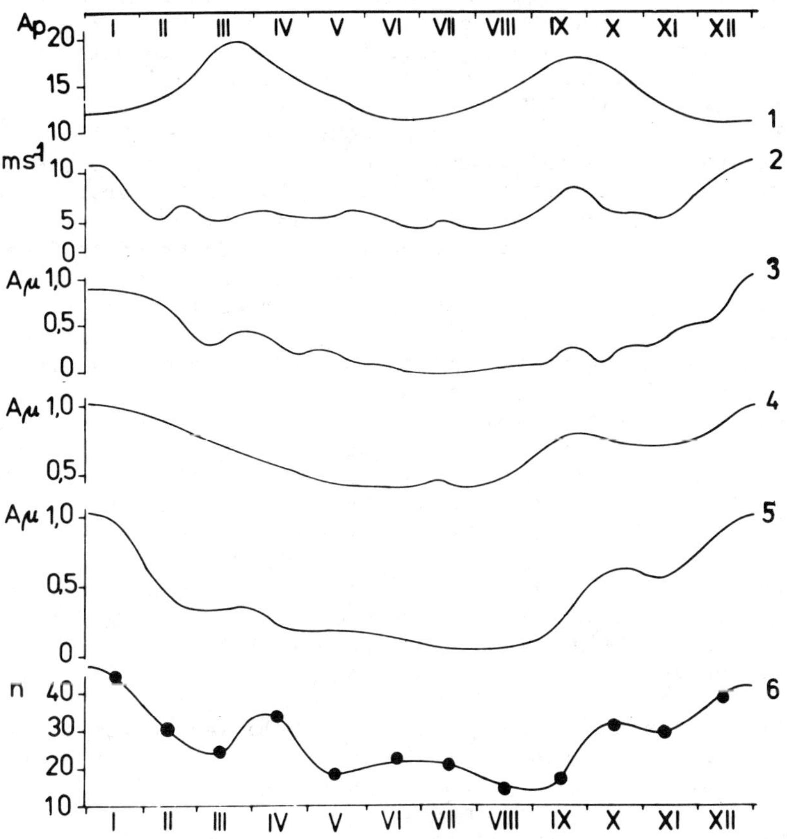

Figure 19. Half-year period of geomagnetic activity (curve 1), average wind velocity in 1975 on the coast of Europe (curve 2 - fifteen-day averages), microseismic activity in 1973 and 1975 in Europe (curve 3), microseisms for the territory of Japan (Zátopek, 1970 - curve 4) and Europe (Zátopek, 1970 - curve 5), earthquake occurrence (monthly totals) for the territory of Bohemia and Moravia (data from Schenková and Kárník, 1981 - curve 6).

ned by the model of the dynamic system of the world-wide plate tectonics and of its developments (Bucha, 1980 c). The distribution of continents in the subsequent geological periods investigated by palaeomagnetic methods agrees well with palaeontological facts about the occurrence of warm climate (when nearly all continents were either on the equator or close to it, e.g. in the Ordovician and Silurian) and about the glaciations (when several continents were close to the pole, e.g. in the lower Cambrian and Carboniferous) (Bucha, 1980 c).

The second group of climatic oscillations includes the dominant cycle of 100 thousand years due to variations in eccentricity (according to the astronomical theory) and three other smaller peaks - 43 thousand years corresponding to axial tilt and 23 and 19 thousand years corresponding to variations in precession (Hays et al, 1976).

It is, however, unknown how the triggering mechanism could operate and why the 100 thousand year cycle appeared to be so strong. We suggest that the changes in the earth's orbital geometry have an effect on the instabilities in the earth's magnetic field, i.e. mainly on the changes in the position of the geomagnetic pole as a centre of the auroral oval which, as we have shown in this paper, significantly influences the weather and climate due to the processes occurring in the oval. That is why we concentrated our attention to the fluctuations of geomagnetic dip and intensity. As shown by Bucha (1980 b), the decreased values of geomagnetic dip in Europe correspond to coolings in Europe as given by the average global air temperatures and to warmings in the Pacific. This means that the position of the geomagnetic pole was closer to the Pacific at that time. The spectral analysis of palaeomagnetic data from the Stirone river (Fig. 20) showed periods 100, 140 and 20 - 30 thousand years (Bucha, 1980b).

For the time interval of the past 20 thousand years, more precise palaeomagnetic results were obtained, showing for three territories on the northern hemisphere that the geomagnetic pole was located several times in the Pacific during the interval 18 - 12 thousand years ago (Bucha, 1980 a,b,c, 1983) (Fig. 21). Also the intensity of the geomagnetic field was decreased. This pole moved from the Pacific to the eastern part of North America in the interval 12 to 10 thousand years ago when also the last period of glaciation terminated.

This correlation seems to provide one possibility of explaining the alternation of glacial and interglacial periods. But it does not answer the question what is the role of the geomagnetic pole and what is the physical mechanism leading to such expressive changes in temperature between glacial and interglacial periods.

The next spectrum of climatic changes includes 1000 - 600 years. The approximate position and wandering of the virtual geomagnetic pole between 200 B. C. and the present as well as variations of the geomagnetic inclination in the Caucasus and Bulgaria between 1,200 B. C. and the present(Fig. 20), r.h.s. (Burlackaya, 1965; Kovacheva, 1969) closely indicate that the approach of the geomagnetic pole in the direction to Europe and the dip maxima are practically identical with the temperature maxima and vice versa (Bucha, 1980 a,b).

If we want to understand the changes in the past, we must be able to explain the present day climatic and weather changes, which are documented in the more reliable direct observations

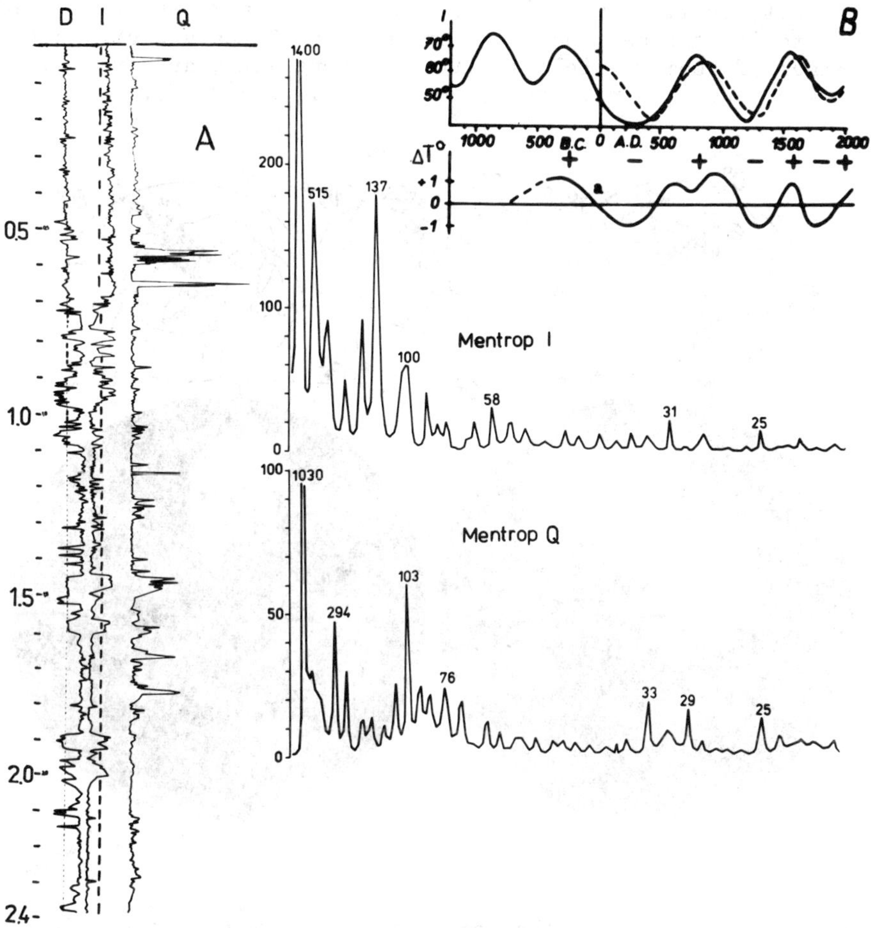

Figure 20. A) Variations of the geomagnetic field (declination, dip, Q coefficient) determined by investigating the marine sediments along the River Stirone (Italy) (Bucha, 1980 b) for the period 2.4 million years ago and the spectral functions expressing marked maxima for periods of 140, 100 and 25 thousands years.
B) Variations of I (— Caucasus --- Bulgaria) and T in Europe.

than the long-term changes in the geological past represented by sporadic and indirect data only.

In addition to the above-mentioned effects there is another factor - solar activity - which seems to influence climatic changes considerably (Bucha, 1980 a). Changes in geomagnetic activity, solar activity - R numbers (Eddy, 1976) and in temperature defined by changes in the distribution of Alpine glaciers show periods of 200 - 250 and 450 - 600 years (Le Roy Ladurie, 1969) which are according to Fairbridge (1983) chronologically synchronous with orbital variations of the planets and the sun. In addition to it, the periods of 80 and mainly 22 years occur in the changes of geomagnetic activity, in temperature and in agricultural production in Europe (see Figs 10, 11 and 17).

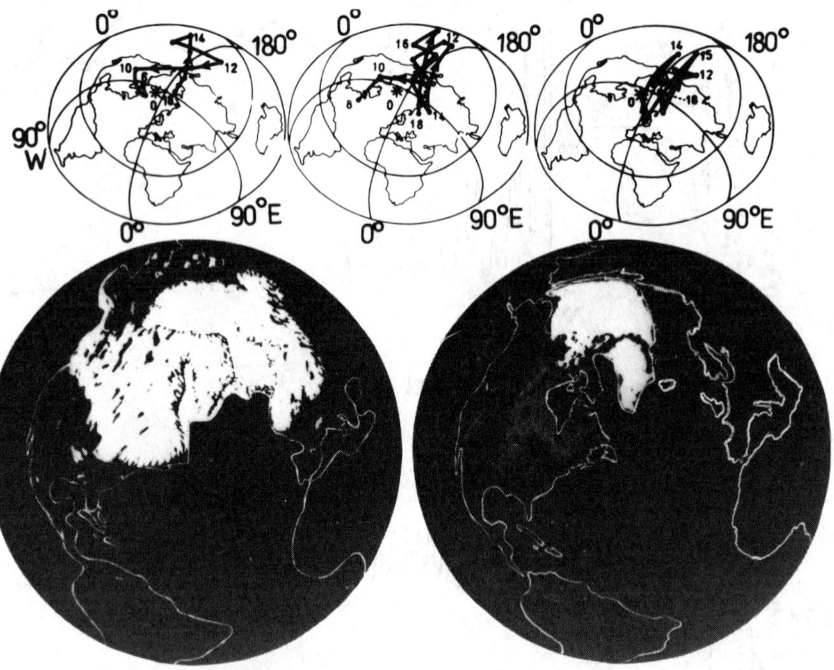

Figure 21. Changes in the position of the geomagnetic pole over the last 18 000 years determined from palaeomagnetic investigations in Czechoslovakia (left top) (Bucha, 1976), in Japan (middle top) (Kawai et al, 1973), and in the USA (right top) (Easterbrook & Othberg, 1976). The glaciation of the northern hemisphere during the last glacial period approximately 20 000 - 13 000 years ago (left bottom) and the present distribution of glaciers (right bottom - Imbrie, 1979).

That is why we started with the comparison of the changes of geomagnetic activity indicating the corpuscular radiation with changes in temperature in winter when the solar electromagnetic radiation is decreased in the northern hemisphere. This gave us very interesting correlations. It was found (Fig. 10) that mild winters occurred in central Europe when the geomagnetic activity was increased; in the western part of North America and in Alaska, however, severe winters occurred simultaneously. When the geomagnetic activity was decreased, the effect was opposite. The correlation coefficient between C_i and T_i in Prague for the period 1949 - 76 is equal to + 0.63 and for Havre for the period 1949 - 64 is - 0.457. We found that these relatively high differences, e.g. between 1962 - 63 (low geomagnetic activity) and 1974 - 75 (high geomagnetic activity) winters represented more than 7°C in Czechoslovakia and seem to be connected with different types of atmospheric circulation as shown in this paper (Figs. 1 - 19).

CONCLUSION

As our results of palaeomagnetic research into sedimentary rocks have shown (Fig. 21, Bucha, 1976), during the interval 18 - 12 thousand years ago the geomagnetic pole was located several times for longer periods in the Pacific (Bucha, 1976, 1980 a,b, Easterbrook, 1976, Kawai et al, 1975, Mörner, 1978). This would mean that also the position of the auroral oval, the centre of which is the geomagnetic pole, was displaced towards the Pacific, and the processes generated by the penetration of electrons into the auroral oval on commencement of geomagnetic storms were unable to affect the changes of the atmospheric circulation over the Atlantic as much as they do now. That is probably why cold meridional flow into Europe and the eastern part of the North American continent predominated; this flow is responsible for the substantial decrease in temperature in these regions for short intervals at the time of low corpuscular radiation now, but was apparently responsible for the decrease over long intervals also during the past periods of glaciation. This then means that the meridional flow which prevailed for long intervals during glaciations when the geomagnetic pole was closer to the Pacific (Bucha, 1980 a,b,c, 1983), caused extensive glaciations in the northern hemisphere, particularly in its western part, i.e. in North America and Europe. A comparison with the present distribution of the regions with negative deviations of temperature from normal at low geomagnetic activity (compare l.h.s. and r.h.s. of Fig. 9 and Fig. 21) shows that the North American and European continents were affected most, whereas the North Pacific and East Siberia experienced no pronounced glaciation.

This is in agreement with our hypothesis about the ffect of the position of the geomagnetic pole and, therefore, also of the auroral oval on the increase of warmer zonal atmospheric flow under enhanced geomagnetic activity and, conversely, on the formation of a cold meridional flow in Europe and the eastern part of North America under very low geomagnetic activity when only weak processes take place in the auroral oval (Figs. 1 - 11).

REFERENCES

Bucha, V., 1976. Variations of the geomagnetic field, the climate and weather, Studia geoph. et geod., 20, 149 - 167.

Bucha, V., 1980 a. Weather and climate predictions in the northern hemisphere based on solar-terrestrial relations. Solar-Terrestrial Predictions Proc., 4, NOAA, F-18, Boulder, F-18 - 43.

Bucha, V., 1980 b. Mechanism of the relations between the changes of the geomagnetic field, solar corpuscular radiation, atmospheric circulation and climate. J. of Geomagnetism and Geoelectricity, 32, 217 - 264.

Bucha, V., 1980 c. Model of the world plate tectonics since the early Palaeozoic. J. Geomag. Geoelectr., 32, Suppl. III, S III 47 - 60.

Bucha, V., 1981. Changes in the corpuscular (geomagnetic) activity, atmospheric circulation and seismicity in Europe. Proc. 2nd Int. Symp. on the Analysis of Seismicity and on Seismic Hazard, Czechoslovakia, May 18 - 23, 351 - 360.

Bucha, V., 1983. Direct relations between solar activity and atmospheric circulation, its effect on changes of weather and climate. Studia geoph. et geod., 27, 19 - 45.

Budyko, M.I., 1969. The effect of solar radiation variations on the climate of the earth. Tellus 21, 611 - 619.

Burlackaya, S.P., 1965. Archeomagnetizm. Izd. Nauka, Moskva.

Charney, J.G. & Strauss, D.M., 1980. Form-drag instability, multiple equilibria and propagating planetary waves in baroclinic, orographically forced, planetary wave systems. J. Atmos. Sci., 37, 1157 - 1176.

Chromov, S.P., 1968. Meteorológia, Vyd. SAV, Bratislava.

Clauer, C.R., 1980. Notes from the Stanford sun-weather workshop, SUIPR Rep. NO. 825, Stanford, California, Dec.

Dessler, A.J., 1974. Some problems in coupling solar activity to meteorological phenomena, Symp. Possible Relations between Solar Activity and Meteorological Phenomena, Nov., 1973 Aeronautics and Space Admin., 299 - 310.

Diaz, H.F. & Quayle, R.G., 1978. The 1976-77 Winter in the contiguous Unites States in comparison with past records. Mo. Wea. Rev. 106, No 10, 1393 -1421.

Easterbrook, D.J. & Othberg, K., 1976. Paleomagnetism of pleistocene sediments in the Puget Lowland, Washington. Quat. Glaciations in the Northern Hemisphere, Project IUGS,UNESCO, 73/1/24, Bellingham, Prague, 189 - 207.

Eddy, J.A., 1976. The Maunder Minimum. Science, 198, 824 - 829.
Fairbridge, R.W., 1983. Planetary periodicities and terrestrial climate stress. Second Nordic Symp. on Climate Changes and Related Problems, Abstracts, May 16 - 20, 29 - 32.
Faure, H., Gac, J.Y., Hillaire-Marcel, C., Lezine, A.M., Monteillet, J., Ngom, P.M., Pezeril, G. & Saos, J.L., 1983. Tropical peats: hydrogeologic and climatic control. Second Nordic Symp. on Climate Changes and Related Problems, Abstracts, May 16 - 20, 33 - 34.
Fischer, G., 1983. Some aspects of blocking formation as a general circulation model. Symp. on Blocking and Atm. Prediction. IAMAP Abstracts, IUGG Ass., Hamburg.
Hays, J.D., Imbrie, J. & Shackleton, N.J., 1976. Variations in the Earth's orbit: Pacemaker of the Ice Ages, Science, 194, 1121 - 1132.
Hydromet. ústav, 1976. Meteorologická pozorování v Praze - Klementinu 1775 - 1975, Praha.
Imbrie, J. & Imbrie, K., 1979. Ice ages solving the mystery. Enslow Publ. New Jersey.
Jacchia, L.G., Slowey, J. & Verniani, F., 1967. Geomagnetic perturbations and upper-atmosphere heating. J. Geoph. Res., 72, 1423 - 1438.
Kawai, N., Nakajima, T., Yaskawa, K., Torii, M. & Natsuhara, N., 1975. Paleomagnetism of Lake Biwa sediments. Rock Magn. Paleogeophys., 3, 24 - 37.
Kelly, P.M., Jones, P.D., Sear, C.B., Cherry, B.S.G. & Tavakol, R.K., 1982. Variations in surface air temperatures: Part 2. Arctic Regions, 1881 - 1980. Mo. Wea. Rev. 110, No 2, 71-83.
Kovacheva, M., 1969. Inclination of the Earth's magnetic field during the last 2000 years in Bulgaria. J. Geomagn. Geoel., 21, 573 - 578.
Labitzke, K., 1979. A survey over the PMP-1 winters 1955/56 - 1978/79. J. Japan Met. Soc., 60, 60 - 69.
Le Roy Ladurie, E., 1969. Histoire du climat depuis l'an mil. Flammarion, Paris.
Lejenäs, H. & Okland, H., 1983. Characteristics of northern hemisphere blocking as determined from a long time series of observational data. Tellus, 35A, 350 - 362.
Mörner, N.A., 1978. Paleoclimatic, palecmagnetic and paleogeoidal changes: Complexity and interaction. Proc. of the Nice 1978 conference. Evolution des atmospheres planetaires et climatologie de la Terre. Contribution No. 142.
Nicholson, S.E., 1979 a. Revised rainfall series for the west African subtropics. Mo. Wea. Rev. 107, No 5, 620 - 623.
Nicholson, S.E., 1979 b. Statistical typing of rainfall anomalies in Subsaharan Africa. Erdkunde, Archiv für Wiss. Geogr. 33, 2, 95 - 103.
Rex, D.F., 1950. Blocking action in the middle troposphere and its effect upon regional climate. Tellus 2, No 4, 275 - 301.

Schenková, Z. & Kárník, V., 1981. Catalogue of earthquakes for central and eastern Europe, Geophys. Inst., Czechosl. Acad. of Sci., Prague.

Voskresenskij, A.I., 1977. Meteorologičeskaja izučenost Antarktidy za 20 let. Informacionyj bjuletin sovětskoj expedicii, 95, 63 -81.

Winstanley, D., 1973. Rainfall patterns and general atmospheric circulation. Nature, 245, Dept. 28, 190 - 194.

Zacharov, V.F., 1976. Pocholodanie Arktiki i ledjanoj pokrov arktičeskich morej. Trudy, 337, Gidrometeoizdat, Leningrad.

Zátopek, A., 1970. Generalized Amplitudes of IGY Microseisms in Europe and in Japan, Bull. Int. Inst. of Seismology and Earthquake Eng., 7, 79 - 92.

SOLAR ULTRAVIOLET IRRADIANCE VARIATIONS AND THE EARTH'S ATMOSPHERE

J. L. Lean

CIRES, University of Colorado/NOAA, Boulder,
CO 80309, USA.

Solar ultraviolet (UV) irradiance variations are frequently invoked as a source of variance in atmospheric parameters, and as a possible cause of climate change. Recent observations have established that the solar ultraviolet irradiance does indeed vary with time. However, the extent to which these variations modulate the earth's middle and lower atmosphere is not well known. Not all correlations observed between solar activity and atmospheric variability are consistent with model calculations of the atmosphere's response to solar UV irradiance variations. Physical mechanisms postulated to explain a UV-climate connection require that solar induced variations in the stratosphere be coupled to the troposphere. To date, such coupling mechanisms have been explored only qualitatively. One difficulty in establishing the magnitude of any solar UV-climate link is the lack of detailed information about the temporal behaviour of the solar UV fluxes. In this paper, the interaction of solar UV radiation with the earth's atmosphere is described, solar UV irradiance variability is examined, and possible links between solar UV irradiance variations and atmospheric variability are discussed.

1. INTRODUCTION

Solar UV radiation at wavelengths less than 300 nm represents 1.05% of the total photon energy emitted from the sun. During the past decade (primarily in the past 5 years), satellite and rocket observations of the solar irradiance have confirmed that this radiation varies over the 27 day period of solar rotation as well as the longer, 11 year solar cycle. Simon (1981)

N.-A. Mörner and W. Karlén (eds.), Climatic Changes on a Yearly to Millennial Basis, 449–471.

has reviewed the UV irradiance data. The magnitude of UV irradiance variations decreases with increasing wavelength, and, over the solar cycle, is approximately a factor of two for Lyman alpha radiation (Lean and Skumanich, 1983), decreasing to about 25% at 200 nm and to, at most, a few percent at 300 nm (Lean, 1984). Unfortunately, the detailed temporal behaviour of the solar UV irradiance throughout past solar cycles is not observationally well defined, and remains poorly understood. This presents a serious obstacle for investigating the possible correlations of climate with UV flux variations.

That the earth's atmosphere at altitudes above 100 km responds to changes in solar activity is well recognized (Jacchia, 1975; Roble et al., 1977; Kockarts, 1981). Both the exospheric temperature and the atmosphere's density increase during times of maximum solar activity, in response to changes in both extreme ultraviolet (EUV) radiation, and solar geomagnetic activity. At decreasing altitudes, the influence of solar variability on the atmospheric system becomes increasingly more obscure. Solar-related variations in stratospheric parameters have been investigated numerically and observed experimentally (see Section 5, below). However, the reality of any influence of solar UV flux variability on the troposphere, where weather and climate are experienced, remains, at the present time, unconfirmed. Energetically, the total internal energy of the troposphere is much greater than the energy changes associated with solar UV irradiance variations. Thus, if solar UV irradiance variations do influence the troposphere, their effects must occur via complex and indirect coupling and/or triggering mechansisms by which a perturbation of very little energy produces an amplified response (Studies in Geophysics: Solar Variability, Weather and Climate, 1982). Typical coupling mechanisms (see Section 4, below) invoke solar induced variations in the stratosphere, followed by stratosphere-troposphere coupling, either dynamical (Bates, 1981), chemical or radiative (Chamberlain, 1982; Ramanathan, 1982).

Records of atmospheric parameters reflect the integrated effects of all atmospheric variability, both internal and external. Only a small part of the observed atmospheric variance can be attributed to external solar forcing (Prohaska and Willett, 1983; Pittock and Schapiro, 1982). As discussed by Holton (1982a), even the internal variability of the atmosphere is not yet completely understood. Solar UV forcing is one of many external influences on a climate system which responds to, for example, changes in global carbon dioxide, aerosols, volcanoes, the earth's orbital parameters and solar outputs other than the UV flux. Furthermore, the processes of both solar and climate variations are non-stationary so that establishing the statistical validity of an observed correlation between them requires statistical techniques more sophisticated than are currently employed (Pittock, 1978). Historically, the identification of any real sun-weather link is hindered by the inherent variance of the records, which are themselves often of poor quality and insufficient time coverage, and by a poor understanding and paramaterization of the variability of the solar outputs most appropriate for determining a solar influence.

2. ABSORPTION OF SOLAR ULTRAVIOLET RADIATION BY THE EARTH'S ATMOSPHERE

Although the UV flux is only a small fraction of the total solar irradiance, its variability is extremely important for understanding the behaviour of the earth's atmosphere, where it is almost completely absorbed, and plays a crucial role in controlling the atmosphere's radiation, photochemistry, and dynamics. Figure 1 summarizes the primary interactions of the UV flux with the earth's atmosphere. In the top panel is the solar UV spectrum from 0 to 300 nm. The altitude at which solar UV radiation is attenuated to 1/e of its incident value (shown in the second panel) depends on the absorbing properties of the atmospheric constituents, and is a strong function of wavelength. Extreme ultraviolet radiation, at wavelengths between 10 and 100 nm, is absorbed by O_2, N_2, O and N in the thermosphere (above 120 km). This absorption maintains the earth's ionosphere but has little, if any, influence on the stratosphere and troposphere. The Hydrogen Lyman alpha emission at 121.57 nm is the singularly most important line in the solar ultraviolet spectrum. Its flux is equivalent to the total flux at all wavelengths less than 150 nm. Because of a window in the O_2 absorption cross section near 121.6 nm, Lyman alpha radiation penetrates into the mesosphere (60 - 80 km) where it ionizes NO and dissociates H_2O to produce OH. Solar radiation at wavelengths between 140 and 170 nm is absorbed in the lower thermosphere (90 - 150 km) by the strong Schumann Runge continuum of O_2. Molecular oxygen in the mesosphere and stratopshere (30-70 km) absorbs solar radiation at wavelengths between 170 and 242 nm, in its predissociated Schumann Runge Band system. UV radiation from 220 to 300 nm is absorbed by the strong Hartley band of ozone, primarily in the stratosphere (20-50 km). The earth's atmosphere is relatively transparent to solar radiation at wavlengths longer than 300 nm.

Absorption of solar UV radiation heats the earth's atmosphere. Identification of the primary atmospheric regions is derived from the variation of temperature with altitude. This is illustrated in the temperature-altitude profile in Fig. 1. At the earth's surface, the temperature is determined as a balance between the incoming solar radiation, which has the spectral

1. Radiation

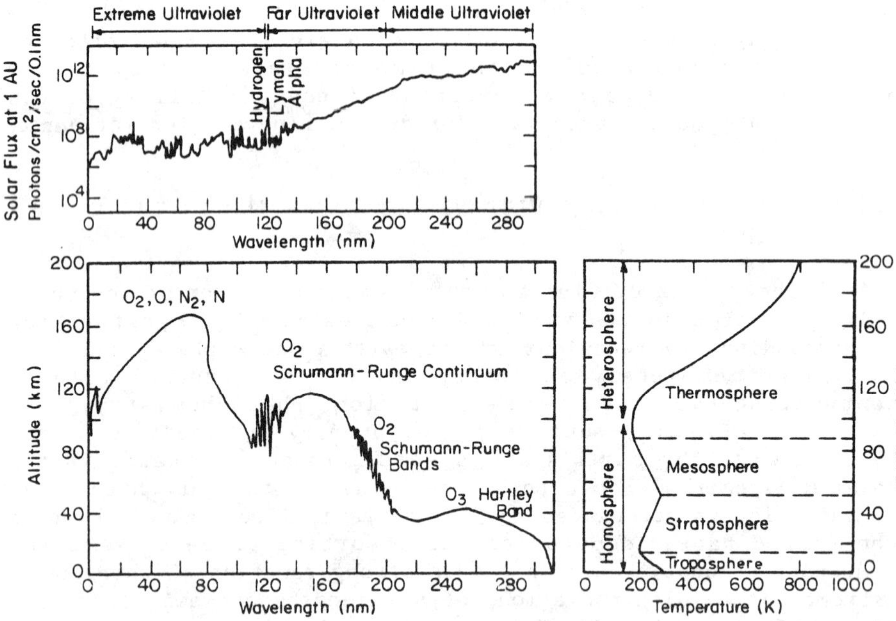

2. Chemistry

O_x Photochemistry

$O_2 + h\nu \rightarrow O + O \quad (h\nu < 242\ \text{nm})$

$O + O_2 + m \rightarrow O_3 + m$

$O_3 + h\nu \rightarrow O + O_2 \quad (h\nu < 1.08\ \mu\text{m})$

$O + O + m \rightarrow O_2 + m$

$O + O_3 \rightarrow 2O_2$

Catalytic Ozone Destruction

$X + O_3 \rightarrow XO + O_2$

$XO + O \rightarrow X + O_2$

$O + O_3 \rightarrow 2O_2$

$X:\ HO_X,\ NO_X,\ Cl_X$

3. Dynamics

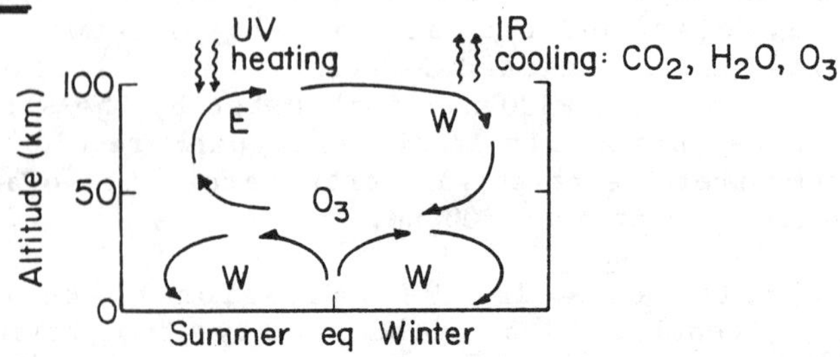

Figure 1. The interaction of the solar ultraviolet radiation from 100 to 300 nm with the earth's atmosphere, showing the primary effects for radiatin, chemistry and dynamics.

distribution of a black body at 5800 K, and the outgoing terrestrial radiation which is characteristic of a black body at 250 K. In the troposphere (below 15 km) there is a negative temperature gradient with altitude because the infrared (IR) radiation reemitted by the earth is absorbed by H_2O and CO_2, both of which have densities which decrease with altitude. In the stratosphere (15-50 km), the temperature gradient becomes positive because of heating by ozone absorption of solar UV radiation. By 50 km the ozone concentration has decreased and there is additional cooling because H_2O and CO_2 reradiate IR energy to space. Near 90 km the atomic oxygen density begins to increase and this enables heating by EUV absorption. Additional heating arises from O_2 absorption in the Schumann Runge continuum, and, at higher altitudes, by N_2, O_2 and O ionization. The thermospheric temperature profile becomes quasi-stable (at the "exospheric" temperature) at altitudes above about 200 km.

Molecular oxygen is dissociated by absorption of radiation at wavelengths less than 242 nm. The resultant oxygen atoms initiate the chemistry of the earth's atmosphere. The five O_x reactions listed in Fig. 1 are the primary reactions within a pure oxygen atmosphere and are called collectively "Chapman chemistry". Atomic oxygen recombination with molecular oxygen produces ozone which is then destroyed by either recombination with atomic oxygen or by photodissociation. Since both the reactions $O + O_2$ and $O + O_3$ are temperature dependent, the ozone concentration is strongly coupled to the local temperature. For example, increasing the stratospheric temperature by 1% causes the ozone concentration to decrease by 3-4%. Recall, however, that the ozone concentration is an important factor in determining the stratospheric temperature, through UV absorption. The feedback between temperature and ozone concentration is one example of numerous feedback mechanisms operating in the earth's atmosphere.

In addition to the oxygen family of species (O_x) are many other atmospheric constituents, the most important of which comprise the hydrogen (HO_x), nitrogen (NO_x) and chlorine (Cl_x) families (see Thrush, 1979 for a review of stratospheric chemistry). Radical species within these families (for example, OH, NO, Cl and ClO) participate in catalytic cycles which destroy ozone but which leave the radical concentration unaltered (Fig. 1). These radicals are formed either by photodissociation of more stable species, or as the product of photochemical reactions. For example, the photodissociation of mesospheric H_2O vapour by Lyman alpha radiation produces the hydroxyl radical (OH); dissociation of O_3 by radiation at wavelengths shorter than 300 nm yields the important excited atomic oxygen radical $O(^1D)$, which then oxidizes N_2O and H_2O to yield, respectively, additional radicals nitrous oxide (NO) and hydroxyl (OH). An example of UV photodissocation of particular importance for understanding atmospheric pollution is the dissociation of chlorofluoromethanes (CFMs) by solar radiation near 200 nm, to produce the chlorine (Cl) radical. In summary, the UV radiation is responsible not only for the production of ozone, but also for its photodissociation, as well as the production, either directly or indirectly, of many of the radical species which destroy it.

The earth's atmosphere is not static. Its constituents are continually mixed and circulated by a spectrum of motions extending from small scale eddy diffusion to global planetary waves (see Geller, 1983 for a tutorial on middle atmosphere climatology). Absorption of UV radiation by ozone fuels the atmospheric circulation. As depicted schematically at the bottom of Fig. 1, the tropospheric circulation is from the heat source, at the equator, to the heat sinks, at the poles. However, in the stratosphere and mesosphere, the meridional circulation is from the warm hemisphere to the cold hemisphere, with the mean zonal flow eastward in the summer hemisphere and westward in the winter hemisphere. As a result of the meridional circulation, a parcel of air at 90 km over the equator, for example, will be transported to a height of 30 km above the poles, in a period of about 3 months. Thus these different altitude regions of the middle atmosphere are coupled on dynamical time scales, so that effects of solar variability in the upper atmosphere may be manifested at lower altitudes, and different latitudes.

3. SOLAR ULTRAVIOLET IRRADIANCE VARIATIONS

Electromagnetic radiation is one form of energy output from the sun. Other manifestations of solar energy output include the solar wind, solar corpuscular radiation, cosmic rays and geomagnetic activity (see Newkirk's, 1982 review of solar variability). Each of these outputs is modulated by solar activity, and together they induce a wide variety of responses in the earth's atmosphere. Only solar UV irradiance variations are discussed here.

Recent satellite observations have confirmed, beyond doubt, that the electromagnetic radiation from the sun varies. In particular, the Nimbus 7 (Heath, 1980; Heath et al., 1983), and the Solar Mesosphere Explorer (Rottman, 1983) satellites have provided excellent data characterizing the short term variability (caused by solar rotation) of radiation at wavelengths from Lyman alpha to 300 nm. Solar ultraviolet irradiance variability is associated with the evolution and rotation, on the solar disc, of magnetically active features, called plage. Plages are regions of the solar atmosphere which extend throughout the photosphere and chromosphere, and which are enhanced in both UV emission and (to a lesser extent) visible continuum emission at longer wavelengths. Examples of solar UV irradiance variations observed during 1982 by the Solar Mesosphere Explorer (SME) are illustrated in Fig. 2. The temporal variation of the projected plage area, recorded by the World Data Center of the National Oceanic and Atmospheric Administration (WDC/NOAA), is included for comparison. Variations in the UV irradiance can be seen, in Fig. 2, to be closely related to variations in enhanced emission from plage regions.

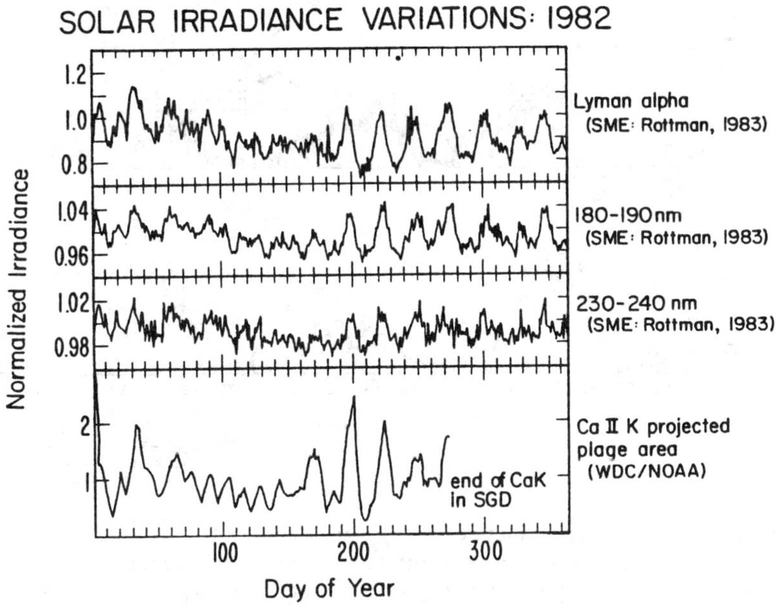

Figure 2. Solar ultraviolet irradiance variability observed by the Solar Mesosphere Explorer (Rottman, 1983) during 1982. The projected CaII K plage area is included for comparison.

It is well established, from ground based observations, that the occurrence of plages on the solar disc (like that of sunspots) varies with the 11- year solar activity cycle and the 27-day solar rotation. This is demonstrated for the past three solar cycles in Fig. 3. Times of minimum solar activity occurred during 1964 and 1976. Information about solar plage is obtained daily from spatially resolved photographs of the solar disc (such as the photographs of the chromosphere shown in Fig. 3), taken in the core of the CaII K solar absorption feature at 393 nm. This line is formed in approximately the same region of the solar atmopshere (i.e, the photosphere and chromosphere) as is the UV emission. Since plage regions observed in CaII K emission are also enhanced in UV emission, solar UV irradiance variations are expected to occur concurrently with CaII K plage variations.

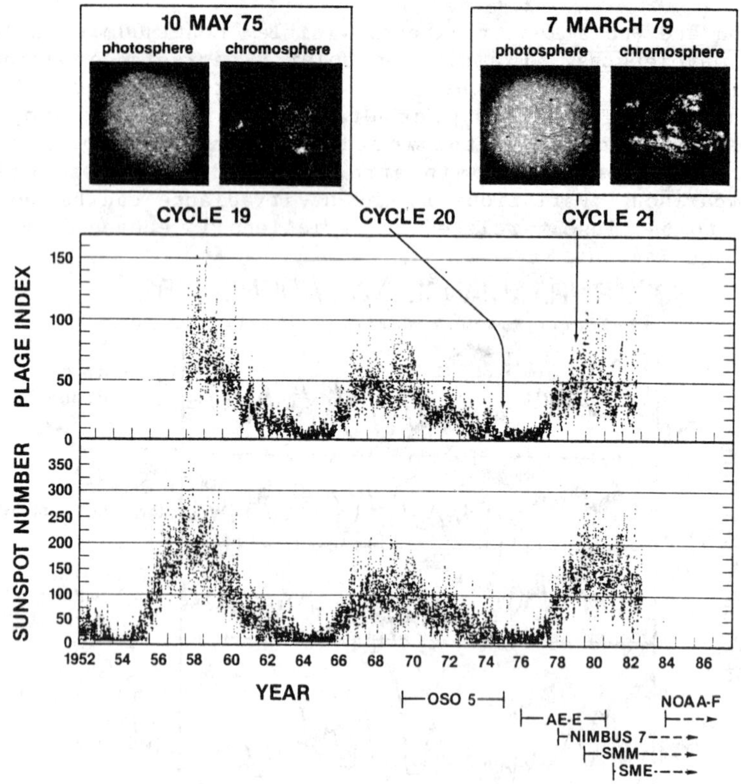

Figure 3. Ground based observations of two solar activity indices (plage index and sunspot number) during the past three solar cycles. The photographs of the photosphere illustrate the presence of sunspots on the solar disc, while the photographs of the chromosphere show the CaII K plage areas.

However, the magnitude of the related UV irradiance variations is not observationally well established. Measurements of the solar UV irradiance and its variations are difficult. They must be made from rocket or satellite platforms, launched above the earth's absorbing atmosphere, and have, typically, uncertainties of the magnitude of the expected solar variability (Lean et al., 1982; Simon and Brasseur, 1983; Lean, 1984).

An alternative approach for investigating the magnitude of solar irradiance variations is to model the solar UV flux by using the ground based CaII K observations of solar plage. Comparison, in Fig. 4, of the solar disc when viewed at CaII K (393 nm) and at Lyman alpha (121.57 nm), illustrates the similarities between the distribution of enhanced CaII K and Lyman alpha emission on the disc. A more detailed discussion of this figure is provided by Lean and Skumanich (1983). Knowing the contrast for plage emission to that from the quiet sun (i.e, at solar minimum) the total solar flux at a particular UV wavelength, emitted from both quiet and active regions on the disc, can be calculated. The plage regions recorded by the WDC/NOAA are the brightest, most compact regions of enhanced emission on the solar disc, and their evolution and rotation generates the observed 27 day variability of the UV irradiance. Over the longer time scale of the eleven year solar cycle, an increase in smaller regions of enhanced activity (called active network), in addition to an increase in plage, appears to contribute to the total flux (Skumanich et al., 1984). Thus a three-component (3C) model of solar active features is required to model the solar UV flux variability over both the 27 day period of solar rotation, and the 11 year solar cycle (Lean et al., 1982). An example of the 3C model calculations, showing the contribution to the full disc Lyman alpha flux from plage and active network emission

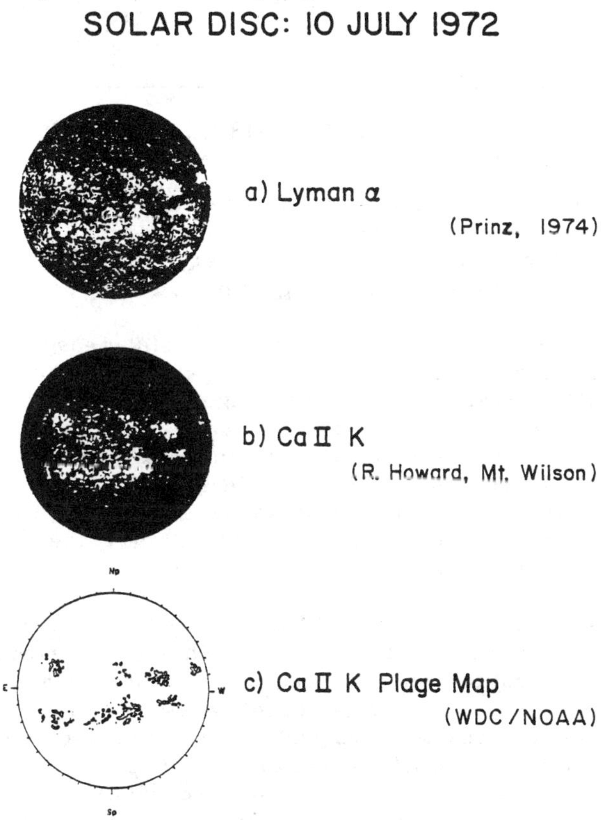

Figure 4. Comparison of the distribution of enhanced emission on the solar disc, viewed at Lyman alpha (Prinz, 1974) and CaII K. The plage areas recorded by the WDC/NOAA are included.

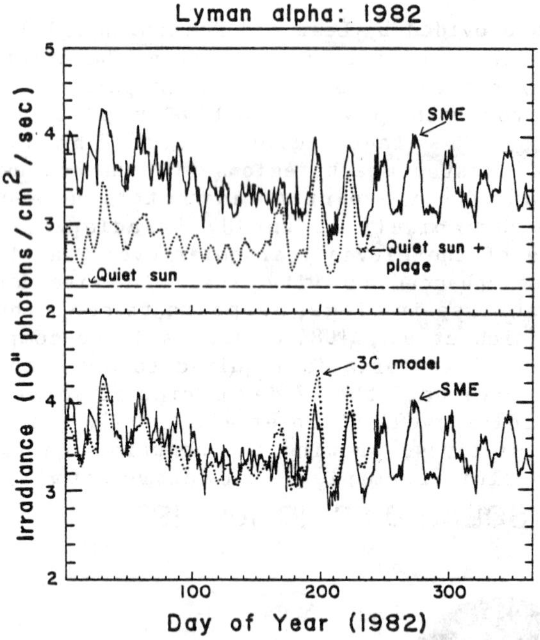

Figure 5. Model calculations of the variability of the Lyman alpha flux during 1982, due to plage areas alone (upper), and to plage and active network emission (lower), compared with measurements by the Solar Mesosphere Explorer.

during 1982, is provided in Fig. 5. The SME measurements of the Lyman alpha flux are included for comparison. The magnitude of the UV flux variation over the longer time interval of the 11 year cycle is not known as accurately as the short term variations. This is especially true for the longer UV wavelengths, which are less variable. In Fig. 6, model calculations of the variability, during solar cycle 21, of the UV flux in two wavelength intervals (180-190 nm, 230-240 nm) are compared with available observations.

Figure 7 summarizes the 3C model predictions of the UV flux variability during solar cycle 21. These calculations can be considered to provide an estimate of the upper limit of the flux variability, from a solar point of view (see Lean, 1984 for a discussion of the assumptions and uncertainties of the 3C model). Thus, for solar Lyman alpha radiation, the variability during solar cycle 21 is about a factor of two, decreasing to no more than 25% or so at 200 nm, 10% at 250 nm and 2% at 300 nm. Also included in Fig. 7 are estimates of solar UV flux variability assumed in various investigations of solar-related atmospheric variability. It is important to note in Fig. 7 that the UV flux

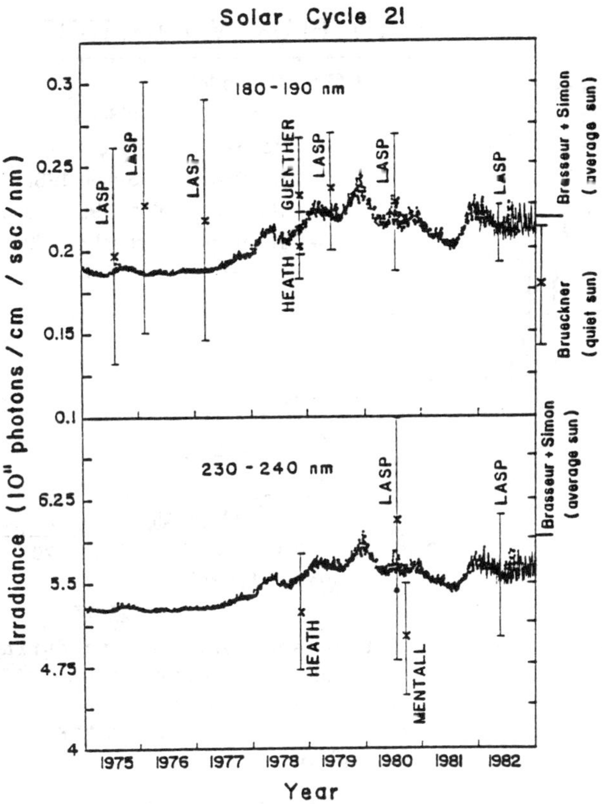

Figure 6. Model calculations of the solar irradiance variability at 180-190 nm and at 230-240 nm, compared with rocket and satellite observations during solar cycle 21.

variability calculated by the 3C model is smaller than has been assumed by past numerical investigations of the effects of solar variability on the atmosphere. Only in the most recent studies of Brasseur and Simon (1982) and Garcia et al. (1983) is the assumed variability comparable, and even then, the spectrum of the variability is different in each case. Some results of numerical investigations of the consequences of solar UV variability for the earth's atmosphere are provided in Section 5.

In terms of energy, the total solar UV radiation between 200 and 300 nm is calculated by the model to vary from 14.6 Watt/m^2 (July, 1976) to 15.5 Watt/m^2 (December, 1979). This is equivalent to a change in the total solar iradiance of about 0.05%. Thus the energy change associated with UV variations over the solar activity cycle is small compared with, for example, the 6% change in solar irradiance (at all wavelengths) caused by the annual sun-earth distance variation (eg., The American Ephemeris

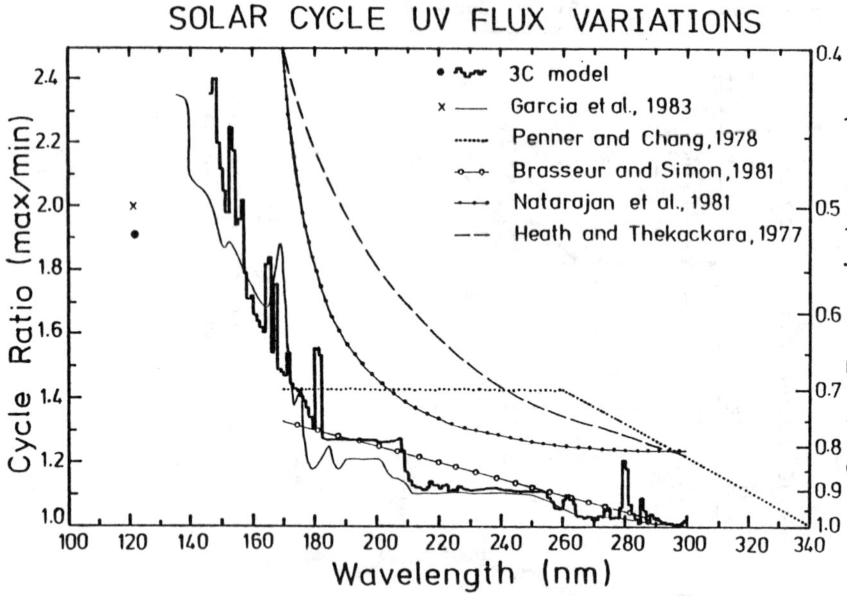

Figure 7. Solar ultraviolet irradiance variability during solar cycle 21. Model calculations of the ratio of the flux for December, 1979 to July, 1976 are compared with the solar cycle variability assumed in studies of the effects of solar variability on the earth's atmosphere.

and Nautical Almanac for 1971). This latter variation is, however, spectrally independent, as are variations in solar irradiance due to the earth's orbital parameters. The potential importance of UV flux variability for the earth's atmosphere lies in the spectral dependence of the variations, which may induce atmospheric variations that are different at different altitudes (Callis and Nealy, 1978; Natarayan et al., 1981; Ramanathan, 1982; Chamberlain, 1982).

4. PHYSICAL MECHANISMS FOR A SOLAR UV-CLIMATE CONNECTION

Since the photochemistry and dynamics of the earth's middle atmosphere is initiated and maintained by the absorption of solar UV radiation (as summarized in Section 2), any changes in this radiation can be expected to have some influence on the atmosphere. The most important solar UV wavelengths for the stratosphere and troposphere are the intervals 175 - 242 nm, which creates ozone (through the production of atomic oxygen from molecular oxygen dissociation) and 240 - 300 nm which destroys ozone by photodissociation and through atmospheric heating. Solar UV flux variability has been reported to be the cause of changes in a variety of atmospheric parameters such as the ozone concentration (Angell and Korshover, 1978, 1983; Hood, 1982; Gille and Smythe, 1983; Dutsch, 1983), hydroxl concentrations (Burnett and Burnett, 1981), zonal wind speeds (Gregory, 1981; Gregory et al., 1983) and the geopotential and temperature fields (Quiroz, 1979; Ebel et al., 1981; Ebel and Schwister, 1981, 1983). Although UV irradiance variability appears to directly impact the climatology of the middle atmosphere, its importance for understanding the behaviour of the troposphere remains unconfirmed, but has been postulated to occur through stratosphere-troposphere coupling of a radiative, chemical and dynamical nature.

4.1. Radiative and chemical coupling between the stratopshere and troposphere.

The radiant solar energy incident on the troposphere is determined by the stratospheric ozone concentration, which depends on both the solar UV radiation and the atmospheric temperature. The temperature of the atmosphere at a particular latitude, longitude and altitude results from a balance between heating by absorption of solar UV energy and cooling by radiation of IR energy. The IR radiation emitted by stratospheric gases (O_3, H_2O, and CO_2) is itself dependent on the temperature, and is enhanced at higher temperatures. The net heating depends on the vertical distribution of ozone and thus on the spectral distribution of the UV flux. Since the solar UV irradiance affects the ozone concentrations through both photochemistry and temperature, solar UV flux variations may modulate the solar energy received in the troposphere, and hence influence the climate. For example, model calculations reported by Ramanathan (1982) showed that a 10% uniform increase (decrease) in stratospheric ozone warms (cools) the surface and troposphere by about 0.1 K. More detailed discussions of radiative and chemical coupling between the stratosphere and troposphere are provided by Chamberlain (1982) and Ramanathan (1982).

4.2. Dynamical coupling between the stratosphere and troposphere.

The atmospheric circulation in the middle atmosphere transfers heat from the summer pole to the winter pole (see Fig. 1). The strength of this circulation depends on the latitudinal gradient of ozone and the geometry of solar radiation (the IR cooling is distributed reasonably uniformly with latitude). Thus changes in the ozone concentration in response to solar UV irradiance variations may affect the general circulation of the middle atmosphere. In the winter hemisphere (where the mean zonal winds are westerly) the troposphere and stratosphere are coupled by planetary scale waves. These waves are generated by landforms and longitudinal variations in heating, and propagate energy and momentum vertically over many scale heights. Reflection of planetary waves by the stratosphere has been postulated as a mechanism through which solar-related changes in the stratosphere may couple with the troposphere wind pattern (Geller and Alpert, 1980; Bates, 1981). The transmission-reflection properties of the waves is determined by the state of the stratosphere, primarily the wind structure, vertical temperature gradient and radiative dissipative processes (Ramanathan, 1982), all of which may be modulated by changes in UV radiation. For example, using a theoretical planetary wave model, Geller and Alpert (1980) concluded that changing the mean zonal flow by 20% in the region of 35 km or below would give rise to changes in the tropospheric planetary wave pattern of the same order as the observed inter-annual variability in the tropospheric wave pattern at middle and high latitudes. Holton (1982b) has reviewed dynamical coupling as a possible mechanism by which solar variability may influence climate.

5. THE RESPONSE OF THE STRATOSPHERE TO SOLAR ULTRAVIOLET IRRADIANCE VARIATIONS

Mechanisms for a solar UV-climate connection rely on strato-sphere variability, primarily in ozone concentrations and temperature, coupling to the troposphere. Hence, a prerequisite for establishing the reality of any solar UV-climate connection is the verification and quantification of solar-related changes in the stratosphere. This question has been investigated with the use of a variety of numerical models of the middle atmosphere, as well as through observations of correlations between solar UV and atmospheric variability. The extent to which solar UV variations perturb the state of the stratosphere is, at present, controversial.

5.1. Observations

Apparent correlations between the variations in a number of stratospheric parameters with indices of solar activity (such as the 10.7 cm radio flux or the sunspot number) have been observed over the time scale of one solar cycle (11 years). Quiroz (1979) reported a decrease in stratospheric temperatures of about 5°C from 1970 (near the peak of solar cycle 20) to 1976 (a period of minimum solar activity). Burnett and Burnett (1981) observed the OH content in a vertical column increase by a factor of two from 1976 (solar minimum) to 1980 (solar maximum). Angell and Korshover (1978) reported an increase in upper stratospheric ozone from 1965 (a period of minimum solar activity) to 1970 of as much as 12%. However, subsequent observations (Angell and Korshover, 1983) have not confirmed this apparent solar cycle variation. Rather, Dutsch (1983) has attributed the strong positive trend in the sixties to changes in the general circulation, but reports the presence of a solar cycle influence on the ozone content of the upper stratosphere, after the removal of volcanic disturbances (Agung and Fuego). As emphasised by Pittock (1978) and Ramanathan (1982), it is not possible to derive statistically significant correlations between two variables by using data extending over only one cycle. Furthermore, stratospheric variations arising from other external forcing, such as changes in CO_2, CFMs and volcanic aerosols, are not well understood and may obscure, or compensate, solar UV influences (see Callis et al., 1983 for a discussion of coupling and interactive mechanisms which may provide compensating effects within the atmosphere).

An alternative approach for identifying stratospheric variations that may be related to solar UV flux variability is to search for short term changes associated with the (smaller) solar rotation variations. Using this approach, Ebel et al. (1981) and Ebel and Schwister (1981,1983) have identified statistically significant oscillations in stratospheric (20-30 km) geopotential and temperature fields, apparently connected with the sun's rotation. Fluctuations in the low latitude stratospheric ozone densities, also apparently associated with solar rotation modulation of the UV fluxes, have been recently reported by Hood (1983) and Gille and Smythe (1983). The observed ozone variations are a few percent.

5.2. Model Calculations

One-, two- and three dimensional numerical models of the earth's middle atmosphere have been employed to investigate the effects of long term (i.e., 11 year) solar UV irradiance variability on stratospheric thermal, chemical and dynamical structure. Results of studies using one dimensional models have been reported by Penner and Chang (1978, 1980), Callis and Nealy (1978), Callis et al. (1979) and Natarajan et al. (1981). Calculations with two dimensional models have been described by Brasseur and Simon (1981) and Garcia et al. (1983). None of the current models has yet been used to investigate daily changes, such as those associated with solar rotation modulation of the UV fluxes. The UV flux variations assumed in the various model investigations, shown together in Fig. 7, are quite different from each other, so that a quantitative comparison of the results of different model calculations is difficult. However, the typical findings are that both the stratospheric temperature and the ozone concentrations are increased during times of maximum solar activity. Temperatures in the mid-latitude stratosphere are predicted to increase by 2 to 5°C. The ozone concentrations between 30 and 40 km, where the solar effect is a maximum, may increase by between 5 and 10%. Total ozone (in a vertical column) is also predicted to increase, by 3 to 5%, from solar minimum to solar maximum.

Figure 8 illustrates the solar cycle changes in the mean zonal wind speed (U) and the ozone concentration, calculated by Garcia et al. (1983). A solar cycle change in the mean zonal wind speed of 1 m/sec at 35 km (about 2%) is much less than the 20% perturbation assumed by Geller and Alpert (1980) in their study of how changes in the mean zonal wind speeds may influence the propagation of planetary waves (see Section 4). The stratospheric ozone concentration, in Fig. 8, at mid to low latitudes increases, in agreement with one dimensional model studies. However, large decreases in high latitude ozone concentrations are predicted to result from increased thermospheric production of NO at solar maximum, caused by increased EUV radiation and auroral particles. Recall (Fig. 1) that the NO radical destroys ozone.

The latitude-altitude plots in Fig. 8 illustrate the spatial morphology of the effects of solar variability. Geller and Alpert (1980) also noted that the tropospheric planetary wave response to solar-induced changes occurs regionally. As well, the phase between solar variability and the atmosphere's response is predicted by the models to depend on altitude. Thus the task of identifying solar related perturbations, using a ground based global monitoring network, may be difficult, and even global satellite observations will require careful interpretation.

5.3. Comparison of observations and model calculations

Perhaps the singularly most important parameter for understanding atmospheric variability is the temperature (its vertical profile and latitudinal gradient). Figure 9, which compares the measured solar cycle variability of stratospheric temperature with corresponding model calculations, illustrates that temperature variations in the stratosphere are not well understood. Both Brasseur and Simon (1981) and Garcia et al. (1983) adopted a flux variability similar to that predicted by the 3C model (see Fig. 7 for the solar UV flux variability used in each of the model calculations). Both studies predict an increase in the mid latitude temperature at 35 Km of no more than 2°, which is a factor of two smaller than the solar cycle variations reported by Quiroz (1979). This suggests that the model calculations are not sufficiently sensitive to solar perturbations. This is also apparent in Fig. 10, which shows the factor of two increase in column hydroxl measured by Burnett and Burnett (1981), compared with an increase of no more than 20%, predicted by the models.

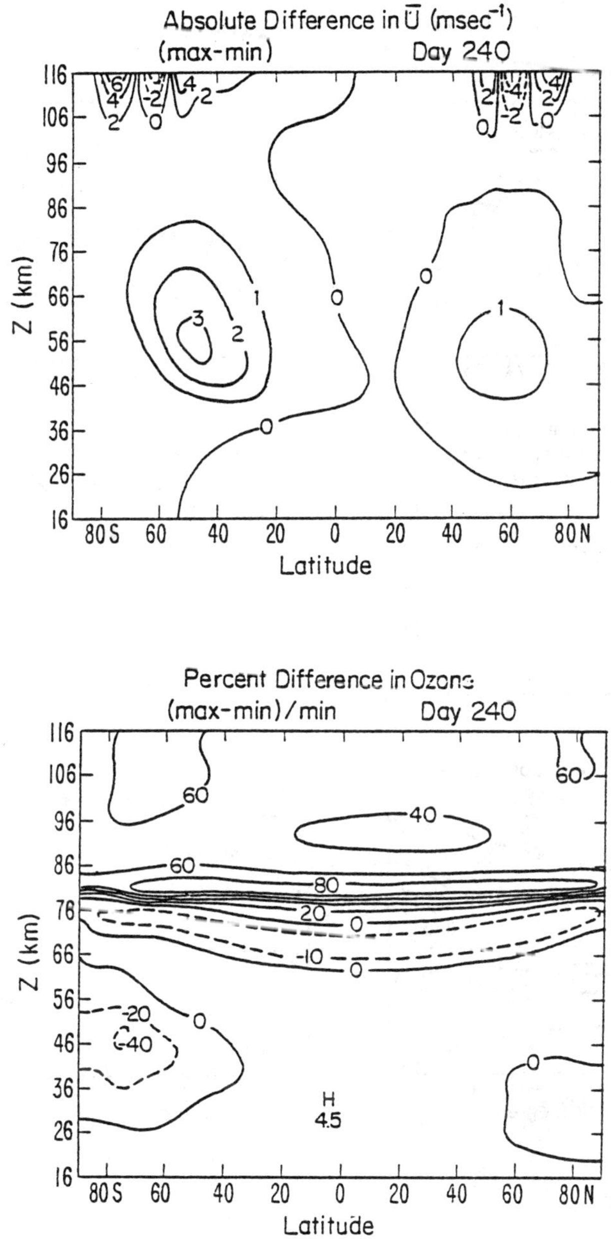

Figure 8. Model calculations by Garcia et al. (1983) of the solar cycle change in the mean zonal wind speed (U) and the ozone concentration.

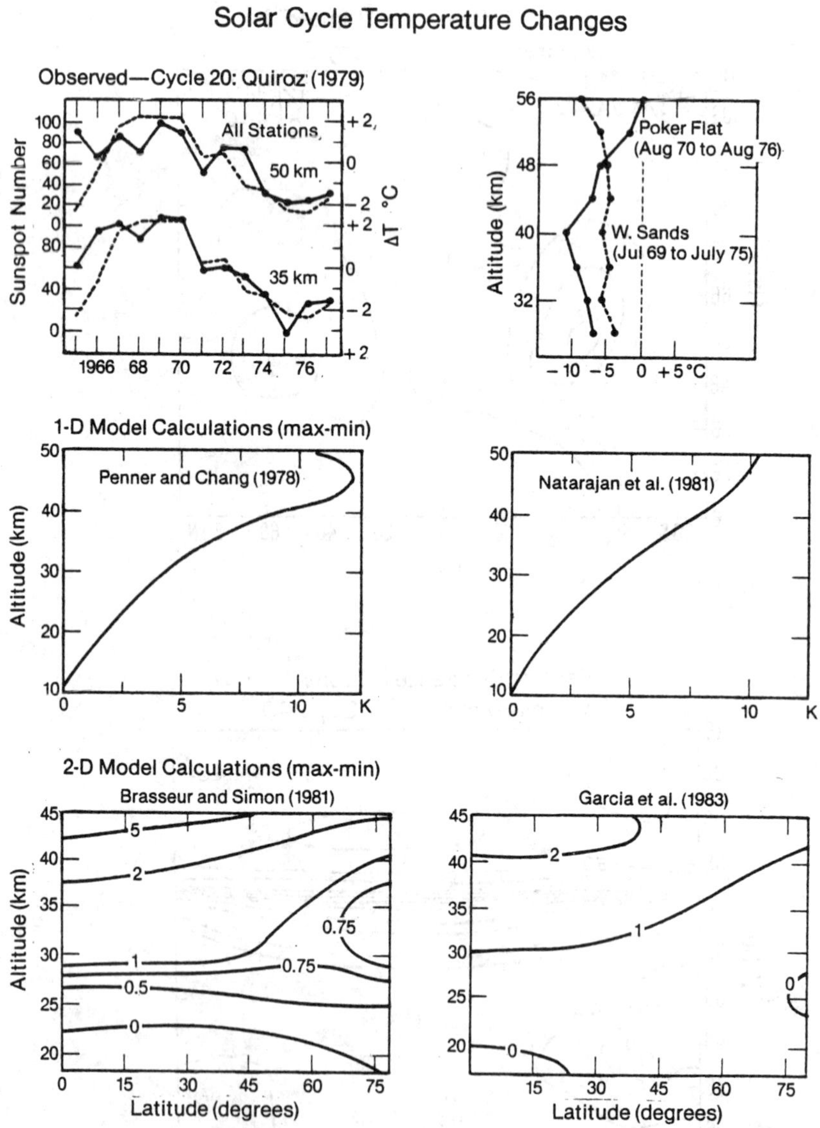

Figure 9. Solar cycle temperature changes, observed and calculated.

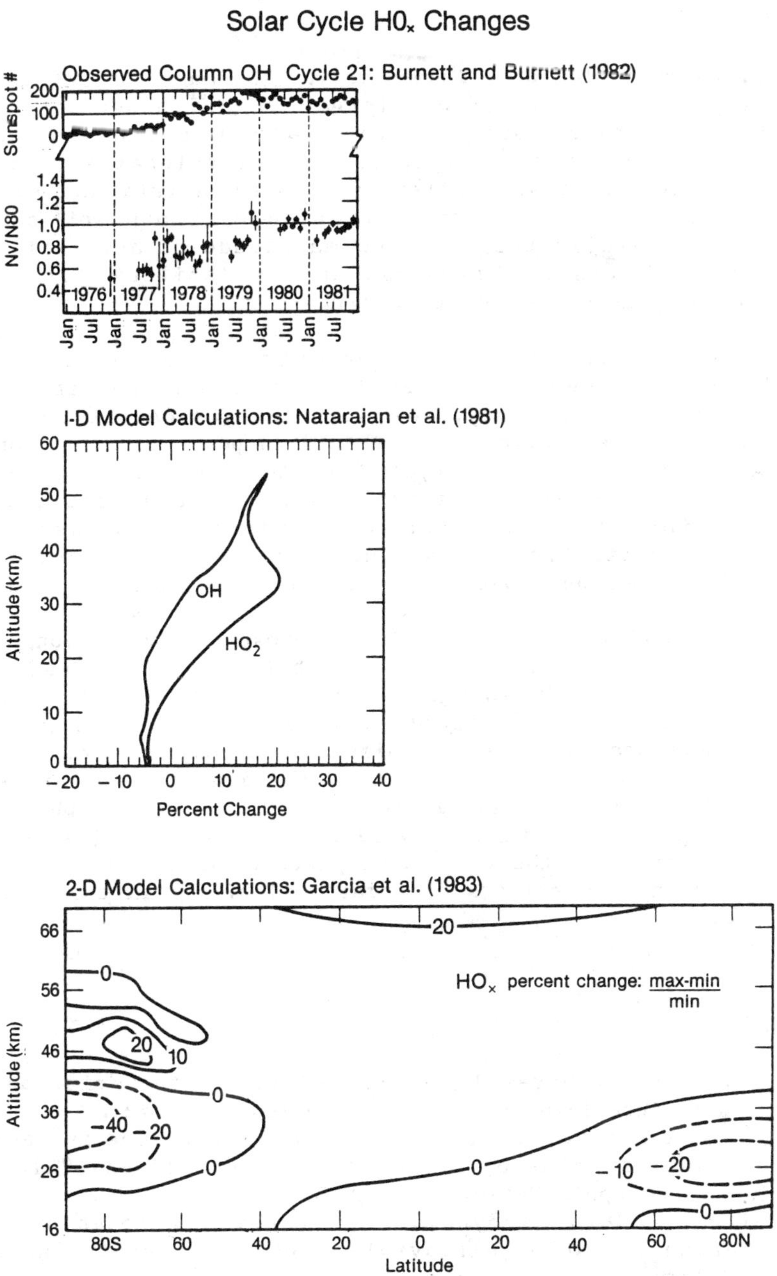

Figure 10. Solar cycle odd hydrogen changes, observed and calculated.

6. SUMMARY

Solar UV radiation does indeed vary. The magnitude of this variability is a function of wavelength and, although not yet known with complete certainty, appears to be less than has been assumed previously. There is increasing observational evidence to suggest that the stratosphere responds to variations in the solar UV flux. However the interpretation of atmospheric data records is so complicated that the magnitude of any solar-related signal in observed stratospheric variability (for eg., temperature and ozone) is not yet known.

Only when the effects of solar UV variations on the stratosphere have been accurately determined will it be possible to estimate with any certainty, how such changes couple to the troposphere and affect climate. For this purpose, accurate long term records (over many solar cycles) of the temporal behaviour of many solar and atmospheric parameters are essential. Interpretation of these data will require the use of sophisticated statistical techniques, together with numerical models of increased sensitivity to solar perturbations.

Physical mechanisms for coupling stratospheric variations to the troposphere have been postulated but not yet proven. A realistic numerical investigation of the effects of solar irradiance variations on the atmospheric system, employing all known feedbacks and coupling mechanisms, is a complex task which has not yet been attempted. This will require, in the future, the combined inputs from many disciplines of atmospheric physics and meteorology. As emphasised by Herman and Goldberg (1978), a realistic treatment of the whole atmosphere is necessary in order to understand the end effects of solar variability in the lower troposphere.

REFERENCES

Angell, J. K. & Korshover J., 1978. Global ozone variations: An update into 1976. Mon. Weather Review, 106, 725.

Angell, J. K. & Korshover J., 1983. Global variations in total ozone and layer-mean ozone: An update through 1981. J. Climate and Appl. Meteorol., 22, 1611.

Bates, J. R., 1981. A dynamical mechanism through which variations in solar ultraviolet radiation can influence tropospheric climate. Solar Phys., 74, 399.

Brasseur, G. & Simon, P. C., 1981. Stratospheric chemical and thermal response to long term variability in solar UV irradiance. J. Geophys. Res., 86, 7343.

Burnett, C. R. & Burnett, E. B., 1981. Spectroscopic measurements of the vertical column abundance of hydroxyl (OH) in the earth's atmosphere. J. Geophys. Res., 86, 5185.

Callis, L. B. & Nealy, J. E., 1978. Solar UV variability and its effect on stratospheric thermal structure and trace constituents. Geophys. Res. Lett., 5, 249.

Callis, L. B., Natarajan, M. & Nealy, J. E., 1979. Ozone and temperature trends associated with the 11-year solar cycle. Science, 204, 1303.

Callis, L. B., Natarajan, M. & Boughner, R. E., 1983. On the relationship between the greenhouse effect, atmospheric photochemistry, and species distribution. J. Geophys. Res., 88, 1401.

Chamberlain, J. W., 1982. The influence of solar ultraviolet variability on climate. Planet. Space Sci., 30, 147.

Dutsch, H. U., 1983. Ozone variability. Planet. Space Sci., 31, 1053.

Ebel, A., Schwister, B. & Labitzke, K., 1981. Planetary waves and solar activity in the stratosphere between 50 and 10 mbar. J. Geophys. Res., 86, 9729.

Ebel, A. & Schwister, B., 1981. The sun's rotation and perturbations of geopotential height and temperature fields in the stratosphere. Solar Phys., 74, 385.

Ebel, A. & Schwister, B., 1983. Relations between stratospheric and tropospheric oscillations correlated with solar activity at periods between 13 and 27 days. In: Weather and Climate Responses to Solar Variations (ed. B. M. McCormac), pp. 169-177, Colorado Associated University Press, Boulder, Colorado.

Garcia, R. R., Solomon, S., Roble, R. G. & Rusch, D. W., 1983. A numerical study of the response of the middle atmosphere to the 11-year solar cycle. Planet. Space Sci., in press.

Geller, M. A., 1983. Dynamics of the middle atmosphere. Space Sci. Rev., 34, 359.

Geller, M. A. & Alpert, J. C., 1980. Planetary wave coupling between the troposphere and middle atmosphere as a possible sun-weather mechanism. J. Atmos. Sci., 37, 1197.

Gille, J. C. & Smythe, C. M., 1983. Response of stratospheric ozone to solar UV variations. IUGG XVIII General Assembley (IAMAP), Hamburg, 15-27 August.

Gregory, J. B., 1981. Wind measuring techniques for studies of middle atmosphere dynamics. 4th IAGA Scientific Assembley, Edinburgh, August 3-15.

Gregory, J. B., Meek, C. W. & Manson, A. H., 1983. Solar cycle variations of mesospheric winds at 52N. In: Weather and Climate Responses to Solar Variations (ed. B. M. McCormac), pp. 161-167, Colorado Associated University Press, Boulder, Colorado.

Heath, D. F., 1980. A review of observational evidence for short and long term ultraviolet flux variability of the sun. In:

Sun and Climate, p. 163, International Conference, Toulouse, France, 30 Sept - 3 Oct.

Heath, D. F. & Thekaekara, M. P., 1977. The solar spectrum between 1200 and 3000 A. In: The Solar Output and its Variation (ed. O. R. White), pp. 193-212, Colorado Associated University Press, Boulder, Colorado.

Heath, D. F., Donnelly, R. F. & Merrill, R. G., 1983. Nimbus-7 SBUV observations of solar UV spectral irradiance variations caused by solar rotation and active region evolution for the period November 7, 1978 - October 26, 1979. NOAA Technical Report ERL 424-ARL 7.

Herman, J. R. & Goldberg, R. A., 1978. Sun, Weather and Climate. NASA SP-426.

Holton, J. R.,1982a. Weather Variability. In: Studies in Geophysics: Solar Variability, Weather and Climate, pp. 48-52, National Academy Press, Washington D.C.

Holton, J. R., 1982b. Possible Physical Mechanisms: Dynamic Coupling. In: Studies in Geophysics: Solar Variability, Weather and Climate, pp. 79-82, National Academy Press, Washington D. C.

Hood, L. L., 1982. Long-term trends in Nimbus IV BUV ozone data. EOS, 63, 895.

Hood, L. L., 1983. The temporal variability of upper stratosphere ozone at equatorial latitudes, Nov. 1970 - Nov. 1972. Submitted to J. Geophys. Res.

Jacchia, L. G., 1975. The Earth's upper atmosphere - I. Sky and Telescope, 155, March 1975; The Earth's upper atmosphere - II. Sky and Telescope, 229, April, 1975; The Earth's upper atmosphere - III. Sky and Telescope, 294, May, 1975.

Kockarts, G., 1981. Effects of solar variations on the upper atmosphere. Aeronomica Acta, 229.

Lean, J. L., White, O. R., Livingston, W. C., Heath, D. F., Donnelly, R. F. & Skumanich, A., 1982. A three component model of the variability of the solar ultraviolet flux: 145-200 nm. J. Geophys. Res., 87, 10307.

Lean, J. L. & Skumanich, A., 1983. Variability of the Lyman alpha flux with solar activity. J. Geophys. Res., 88, 5751.

Lean, J. L., 1984. Estimating the variability of the solar flux between 200 and 300 nm. J. Geophys. Res., 89, 1.

Natarajan, M., Callis, L. B. & Nealy, J. E., 1981. Solar UV variability: Effects on stratospheric ozone, trace constituents and thermal structure. Pure Appl. Geophys., 119, 750.

Newkirk Jr., G. A., Solar Variability. In: Studies in Geophysics: Solar Variability, Weather and Climate, pp. 33-47, National Academy Press, Washington D. C.

Penner, J. E. & Chang, J. S., 1978. Possible variations in atmospheric ozone related to the eleven-year solar cycle. Geophys. Res. Lett., 5, 817.

Penner, J. E. & Chang, J. S., 1980. The relation between

atmospheric trace species variabilities and solar UV variability. J. Geophys. Res., 85, 5523.
Pittock, A. B., 1978. A critical look at long-term sun-weather relationships. Rev. Geophys. Space Phys., 16, 400.
Pittock, A. B. & Schapiro, R., 1982. Assessment of evidence of the effect of solar variations on weather and climate. In: Studies in Geophysics: Solar Variability, Weather and Climate, pp. 64-75, National Academy Press, Washington D.C.
Prinz, D. K., 1974. The spatial distribution of Lyman alpha on the sun. Astrophys. J., 187, 369.
Prohaska, J. T. & Willett, H. C., 1983. Dominant modes of relationship between U.S. temperature and geomagnetic activity. In: Weather and Climate Responses to Solar Variability (ed. B. M. McCormac), pp. 489-494, Colorado Associated University Press, Boulder, Colorado.
Quiroz, R. S., 1979. Stratospheric temperatures during solar cycle 20. J. Geophys. Res., 84, 2415.
Ramanathan, V., 1982. Coupling through radiative and chemical processes. In: Solar Variability, Weather and Climate, pp. 83-91, National Academy Press, Washington D. C.
Roble, R. G., Dickinson, R. E., & Ridley, E. C., 1977. Seasonal and solar cycle variations of the zonal mean circulation in the thermosphere, J. Geophys. Res., 82, 5493.
Rottman, G. J., 1983. 27-day variations observed in solar UV (120-300 nm) irradiance. Planet. Space Sci., 31, 1001.
Simon, P. C., 1981. Solar irradiance between 120 and 400 nm and its variations. Solar Phys.,74, 273.
Simon, P. C. & Brasseur, G., 1983. Photodissociation effects of solar UV radiation. Planet. Space Sci., 31, 987.
Skumanich, A., Lean, J. L., White, O. R., & Livingston, W. C., 1984. The sun as a star: Three component analysis of chromospheric variability in the Calcium K line. Astrophys. J., in press.
Studies in Geophysics: Solar Variability, Weather and Climate, 1982. National Academy Press, Washington, D.C.
The American Ephemeris and Nautical Almanac for the year 1971, 1969. U.S. Government Printing Office, Washington D. C.
Thrush, B. A., 1980. The chemistry of the stratosphere. Phil. Trans. R. Soc. Lond. A, 296, 149.

CYCLES OF SOLAR FLARES AND WEATHER

Theodor Landscheidt

Schroeter Inst. Res. Cycles Solar Activity

ABSTRACT. Sunspots only constitute potentials of solar activity which are actually released by solar eruptions. Single energetic flares and periods of enhanced eruptional activity seem to be related to weather. This is valid for the quality of weather forecasts (Scherhag, Reiter), atmospheric circulation changes (Schuurmans), rainfall (Markson), and thunderstorm incidence (Bossolasco et al.). There are models that explain this effect (e.g. Roberts and Olson, Markson, Neubauer, and Bucha). This poses the problem of the prediction of solar flares. Such eruptions seem to be distributed in a stochastic manner. But closer examination reveals cycles of solar flares with mean periods of 9 years, 2.25 years, and 3 months. They are accessible to forecasts, because they run parallel with special phases in the Sun's motion about the center of mass of the solar system, and with a cyclic pattern formed by the change in the angular acceleration of the vector of the tidal forces of the planets Venus, Earth, and Jupiter.

RELATIONS BETWEEN SOLAR ERUPTIONS AND WEATHER

A large flare can release energy equivalent to the explosion of more than ten million hydrogen bombs in a few minutes' time. Momentary temperatures exceed twenty million ^{o}K, hotter than in the dense core of the Sun (Eddy, 1979). The small fraction of the energy of such flares, that the Earth intercepts, is adequate to gene-

N.-A. Mörner and W. Karlén (eds.), Climatic Changes on a Yearly to Millennial Basis, 473–481.

rate striking effects in the upper atmosphere. These effects are produced by the ultraviolet, x-ray, and energetic particle emissions from flares and other eruptions, and not by an appreciable change in the overall output of the Sun, which at most reaches one thousandth of 1%. None the less, single energetic eruptions and periods of enhanced flare activity seem to be related to weather. Roberts (1979) stated: "We now have, I venture, pretty solid evidence that there are some bona fide influences of variable solar activity on the troposphere that operate at the weather lead time of a few days". Indirect evidence of this is the observed change in the quality of weather forecasts after flares. Scherhag (1952) observed a sharp degradation of the precision of weather predictions in Great Britain and Germany after the proton flare on February 23, 1952. This connection was corroborated by Reiter (1979a), who investigated a sample of 79 flares observed January 1975 to January 1977. He found a 14% increase in weather forecasts of poor quality on the key day, marked by stratospheric intrusions of Be7. According to Reiter (1979b) the Be7-concentration in the troposphere is a measure of the intensity of stratospheric-tropospheric exchange caused by flares.

As observed by Reiter (1979c) and Cobb (1967), a sharp increase in air-Earth current density occurs on flare days. The departure from normal lasts three to five days. Accordingly, Bossolasco et al. (1972) and Muir (1979) found a 70% variation of thunderstorm incidence with the occurrence of solar flares. Markson (1979) developed a model that shows how flare induced variations in electric fields and ionizing radiation can control cloud electrification, rainfall, and circulation by the redistribution of energy already present within the atmosphere. Schuurmans (1979) could demonstrate that strong flares cause atmospheric circulation changes or variations in mass and temperature distribution at middle and high latitudes, starting within 6 to 12 hours after the flare and lasting approximately one day. Zerefos (1975) obtained similar results. Neubauer (1983) and Bucha (1983) have developed models that connect changes in weather with sudden commencements of geomagnetic storms, generated by solar eruptions. Roberts and Olson (1973) hold that solar particle invasions after flares with increased ionization at stratospheric and probably upper-tropospheric levels could generate increases in high-level cirrus by augmentation of the efficiency or number of sublimation nuclei. An increase in cirrus could affect the latitudinal gradient of the radiation balance and the distribution of wind fields.

9-YEAR AND 2.25-YEAR CYCLES OF ENERGETIC FLARES

The Sun, rotating on its axis, and the Sun, revolving around the center of mass (CM), could be looked at as coupled oscillators capable of internal resonance, resulting in slight positive or negative accelerations in the Sun's spin. Such accelerations are actually observed. Speeding up or slowing down of the Sun's rotation rate is liable to influence the Sun's activity. Slower rotation seems to be linked to enhanced activity and faster rotation to weak activity. According to investigations by Eddy (1977), based on historical observations by Scheiner and Hevelius, the rotation of the Sun's equator sped up just before the Maunder minimum, while its rotation rate about 1620, near a secular maximum of sunspots, was much the same as it is today. Modern observations confirmed this relation.

Mt. Wilson observations (Howard, 1975), presented in Fig. 1, show two striking jumps in the Sun's rate of rotation in 1967 and 1970. These jumps occurred when the Sun's center, the center of mass of the solar system, and the planet Jupiter were on a line. These heliocentric constellations are indicated in Fig. 1 by arrows. Jupiter plays a dominant role even among the giant planets, that regulate the Sun's oscillations about CM. Jupiter holds 71% of the total mass of the planets and 61% of the total angular momentum of the system, while the Sun governs less than 1% of the angular momentum. This seems to be indicative of a case of spin-orbit coupling of the spinning Sun and the Sun revolving about CM, involving transfer of angular momentum from Jupiter to the revolving Sun and eventually to the spinning Sun. With respect to the angular momentum conservation law, it makes sense that the observed slowing down in the Sun's spin went along with an increase in its orbital angular momentum.

Conjunctions of Jupiter and CM relative to the Sun's center constitute a cycle with a mean period of 9 years and a range of variation from 3 to 15 years. Since the spectroheliosocope by Hale was not fully developed until 1931, the 17th 11-yr cycle of the late thirties was the first during which flares were observed systematically. Since that period, conjunctions occurred in 1942, 1951, 1959, 1967, 1970, 1974, and 1982. There was always very energetic eruptional activity around these events (Landscheidt, 1976). As to the activity in 1982, the reliability of the relation was tested by a long range forecast. X-ray bursts are a good measure of the energy released by flares. They were listed since 1970. Bursts greater than X9 are rare events. Only two of this category occurred 1970 to 1981. According to a forecast issued January 1982, events greater than X9 were expexted 29 April to 5 May, 22 May to 9 June, and 27 September to the middle of December 1982. The observations match the prediction. Bursts X12, X12.9, and X10 occurred on 6 June, 15 December, and 17 December 1982. In addition several X7, X8, and X9-events were observed. The forecast was checked by the Space Environment Services Center, Boulder, and the astronomers Gleissberg, Pfleiderer, and Wöhl.

If the interval from one conjunction or opposition of Jupiter and CM to the next one, from one jump in the Sun's rotational velocity to the next one, is looked at as a cycle and subjected to Fourier analysis, the fourth, fifth, and sixth harmonic prove to be strong. A superposition of these harmonics forms a new cycle, the mean period of which is 2.25 years, one fourth of the fundamental. A Pearson-test of x-ray bursts $\geq$ X1, related to this shorter cycle, yields the value 40 for 1 degree of freedom ($P < 0.00001$) for the years 1970 to 1974. A replication for the fundamental period 1974 to 1982 gives out the value 79 for 1 degree of freedom ($P \ll 0.00001$).

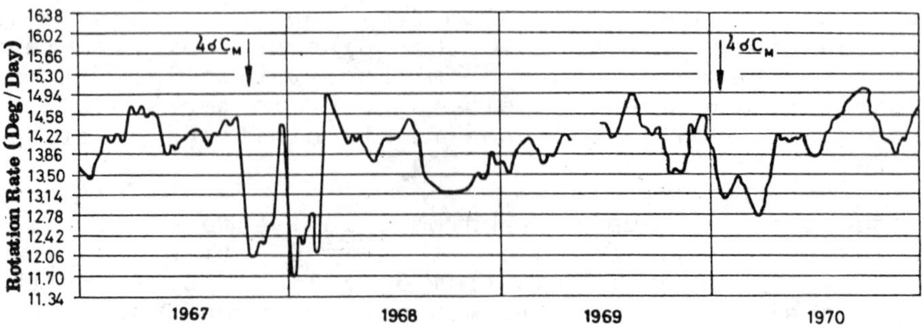

Fig. 1. Sudden decrease in the Sun's rate of rotation (from Howard, 1975) related to heliocentric conjunctions of the planet Jupiter and the center of mass.

ANGULAR ACCELERATION OF THE VECTOR OF TIDAL FORCES AND 4-MONTH CYCLE OF X-RAY BURSTS

The Sun's oscillations around CM are regulated by changing constellations and variations in the distance of the four giant planets. Another cycle is generated by the tidal planets Venus, Earth, and Jupiter. There is a wealth of literature dealing with a presumed influence of tidal planets on sunspots. Critical authors stress that all tidal planets, when conjunct, could only raise a tide of a few millimeters on the Sun. But it should be taken into consideration, that the horizontal component is not negligible, since the Sun's gravity acceleration is 28 times that of the Earth's. Öpik (1972) has shown that the mean velocity of tidal currents on the Sun reaches about one third of tidal currents generated by the Moon on Earth. As to relations of tidal planets with flares, there are only few papers (Blizard, 1965; Svestka, 1968; Landscheidt, 1976, 1981, 1983). This disproportion is not justified. We know from Skylab observations that flares are set off by initial disruptions in hot coronal loops over active regions (Eddy, 1979). It is easier to imagine that tidal disturbances may trigger such events in an unstable zone of the Sun's atmosphere than to concede that tide generating forces could act on the strong magnetic fields constituting sunspots. Calculations of the relative tidal force of the planets Mercury to Saturn show that the latter is negligible as well as Mars. Venus is sometimes stronger than Jupiter. Comparison of the combined vector of the tidal forces of Venus, Earth, and Jupiter, including or excluding Mercury, shows that the vector including Mercury oscillates around the vector of Venus, Earth, and Jupiter. Therefore, only the latter has been investigated in its relation to energetic flares marked by x-ray bursts.

Figure 2 presents the result, revealing another cycle of flares. Unexpectedly, no interesting correlations with variations in the magnitude of the vector emerged. But the change in direction proved to be crucial. The angular acceleration $d\omega/dt = d^2\varphi/dt^2$ of the vector of the tidal forces of Venus, Earth, and Jupiter forms a cyclic pattern, which shows a strong relation to x-ray bursts observed since 1970. Figure 2 shows the course of the cycle in 1982. The axis of abscissae designates the days of this year. The ordinates represent the time rate of change of the angular velocity of the vector. The active phase of a flare cycle begins when the curve crosses the time axis. This occurs when $d\omega/dt$ changes from positive to negative or from negative to positive

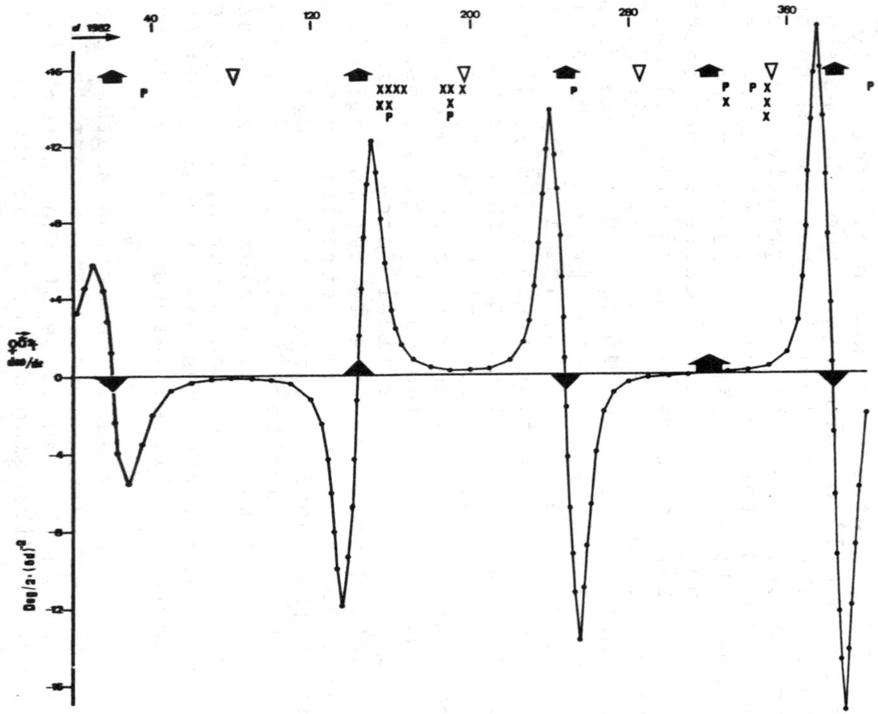

Fig. 2. The change in the direction of the vector of the tidal forces of Venus, Earth, and Jupiter, measured by the angular acceleration d /dt, forms a cyclic pattern which correlates with energetic solar eruptions. The plot shows the course of d /dt and the distribution of energetic x-ray bursts in 1982.

values. These zero phases are indicated in Fig. 2 by fat triangles, and in one case by a fat arrow. The effect is stronger when the curve ascends than when it descends. The strength of the effect on flares is inversely proportional to the steepness of the ascent. Slow ascent releases prolonged flare activity reaching a high level of energy display. The fat arrow in Fig. 2 marks such a situation. In 1982 the ascent, marked by a triangle pointing upwards, had as strong effects as the second case with slow ascent. But this category showed a stronger overall effect since 1970. The lull begins always in the middle between two zero values. On top of Fig. 2 the active and lull phases of the flare cycle are marked by arrows pointing upwards and open triangles pointing downwards. Observed proton events (P) and x-ray bursts ≧ X3 (X) match the phases of activity without exception. If all of those 109 x-ray bursts ≧ X2 observed since 1970 are tested, 91 of them fit active phases in the cycle and only 15 inactive ones. A Pearson test yields the value 52 for 1 degree of freedom ($P < 0.00001$). All events ≧ X6 fell into the active periods. As the sample covers 36 cycles and shows a homogeneous distribution, the result points to a dependable relation. The high level of significance breaks down when less energetic flares are analysed, even if they belong to an upper optical category. Such optical flares are no reliable indicator of solar-terrestrial relations. Unfortunately, many interdisciplinary papers dealing with solar-terrestrial interaction do not consider this, even if the author investigates flares because he knows that flares are a better criterion than sunspots.

EXPERIMENTUM CRUCIS

Beside the prediction of x-ray bursts >X9, mentioned already, long range forecasts, based on the flare cycles presented and on additional experience, showed significant reliability. Predictions of periods of energetic x-ray bursts and proton events were published one year in advance and checked by the Space Environment Services Center, Boulder, and the astronomers Gleissberg, Pfleiderer, and Wöhl. Out of 29 events observed 1979 to 1981, 27 hit the predicted periods of activity (Landscheidt, 1983). A Pearson-test yields the value 20.9 for 1 degree of freedom ($P < 0.00001$). A new forecast, issued January 1983, covers the period 1983 to 1990. In 1983 all events ≧ X1 observed match the predicted periods of activity.

These results, when judged with respect to the relations between energetic flares and weather discussed above, should encourage interdisciplinary research, which could lead to an improved understanding of the phenomena involved.

REFERENCES

Blizard, J.B., 1965. Predictions of solar flares months in advance. Astron. J. 70, 667.
Bossolasco, M., Dagnino, I., Elena, A., and Flocchini, G., 1972. Instituto Universitario Navale di Napoli, Meteorol. Oceanog. 1, 213.
Bucha, V., 1983, Direct relations between solar activity and atmospheric circulation, its effect on changes of weather and climate. Studia geoph. et geod. 27, 19-45.
Cobb, W.E., 1967. Evidence of a solar influence on the atmospheric electric elements at Mauna Loa Observatory. Mon. Weather Rev. 95, 905 - 911.
Eddy, J.E., 1977. The case of the missing sunspots. Scient. American 236, 80 - 92.
Eddy, J.E., 1979. A new Sun. The solar results from Skylab. NASA SP-402. Washington D.C.
Howard, R., 1975. The rotation of the Sun. Scientific American 232 106 - 115.
Landscheidt, T., 1976. Beziehungen zwischen der Sonnenaktivität und dem Massenzentrum des Sonnensystems. Nachr. d. Olbers-Ges. Bremen, No. 100, 2 - 19.
Landscheidt, T., 1981. Cycles of solar flares. In: Abstract Volume 9th International Congress of Biometeorology, Osnabrück 1981 (eds. Overdieck, D., Müller, J., Lieth, H.), p. 103.
Landscheidt, T., 1983. Solar oscillations, sunspot cycles, and climatic change. In: Weather and climate responses to solar variations (ed. Billy M. MacCormac), pp. 293 - 308. Colorado Associated University Press, Boulder, Colorado.
Markson, R., 1979. Atmospheric electricity and the Sun-weather problem. In: Solar-terrestrial influences on weather and climate (eds. Billy M. MacCormac and T. A. Seliga), pp. 215 - 232. Reidel Publishing Company.
Muir, M.S., 1979. The role of atmospheric electricity in Sun-weather relationship. In: McCormac and Seliga, 1979, pp. 259 - 262.
Neubauer, L., 1983. The Sun weather connection. In: McCormac, 1983, pp. 395 - 397.
Öpik, E., 1972. Planetary tides and sunspots. Irish Astron. J. 10, 298 - 301.

Reiter, R., 1979a. Influences of solar activity on the electric potential between ionosphere and Earth. In: McCormac and Seliga, 1979, pp. 243 - 251.
Reiter, R., 1979b. Influences of solar activity on the exchange intensity between stratosphere and troposphere. In McCormac and Seliga, 1979, pp. 289 - 296.
Reiter, R., 1979c. Influences of solar activity on the electric potential between ionosphere and Earth. In: McCormac and Seliga, 1979, p. 245.
Roberts, W. O., and Olson, R.H., 1973. New evidence for effects of variable solar corpuscular emission on the weather. Rev. Geophys. Space Phys. 11, 731 - 740.
Roberts, W. O., 1979. Introductory review of solar-terrestrial weather and climate relationships. In: McCormac and Seliga, 1979, pp. 29 - 39.
Scherhag, R., 1952. Die explosionsartigen Stratospären-erwärmungen des Spätwinters 1951/52. Berichte d. Dt. Wetterd. US-Zone No. 38, 51 - 63.
Schuurmans, C.J.E., 1979. Effects of solar flares on the atmospheric circulation. In: McCormac and Seliga, 1979, pp. 105 - 118.
Svestka, Z., 1968. On long-term forecasting of proton flares. So. Phys. 4, 18 - 29.
Zerefos, C.S., 1975. Planet. Space Sci. 23, 1035.

PLANETARY, SOLAR, ATMOSPHERIC, HYDROSPHERIC AND ENDOGENE PROCESSES AS ORIGIN OF CLIMATIC CHANGES ON THE EARTH

Nils-Axel Mörner

Stockholm University, Sweden

The spectrum of climatic cycles and changes given in the introduction (Mörner, this volume, Fig. 1) can be divided into four main groups:

(1) the Milankovitch variables that have a clear origin in the Sun-Earth constellation but still unknown mechanisms by which they affect the Earth's climate.
(2) the spectrum in the order of some decades to some millennia that has an unknown origin.
(3) the sunspot cycle and multiples of this cycle that have a clear origin in the Sun (but a fairly unknown mechanism of operation),
(4) the annual cycle that originates from the tilt of the Earth's rotational axis.

In this paper I will try to shed light on the decades to millennial cycles. In doing so, I must, however, also analyze the possible mechanisms involved in the solar and Milankovitch cycles.

Before going into this discussion, I want to recall that the Earth's climatic changes may originate from a large number of different mechanisms (see Table 1 in the concluding chapter; Mörner this volume) and are the function of a highly integrated and multidirectionl system (e.g. Fig. 1).

1. THE SUNSPOT CYCLES AND TERRESTRIAL EFFECTS

In a multibody system like our planetary system, the bodies in motion cannot have fixed, strictly Keplerian, motions, but must be constantly "vibrating". This must give rise to changing distances to the Sun, changing orbital radii, changing speeds of the celestial bodies in their orbits, changing rates of rotation

N.-A. Mörner and W. Karlén (eds.), Climatic Changes on a Yearly to Millennial Basis, 483–507.

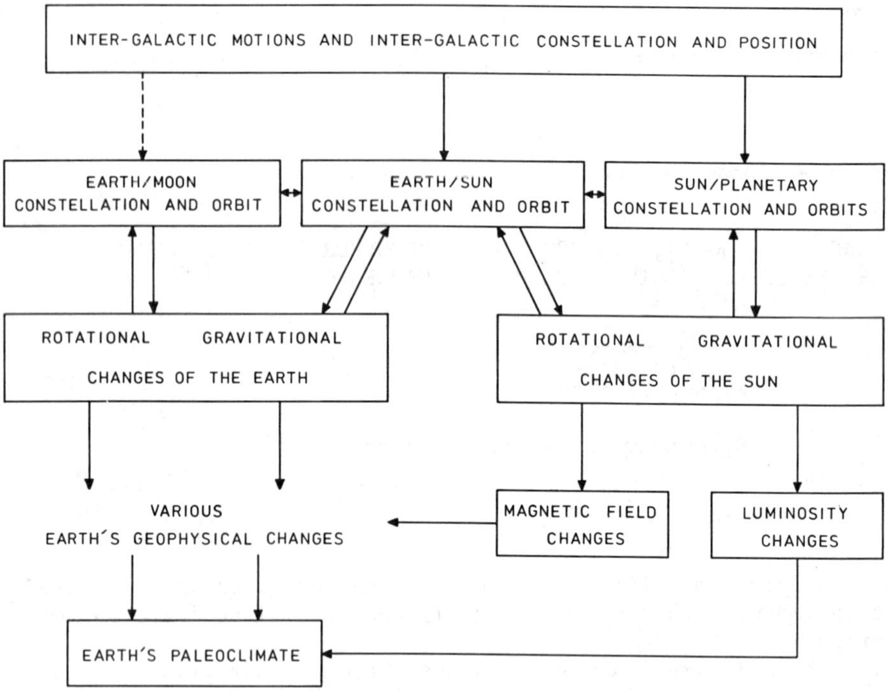

Fig. 1. Integrated planetary gravitational and rotational effects (transfer of angular momentum) on the Earth and the Sun and independent lines of affecting Earth's geophysical variables and paleoclimate.

of these bodies around their spin axes, changing gravity forces and tides, etc.

The 11-year sunspot cycle and the 22-year (Hale) cycle of magnetic polarity of the sunspots (and multiples of these cycles) can, a priori, be driven by several different mechanisms; viz.

(1) Independent processes within the Sun (a hidden "chronometer"; Dicke, 1987).
(2) The gravitational attraction by the planets. The force is very small even at a 9-planets conjunction. The tide-raising planets (Mercury, Venus, Earth, Jupiter) produce periodicities that do not match the sunspot cycles.
(3) The displacement of the Sun with respect to the center of mass (CM) in the solar system and its orbital motions around this center.
(4) Changes of the Sun's angular momentum (orbital and rotational) due to a transfer of orbital angular momentum within the planetary system due to their mutual gravitational interaction.

The Sun can, of course, not remain fixed in this planetary multibody system in motions, but has constantly to adjust its position with respect to the center of mass (CM). The revolution of the Sun around the CM varies between 9 and 14 years, with the average at about 11 years. The Sun may be displaced with respect to the CM by up to 2.19 times the solar radius. These motions of the Sun with a periodicity of about 11 years may well be the triggering and/or causation mechanism of the well-known 11-year sunspot cycle.

The orbital angular momentum in the solar system is to 99% caused by the four giant planets, viz. Jupiter (61.2%), Saturn (24.7%), Uranus (5.5%) and Neptun (8.0%). In pairs, these four planets produce synodic periods of 19.85, 45.36 and 171.39 tropical years. The multibody gravitational interaction gives rise to accelerations and decelerations of the planet's orbital motions. Mörth & Schlamminger (1979) showed that the synodic half-period (i.e. between opposition and conjunction) of Jupiter and Saturn was 9.93 years, of Jupiter-Saturn versus Uranus-Neptun was 11.23 years (or 22.46 of the same latitudinal separation) and of Uranus and Neptun was 85.7 years.

The Sun's motions around the CM and changes in the Sun's angular momentum may cause structural changes of the superadiabatic zone in the Sun and hence give rise to luminosity changes as proposed by Dearborn & Blake (1980) and discussed by Mörner (1983a, Fig. 1; 1984a). Furthermore, Gough (1980) stated that the luminosity "easier gets out" at sunspots and that "bursts of the magnetic field increase luminosity". These possible luminosity effects are illustrated in Fig. 2 (line 4).

Solar activity (solar wind) has a direct effect on the ^{14}C-production in the Earth's atmosphere (e.g. Stuiver, 1980; Stuiver & Quay, 1980). This mechanism seems to be responsible for the ^{14}C-fluctuations in the order of up to about 200 years (Stuiver, 1980, Fig. 2A). As demonstrated by Stuiver (1980), however, there is "a complete absence of a direct relationship between the solar Q and climatic changes". Similarly, it has not been possible (Mörner, 1973a, 1978, Tab. 1) to find any clear correlation between ^{14}C-production fluctuations on one hand and eustatic, climatic and glacial fluctuations on the other hand when it concerns variations in the order of centuries-to-decades (cf. Fig. 6). Even the correlation claimed (e.g. Eddy, 1977) between minima in the sunspot activity and the Little Ice Ages is not without problems as illustrated in Fig. 4 in the concluding paper (Mörner, this volume).

The main cycle of atmospheric ^{14}C-production (Fig. 5) has such an amplitude that it hardly can be explained in terms of solar wind variations. In this case, it seems to be changes in the Earth's own geomagnetic field that is the controlling variable. This "double origin" of atmospheric ^{14}C-production has been discussed by Mörner (1978, 1984a) and is illustrated in Fig. 2 (line 1).

Undoubtedly, however, there are many terrestrial records that

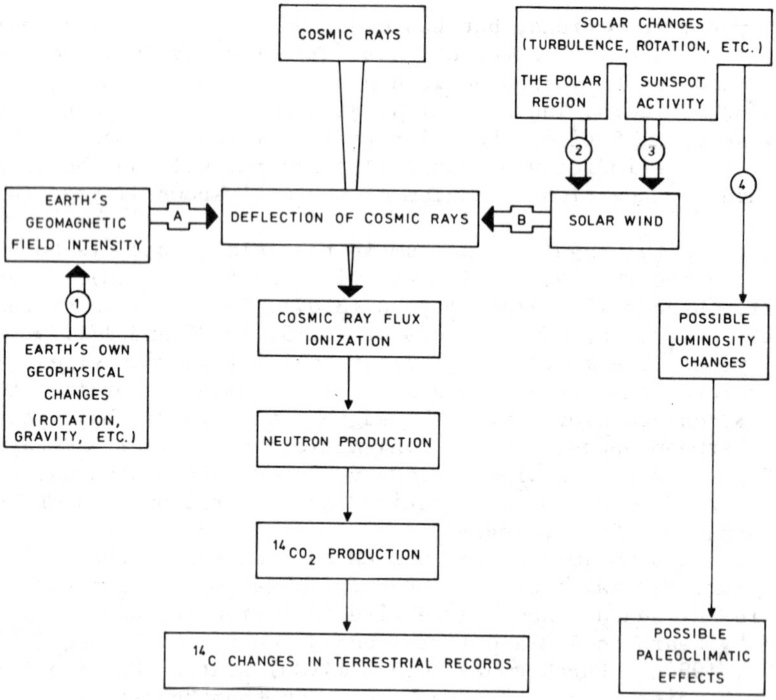

Fig. 2. "Internal" (A) and "external" (B) ways of controlling the incoming cosmic rays and hence the atmospheric ^{14}C-production.

record an 11-year (sunspot) cycle and a 22-year (Hale) cycle (and their multiples). This could bear witness of solar luminosity variations (line 4 in Fig. 2) or solar wind variations (lines 2-3 in Fig. 2) that affect the Earth's atmosphere (see paper by Lean and Bucha, both in this volume) with secondary effects on weather and climate. (The effect of the solar wind on the Earth's geomagnetic field is well-recorded). This may be used to explain various paleoclimatic records of the 11- and 22-year cycles.

The problem is not that simple, however. The 11-year cycle is seen in several geophysical variables that seem hard to connect with solar variations; viz. changes in the Earth's rate of rotation, sea level and ocean tides, Earth tides, subglacial melting (i.e. where the heat variations seem to be endogene). This seems to indicate that the planet Earth is subjected to a similar 11--year beat as is the Sun. If this is true, it gives evidence of a planetary (or galactic) beat that affects the Sun and the Earth simultaneously and sets up independent effects in both celestial bodies, where the Sun's solar wind changes have secondary effects on the Earth, too. This is illustrated in Fig. 1 (cf. Fig. 13).

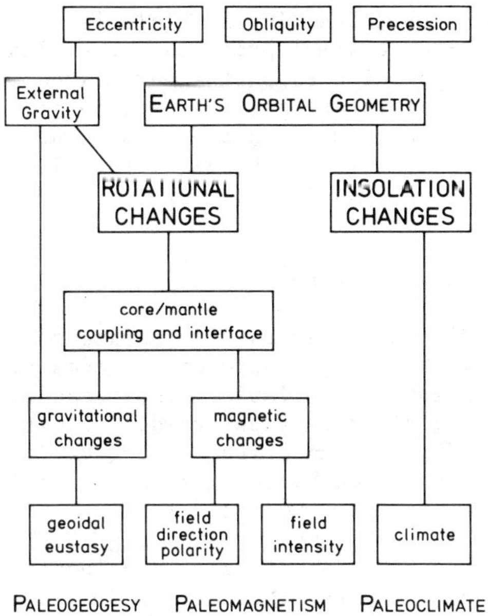

Fig. 3. Multiple effects of the so-called Milankovitch variables (Mörner, 1978, 1981a).

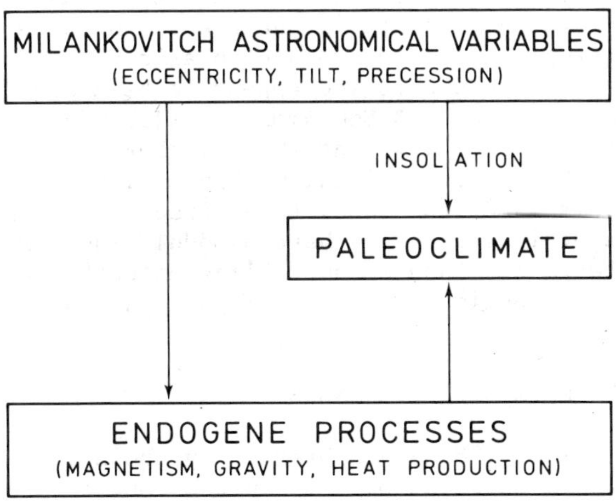

Fig. 4. The effects on paleoclimate of the Milankovitch variables may go via insolation (as generally assumed) and/or via various endogene processes, here proposed to have had a significant, maybe dominant, influence (cf. Fig. 13).

This would mean that the 11-year cycle is very complex and includes (1) extra-terrestrial processes (the planetary beat, the solar changes, the effects on the earth's geomagnetic field, atmosphere and rate of rotation), (2) exogene processes (the responses in weather and climate) and (3) endogene processes (the internal responses in gravity, the differential rotation, the heat production, the internal stress, etc.).

2. THE MILANKOVITCH VARIABLES

The Earth's paleoclimatic changes were apparently to a great extent controlled by changes in the orbital geometry according to the so-called Milankovitch variables. The mechanisms by which the orbital geometry signals are transferred into paleoclimatic changes are obscure and poorly understood, however. It is generally assumed that this is caused by changes in insolation. Changes in the Earth's orbital geometry must simultaneously affect several terrestrial variables as illustrated in Fig. 3. This may explain why there is a correlation in changes in paleoclimate, in paleomagnetism and in paleogeodesy (Mörner, 1978, 1981a). It is therefore quite possible that the main power behind the paleoclimatic changes of cyclicities in the order of the Milankovitch variable (e.g. Hays et al., 1976; Imbrie, 1982) comes via endogene processes, and not at all via insolation, as indicated in Fig. 4 and further discussed below.

Broecker (1978) concluded: "It is difficult to see how insolation cycles which are dominated by frequencies near 20,000 and 40,000 years and which are not in phase between the northern and southern polar regions, could produce polar cycles which have the same amplitude, are in phase, and appear to be dominated by the 100,000 year cycle". Mörth & Schlamminger (1979) noted that: "Thus arises the possibility that the Milankovitch variations influence climate by a gravitational mechanism". Mörner (1981a) stated: "It seems as if the insolation effects from the changes in the Earth's orbital geometry have a modulating effect on the Earth's climate rather than a controlling effect". This is in agreement with the philosophy behind Figs. 3 and 4.

3. THE HOLOCENE MAIN CYCLIC CHANGES

Mörner (1976a, 1978, 1980a) has previously shown that there seems to be a remarkable correlation between the long-term Holocene changes in global (eustatic) geoid configuration, archaeomagnetism and atmospheric ^{14}C-production, and proposed a mutual causal mechanism. This is illustrated in Fig. 5. The scenario is as follows; core/mantle processes (for example, differential movements due to rotational changes; cf. Fig. 10) give rise partly to geodynamic processes in the mantle that affect phase-transitional

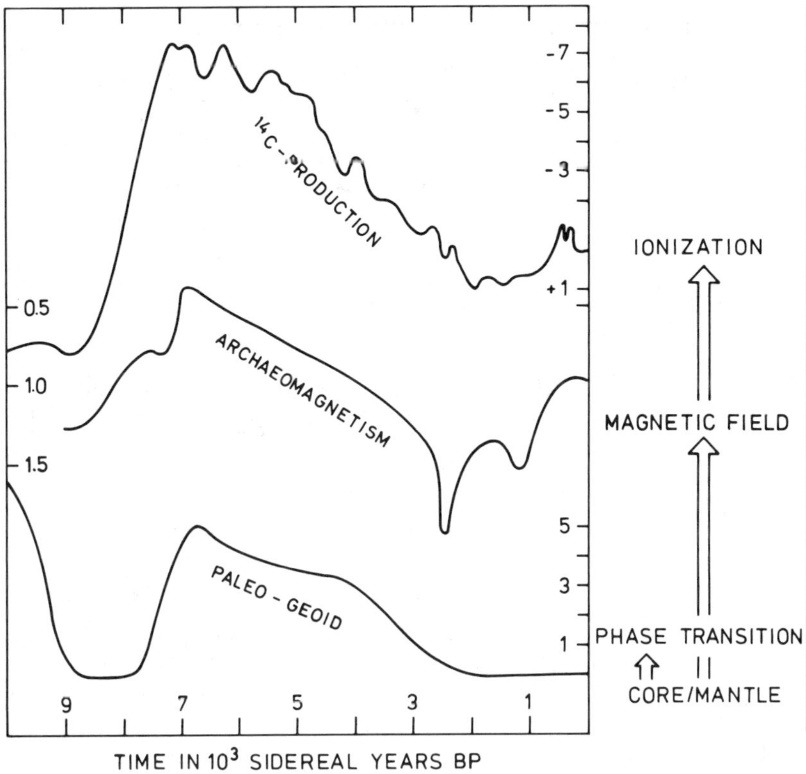

Fig. 5. Holocene main cyclic changes in ^{14}C-production, archaeomagnetism and paleogeoid configuration (Mörner, 1978, 1980a).

boundaries causing geoid deformations at the surface (cf. Figs. 8-9; Mörner, 1983b, 1984b) and partly to deformations of the Earth's geomagnetic field and hence the ionization controlling the atmospheric ^{14}C-production (as illustrated in Fig. 2).

The endogene origin seems obvious, though the ultimate triggering mechanism may be internal (terrestrial) or external planetary effects on the rotation. The origin of these cyclic changes could well be rotational changes caused by phasing of the Milankovitch variables and possible free oscillation (e.g. Ghil & Treut, 1981).

4. CHANGES IN THE ORDER OF CENTURIES AND DECADES AND INTEGRATION OF SEVERAL TERRESTRIAL VARIABLES

Mörner (1978, 1980a) showed that there was a strong correlation

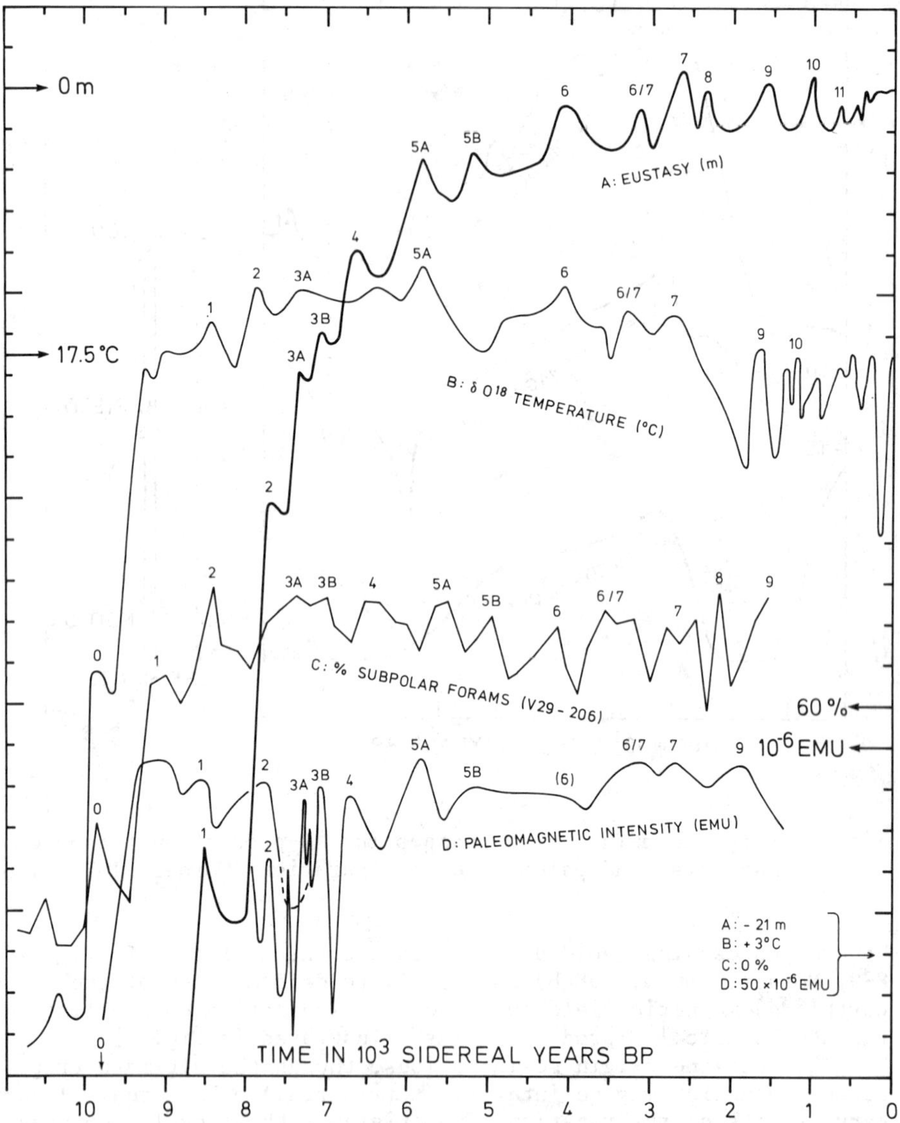

Fig. 6. Holocene short-term fluctuations in eustasy (A) in northwestern Europe (Mörner, 1969, 1980b), temperature (B) on the island of Gotland (Mörner & Wallin, 1976; Mörner, 1980c), percent subpolar forams (C) in the Denmark Strait (Kellogg, this volume) and paleomagnetic intensity (D) in two South Scandinavian cores (Mörner, 1978, 1980a). Vertical scales in meters (A), centigrades (B), 10% (C) and 10×10^{-6} EMU (D). Geographical locations in Fig. 11.

between Holocene short-term changes in climate, eustasy and paleomagnetic intensity (and, maybe, minor directional changes, too). This seemed to call for an endogene origin like that discussed for the main Holocene changes (Fig. 5).

In Fig. 6, four different records are compared. Curve A is the eustatic curve of northwestern Europe (Mörner, e.g. 1969, 1980b). It has a very firm dating control. Curve B is a temperature curve from the island of Gotland in the Baltic based on δ 018 in lake marl (Mörner & Wallin, 1976; Mörner, 1977a, 1980c). It seems to have a good dating control. Curve C is a paleoclimatic curve from a North Atlantic deep-sea core taken in the Denmark Strait west of Iceland (Kellogg, this volume). The dating control is not strict, rather floating. Curve C is a paleomagnetic intensity curve (based on two overlapping cores) from southern Scandinavia (Mörner, 1977a, 1978, 1980a). It has a firm dating control. The curves have been supplied with the eustatic transgression-peak numbers applied by Mörner (1969) to the curve A basic records. The number of peaks and bottoms and their ages are very similar (not to say almost identical) in curves A, B and D. Curve C exhibits a very similar picture, only the ages of the first 3 peaks are too old, which, however, fully lies within the uncertainties in the dating control of curve C.

This means that there in the northwest European region were some 16 low-amplitude eustatic oscillations during the last 10,000 years and that these changes were linked to similar changes in temperature and probably geomagnetic intensity, too. This could be explained in terms of the models explained in Figs. 8 and 9, i.e. regional changes of phase-boundaries within the Earth (e.g. the eclogite/granulite boundary) causing a rise of the geoid level, a geomagnetic lowering and a heat-flow rise, or vice versa.

The possible correlation with paleoclimatic changes west of Iceland opens new possibilities, however, as discussed below (Figs. 11 and 12). The rises and falls in the abundance of subpolar forams seem to record pulsating changes in the Gulf Stream activity (Fig. 12). Such changes could easily explain the changes in curve B and perhaps also explain the oscillations in curve A, though this seems hardly likely.

The 16 oscillations recorded in Fig. 6 represent changes in northwestern Europe and the bordering parts of the North Atlantic. In other regions there should be other numbers and amplitudes. This is fully in line with the long-term meteorological analysis of the geographic distribution of certain climatic events (Lamb, e.g. 1969).

It is, therefore, interesting to note that there in the Brazilian region seems to be a quite different situation. The eustatic curve shows 3 transgressions separated by two 3-4 m deep, rapid regressions during the last 7000 years (Mörner, 1981a, 1981b). In the Amazonas, Servant et al. (1981) have recorded sedimentary sequences that reveal the existence of 3 humidity/aridity cycles during the last 7000 years.

The 16 northwest Eupean low-amplitude oscillations and the 3 Brazilian high-amplitude oscillations seem to represent different regional expressions of the same intra-regional relationships and causal mechanisms.

5. HEAT FLOW VARIATIONS WITH TIME

Warm-base glaciers and subglacial meltwater tunnels with deposition of glacifluvial eskers and deltas (subaquareously) and various erosion features from water under high hydrostatic pressure give evidence of basal melting and addition of heat from the lithosphere.

There is no reason why this heat flow should have remained constant with time; even as short a time as the last glaciation or shorter parts of this period. On the contrary, there are reasons to suspect that some of the changes in the rate of deglaciation were controlled by lithospheric heat flow variations. The rapid melting of the Fennoscandian ice cap which started fairly abruptly at about 10,000 BP and ended by the disappearance of the ice cap at about 8,800 BP (with an annual rate of recession in the order of 200 m/year), cannot be explained in terms of surface and marginal melting but requires an additional heat source. This additional heat source is here proposed to be the lithospheric heat flow.

It seems significant that uplift - and subcrustal horizontal inwards migration of asthenospheric mass - peaked just at about 10,000 BP (Mörner, 1979, 1980d). This means that the uplift acceleration phase pre-dated the period of rapid deglaciation and that the latter instead coincided with the deceleration phase. The rapid uplift, therefore, seems to lead - not lag - the period of rapid deglaciation. This agrees with the theory of an origin of the rapid ice recession in intensified lithoshperic heat flow.

In a study of the deglaciation in the area south of Stockholm (e.g. Mörner, 1977b), it was possible to demonstrate that intensified accumulation of glacifluvial deposits (at Jordbro and Pålamalm) correlated with intensified melting and ice recession; not a retardation in the ice recession as previously assumed. This intensified melting must be subglacial becuase the meltwater that debouched some 200 m below sea level was under a very high hydrostatic pressure which, of course, excludes any connection with the surface-melting via fissures and fractures.

Cyclic variations in this basal melting are indicated by the fact that the succession of esker-centers sometimes exhibits an 11-year cyclicity (Möller, 1962). This cyclicity cannot originate from surface-melting in response to the sunspot cycle, but must lead its origin in rhythmic variations in heat flow. This gives evidence of an independent effect of a planetary beat on the Earth itself (as discussed above).

If this is true for the 11-year cycle, it may also hold for

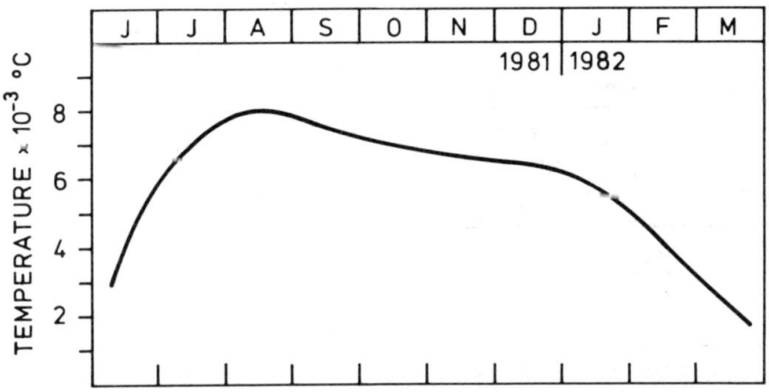

Fig. 7. Annual temperature variations at 600 m in the Dannemora mine.

the long-term Sun-Earth constellation changes as indicated in Fig. 4.

In order to investigate the heat flow problem further and to integrate it with geomagnetic and Earth tide changes, we installed a high-sensitivity temperature recorder 600 m down (in a drill hole) in the Dannemora mine. Fig. 7 gives a 10-months record. It exhibits a clear annual cyclicity of about 0.006^{o} C (dayly and monthly cycles are also recorded). This indicates that the heat flow really fluctuates with time.

Mörner (1982) suggested that the dayly Earth tides and the dayly heat flow changes were expressions of the dayly deformation of the equipotential surface due to tidal forces.

In deep wells in Hungary, Ronai (personal communications 1982) has been able to trace an annual cycle (in deep groundwater) that is likely to be an annual deformation cycle of the equipotential (geoid) surface.

6. TERRESTRIAL EFFECTS OF CHANGES IN GRAVITY (PALEOGEODESY)

In several papers, I have presented observational data indicating that the Earth's geoid surface is constantly changing and deforming, in the long as well as in the short time range (Mörner, 1976a, 1980a, 1981a, 1981b, 1981c, 1981d, 1983c, 1984b, 1984c). Finally, it has been possible to see the type of deformations that are involved (Mörner, 1984b).

Figure 8 summarizes the input and output variables regarding geoid deformations. All processes that cause a redistribution of mass or a change in density lead to a deformation of the gravitational equipotential surfaces.

The equipotential surface that corresponds to the density of oceanic water is called the geoid and determines the ocean's mean eustatic sea level (e.g. Mörner, 1976a, 1980a). It is highly

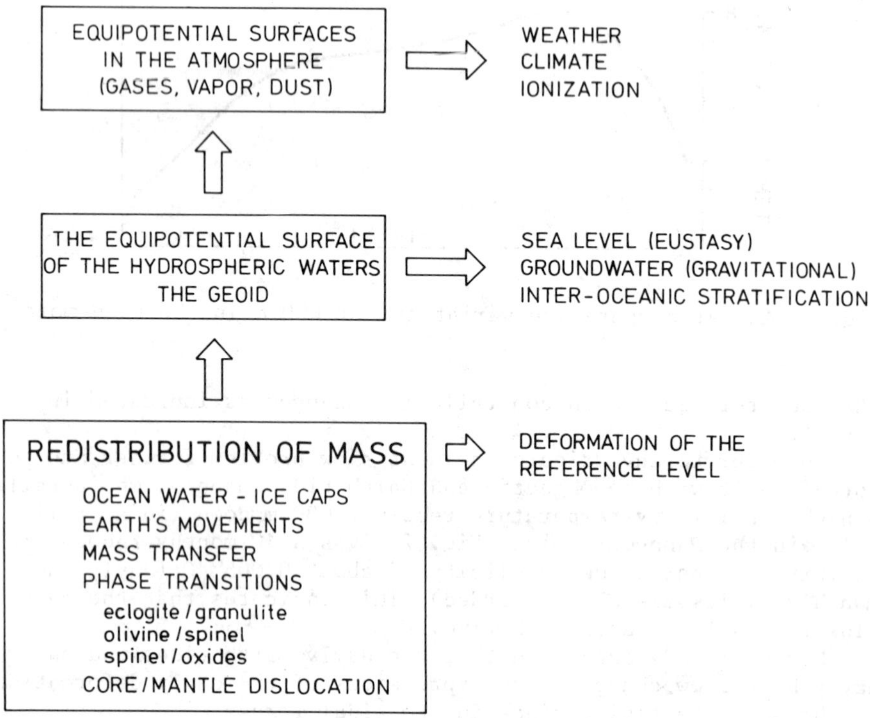

Fig. 8. Origin of geoid changes and their multiple effects via deformations in the hydrosphere and asthenosphere.

irregular with a present relief of up to 180 m. This equipotential surface also determines the gravitational groundwater level (Mörner, 1982). Similarly, it controls and/or affects the intraoceanic vertical stratification; e.g. the CCD level.

Any mass/density changes will lead to corresponding deformations and changes of these hydrospheric variables.

The atmosphere contains many equipotential surfaces, or mb--level, corresponding to different densities of the atmosphere caused by vapor, dust and gases. The dynamic forces are quite strong in the atmosphere, however, which in general leads to heavily irregular atmospheric dynamic surfaces. In the case of the middle atmosphere, Ebel (1983) concluded that "gravity induced fluctuations represent an important, in many cases the dominant, source of eddy kinetic energy" which controls the heat. Very short (hours and minutes) gravitational oscillations are recorded in airglow data (e.g. Tuan et al., 1979).

Any terrestrial change in mass/density will, however, lead to corresponding deformations and changes of the atmospheric equipotential surfaces (Mörner, 1983d). This affects weather and climate and probably the ionization, too (as shown in Fig. 8).

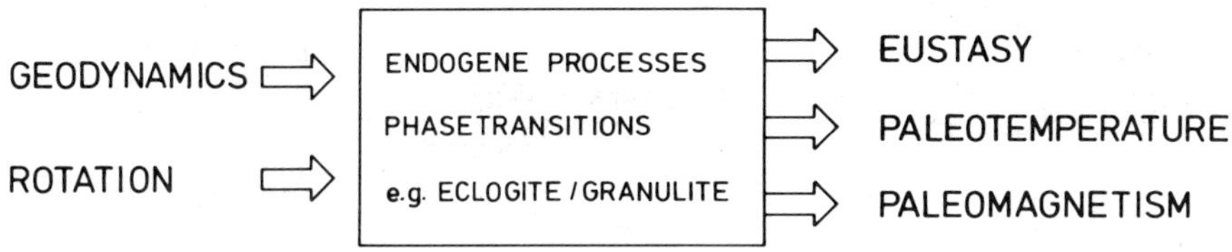

Fig. 9. Possible background to the recorded correlations in eustasy (paleogeoid), paleoclimate and paleomagnetism.

The MAGSAT data (e.g. Langel & Estes, 1982) contain effects from the main dipole field, from the non-dipole core-field and from the non-dipole "standing" lithosperic field. Mörner (1983e) argued that the power of the lithospheric field was also present in the low harmonics. At least a part of the "standing" field is, in fact, changing with time due to geodynamic processes (cf. Fig. 9). The phase-transitional boundary between eclogite and granulite is a boundary between denser, virtually non-magnetic and hotter mass and less dense, magnetic and cooler mass. This could explain regional changes in eustasy (geoid), magnetism and temperature (cf. Fig. 9) such as those recorded in Figs. 5 and 6. Even the very large-scale geomagnetic changes restricted to the northern hemisphere at about 1900 and 1967 (Courtillot at al., 1978) might be explained in a similar way. It is a general fact (e.g. von Fresen et al., 1983) that a positive gravity anomaly correlates with a negative magnetic anomaly and vice versa.

These possible effects are illustrated in Fig. 9. Rotational stress and other geodynamical forces that cause displacements of phase-transitional boundaries - such as that of granulite/eclogite - will simultaneously cause (1) eustatic changes via the deformation of the geoid, (2) geomagnetic changes due to deformation of the "standing" lithospheric field and, maybe, also due to deformation of the main field either via the rotational effects on the core/mantle interface or via deformations of the equipotential surface related to the ionosphere, and (3) climatic changes due to deformations of the atmospheric equipotential surfaces and/or due to changes in the lithospheric heat flow.

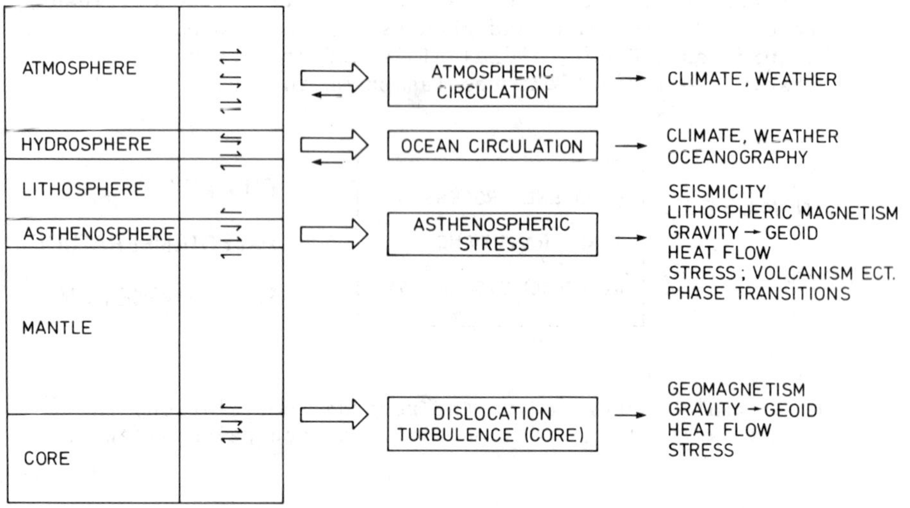

Fig. 10. Places where compensational differential rotation may take place and their multiple causation effects.

7. TERRESTRIAL EFFECTS OF CHANGES IN ROTATION

The Earth's rate of rotation seems constantly to change in the long as well as in the short-term range. If these accelerations and decelerations are not compensated or caused by changes in the Earth's radius (e.g. via sea level rises and falls), they must be compensated - to keep the angular momentum constant - by differential movements (dislocations).

Figure 10 summarizes the possible regions of differential movements (dislocations) and their corresponding effects.

A core/mantle dislocation will generate geomagnetic field changes, gravity effects, heat flow changes and changes in the stress field.

In the low-viscosity asthenosphere, there cannot be any rapid dislocation but only altered stress, which, however, may be very important and set up a lot of geodynamic effects, including changes in seismicity, volcanism, lithospheric magnetism, gravity, heat flow, etc. (cf. Fig. 9).

In the hydrosphere, dislocation and differential movements may be set up between the lithospere and the hydrosphere or, rather, within the hydrosphere as changes in the oceanic circulation. This will affect weather and climate and various oceanographic variables, including the upwelling, so important in biologic productivity, carbondioxide mixing and air/ocean heat interchange.

In the asthenosphere, differential movements are easily set up. They can take place at different altitudes. These changes will, of course, directly affect weather and climate (and, maybe, ionization, too). Barnes et al. (1982) has shown that the rotational changes wihtin a year are completely compensated within the atmosphere.

A rotational change may be compensated at anyone of these levels (Fig. 10).

The rotational changes at 1900-1910 and 1967 (Courtillot et al., 1978) are linked to significant geomagnetic field changes and must hence be compensated at the core/mantle interface or within the asthenosphere or, rather, in both. Both of these changes exhibit long wave-lengths geomagnetic changes that are limited to the northern hemisphere (Courtillot et al., 1978), which rather resembles a large scale deformation of the asthenosphere/lithosphere region than a non-dipole core field deformation.

The rotational change at 1900-1910 is linked to an increase in volcanism and seismicity (Anderson, 1974). Mörner (1981d, 1981e) stressed that this event also coincided with a marked change in the energy release at earthquakes in Fennoscandia and, maybe, in the rate of uplift, too. This calls for a considerable effect on the asthenospheric stress (cf. Fig. 10). New data from Venezuela (Aggarwal, personal communications 1983) indicate that there was a very distinct change in the seismic activity at about 1900-1910. Together with the Fennoscandian records, this indicates that the Earth's entire plate system was affected by this rotational change, i.e. that the asthenospheric zone was affected (Fig. 10).

8. EFFECTS OF CHANGES IN OCEANIC CIRCULATION

Mörner (1973a) demonstrated that the high-sedimentation-rate cores of Wollin et al. (1971) correlated very well with the eustatic changes established in southwest Sweden (Mörner, 1969) and - at that time - concluded that they represented different expressions of global climatic changes. With the introduction of the geoidal eustatic or paleogeoid theory (Mörner, 1976a), the globality could no longer be tenable.

The detailed climatic record of Kellogg (this volume) calls for a new analysis of the entire problem.

In Fig. 6, I have shown that there seems to be a very good correlation even with Kellogg's cores from the Denmark Strait. This suggests that the Gulf Stream has really been pulsating or vibrating with time (in this case through the Holocene).

Figure 11 gives a map of the Gulf Stream and the location of 3 cores studied by Wollin et al. (1971) and Mörner (1973a), of the 2 cores of Kellogg (this volume), of South Carolina where Calquhoun et al. (1983) have claimed that there is a correlation between sea level oscillations and climatic changes and that these

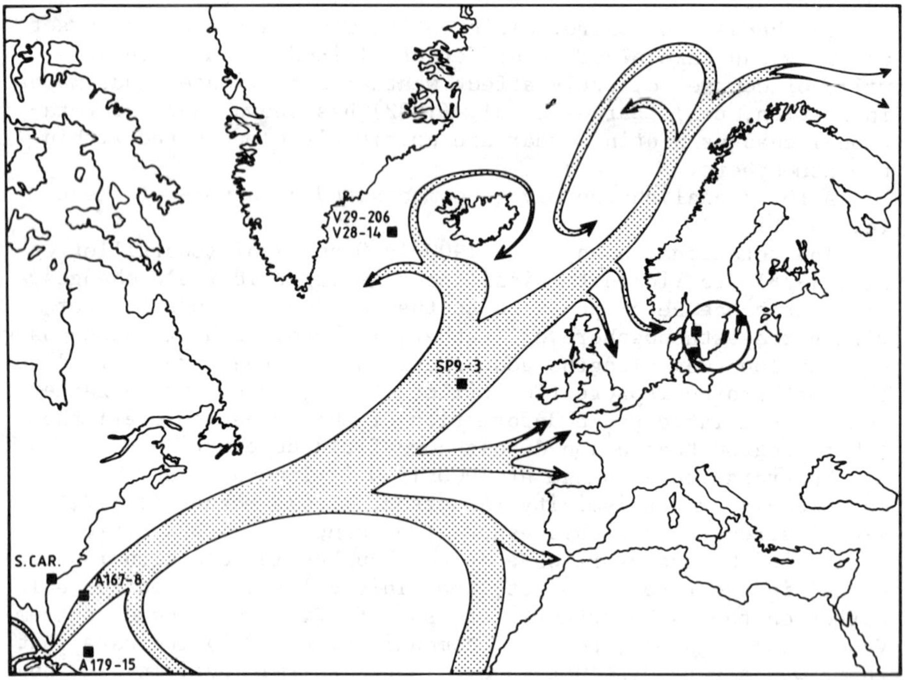

Fig. 11. The North Atlantic Gulf Stream pattern and location of places and cores discussed in the text. S.Car. = South Carolina; A179-15, A165-8 and SP9-3 are cores discussed by Wollin et al. (1971) and Mörner (1973a); V29-206 and V28-14 are cores discussed by Kellogg (this volume); circle denotes Southern Scandinavia where the eustatic, temperature and paleomagnetic records in, for example, Fig. 6 were obtained (Mörner, 1978, 1980a).
The Gulf Stream activity and the Polar Front movements with time are supposed to have played a significant role in paleoclimatology; in the very long time range at, for example, the 3.1 MA event (e.g. Mörner, 1978), in relation to the Milankovitch variables (Figs. 3-4), for example, the 23,000-cycle discussed by Ruddiman & McIntyre (1981), the short-term changes in climate during the last 35,000 years here discussed and historical changes (Lamb, 1979).

changes correlate with those in Europe (questioned by Mörner, 1981f, however), and of the basic data for curves A, B and D in Fig. 6 in Southern Scandinavia.

The scenario here offered is illustrated in Fig. 12. Increases and dedreases in the Gulf Stream activity affect the local climate and sea level in South Carolina (and surrounding areas),

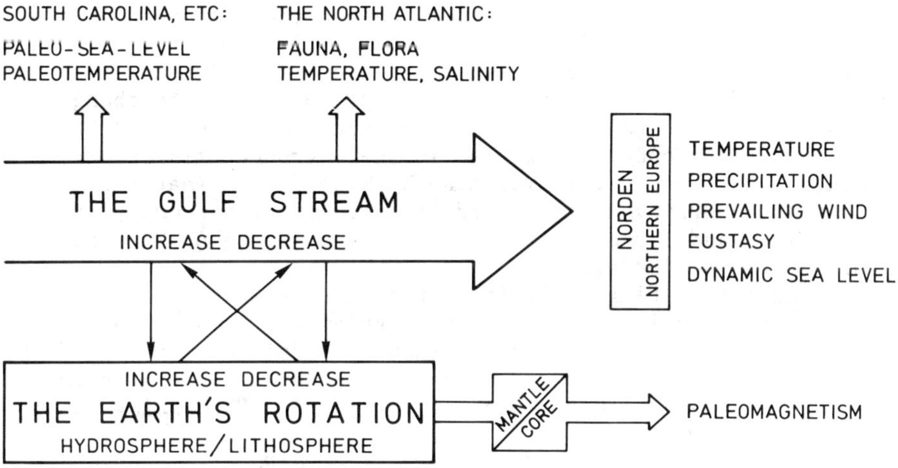

Fig. 12. Proposed model of Gulf Stream "vibration" and rotational changes explaining recorded Holocene short-term changes (e.g. Fig. 6) but probably having much wider applications.

cause the fluctuations recorded by Kellogg in the Denmark Strait and affect tempereature, precipitation, wind direction, dynamic sea level (hardly eustasy), etc. in northwestern Europe including the Nordic countries (i.e. "Norden"). Changes in the Gulf Stream activity will affect the Earth's rotation and vice versa; rotational changes will generate changes in the Gulf Stream activity (Fig. 10). Rotational changes will generate geomagnetic field changes. They may also generate many other geodynamic effects via changes in the asthenospheric stress (Figs. 9 and 10).

The balance between Gulf Streem activity and Earth's rate of rotation is a feed-back coupling (Fig. 12). Increased Gulf Stream activity brings more water to high latitudes which increases the rate of rotation, which in its turn brings water to the equator and decreases the Gulf Stream activity which both decrease the rate of rotation so that the Gulf Stream activity could increase again. This mechanism may well explain the "vibrations" recorded.

It may be significant that the combined eustatic record of Mörner (1969) and the Atlantic paleotemperature records of Wollin et al. (1971) give cycles that are frequency changing with time (Mörner, 1973a, 1973b); the longer cycle has a range of 1000--3600 years and the shorter a range of 230-1000 years. A feed--back mechanism of the type proposed in Fig. 12 would be expected to behave precisely in this way and to give rise to irregular "cycles". Mörner (1973a, 1973b) identified these cycles for the last 35,000 and 20,000 years, respectively. This means that the mechanism proposed on the basis of Holocene data may have operated also in pre-Holocene time and hence may be responsible also for fluctuations in Late Weichselian time. So, for example, do the two major warming pulses in the continental records at around 13,000-12,700 BP and at 10,000 BP correspond to changes in the Atlantic circulation bringing more temperate water to northwestern Europe as discussed in the introductorial paper (Mörner, this volume). It may also be significant that the 13,000-12,700 BP event is linked to a geomagnetic anomaly (Mörner, 1976b) and large scale deformations of the global geoid (Mörner, 1976a, 1980a).

In an interesting analysis of the changes in ocean and wind circulation in the northeastern Atlantic, Lamb (1979) showed that "sea surface temperature is liable to fall by 3 to 5° C" (or even more) in times as short as years and decades. This is in perfect agreement with the present theory of a "vibrating" Gulf Stream.

The Kuro Siwo stream in the northwestern Pacific should exhibit a somewhat similar pattern if the Fig. 12 mechanism would really operate. It is, therefore, very interesting that Taira (1981) finds a high degree of agreement between his Japanese results and my northwest European results. It is also significant that he records a correlation between sea level changes and seismo-tectonic events which is just what one would expect with the model here presented. Furthermore, Wollin et al. (1980) showed that there is a high correlation (with a 3 day lag) between changes in North Pacific seasonal sea surface temperature (1950-1975) and seasonal magnetic intensity changes especially in Japan but also in Alaska and Arizona. This is in full agreement with the model here proposed. (The explanation of Wollin et al. is quite different, however).

The mechanism offered is a rotational-gravitational-oceanographic model that includes 3 main basic variables; (1) the Earth's rate of rotation (Fig. 10), (2) the gravitational changes (Figs. 8 and 9), and (3) the Gulf Stream activity (Fig. 12). Furthermore, if rotation is the primary source (which we do not know), it should be caused by planetary forces (cf. Figs. 1 and 13).

Changes in the oceanic circulation should, of course, also apply on the southern hemisphere. Because the main circulation pattern is here quite different from that on the northern hemisphere, the effects would be quite different (in time and amplitudes) to those on the northern hemisphere. The Brazilian records of the last 7000 years may be a record of this.

The circum-Antarctic West Wind Drift must be very sensitive to rotation (and vice versa). As it is the main driving source for the Peruvian, Benguela and West Australian cold currents, changes in this circulation system will have drastic effects on these three cold coastal currents and hence the upwelling and biologic productivity that take place here. Fluctuations in the biologic productivity in these regions may explain (cf. McElroy, 1983; Newell & Hsiung, this volume) the rapid high-amplitudo changes in atmospheric CO_2 that have been recorded (e.g. Stuiver et al. in Bojkov, 1983, Fig. 5). The analogy with present day, short-term, El Niño events (see Newell & Hsiung, this volume) seems obvious, only the duration and amplitudes are larger. This is proposed as an important mechanism for regional paleoclimatic events in the order pf centuries and decades (cf. the concluding chapter of Mörner, this volume).

9. PLANETARY INTERCHANGE OF ANGULAR MOMENTUM

It is clear that the planetary multibody system in motion must generate changes in orbital rates and radius of individual planets, including the Sun. It is not equally clear if this would also lead to corresponding changes of individual planets' spin rates and radii or compensational internal differential movements.

In principle, a change in the orbital radius should also lead to a change in the Earth's rate of rotation. This opens two possibilities of interest in paleoclimatology and related problems.

(1) A planetary induced change in the orbital motions leads to a change in the Earth's rate of rotation. This rotational change is compensated (to keep the angular momemtum constant) at some level according to the Fig. 13 model. This agrees with the fact that the Earth's rate of rotation exhibits an 11-year periodicity which should originate from the Jupiter-Saturn beat (cf. above) in line with the Figs. 1 and 13 models.

(2) A change in the Earth's radius - i.e. by a major sea level rise or fall - would change the Earth's spin rate leading to a change of the orbital radius according to Fig. 4 in Mörner (1984c) which changes the distance to the Sun and hence the incoming solar radiation. This mechanism may operate at major glacial/interglacial alternations. (A sea level rise causes a deceleration that increases the Earth-Sun distance causing less insolation, glaciation and a sea level drop, which in its turn give rise to an acceleration that decreases the Earth-Sun distance causing increased insolation, deglaciation and a sea level rise again).

10. INTEGRATION

Figure 13 illustrates how planetary motions simultaneously and independently can affect the Sun and the Earth, how this can change the orbital distances and rotations (with effects on the insolation and Earth-Moon distance), how this affects the Sun's lumino-

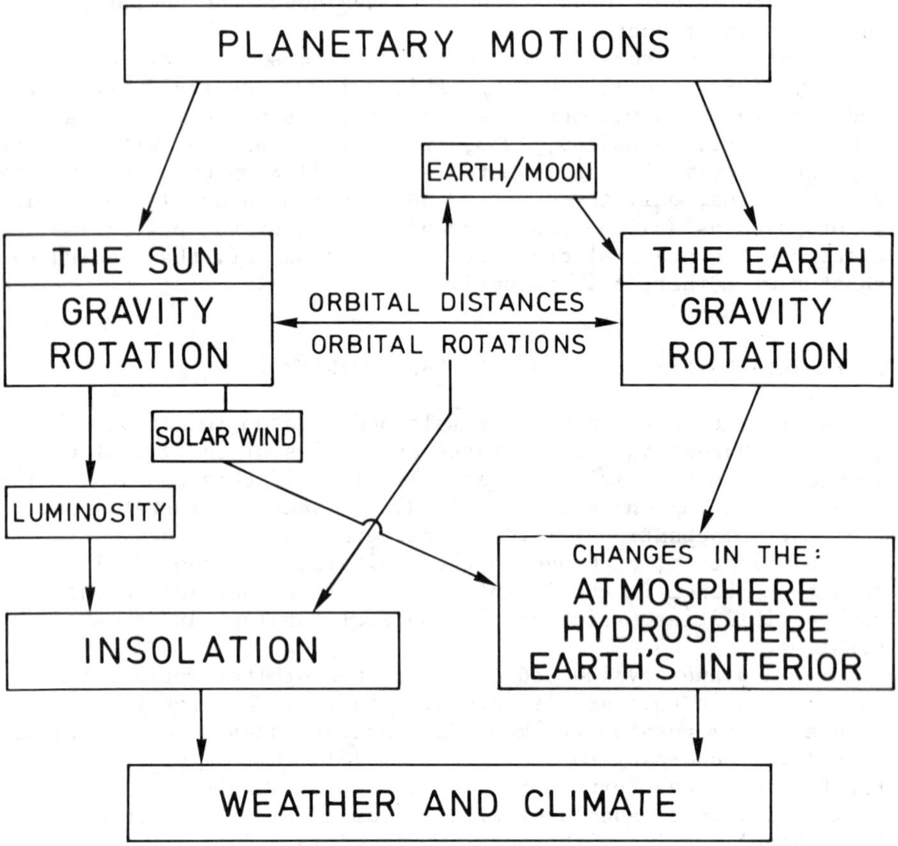

Fig. 13. Independent effects of planetary beat on the Sun and the Earth and various lines by which the Earth's weather and climate may be controlled. The terrestrial mechanisms involved are given in Figs. 8, 10 and 12. The solar wind and atmosphere interaction is discussed by Bucha (this volume). The Fig. 13 system implies that the Earth's climate and paleoclimate are controlled by planetary, solar, atmospheric, hydrospheric and internal (endogene) processes in a complex interaction.

sity and solar wind emission, how this can affect the Earth and cause changes in the atmosphere (circulation, equipotential surfaces, geomagnetic field, concentrations of gasses and vapor, etc.), in the hydrosphere (circulation and geoid surface) and in the Earth's interior (heat flow, gravity potential, geomagnetism, stress, motions, etc.) and finally how both these systems affect and control weather and climate.

Brilliantly, has Lean (this volume) discussed the effects of UV radiation on the Earth's atmosphere, and has Bucha (this volume) discussed a mechanism for how the solar wind should affect the Earth's geomagnetic field and atmosphere causing changes in weather and climate. In this paper, I have shown that an 11-year rhythm is independently set up in the Sun and the Earth. I have argued (cf. Mörner 1978, 1981a) that the Milankovitch variables are not affecting Earth's climate only - and, maybe, not even dominantly - via insolation but also, or rather, via internal processes. I have shown that the short-term Holocene changes (in different variables and at different places) call for variations in the geoid (Fig. 8), in the rotation (Fig. 10) and in ocean circulation (Fig. 12), and that all these variables control several second-order variables (Fig. 8-12). I have proposed a rotational-gravitational-oceanographic model to explain the short--term Holocene fluctuations in different variables. This model seems also to apply for the short-term fluctuations identified prior to the Holocene. The basic principles of the model may even hold for fluctuations of larger time ranges such as the main Holocene cyclic changes (Fig. 5) and the so-called Milankovitch variables (Fig. 3-4). Simultaneous changes in gravity, rotation and ocean circulation may explain the correlation found between geoid deformations, geomagnetic field disturbances and paleoclimatic variations (Mörner, 1978, 1980a). An implication of this is that oceanographic records of paleoclimatic fluctuations may represent ocean circulation variations and not strictly paleoglaciation changes. This could solve the dilemma with the isotopic substage 5d (and 5b) which from the oceanographic records (especially oxygen isotope records) should represent a full glaciation, but in the land records is seen as a very short and severe cooling without any traceable glaciation (cf. the concluding chapter; Mörner, this volume).

Figure 13 is proposed as the integrated large-scale causation system behind changes in weather and climate, with the special "changes in the: atmosphere, hydrosphere and Earth's interior" specified in Figs. 5-12).

11. CONCLUSIONS

(1) The sunspot cycles seem to be driven by a planetary beat that affects the Earth simultaneously and independently. The solar wind and possible luminosity changes affect the Earth. As a whole, this gives evidence of extra-terrestrial, endogene and exogene processes operating at the same time.

(2) The Milankovitch variables may operate via endogene processes. Changes of the Earth's radius due to glacial eustasy may affect the Earth-Sun distance in a feed-back mechanism.

(3) The main Holocene cyclic changes originate from internal changes.

(4) The 16 northwest European short-term fluctuations seem to lead their origin in an interaction of rotational changes, gravitational changes and "vibrations" of the Gulf Stream; a rotational--gravitational-oceanographic model is proposed.

(5) Heat flow varies with time. Dayly, annual and 11-year cycles are recorded, and longer period changes that may have controlled the variations in ice recession. The rapid ice recession from 10,000 to 8,800 BP was, at least partly, caused by increased heat flow.

(6) Gravitational changes have multiple effects via deformations of the geoid and other, atmospheric equipotential surfaces.

(7) Rotational changes have multiple effects via internal geodynamic processes, changes in ocean circulation and changes in atmospheric circulation.

(8) Oceanic circulation changes are important for short-term climatic changes in the North Atlantic (and the north Pacific, too). In the southern hemispere, similar processes may generate El Niño like changes in the order of centuries and decades.

(9) Planetary interchange of angular momentum affects the orbital and rotational motions of the Sun and the Earth, in both of which processes are set up that affect and control the Earth's weather and climate, and many other processes, too.

(10) Climate and paleoclimate are controlled by planetary, solar, atmospheric, hydrospheric and internal (endogene) processes in a complex system of interaction.

REFERENCES

Anderson, L.D., 1974, Earthquakes and the rotation of the Earth. Science, 186,49-50.

Barnes, R.T.H., Hide, R., White, A.A. & Wilson, C.A., 1982. Length-of-day changes and the polar motion correlated with atmospheric angular momentum fluctuations. Abstracts, EGS/ ESG, Leeds 1982, p. 78.

Bojkov, R.D., 1983. Report of the WMO (CAS) meeting of experts on the CO_2 concentrations from pre-industrial times to I.G.Y. WMO Proj. Res. Monitor. Atm. CO_2, Rep. 10, 1-34.

Broecker, W.S., 1978. The cause of glacial to interglacial climatic change. In: Evolution of planetary atmosphers and climatology of the Earth. CNES, Toulouse, Colloq. Intern., 165-190.

Bucha, V., this volume. Mechanism for linking solar activity to weather-scale effects, climatic changes and glaciations in the northern hemisphere.

Colquhoun, D.J., Brooks, M.J., Brown, J.G. & Stone, P.A., 1983. Correlation between sea level and climatic change in the middle Holocene of the southeastern United States. IUGG Hamburg 1983, Abstracts ICL, p. 89.

Courtillot, V., Ducruix, J. & Le Mouel, J.-L., 1978. Sur une accélération récente de la variation séculaire du champ magnétique terrestre. C.R. Acad. Sc. Paris 287:D, 1095-1098.
Dearborn, D.S.P. & Blake, J.B., 1980. Magnetic fields and the solar constant. Nature 287, 365-366.
Dicke, R.H., 1978. Is there a chronometer hidden in the Sun? Nature, 276, 676-680.
Ebel, A., 1983. Gravity waves - sources of momentum, heat and turbulent kinetic energy in the middle atmosphere. IUGG Hamburg 1983, Abstracts IAMAP, p. 247.
Eddy, J.A., 1977. Climate and the changing Sun. Climatic Change 1, 173-190.
von Fresen, R.R.B., Olivier, R. & Hinze W.J., 1983. Long-wavelength magnetic and gravity anomaly correlations of Africa and Europe. IUGG Hamburg 1983, Abstracts IAGA, Bull. 48, 140.
Ghil, M. & Le Treut, H., 1981. A climatic model with cryodynamics and geodynamics. J. Geophys. Res. 86, 5262-5270.
Gough, D., 1980. Solar interior structure and luminosity variations. Abstracts, 14th ESLAB Symp., Scheveningen 1980.
Hays, J.D., Imbrie, J. & Shackleton, N.J., 1976. Variations in the Earth's orbit: pacemaker of the Ice Ages. Sciences 194, 1121-1132.
Imbrie, J., 1982. Astronomical theory of the Pleistocene Ice Ages. A brief historical review. Icarus 50, 408-422.
Kellogg, T.B., this volume. Late-Glacial - Holocene high-frequency climatic changes in deep-sea cores from the Denmark Strait.
Lamb, H.H., 1969. Investigation of the climatic sequence: A meteorological-empirical approach. Palaeoecology of Africa 5, 21-63.
Lamb, H.H., 1979. Climatic variation and changes in the wind and ocean circulation: The Little Ice Age in the Northeast Atlantic. Quatern. Res. 11, 1-20.
Langel, R.A. & Esters, R.H., 1982. A geomagnetic field spectrum. Geophys. Res. Letters 9, 250-253.
Lean, J.L., this volume. Solar ultraviolet irradiance variations and the Earth's atmosphere.
McElroy, M., 1983. Marine biological controls on atmospheric CO_2 and climate. Nature 302, 328-329.
Möller, H., 1962. Om periodiska växlingar i isälvarnas intensitet och solfläcksperiodern . Geol. Fören. Stockh. Förh. 84, 119-133.
Mörner, N.-A., 1969. The Late Quaternary history of the Kattegatt Sea and the Swedish West Coast; deglaciation, shorelevel displacement, chronology, isostasy and eustasy. Sveriges Geol. Undersökn. C-640, 1-487.
Mörner, N.-A., 1973a. Climatic changes during the last 35,000 years as indicated by land, sea and air data. Boreas 2, 33-53.
Mörner, N.-A., 1973b. Climatic cycles during the last 35,000 years. J. interdiscipl. Cycle Res. 4, 189-192.

Mörner, N.-A., 1976a. Eustasy and geoid changes. J. Geol. 84, 123-151.
Mörner, N.-A., 1976b. The Gothenburg Magnetic Excursion. Quatern. Res. 7, 413-427.
Mörner, N.-A., 1977a. Climatic framework of the end of the Pleistocene and Holocene. Paleoclimatic variations during the last 35,000 years. Géogr. Phys. Quatern. 31, 23-35.
Mörner, N.-A., 1977b. Varve chronology on Södertörn: recording and dating of the "drainage" of the Baltic Ice Lake and correlation with the Finnish varve chronology. Geol. Fören. Stockh. Förh. 99, 423-425.
Mörner, N.-A., 1978. Paleoclimatic, paleomagnetic and paleogeoidal changes: interaction and complexity. In: Evolution of planetary atmospheres and climatology of the Earth, CNES Toulouse, Coll. Intern., 221-232.
Mörner, N.-A., 1979, The Fennoscandian uplift and Late Cenozoic geodynamics: geological evidence. GeoJournal 3, 287-381.
Mörner, N.-A., 1980a. Eustasy and geoid changes as a function of core/mantle changes. In: Earth Rheology, Isostasy and Eustasy (N.-A. Mörner, Ed.), Wiley, 535-553.
Mörner, N.-A., 1980b. The nortwest European "sea-level laboratory" and regional Holocene eustasy. Palaeogeogr. Palaeoclim. Palaeoecol. 29, 281-300.
Mörner, N.-A., 1980c. A 10,700 years' paleotemperature record from Gotland and the Pleistocene/Holocene boundary event. Boreas 9, 283-287.
Mörner, N.-A., 1980d. The Fennoscandian uplift: geological data and their geodynamical implication. In: Earth Rheology, Isostasy and Eustasy (N.-A. Mörner, Ed.), Wiley, 251-284.
Mörner, N.-A., 1981a. Eustasy, paleoglaciation and paleoclimatology. Geol. Rundschau 70, 691-702.
Mörner, N.-A., 1981b. Eustasy, paleogeodesy and glacial volume changes. IAHS Publ. 131, 277-280.
Mörner, N.-A., 1981c. Revolution in Cretaceous sea level analysis. Geology 9, 344-346.
Mörner, N.-A., 1981d. Space geodesy, paleogeodesy and paleogeophysics. Ann. Geophysique 37, 69-76.
Mörner, N.-A., 1981e. Crustal movements and geodynamics in Fennoscandia. Tectonophysics 71, 241-251.
Mörner, N.-A., 1981f. Holocene sea level fluctuations on a global scale: Evidence for geoidal eustasy. Bull. INQUA Neotectonics Comm. 4, 100.
Mörner, N.-A., 1982. Water and water. Proc. LANDPLAN-1 Symp. Bangkok 1982 (P. Nutalaya, Ed.), D4:1-5. (Also: Bull. INQUA Neotectonics Comm. 6, 72-76).
Mörner, N.-A. 1983a. Internal and/or external origin of fluctuations in terrestrial variables of the order of 10, 10^2, 10^3, and 10^4 years time range. Abstracts, 2nd Nordic Symp. Climatic Changes Related Problems, Dansk Meteorol. Inst., 88-92.
Mörner, N.-A., 1983b. Geoid topography: origin and time consis-

tency. IUGG Hamburg 1983, Abstracts I, 169.
Mörner, N.-A., 1983c. Sea Levels. In: Large-scale Geomorphology (R.A.M. Gardner & H. Scoging, Eds.), Oxford Univ. Press 1983, 73-91.
Mörner, N.-A., 1983d. Illusions and problems in water-budget synthesis. In: Variations in the global water budget (A. Street--Perrott, M. Beran, R. Ratcliffe, Eds.), Reidel, 419-423.
Mörner, N.-A., 1983e. The lithospheric field even in the lower harmonics? IUGG Hamburg 1983, Abstracts II, p. 621.
The Lithospheric field: geomagnetic anomalies and deformations. IUGG Hamburg 1983, Abstracts IAGA, Bull. 48, 138.
Mörner, N.-A., 1984a. Terrestrial, solar and galactic origin of the Earth's geophysical variables. G ogr. Ann. 66A: 1/2, in press.
Mörner, N.-A., 1984b. Geoidal topography: origin and time consistency. Marine Geophys. Res., in press.
Mörner, N.-A., 1984c. Paleogeoid changes: Rates, amplitudes and origin. Proc. IAG symp. IUGG Hamburg 1983, in press.
Mörner, N.-A. & Wallin, B., 1976. A 10,000 year temperature record from Gotland, Sweden. Palaeogeogr. Palaeoclim. Palaeoecol. 21, 113-138.
Mörth, H.T. & Schlamminger, L., 1979. Planetary motion, sunspots and climate. In: Solar-terrestrial influence on weather and climate (B.M. McCormac & T.A. Seliga, Eds.), Reidel, 193-207.
Newell, R.E. & Hsiung, J., this volume. Sea surface temperature, atmospheric CO_2 and the global energy budget: Some comparisons between the past and the present.
Ruddiman, W.F. & McIntyre, A., 1981. Oceanic mechanisms for amplification of the 23,000-year ice-volume cycle. Science 212, 617-627.
Servant, M., Fontes, J.-C., Rieu, M. & Saliège, J.-F., 1981. Phases climatiques arides holocènes dans le sud-ouest de l'Amazonie (Bolivie). C.R. Acad. Sc. Paris 292, II, 1295-1297.
Stuiver, M., 1980. Solar variability and climatic changes during the current millennium. Nature 286, 868-871.
Stuiver, M. & Quay, P.D., 1980. Changes in atmospheric carbon-14 attributed to a variable Sun. Science 207, 11-19.
Taira, K., 1981. Holocene tectonism in eastern Asia and geoidal changes. Palaeogeogr. Palaeoclim. Palaeoecol. 36. 75-85.
Tuan, T.F., Hedinger, R., Silverman, S.M. & Okuda, M., 1979. On gravity wave induced Brunt-Vaisala oscillations. J. Geophys. Res. 84, 393-398.
Wollin, G., Ericson, D.B. & Ewing, M., 1971. Late Pleistocene climates recorded in Atlantic and Pacific deep-sea sediments. In: The Late Cenozoic Glacial Ages (K.T. Turekian, Ed.), Wiley, 199-214.
Wollin, G., Ryan, W.B.F., Ericson, D.B., 1980. Short period climatic variations as a geomagnetism-ocean-atmosphere problem. Manuscript.

PLANETARY PERIODICITIES AND TERRESTRIAL CLIMATE STRESS

Rhodes W. Fairbridge

Columbia University, New York, N.Y. 10027

ABSTRACT

Extraterrestrial climatic stress is applied to the planet Earth by four deterministic processes:

1. Planetary orbital motions, dominated by Jupiter and Saturn, that transmit momentum by gravitational torques, causing changes in velocity and spin rate to successive planets and to the Sun itself. On Earth, spit rate change appear to trigger seismicity and volcanicity (and therefore dust veils).

2. The Sun accordingly develops its own mini-orbit around the systemic barycenter,with abrupt changes in its acceleration and turning angle that are expressed in the 11 and 22 yr solar cycle of sunspots, electromagnetic radiation and particulate emissions that reach the Earth and beyond as the "Solar Wind."

3. The Earth's geomagnetic field is modulated by the solar wind, which triggers geochemical reactions within the gases of the upper atmosphere. In turn the latter control the stratospheric greenhouse effect, high-altitude clouds and other factors that influence the general circulation. Insolation is further modulated by long-term (10,000 to 100,000 yr) components of orbital motion, eccentricity, tilt and precession, commensurable with category 1 periodicities. The geologic spacing of the great ice ages probably reflects the Galactic Cycle.

4. Lunar tidal cycles, identified in many terrestrial climate series, develop standing waves in the atmosphere and help to trigger major seismic and volcanic events with contribution to the dust veil. The 18.6 yr nodal periodicity also corresponds to a nutation of the precession parameter and is commensurable in turn with the basic cycles of category 1.

N.-A. Mörner and W. Karlén (eds.), Climatic Changes on a Yearly to Millennial Basis, 509–520.

INTRODUCTION

The history of climatology in its relation to solar-terrestrial processes has undergone a series of quite basic changes in scientific attitude during the last two centuries. During the 19th century numerous correlations (for example, Alexander von Humboldt, transl. 1871) between astronomically observed sunspots and terrestrial climatic fluctuations were widely accepted as positive evidence for a cause-and-effect relationship, although the mechanism was obscure. With the 20th century came improved statistical methods, more data, and a growing scepticism. Attempts to forecast the solar cycle failed repeatedly, and by the mid-20th century it became a virtual paradigm of both meteorology and climatology that no solar forcing function could be recognized.

In the last decade or so, however, with the growth of observational aeronomy and astrophysics, and the evolution of computer science which paralleled the dramatic technological development of rocketry and space travel, an absolute deluge of papers have made it perfectly clear that an exogenetic role must be accepted in many terrestrial processes. So far the data have been scattered and largely phenomenological. It is not enough to be quantitative and accurate if we do not know what it all means, but some of the conceptual patterns are now beginning to take shape.

In this paper no revolutionary discoveries are reported, but I present rather an integration of numbers of partly well-known and partly little-known processes and phenomena, assembled from interdisciplinary fields, notably geology, geophysics, meteorology, climatology, astronomy and astrophysics.

THE EXTERNAL STRESSES

To begin with, it must be appreciated that external stresses are applied to the planet Earth, reacting with its atmosphere, hydrosphere and lithosphere in various ways. The stresses are generated within the Solar System by two mechanisms: (a) gravitational tides and torques, and (b) radiation, electromagnetic and particulate. The delivery of these energy sources is modulated by the overall motions and alignments of the planets around the Sun, and secondarily but very importantly modulated by the Earth-Moon motions.

Two, non-independent barycentric systems are involved: (a) the Solar System, which is in orbit NOT around the Sun, but around the systemic barycenter; the heliocentric point follows an elliptical, looping orbit with a constantly variable radius and may reach more than 1.5 million km away from the barycenter; (b) the Earth-Moon System, where the E-M barycenter lies always beneath the Earth's surface, but likewise at a variable radius. These two barycentric systems are interlocked chronologically in a "commensurability" relationship (Stacey, 1963, 1967; Roy, 1977; Mörth and Schlamminger, 1979). This means that every fundamental cycle in the Solar System is related to every other one in ratios of small integers, although at any given moment in time (due to ellipticity, etc.) the relationships will be merely approximate.

I have recently discovered that the principal cycle generated by the planet's orbital system is the Saturn-Jupiter Lap (19.857 yr), those two planets possessing 86% of the angular momentum involved. The major orbital loops of the Sun around the barycenter ("SOBs") are totally locked-in to this periodicity, although the time from one orbital maximum to the next may vary from 15 to 26 yr. The SJL (Saturn/Jupiter Lap) therefore constitutes the fundamental pulse of the solar system. According to calculations by Stacey (1963, 1967) all of the outer planets realign themselves in the same pattern ("OPR") once every 4448 yr (224 x 19.857), and the "all planets restart" ("APR") cycle is 93,408 yr, to which is evidently related the Earth's eccentricity period (and the principal Glacial Cycle).

The way in which the Sun interacts with the terrestrial climate system is primarily by means of its energetic emissions which, both in the form of electromagnetic (mainly in optical wavelengths) and particulate radiation, vary with the so-called "Solar Cycle" of about 11 yr and the "Hale Cycle" (solar magnetism) of about 22 yr. Like the SOB period, these vary considerably in magnitude and in length, the 11 yr cycle maxima ranging from 6 to 17 yr. The 11 yr periodicity, expressed visually as the Sunspot Cycle (Schove, 1983), is generally attributed to a magnetohydrodynamic process. This dynamo effect is generated by solar rotation and convection. Dicke (1978) spoke of an internal chronometer, but several indicators suggest an external triggering. First, an analysis of the phase lag of the sunspot cycles over two 178-yr periods show a consistent cycle (Fairbridge and Hameed, 1983), which with the aid of auroral proxies I have no extended to some 2600 yr. Landscheidt (1979) has shown that this 178 yr periodicity, and I have confirmed it with additional data, breaks down into a 39-39-100 yr pattern, earlier recognized in part from auroral evidence as the 78-yr or 80-yr "Gleissberg Cycle" (Gleissberg, 1965).

The second indication of external triggering is the fact that the 39-39-100 yr arrangement exactly fits the SOB record which in turn is locked to the Saturn-Jupiter Lap. I have therefore no hesitation in regarding the solar "chronometer" as externally triggered. This still does not explain exactly how the trigger-mechanism works. The average timing of the sunspot cycle coincides with a mean beat-frequency for the outer planets (Mörth and Schlamminger, 1979; Bagby, pers. comm.), and the peaks are observed to match the most abrupt changes in acceleration and orbital curvature of the SOB. The long-term average length of the cycle is not 11.14 yr as calculated by Dicke (1978), but 11.12 yr (Stacey, 1963), which is precisely commensurable with the outer planetary alignments and quadratures (100 per 1112 yr, and 400 per 4448 yr OPR).

THE EARTH'S RESPONSES

In response to the exogenetic stress mechanisms, two distinctive climate-affecting processes seem to be paramount.

First, the Earth is the recipient of two types of solar emission, the amplitude and periodicity of which seem to be directly related to the torques set up by the planetary alignments and quadratures outlined above. The electromagnetic radiation, with most energy in the optical wavelengths, varies only a little, so little in fact that for a long time one spoke of the "solar constant." A large variation, however, occurs in the far UV wavelengths during flares and especially near peaks in the sunspot cycle (Cook et al., 1980). And incoming UV reacts with stratospheric oxygen to produce ozone, which in turn is the primary control of the stratospheric greenhouse effect. The other climate-related radiation is by high-energy subatomic particles which reach the Earth in the "Solar Wind." It is also strongly affected by the sunspot cycle, and reacting with the Earth's magnetic field is responsible for further geochemical reactions, mainly in the stratosphere, and for pressure disturbances in the general atmospheric circulation.

For its part, the Earth is also involved in the type of orbital variations mentioned above for the outer planets. The three long-term variables concern the eccentricity, tilt and precession, the so-called "Milankovitch parameters," which Milankovitch (1941) recognized as controlling the "insolation," or effective magnitude and distribution of solar radiation reaching the Earth. Because of the asymmetric and irregular distribution of continents and oceans on the Earth at the present time, the role of the Milankovitch insolation is very great and is responsible for climatic variations in the range of 10,000 to 100,000 yr, which is the "Glacial Cycle" (Imbrie and Imbrie, 1979). From recent studies it seems that the Milankovitch parameters are sometimes in phase with one another and may also be reinforced by the short-term radiation variables to create brief but very extreme climatic fluctuations. At other epochs they will be variously out-of-phase, so that very little cyclicity is recorded (Fairbridge, 1984).

The second type of response is controlled by the Moon. The orbital motions of the Earth-Moon pair around our local barycenter are well known to students of oceanic tides. Similar, but much larger tides are recorded in the atmosphere, and smaller ones with the lithosphere. The Moon's satellite orbit is, like the others, also variable and is tied in to the larger periodicities of the Earth and the other planets (Brown, 1896; Roy, 1977). Apart from the daily and fortnightly tides, the 18.6 yr lunar nodal precession (that is locked into a nutation of the Earth's precession) is now identified in MSL, temperature, precipitation and atmospheric pressure (Currie, 1979, 1981, 1984). A longer earth-moon cycle is the Progression of the Lunar Perigee (556 yr) that is also represented in the geological (sedimentary and geomorphic) time series (Stacey, 1963). The maximum fluctation in the spin rate of the Earth-Moon pair is 93, 408 yr, which has the same duration as the APR cycle and the Earth's Perihelion progression Cycle.

As noted by Berger (1981, p. 511), in the high latitudes the terrestrial insolation variations caused by change in the earth's obliquity (c. 40,000 yr) are much greater than those generated by the precession parameter. The precession signal in contrast is much more powerful in low latitudes (as seen clearly in the oxygen-isotope data in deep-sea cores: Hays, et al., 1976). Short-term (18.6 yr) nutation of the precession variable is minor and hardly to be expected in detailed data series, but it is astronomically locked into the nodal cycle of the moon, which controls the maximal tilt of the moon's fortnightly period, and therefore governs the latitudinal extremes of the lunar gravitational tide ("spring" tides). The farthest departure that this zenith effect is felt will be shifted by approximately 1100 km (10° of latitude) north (and south) of the equator over the course of one 18.6 yr nodal cycle. It is to be anticipated, therefore, that in climatic time-series covering these critical latitudes, the tidal component will be of varying amplitude and in certain epochs may go out-of-phase.

The Earth's Climate

Before discussing planetary-solar activities, it seems desirable to underline certain points about the terrestrial climate record that emphatically suggest extraterrestrial forcing.

a) Over the geological record of the planet Earth we possess about 4 billion (4×10^9) yr of continuous sedimentary records that show a long-term steady state situation. There has always been liquid water covering most (but not necessarily all) of the surface, with a mean atmospheric temperature generally in the order of 20 ± 5°C. This fact requires that the Earth's relationship to the Sun and other planetary bodies must have remained relatively constant. A serious orbital deviation would bring us in the regime of Mars on one side and Venus on the other. The same fact implies that although the Sun's radiant energy may be slowly increasing (as cosmologists argue), there are sufficiently effective feed-back and equilibrium controls in the Earth's atmospheric-hydrologic systems to maintain an almost uniform environmental pattern.

b) Over nearly this same time span (at least 3.5 x 10^9 yr) organic life has been in existence. The "Law of Biologic Uniformity" requires that at no time, not for one instant, has the biosphere been subjected to a catastrophic condition of such magnitude that all life has been wiped out (Fairbridge, 1981). The continuous, although undoubtedly "punctuated" course of evolution has never once been terminated, to start once again from the primitive molecular building blocks. The effects of meteoritic or asteroid impacts (that have demonstrably occurred) or of not-too-far-off nova or galactic explosions (raining down neutron showers) would no doubt have played a role in selective extinction of less well protected or adapted organisms. But the "bottom line" is this: no major phylum of either plant or animal kingdoms has ever become extinct. Such biological durability demands a highly stable environmental setting. Total cataclysms and "Velikovskian" solutions are absolutely ruled out.

c) Against this double record of stability through geological times, we have equally well-demonstrated evidence of cyclic ice ages. However, the geological explorations of the present century have shown that although the "Uniformitarians" of the last century thought that ice ages would logically be frequently recurring, in fact they are very infrequent, occupying less than 5% of all geologic time, and appearing at intervals of the order of around 200 million yr. Their regularity has suggested a coincidence with the galactic cycle, the length of which and the mechanisms (intergalactic dust clouds?) are still uncertain. While paleogeographic changes within the Earth are undoubtedly essential in providing the right setting for an ice age, the periodicities cannot be easily explained from within the planet and the galactic period would be most appropriate for an extraterrestrial forcing.

d) Within any ice age, we now recognize a fundamental "glacial cycle" of about 100,000 yr, with shorter cyclic recurrences of glacial advances and retreats. The timing of those cycles was explained by Milankovitch (1941) as explicable in terms of his insolation theory based upon orbital variations of the planet Earth, but with the assumption of a constant solar radiation. At first Milankovitch was treated by most of the "establishment" with total contempt, but it is satisfactory to note that at a meeting convened at Columbia University late in 1982 members of his family were invited to a large symposium that could leave no doubt whatever that in principle, Milankovitch was correct, and at least the glacial cycle was due to extraterrestrial forcing.

e) From the billion yr, to the million yr, to the 100,000 - 10,000 yr level, the Earth's surface processes are found to be subjected to extraterrestrial forcing. Now we come to the short period processes that, curiously enough, have generated the most controversy.

When it comes to the shorter cycle patterns of climate, there have been innumerable attempts to relate them to the 11 year sunspot cycle, but with mixed results. Many of the attempts have been foredoomed by the selection of statistically invalid samples, for example, using most of the data from stations that are clustered in certain geographic areas (often political units) or mixing the data from different climatic systems. In general, low latitude, high-pressure tropics in continental situations show good correlations with the solar cycle, whereas mid-latitude maritime locations naturally are greatly influenced by temperatures of sea-water which has a much slower exchange rate than the atmosphere. Mean annual figures often smooth over opposing trends in summer and winter. Compounding the difficulties for the early investigators is the now well-recognized mercurial behavior of the solar cycle itself, that has still to be explained (a comprehensive "benchmark" collection has just been published: Schove, 1983). When the length of the fluctuations, from peak to peak, varies by as much as 7 to 17 years, it is not surprising that it has been a despair for meteorologists. For geologists, in contrast, being accustomed to vaguely established patterns, certain solar-climatic reactions seem at least reasonable, and it seems likely that the solar cycle will soon become predictable (Fairbridge and Hameed, 1983; Landscheidt, this volume). At the present state of knowledge, nevertheless, the longer cycles are more informative than the so-called "11 yr" cycle.

The most generally accepted of the longer climate periodicities is the 22-year cycle of the continental mid-latitudes to subtropics (Schuurmans, 1981; Schove, 1983). This corresponds to the so-called "double sunspot" or "Hale" cycle, or heliomagnetic cycle, of the solar specialists, being characterized by the reversal of the sun's magnetic fields, and sunspot orientation, and variations (possibly) of the solar diameter.

The linking mechanisms between solar emissions and climate appear to the writer to lie largely in the atmospheric chemistry of the stratosphere and the modulation of the geomagnetic field by solar "wind." Solar short-wave radiation, essentially in the far-UV spectrum, has been shown to generate ozone in the upper stratosphere, where the ozone plays an important "greenhouse" role (Kondratyev, 1969). During magnetic lows, there is enhanced cosmic ray flux with production of NOx, thus reducing the O3 concentration and leading to stratospheric cooling. With variations of the ozone content in the stratosphere of 25% or more over both decadal and shorter periods, the ozone layer acts as a "shutter" to incoming solar energy to the troposphere.

As is well established, terrestrial magnetic activity varies with the 22-year solar magnetic cycle (Chernosky, 1966), as does the "Forbush Effect" which refers to the cosmic ray flux. The so-called 22-year cycle is no more regular than is the 11-year sunspot cycle, so it is not reliable for short-term forecasts, unless handled within a long-term framework. The pattern is then easier to discern when measured from minimum to minimum. Barta (1956) has already suggested a 40 to 50-year geomagnetic cycle. Gleissberg (1965) found a 78 to 80-year period that now bears his name and Landscheidt (1979; this volume) places these in a 39-39-100 year sequence, which the writer has been able to confirm in a 1300-yr time series (in preparation). The 39-39-100 yr pattern has small variations but fits into the 178-yr framework of solar phase lag (Fairbridge and Hameed, 1983). A 20-60 yr lag time is generally assumed but at various latitudes and at various solar periods it is likely that the 14-C values will show very different lag times.

In the search for periodicities in climate records that exceed the interval covered by instrumental data, an apparently insuperable problem presented by the limited time series can in part be solved by the use of dendrochronology, which now has a year-by-year accuracy, reaching back for 8000 years, with also some "floating" segments for several thousand years more (Suess, 1970; Damon, 1977). Stuiver and Quay (1980) have shown that the 14-C values of tree rings tend to reflect the amplitude of the solar cycle, and in a favorable series (Californian bristlecone pine of Capito Mountain) Sonett and Suess (1984) correlate the solar signal (14-C) with climate (mean ring width reflecting temperature).

Other annual layer data series are provided by ice cores (Hammer et al., 1980) and by varves (e.g., in Schove and Fairbridge, 1984). Other historical "proxies" are provided in the form of written records of climatic events (e.g., Lamb, this volume), while auroras (e.g., Schove, this volume; Siscoe, 1980) and visual sightings of sunspots through haze (Wittmann, 1978) provide insights into the solar record that can be cross-checked with geological data on climatic reactions. Nevertheless, the pre-instrumental records are eclectic ones, full of lacunae and of questionable amplitudes.

In recent years the geologists (notably those concerned with the INQUA commissions, respectively, on the Holocene, Shorelines and Neotectonics) have been searching for examples of continuous, uninterrupted sedimentary sequences that contain evidence of the detailed climatic and geodynamic modulations of the last 10,000 years or so. The International Geological Correlation Project also plays a vital role. The deep-sea record has been disappointing on the whole because the rate of abyssal accumulation is generally too low to permit detailed resolution on less than about 3000 years; further, the bioturbation by bottom-living organisms tends to convert even the sharpest boundaries into homogenized "gray zones." Accordingly the search has turned more to littoral and lacustrine sections.

The isostatically uplifted coasts of Hudson Bay have yielded outstandingly useful results (Fairbridge and Hillaire-Marcel, 1977; Fairbridge, 1981) and comparable sequences occur in other high-latitude uplift regions (Sweden, Spitsbergen, Greenland, and Canadian Arctic). A particularly important discovery, supported by over 100 14-C dates, has been made by Filion (1983) who found that the low sea-level phases on the Hudson Bay matched cold climatic fluctuations that led to dry conditions, brush fires and dune growth.

Several other very important geological time-series of the last 10,000 years or so are partially known but have not yet been adequately explored in depth, in large measure because of a lack of appreciation of their potential for world-wide climatic, geodynamic, paleomagnetic and solar activity analyses, and ultimate use for prediction. One could mention specifically the high-latitude annual varve series of Sweden that goes back (potentially) to about 13,000 BP, and is currently under long-term restudy, but which could respond to a much more intensive interdisciplinary effort. There are also mid-latitude varves from lake basins that could provide chronologies of solar-climatic events back for comparable periods, notably the Lake Saki varves of the Crimea, USSR (Schove, 1978); and Lake Van, Turkey (Kempe and Degens, 1979). The marine varves from the Santa Barbara basin off southern California show great promise (Pisias, 1978), and others in the Gulf of California are under investigation (J. Curray, pers. comm.).

On the basis of Dansgaard's Greenland ice core isotopic data, Bray (1972) suggested 1300-year and 2000 to 2800-year periodicities. A 2800-yr periodicity may represent the precession of the non-dipole magnetic field, which the writer would interpret as being modulated by lunar precession (30 x 18.53 yr = 556 yr; the latter x 5 = 2780 yr. An important aspect to this question has been pointed out by Siscoe (1976); as the dipole field weakens, the quadripolar effect becomes stronger and at such times the auroras can reach as far as 45° away from the NMP. Depending on its situation from time to time, this phenomenon could bring auroras to be visible as low as latitude 35°, where they are recorded in historical documents as highly unusual and remarkable events.

Bucha (1977, 1979; this volume) has assumbled some interesting evidence pointing toward a possible mechanism for a direct link between magnetic moment and climate. He showed that within 2-3 days of any violent solar flares, eruption and geomagnetic activity increase, an extraordinarily deep but constricted low pressure system would develop near the magnetic pole that would lead at the 500 mb level to an air temperature rise of up to 30°C; this creates a cyclonic situation that would generate catastrophic storm tracks that would sweep in from Alaska or the Hudson Bay. Following this disturbance a powerful flow of arctic air pours out to reach northwest and central Europe one or two weeks later where a dramatic fall of temperature is recorded within a few hours.

CONCLUSION

Perusing the four fundamental celestial systems that are here regarded as critical in determining the extraterrestrial forcing of terrestrial atmospheric behavior, both astronomers and meteorologists will naturally ask questions. Few of these ideas are new or untested, but the general integration now proposed is original. At the present time it is certainly not possible to provide a satisfying answer to each point. The material presented here can be no more than a working paper that will require searching studies, data collection and future analysis. Experimental tests will then present themselves. As the writer sees it, we stand now at the threshold of a new phase of astronomic-climatic research.

Acknowledgements: I would like to express appreciation to those friends and colleagues who have encouraged me to push into this previously unknown territory: the late Clyde Stacey, who patiently instructed me on the solar system and commensurability; John Sanders, who inspired me to investigate orbital data; Steven Unger, of NASA, who provided ephemeris calculations covering 13 centuries; and Sultan Hameed and Robert Currie who helped with the data analysis.

REFERENCES

Barta, G., 1956. A 40-50 year period in the secular variation of the geomagnetic field. Acta Geol. (Budapest), 4: 15-52.

Berger, A. (ed.), 1981. Climatic Variations and Variability: Facts and Theories. Dordrecht: Reidel.

Bray, J.R., 1972. Cyclic temperature oscillations from 0-20,300 yr BP. Nature, 237: 277-279.

Brown, E.W., 1896. Introductory Treatise on the Lunar Theory. Dover.

Bucha, V., 1977. Mechanism of solar-terrestrial relations and changes of atmospheric circulation. Studia Geophys. et Geodet, 21: 350-360.

Bucha, V., 1979. Connections between geophysical and meteorological processes. Studia Geophys. et Geodet, 23: 55-67, 102a-102d.

Chernosky, E.J., 1966. Double sunspot-cycle variation in terrestrial magnetic activity, 1884-1963. Jour. Geophys. Res., 71(3): 965-974.

Cook, J.W. et al., 1980. Variability of the solar flux in the far ultraviolet 1175-2000 A. Jour. Geophys. Res., 85: 2257-2268.

Currie, R.G., 1979. Distribution of solar cycle signal in surface air temperature over North America. Jour. Geophys. Res., 84: 753-761.

Currie, R.G., 1981. Solar cycle signal in earth rotation: non-stationary behavior. Science, 211: 386-389.

Currie, R.G., 1984. Evidence for 18.6 year lunar nodal drought in western North America during the past millennium. Jour. Geophys. Res. (in press).

Dicke, R.H., 1978. Is there a chronometer hidden deep in the sun? Nature, 276: 676-680.

Fairbridge, R.W., 1980. Prediction of long-term geologic and climatic changes that might affect the isolation of radioactive waste. In: Underground Disposal of Radioactive Wastes, v. 2, pp. 385-405. Internat. Atomic Energy Agency (IAEA-SM-243/43).

Fairbridge, R.W., 1983. The Pleistocene-Holocene boundary. Quaternary Science Reviews, 1: 215-244.

Fairbridge, R.W. & Hameed, S., 1983. Phase coherence of solar cycle minima over two 178-year periods. Astron. Jour., 88: 867-869.

Fairbridge, R.W. & Hillaire-Marcel, C., 1977. An 8000 yr paleoclimatic record of the "Double Hale" 45 yr solar cycle. Nature, 268: 413-416.

Filion, L., 1983. These. Universite Laval (Dep. Geographie), Quebec.

Gleissberg, W., 1965. The eighty-year solar cycle in auroral frequency numbers. Brit. Astron. Assoc. Jour., 75: 227-231.

Hammer, C.U., Clausen, H.G. & Dansgaard, W., 1980. Greenland ice sheet evidence of post-glacial volcanism and its climatic impact. Nature, 288: 230-235.
Hays, J.D., Imbrie, J. & Shackleton, N.J., 1976. Variations in the earth's orbit: pacemaker of the ice ages. Science, 194: 1121-1132.
Imbrie, J. & Imbrie, K.P., 1979. Ice ages: solving the mystery. Enslow Publ.
Kempe, S. & Degens, E.G., 1979. Varves in the Black Sea and in Lake Van (Turkey. In: Moraines and Varves (ed. C. Schlüchter), pp. 309-318. A.A. Balkema.
Kondratyev, K.Y., 1969. Radiation in the Atmosphere. Academic.
Landscheidt, T., 1979. Swinging Sun, 79-year cycle, and climate change. Jour. Interdisc. Cycle Res., 12: 3-19.
Milankovitch, M., 1941. Canon of insolation and the ice-age problem. Kon. Serb. Akad. 132 (33).
Mörth, H.T. & Schlamminger, L., 1979. Planetary motion, sunspots and climate. In: Solar-Terrestrial Influences on Weather and Climate. (ed. B.M. McCormac & T.A. Seliga), pp. 193-207. D. Reidel.
Pisias, N.G., 1978. Paleoceanography of the Santa Barbara Basin during the last 8,000 years. Quaternary Research, 10: 366-384.
Roy, A.E., 1977. Orbital Motion. Wiley.
Schove, D.J., 1978. Tree-ring and varve scales combined, c.13,500 B.C. to A.D. 1977. Pal. Pal. Pal., 25: 209-233.
Schove, D.J., 1983. Sunspot Cycles. Hutchinson & Ross Publ. (Benchmark Vol. 38).
Schove, D.J. & Fairbridge, R.W. (eds.), 1984. Ice-cores, Varves and Tree-rings. Balkema.
Schuurmans, C.J.E., 1981. Solar activity and climate. In: Climatic Variations and Variability: Facts and Theories (ed. A. Berger), pp. 559-575. D. Reidel.
Siscoe, G.L., 1980. Evidence in the auroral record for secular solar variability. Rev. Geophys. and Space Phys., 18: 647-658.
Sonett, C.P. & Suess, H.E., 1984. Correlation of bristlecone pine ring widths with atmospheric 14-C variations: a climate-Sun relation. Nature, 307: 141-143.
Stuiver, M. & Quay, P.D., 1980. Changes in atmospheric carbon-14 attributed to a variable sun. Science, 207: 11-19
Von Humboldt, A., 1871. Kosmos (trans. E.C. Otte & B.H. Paul). Belle, v. 4.
Wittman, A., 1978. The sunspot cycle before the Maunder Minimum. Astron. & Astrophysics, 66: 93-97.

A POSSIBLE MECHANISM OF ABRUPT CLIMATIC CHANGES

H. Flohn

University of Bonn

Abstract
Many investigations have demonstrated the occurrence of "abrupt" climatic changes at a time-scale of 100 years during the transitions between glacial and interglacial periods and (infrequently) during interglacials. They are tentatively interpreted as a consequence of prevailing oceanic upwelling (during cold episodes) and downwelling (during warm episodes) and their effects on the air-sea-exchange of CO_2 and H_2O. In the atmosphere, under natural conditions, both gases are rarified during upwelling and amplified during downwelling, with appropriate changes of their combined greenhouse effect. Quantitative estimates indicate that these processes are in the right order of magnitude to cause abrupt global temperature changes of the order of 2 $^{\circ}$C. Recent measurements of the CO_2 content of air bubbles in a new Greenland ice core have confirmed time-scale and amplitude of such variations.

1. ABRUPT CHANGES

Recent investigations - mainly but not exclusively at continental sites, such as fossil peat-bogs, ice-cores and diatomites - have indicated the occurrence of a number of cases with a remarkable climatic change apparently evolving in a time-span of 100 years or so, but with an amplitude in the order of about half of the difference glacial-interglacial (i.e. about 3 K in annual or summer temperature) (Müller 1965, 1974; Woillard 1979; Eicher-Siegenthaler 1980, 1981; Kukla 1980). A thermal amplitude can be estimated either from vegetational changes (e.g. pollen data), from variations in the relative abundance of planktonic micro-

N.-A. Mörner and W. Karlén (eds.), Climatic Changes on a Yearly to Millennial Basis, 521–531.

Table 1. Abrupt Palaeoclimatic Events (Time-scale ∿100a).

ΔT		Time BP
a) W	Transition Y. Dryas - Preboreal	10.2 ka
b) C	" Alleröd - Y. Dryas	10.8
c) ?	Onset Moist Period (Lat. 15-35°N)	12
d) W	Transition O. Dryas - Bølling	12.8
e) C	" Stages 5a/4	73
f) C	" " 5c/5b	95
g) C	" " 5e/5d	115
h) C	Cold Episode Holstein Interglacial Stages 7, 9?	190?
i) C	" " Stage 19 (Matuyama - Brunhes)	700

fossils or from the $^{16}O/^{18}O$ relationship in fallen snow in ice cores; unfortunately only a few quantitative estimates have as yet been given in the European theatre (Duplessy et al. 1981). The absolute chronology is often subject to several uncertainties and in some cases only found by comparison with well-dated fix-points of the climatic evolution. But the relative chronology is in several cases evidenced by annual layering (Müller 1965, 1974; Woillard 1979) or by a reasonable interpolation from estimated sedimentation rates (Kukla 1980; Duplessy et al. 1981; Dansgaard 1982). At any rate, the time-scale is 2 or 3 orders of magnitude smaller than that of the orbital elements (10^4-10^5a).

Table 1 lists 9 cases of such "abrupt" changes (Müller 1965, 1974; Woillard 1979; Eicher-Siegenthaler 1980, 1981, Kukla 1980; Duplessy et al. 1981; Wijmstra 1978; Grüger 1979). Most cases are evidenced at several far-distant places; cases a), b), d)-g) are well-known short fluctuations observed during Late Glacial or the transition between the last interglacial (Eem) and the first advance of the last glaciation (Stage 4 Shackleton-Emiliani). Cases a), b), d), g) are evidenced also in ocean cores (Kukla 1980; Duplessy et al. 1981), with sea surface temperature changes (SST) up to 12-14°C in the Biscaya. The most dramatic abrupt climatic episode is the Younger Dryas glacial readvance, which lasted, together with both transitions, no more than about 600 years, during which forests disappeared from large areas of central and western Europe. The Atlantic polar water readvanced from the northern Norwegian Sea over 2500 km into the Biscaya (Duplessy et al. 1981). The Scottish highlands developed a regional glaciation (Sissons et al. 1974a, b). Its effect was felt even in the Tropics (van der Hammen et al. 1978), while central and western North America were not affected. Case h) (Fig. 1) is also evidenced in SE-England (time-scale uncertain); case i) remains as yet isolated (Müller 1965), but here the time-scale and the vegetational change

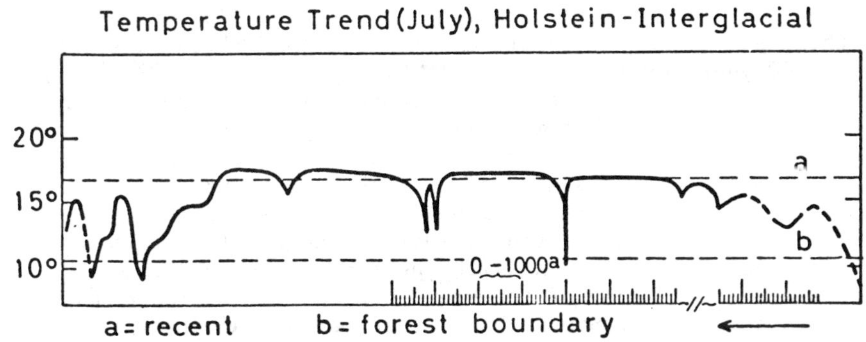

Fig. 1. Abrupt climatic fluctuations in NW Germany (Müller 1974) during the Holstein Interglacial (case h)).

are particularly well represented. Indeed our knowledge of the relationship between spatial and time-scales of climatic change definitely prohibits any purely local interpretation.

Here again it should be realized that due to the limited velocity of the migration of the trees and some other plants, the real time-scale of climatic shift is smoothed and prolonged by this factor, partly similar to the role of bioturbation at the ocean floor (Duplessy et al. 1981). Insect deposits at land confirm the rapidity of climatic change (Osborne 1974; Coope 1977).

Of particular interest is the sudden, apparently catastrophic increase in precipitation near 12 ka BP (cases c)). It is felt in the Tropics and Subtropics (Pastouret et al. 1978; Maley 1980; Nicholson-Flohn 1980; A. Street-Perrott 1983) and caused a short, but intense glaciation in Mexico (Heine 1983), possibly correlated with "Plato's deluge" - a freshwater spike in the Gulf of Mexico (Emiliani et al. 1975). Probably simultaneous is the onset of a very moist period in the Rajasthan desert (Swain et al. 1983) which has been found also in the northern Indian Ocean (Prell et al. 1980; Duplessy 1982), perhaps also a short rapid glaciation in the Hindukush (Haserodt 1982). ^{14}C dates of these events cluster around 11.8 ka BP within the limits of error, which seems to exclude a direct correlation with cases b) or d). A possible coincidence with the reappearance of tropical forests (Shackleton 1977) ought to be checked.

A convincing review of the time-scale of case g) has been given by Kukla (1980), comparing records from 6 quite different European sites including an ice-core from northern Greenland and an ocean core from about 1000 km W Ireland. A detailed study revealed two important shifts of the prevailing ecosystem in Great Britain, eastern France, Germany, west of the Elbe/Oder, Denmark and Sweden up to 62°N: at about 115 ka ago from a rich mixed decideous/coniferous forest (Eem interglacial) to boreal taiga, and

1-3 ka later to a tundra (stage 5d culminating 110 ka ago). Each of these sudden shifts occurred, at several of these sites, in about 100 years or less.

A tentative interpretation has been given for an abrupt cooling (Kukla 1980) but it fails to interpret the geophysical mechanism of the climatic change. It is the intention of this paper to propose such a mechanism (Flohn 1981, 1982; Kellogg 1983) in form of a positive global feedback as a working hypothesis for further discussions. There is no doubt that these abrupt changes are incoherent with the time-scale needed for the accumulation and decay of the great continental ice-sheets of the northern hemisphere. Simple budget estimates indicate that this time-scale should be in the order of something like 8000 years or more. The decay time could be distinctly shorter, if greater parts of them are based below the rising sealevel, calved into the sea (Denton-Hughes 1981). Such a development has been suggested for Hudson Bay, Barents Sea and even parts of the Baltic and perhaps also for the Faroe-Iceland ridge (Stracuh 1973).

The author who has developed the following ideas gradually since 1980, is grateful to many of his colleagues who have discussed this problem critically after his lectures, among them R. Bacastow, C.F. Baes jr., A. Berger, W.H. Berger, B. Bolin, L. Gates, W. Kellogg, H. Oeschger, St. Schneider, R. Wetherald and T. Wigley.

2. OCEANIC UPWELLING AND ATMOSPHERIC COMPOSITION

The phenomenon of upwelling cool waters from the thermocline region (100-400 m depth, temperatures 14-24°C) along some tropical coasts and along the Pacific and Atlantic equator (SST usually 26-28°C) is well-known - here it is impossible to outline its physical mechanism (cf. Wyrtiki 1975; Philander 1983). Its speed is only in the order of 10-50 m/d, but its mass-transport has been estimated (in the equatorial Pacific alone) to $50x10^6m^3/s$ (Wyrtki 1981). Some of its climatic consequences are well understood: stabilisation of air by downward flux of sensible heat, suppression of convective clouds and rainfall. Others (Table 2. Fig. 2) have been found only recently: reduction of evaporation (Henning-Flohn 1980) coinciding with relative humidity in the stable air and high consumption of atmospheric CO_2 by the ocean. This is apparently due to the photosynthese of algae in the nutrient-rich upwelling water from below the thermocline (Newell et al. 1978), in spite of its supersaturation with CO_2-gas and anorganic carbon. The relationship between upwelling and CO_2-exchange rate is a well established fact; its origin is still controversial among chemical oceanographers (Baes 1983; Broecker-Peng 1982). The correlations between the growth rates of atmospheric CO_2 (as produced by fossil fuel, together with man-made destruction of tropical forests and soils, but after removal of the seasonal

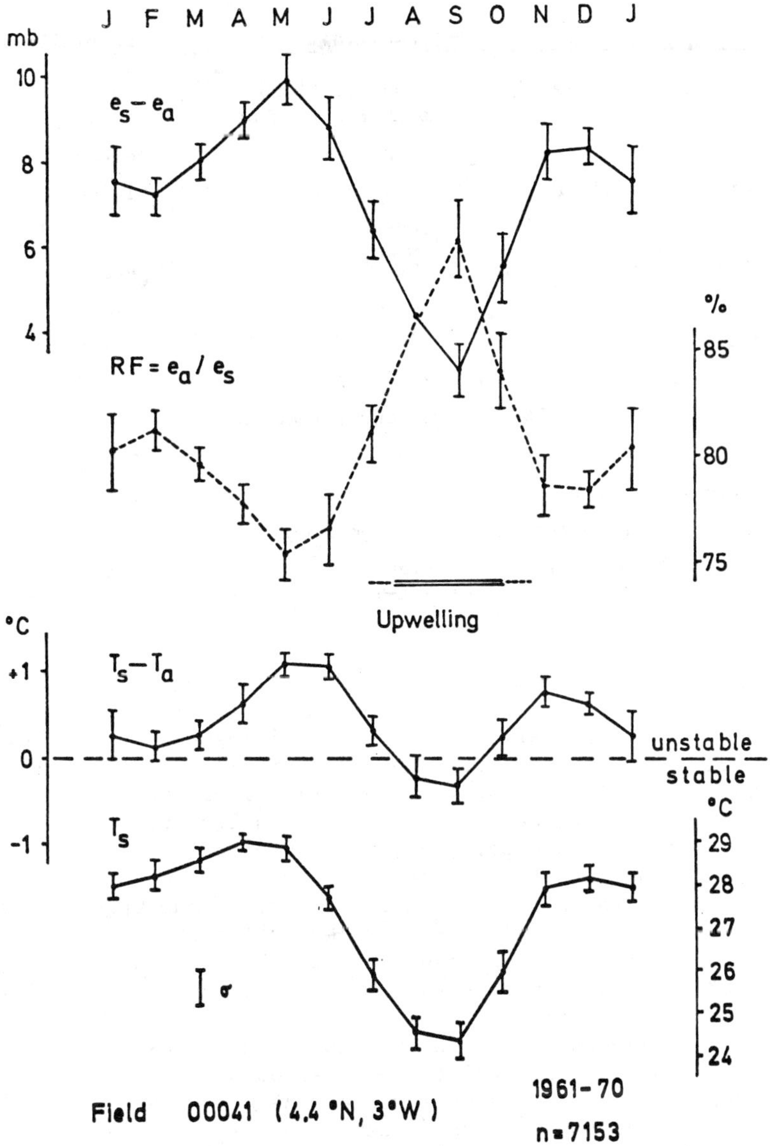

Fig. 2. Seasonal variation of climatic parameters in the upwelling region of the Gulf of Guinea; $e_s(e_a)$ = vapour pressure at sea surface (in air), RF = relative humidity, T_s-T_a = temperatur difference (sea surface - air), σ = standard deviation.

Table 2. Climatic effects of oceanic upwelling/downwelling.

	Upwelling	Downwelling
Sensible Heat (Enthalpy)	downward flux stabilization of air	upward flux destabilization
Convection	suppressed	enhanced
Cloudiness, rainfall	suppressed	large
Vertical momentum exchange	reduced	enlarged
Surface wind speed	low	high
Saturation deficit	low	high
Relative humidity	high (>80%)	low (<75%)
Precipitable water	20-30 l/m^2	40-60 l/m^2
Evaporation	strongly reduced	normal
Nutrient content of ocean	high	very low
Biological CO_2 exchange	air → sea	sea → air
CO_2 growth rate in the air	reduced	enhanced
Heat gain of the ocean	high	negative

cycle) and the intensity of upwelling in the equatorial Pacific are statistically highly significant (Bacastow-Keeling 1981; Angell 1981). A comparison between upwelling and downwelling (warm water) episodes lasting 6 (or even 12) months - made by our working group in Bonn (K.H. Weber) - indicates that the differences between these two samples are equivalent to 1-2 Gt (1 Gigaton = 10^{15}g) carbon per year, i.e. 20-40 percent of the annual release of fossil fuel CO_2 into the air (where only about 55 percent remain). The biological productivity of the upwelling areas (estimated to about $10 x 10^6 km^2$ or 3 percent of the total ocean area) is in the right order of magnitude of 4-5 Gt carbon per year, as measured at several places (Weichert 1980; Lenz 1981; Broecker-Peng 1982, p. 397 f).

This leads to a surprising coincidence of exchange processes: upwelling cold water leads to a depletion of atmospheric CO_2 and H_2O, downwelling warm water enriches the atmosphere with both gases. Both act together as "greenhouse gases" transparent for solar radiation, but absorbing infrared terrestrial radiation and thus warming the atmosphere. Since the downwelling episodes - known from their occurrence along the Peruvian coast as "El Niño" - are correlated with wide-spread weakening (or even a reversal) of the trade winds at the equatorial Pacific, a natural positive feedback process has been suggested (Flohn 1981, 1982) which controls these quite effective air-sea exchange processes (Table 3). Here the relationships between polar cooling, intensity of extra-

Table 3. Natural Hemispheric Climatic Feedback.

Initiation[1] (polar regions)	cooling	warming
Meridional ΔT / Intensity of westerlies	↗	↖
Lat. subtropical high (ϕ_{STA})	equator	pole
Intensity of trades (Hadley cell)	strong	weak
Equatorial Oceans[2]	upwelling	downwelling
SST Equatorial Oceans	cold	warm
Atmospheric content $CO_2 + H_2O$	↘	↗
resulting in further	cooling	warming

[1] triggered e.g. by a cluster (an absence) of volcanic eruptions
[2] except Indian Ocean

tropical westerlies and meriodional shift of subtropical anticyclones are convincingly reflected in the annual cycle; however, for a complete verification a statistical investigation of their anomaly correlations is needed.

3. DISCUSSION OF AIR-SEA INTERACTION AND GLOBAL FEEDBACK MECHANISMS

A comparison between the completely natural "abrupt" events during the late Glacial-Holocene, at the 100^a time-scale, and the model-derived (Flohn 1981; Schlesinger 1983) climatic changes to be expected after a prolonged CO_2-induced climate change at the same time-scale can be made on the base of the proposed feedback mechanism (Table 3). In the cases of a natural cooling (Flohn 1981, 1982, 1983) the natural trigger could be a series of heavy volcanic explosions producing large amounts of stratospheric particles by photochemical processes from injected gases (SO_2, H_2S), occurring more or less simultaneously ("clustering") as e.g. during the early 19th century. This would cause wide-spread cooling accentuated at polar latitudes where solar radiation crosses the aerosol layer under low angles. Vice-versa, a natural warming could be produced by a prolonged lull of severe volcanic events, similar (but often much longer) than the minimum period between 1912 (Katmai) and 1963 (Agung). Such non-random clustering (or absence) of large volcanic events (Flohn 1979) may have been triggered by near-simultaneous changes in the intensity of spreading rates at the world's great rift zones, i.e. by the discontinuous stress of movements in the asthenosphere. Indeed, it would be rather surprising if platetectonic motions are constant in time instead of being discontinuous. What may happen in the next one or two centuries, if we allow the CO_2 content of the air to rise above a certain threshold (Flohn 1981; Schlesinger 1983), can be also predicted by this feedback, if we enter in Table 3 at the second last line by an input of CO_2 from fossil fuel or burning of tropical forests. This will gradually increase surface temperatures (at the ocean after an empirical delay of 20-25 years (Paltridge-Woodruff 1981) and thus evaporation of H_2O. Model results e.g. of the GFDL group at Princeton (Manabe et al. 1980; Manabe 1983) are consistent with Table 3 (Flohn 1983); a similar feedback mechanism has also been discussed by Kellog 1983).

It should be mentioned here that all advanced models of the general circulation of the atmosphere use a positive feedback mechanism which combines the effect of CO_2 with those of H_2O (increased evaporation due to global warming) (Manabe et al. 1980; Hansen et al. 1981; Ramanathan 1981). Negligence of this coupling leads to unrealistic results (Manabe 1983; Schlesinger 1983). Obviously the proposed mechanism adds to this coupling.

The interaction between atmospheric and oceanic circulation is certainly not as simple as outlined here. Since the present oceans are more or less closed basins connected only by the powerful circum-Antarctic current, their circulation is subject to several constraints. Continuous upwelling or downwelling along the equator in a basin like the Pacific would be impossible: permanent zonal wind stress would drive the surface waters towards a meridional coast and create a slope, which forces, above a given threshold of the slope (depending on wind speed u, indeed on u^2), the water to return against the wind. An increase of the zonal winds during the last ice-age - at least between 20 and 14 ka, when the sea-ice advanced from both subpolar belts, to about Lat. 45° (Ruddiman-McIntyre 1981; Hays 1978) - necessarily produced much stronger winds than now. What we can expect is a marked change in the frequency, intensity and/or duration of the individual upwelling or downwelling episodes, which last now between several months and a few years.

Let us assume a predominance of upwelling (of, say 75 percent of a 100 year period), this should easily withdraw 100-200 Gt carbon from the atmosphere, equivalent to 47-94 ppm CO_2. In contrast to this, predominant downwelling - as during the Holocene warm period 12-6 ka ago, when the northern hemisphere received during summer about 7 percent more solar radiation at the top of the atmosphere than now (Kutzbach et al. 1981, 1982) - would increase evaporation and the release of latent heat by precipitation in the tropical belt, thus promoting spreading of tropical forests (Pastouret 1978; Maley 1980; Nicholson-Flohn 1980; Street-Perrott 1983) and the H_2O content of the atmosphere. As suggested by Shackleton (1977), increased productivity and decay of land vegetation would also enhance, by a biological feedback process

(Kellogg 1983) the CO_2 content of the atmosphere. These processes presumably operate more slowly and less frequent - only in periods of favorable orbital geometry - as the cooling process; they could also have contributed to case c) in Table 1.

Careful investigations, with much improved accuracy and time-resolution of measurements, on the CO_2 content of air bubbles in the new ice-core from Southern Greenland (Dye 3: cf. Dansgaard 1982), simultaneous with the temperature-dependent ^{18}O variations, have shown (Oeschger et al. 1983) that the abrupt climate changes during the Late Glacial - cases a), b), d) from Table 1 - have been accompanied by changes in the CO_2 content around 80 ppm occurring at a time-scale near 100 years. These three cases confirm the above-mentioned order of magnitude as deduced from actual observations. These results also elucidate the hitherto enigmatic parallelism of the climatic evolution in high northern and southern latitudes (Oeschger et al. 1983). Since the absolute CO_2 content varied during the Late Glacial between about 200 and 300 ppm, the equivalent variation of the globally averaged surface temperature ought to be in the order of 1.5-2 $^{\circ}C$ (Manabe 1983; Schlesinger 1983); in the European area it was much larger, due to advection of polar water masses (Duplessy et al. 1981; Ruddiman-McIntyre 1981), possibly with seasonal sea-ice. Such observed abrupt climatic fluctuations under absolutely natural conditions, emphasize the possible climatic role of a persisting CO_2 increase for the expected global warming in a foreseeable future, at the same time-scale.

Simultaneous global changes of the H_2O content of the atmosphere are not observable now: the accuracy and comparativity of humidity measurements within our present (not quite homogenous) radiosonde network is hardly sufficient for such task. The geostationary Meteosat satellite measures the H_2O content only at the upper troposphere; perhaps in the future continuous microwave measurements will allow investigations into global balances and their changes. At present the best estimates of the global water budget (evaporation - precipitation) are still uncertain by about 5 (or more) percent. For the glacial-interglacial cycle the variations in the global water budget have been estimated to about 20-25 percent of the present value (Flohn 1983). This difference is partly due to the vertical exchange processes in the atmosphere, but partly also due to changes in the amount of "precipitable water" available in the atmosphere, globally averaged at present very nearly equivalent to 25 mm (= $1/m^2$) rainfall.

What limits the efficiency of the proposed feedback mechanism? Looking into the climatic history of the last million years, there seems to be a bimodal climatic fluctuation between glacial and interglacial periods which are, each group for itself, rather similar. Such an "almost-intransitive" behaviour of the climatic system, with abrupt transitions between the two contrasting modes, has been first suggested by E. Lorenz (1970; cf. also Saltzman et al. 1980, 1982). At present there are two constraints: the upper limit is an ocean temperature of about 30°C (apart from marginal seas as the Persian Gulf, where hot desert air is advected), above which the present solar constant is unable to evaporate sufficient water vapour (Newell et al. 1981). The lower limit is the water temperature just below the thermocline: it is over vast areas, especially in the Tropics and Subtropics, very near to 14°C. In exceptional cases (along the Peruvian coast or even at the western flank of the Galapagos Islands) this value is observed also in upwelling water. But the incoming solar radiation and the absence of high-reaching clouds warms (Weare 1983) this water, after a few days, to 18-22°C.

REFERENCES

Angell, J.K., 1981: Monthly Weather Review 109, 230-243.
Bacastow, R.B., Keeling, C.D., 1981: WMO/ICSU/UNEP Conference Report, Bern.
Baes Jr., C.F. in W.C. Clark (ed.): Carbon Dioxide Review 1982. Oxford Univ. Press 1983, 187-204.
Broecker, W.S., Peng, T.H., 1982: Tracers in the Seas. Lamont-Doherty Geological Observatory. Columbia Univ. Palisades, N.Y., 690.
Coope, G.R.: Philos. Transact. Roy. Soc. London 1977, B280, 313-340.
Dansgaard, W., 1982: Science 218, 1272-1277.
Denton, G., Hughes, T.R. (eds.), 1981: The Last Great Ice Shields J. Wiley, New York, 2 vol.
Duplessy, J.C. et al. 1981: Palaeogeography, Palaeoclimatology, Palaeoecology 35, 121-144.
Duplessy, J.C., 1982: Nature 295, 494-498.
Eicher, U., Siegenthaler, U., 1980: Mitt. Naturf. Ges. Bern 1980, 65-80; Quartern. Res. 15(1981), 160-170.
Emiliani, C. et al., 1975: Science 189, 1083-1088.
Flohn, H., 1979: Quatern. Res. 12, 135-149.
Flohn, H., 1981: Phys. Blätter 37, 184-190; Journ. Meteor. Soc. Japan 60(1982), 268-273.
Flohn, H., 1983 in: F.A. Street-Perrott (ed.): Variations in the Global Water Budget, D. Reidel, Dordrecht, 403-418.
Grüger, E., 1979: Geologica Bavarica 80, 6-64.
Hansen, J. et al., 1981: Science 213, 957-966.
Haserodt, W., 1982: unpublished lecture, Braunschweig.
Hays, J.D., 1978 in E.M. van Zinderen Bakker (ed.): Antarctica Glacial History and World Palaeoenviroments. A. Balkema, Rotterdam, 57-71.
Heine, K., 1983: Catena 10, 1-25; also this volume.
Henning, D., Flohn, H., 1980: Beitr. Phys. Atmos. 53, 430-441; Trempel, U., Weber, K.-H., Becker, C.: unpublished Diploma Theses Bonn 1978-1982.
Kellogg, W.W., 1983: Journ. Geophys. Res. 88, 1263-1269.

Kukla, G., 1980: Paleoecology of Africa 12, 395-408.
Kutzbach, J.E. et al., 1981: Science 214, 59-61; Journ. Atmos. Sci. 39(1982), 1177-1188.
Lenz, J., 1981: Naturwiss. Rundsch. 34. 405-413.
Lorenz, E., 1970: Journ. Appl. Meteor. 9, 325-329.
Maley, J., 1980: Thèse Montpellier, 586
Manabe, S., Stouffer, R.J., 1980: Journ. Geophys. Res. 85, 5529-5554; Manabe, S., Wetherald, R.T., 1980: Journ. Atmos. Sci. 37, 95-118.
Manabe, S., 1983: Advances in Geophysics 25, 39-82.
Müller, H., 1965: Geol. Jahrbuch 83, 327-352; Archiv A21(1974), 107-140.
Newell, R.E. et al., 1978: Pure and Applied Geophysics 116, 351-371.
Newell, R.E., Gould-Stewart, S., 1981: Journ. Atmos. Sci. 38(1981), 2789-2796.
Nicholson, Sh.E., Flohn, H., 1980: Climatic Change 2, 313-348.
Oeschger, H., Dansgaard, W. et al. in: Ghazi (ed.) 1983: Paleoclimatic Research and Models. D. Reidel Dordrecht, 95-107.
Osborne, P.J., 1974: Quartern. Res. 4, 471-486.
Paltridge, G., Woddruff, S., 1981: Monthly Weather Review 109, 2427-2435.
Pastouret, L. et al., 1978: Oceanologica Acta 1, 217-232.
Philander, S.G.H., 1983: Nature 302, 295-301.
Prell, W.L., 1980: Quartern. Res. 14, 309-336.
Ramanathan, V., 1981: Journ. Atmos. Sci. 38, 918-930.
Ruddiman, W.F., McIntyre, A., 1981: Palaeogeography, Palaeoclimatology, Palaeoecology 35, 145-214.
Saltzman, B., Moritz, R.E., 1980: Tellus, 93-118; Journ. Atmos. Sci. 38(1981), 494-503.
Schlesinger, M.E., 1983: Intern. Journ. Environm. Studies 20, 103-114.
Shackleton, N., 1977 in: A. Malahoff (ed.): The Fate of Fossil Fuel CO_2 in the Ocean. Marine Science 6, 401-427.
Sissons, J.B. et al., 1974a: Nature, 822-824; Boreas 1974b: 3, 41-48.
Strauch, F., 1973: Dt. Geol. Ges. 123, 163-177.
Street-Perrott, A., Roberts, N. et al. (eds.), 1983: Variations in the Global Water Budget. D. Reidel, Dordrecht, 331-345.
Swain, A.M. et al., 1983: Quartern. Res. 19, 1-17.
van der Hammen, Th. et al., 1978: Palaeogeography, Palaeoclimatology, Palaeoecology 25, 179-190.
Weare, B.C., 1983: Journ. Phys. Oceanogr. 13, 873-885.
Weichert, D., 1980: Dt. Hydrogr. Zeitschr. 33, 192-198.
Wijmstra, T.A., 1978 in: J. Gribbin (ed.): Climatic Change, Cambridge Univ. Press, 25-45.
Woillard, G., 1979: Nature 281, 558-562.
Wyrtki, K., 1981: Journ. Phys. Oceanogr. 11, 1205-1214.

SEA SURFACE TEMPERATURE, ATMOSPHERIC CO_2 AND THE GLOBAL ENERGY BUDGET: SOME COMPARISONS BETWEEN THE PAST AND PRESENT.

Reginald E. Newell and Jane Hsiung

Department of Earth, Atmospheric and Planetary Sciences
Massachusetts Institute of Technology
Cambridge, Massachusetts, USA

Factors which contribute to air temperature changes within the recent period and between the ice age, the Holocene and the recent period, are considered. The main factors controlling recent low latitude free air temperatures are sea surface temperature changes and volcanic activity. The longer time scale temperature changes at low latitudes are thought to be due to sea surface temperature changes which in turn are driven by changes in the middle latitude circulation. Middle latitude circulation and temperature on this time scale depend upon Milankovitch radiation changes with high latitude radiation controlling ice extent as Milankovitch suggested and middle latitude radiation in summer controlling ocean storage of energy which is transfered from the ocean to the atmosphere in the following winter. This is a new concept of the mechanism by which Milankovitch radiation changes influence climate. The low latitude temperatures were apparently buffered by evaporation to remain below about 30°C during the Holocene although the oceanic area covered by these values would have been larger than normal, as occurred during the 1982-83 El Niño. Atmospheric CO_2 is thought to depend on climate through a modulation of upwelling at low latitudes. More upwelling brings up more nutrients which increase the rate of calcium carbonate formation and so decrease the CO_2 release from the ocean to the atmosphere at low latitudes. Thus atmospheric CO_2 decreased in the ice ages when more upwelling occurred, and increased in the Holocene. There is a saturation on the high atmospheric CO_2 side for continued lack of upwelling cuts off the ultimate CO_2 source of deep water.

N.-A. Mörner and W. Karlén (eds.), Climatic Changes on a Yearly to Millennial Basis, 533–561.

1. Introduction

Stable isotopes in ice cores, distributions of pollen, and distributions of oceanic foraminifera indicate that air and sea temperatures were higher during the general period 8000-3000 BP than at present. Because of the rather small temperature changes involved it is hard to estimate the time of the maximum. Lake levels and the extent of dunes suggest that precipitation was higher and deserts were less extensive in the early Holocene. Lake level changes are thought to have been substantial and the time of occurence of high levels is given as the period 10,000-7000 BP (Street and Grove, 1979; Hastenrath and Kutzbach, 1983), somewhat earlier than the time of possible temperature maximum. Why was the climate warmer and wetter during the Holocene?

Atmospheric carbon dioxide has been measured in ice cores by Delmas et al. (1980), Berner et al. (1980) and Neftel et al. (1982). It was lower than the present during the last ice age and higher than the present during the Holocene. What are the physical and chemical reasons for these differences? Did increased atmospheric CO_2 contribute to higher temperatures during the Holocene or <u>vice versa</u>? Did decreased atmospheric CO_2 result from lower temperatures?

In this paper we take the evidence for the observed features as given and try to give partial answers to the questions; it would be presumptuous to attempt more at this stage. Because we make the case that atmospheric CO_2 is controlled by climate through a series of oceanic processes and that climate itself is strongly dependent on the surface layer of the ocean, we stress here the variability of sea surface temperature, drawing heavily on the recent record including that for the 1982-83 El Niño which may be treated in some respects as a "Mini-Holocene".

2. Global overview of the energy budget

An energy conservation equation for a given latitude belt of the atmosphere-ocean climate system may be written:

$$RN_{EA} = R_S - R_{IR} = S_O + S_L + S_A + S_Q S_I + DIV(F_Q + F_A + F_O) \qquad (1)$$

where RN_{EA} is the net energy available to the earth-atmosphere system, which is the difference between the incoming solar radiation (R_S) and the outgoing long-wave radiation (R_{RI}). S_O, S_L, S_A, S_Q, and S_I represent changes in the storage of heat energy in the ocean, land, atmosphere, moisture content, and ice respectively, and $DIV(F_Q + F_A + F_O)$ represents horizontal divergence of the meridional energy flux carried in the form of moisture

(F_Q), as sensible heat (F_A) and by the ocean (F_O). This energy balance is shown schematically in Fig. 1 and examples of the terms in this equation appear in Table 1 for present day seasonal variations (S_I is not well known and is omitted here). In the northern hemisphere winter, in middle and high latitudes, R_{IR} is greater than R_S and there is a net radiational energy loss which of course results in lower temperatures so that S_A decreases. Some of the energy deficit is made up from decreases in oceanic heat storage S_O and some comes from latent heat liberation as ice forms. Substantial energy is also provided by transport from lower latitudes. For example in the 60°N latitude belt, including Stockholm, the radiational energy deficit in December-February is 146 Wm^{-2}; oceanic cooling provides 29 Wm^{-2} to the air while transport from lower latitudes accounts for 109 Wm^{-2}. Cooling of the air and land together give the remaining 8 Wm^{-2} of the net energy loss to space. There are of course significant variations in these quantities with longitude; these values are averages over all longitudes.

How did this energy balance change as conditions evolved into and out of the last Ice Age and through the Holocene? In the period when ice was increasing at high latitudes, three factors could operate: 1) RN_{EA} could have larger negative, or smaller positive values at high latitudes; both could be brought about by smaller values of R_S. This would be a <u>direct</u> effect of the Milankovitch orbital changes; 2) Oceanic storage change could diminish. Less energy would presumably be lost to the air in winter if less was received by the ocean in summer - an <u>indirect</u> effect of the Milankovitch changes; 3) The energy fluxes from lower latitudes could be diminished but note that it is usual to treat these fluxes as the result of the low-high latitude differences set up by external conditions rather than as their control.

The Milankovitch changes for the past 40,000 years are summarized in Table 2. Following Milankovitch we use 65°N to represent a region of variable ice extent. We quote melting period April-August values. Also summer values for 45°N are taken to represent possible ocean storage changes; when this energy is released in winter it moderates northern hemisphere temperatures as seen from Table 1.

In the transition from the last interglacial to the last glacial maximum 47 x 10^6 km^3 of ice formed north of about 40°N (Flint, 1971). This corresponds to about 45 x 10^{18} kg of water equivalent and for a latent heat of sublimation of 2.83 x 10^6 J kg^{-1} it can be seen that about 127 x 10^{24} J must be lost to space. The period over which the ice accumulated is subject to some debate -- suppose we use here a round number of 20,000 years or 6.3 x 10^{11} seconds. Then the power loss is 2 x 10^{14} W and spread over an area of 0.9 x 10^{14} m^2 (40°N to the pole) this

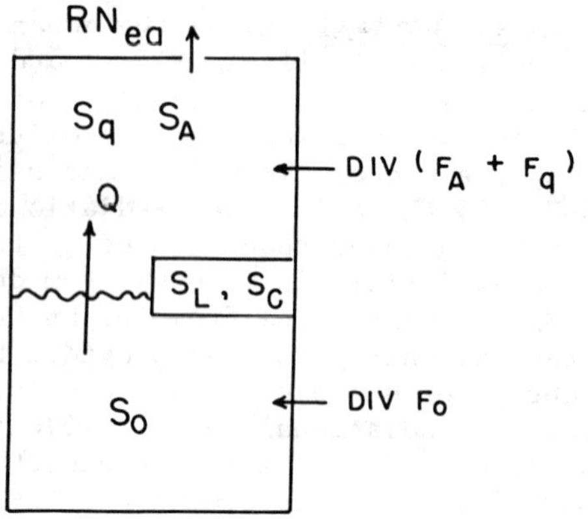

Figure 1
Schematic showing the energy balance of an atmospheric column

Table 1. Components of the energy balance (in watts m^{-2})

	RN_{EA}	S_O	S_L	S_A	S_Q	$DIV(F_Q + F_A + F_O)$
Dec.–Feb.						
80°N	−146	−10	−1	−4	0	−131
60°N	−146	−29	−3	−5	0	−109
40°N	−92	−60	−2	−4	0	−27
20°N	−22	−41	−1	−1	−1	21
0°	41	0	0	0	0	41
20°S	87	49	0	2	2	36
40°S	85	74	0	3	2	8
60°S	19	30	0	3	1	−14
80°S	−36	0	2	7	0	−45
June–August						
80°N	−5	23	1	5	0	−35
60°N	32	43	3	6	1	−22
40°N	68	71	2	4	2	−11
20°N	68	30	1	1	2	35
0°	41	−10	0	0	−1	52
20°S	−24	−54	0	0	−1	32
40°S	−97	−72	0	0	0	−24
60°S	−143	−31	0	−1	0	−111
80°S	−146	0	−2	−1	0	−141

SOURCE: Newell et al. 1974.

corresponds to an energy flux loss of 2.2 Wm^{-2}. The radiational change averaged over the polar cap in the peak of the melting period (May-July) between 32,000 BP and 24,000 BP (a time when ice is increasing) corresponds to a loss of 18 Wm^{-2}. A similar figure for the period between 40,000 and 20,000 BP is 17 Wm^{-2}. These decreases are obviously adequate to account for the observed ice growth. In other words, if less radiation is received over the polar cap in summer, the ice does not melt so that more persists from year to year.

The time average over the year of energy entering the ocean at 45°N where the storage effect is largest does not change - this is an invariant latitude for this particular property. However, for the radiation incident in June-August on the ocean, which provides the energy used to moderate the Northern Hemisphere winter, there is a systematic change with minima between 24 and 16 K and maxima between 12 and 6 K, the difference amounting to about 40 Wm^{-2}, a substantial fraction of S_0 as seen from Table 1. Of course one cannot take radiation on the top of the atmosphere as a real change in storage -- it is just the maximum potential change.

Sometime between 16000 and 12000 BP this large ice volume begins to melt and in fact the melting radiation at 65°N increases from 466 to 483 Wm^{-2} in this period. The average change in radiation over the polar cap is an increase of 16 Wm^{-2} (see Table 2). As much of the ice only has to be converted to water, not re-evaporated, then the energy involved is much less than that in the reverse phase transition from vapor to ice. Thus much less time is taken to melt the ice than to form it and the energy increase over the polar cap can easily take care of the melting.

In Table 2, three parameters are presented: radiation incident at 65°N in summer as a measure of potential ice melting; radiation over the polar cap (40-90°N) also dominated by that in summer and also an indicator of potential ice melting; and radiation at 45°N in summer to represent potential oceanic storage changes which would be reflected in temperatures in the following winter. The first two items are to some extent redundant. It is noteworthy that the physics producing the changes is different. The first two are associated with the obliquity of the ecliptic which changes with a period of about 40,000 years while the oceanic storage variable responds most to the precession of the equinoxes coupled with obliquity changes which yields a net period of about 24,000 years. The two processes are in phase and would tend to produce maximum cooling at about 24,000 BP and maximum heating in the 12-10000 BP period. Prior to these changes the maximum cooling tendency with both processes in phase was at about 72000 BP and the maximum heating at 84-88000 BP.

3. Sea surface temperature factors

An analysis has been made of the modes of variability of global sea surface temperature, defined by empirical orthogonal functions, from the recent (1949-1979) record (Hsiung and Newell, 1983). The first mode shows the following features (Fig. 2a): when the equatorial Pacific is warm (except in the west) the north and south central Pacific is cold; the east Pacific off the American coasts is warm at the same time; the tropical Atlantic is warm; and the middle latitude north Atlantic is cold. The mode alternates between this pattern and one with the temperature anomalies reversed. The time series of this eigenvector pattern (Fig. 2b) is similar to that of SST in the eastern equatorial Pacific. Positive values correspond to the so-called El Niño mode and the sea surface temperature pattern also fluctuates in phase with the Southern Oscillation Index. The second mode shows a cold eastern equatorial Pacific, a cold south Pacific and a cold south Atlantic accompanying a warm north Atlantic and north Pacific. Again, the anomalies alternate with those of opposite sign. During the 1965-79 period this pattern shows a general cooling trend in the north Pacific and north Atlantic. The third mode is characterized by a cold eastern equatorial Pacific, north-east Pacific and north Atlantic and a warm south Atlantic, western north and south Pacific and extreme north-east Atlantic with the anomaly signs again alternating with time. Do any of these three modes encompass events which may be similar to those in the Holocene? If so, then by studying the physics of the recent fluctuations we should be able to shed some light on the causes of the Holocene.

The highest temperatures at low latitudes occur with the extremes of the El Niño pattern or the first mode above (see Fig. 2a). The high temperatures reach up to 30-31°C in the tropics but do not exceed these values as we have stressed previously (Newell et al., 1978). In 1982-83 when the limiting temperatures were achieved the anomalies reached 5-6°C. The limit appears to be set by the balance at the sea surface of incoming solar radiation and evaporation (Newell et al., 1978; Newell, 1979; Newell and Dopplick, 1979). Because of the Clausius-Clapeyron relation additional energy received and consequent surface warming is offset by a rise in saturation vapor pressure that produces an additional energy loss to the air by evaporation. This loss is also a function of the wind speed so that the limiting temperature increases as the wind speed decreases (an example is given by Newell and Dopplick, 1979). The wind plays a double role, for in addition to increasing evaporation and therefore cooling by loss of energy to the air, it mixes water within the oceanic boundary layer and therefore also induces cooling because the water from below is cooler than that at the surface.

Table 2

Incoming Solar Radiation
(Wm^{-2})

(Data from Ledley, 1980)

	Thousands of Years Before Present													
	0	2	4	6	8	10	12	16	20	24	28	32	36	40
April(65°N)	316	310	306	308	315	324	334	339	322	302	299	313	328	331
May	429	424	426	435	448	463	472	465	433	409	400	484	452	432
June	478	482	491	504	517	526	526	503	474	462	473	490	496	496
July	435	445	456	464	466	461	451	431	424	430	438	435	430	433
August	326	337	342	339	330	318	307	303	316	329	325	309	302	310
June (45°N)	484	485	493	504	516	524	525	507	483	476	489	505	507	501
July	462	469	480	489	495	493	485	465	455	462	475	477	469	466
August	402	411	419	421	417	407	397	386	393	407	412	402	390	392
October(65°S)	305	314	327	339	345	342	333	310	298	308	326	309	317	309
November	441	452	462	465	461	452	443	430	429	438	437	428	426	433
December	513	515	513	505	496	489	487	493	501	495	473	462	474	497
Polar cap average flux May to July	461	464	471	481	489	497	497	481	459	454	459	472	476	475

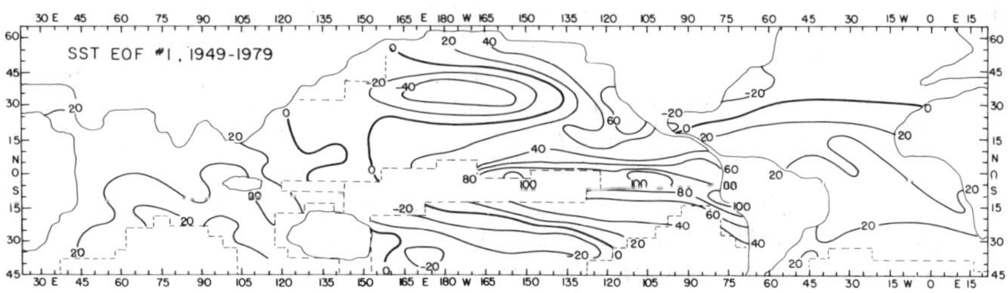

Figure 2a
First eigenvector pattern of global nonseasonal sea surface temperature variations (after Hsiung and Newell, 1983)

The eastern equatorial Pacific sea surface temperature changes occur several months before changes in tropical air temperature and these in turn occur several months before middle latitude air temperature changes (Newell and Weare, 1976a, b; Newell, 1979). The other main factor influencing tropical air temperature is volcanic aerosol and in fact it is possible to write a regression relationship as follows:

$$\Delta T_{air} \doteq 0.5\ \Delta T_{SST} + 20\Delta(\tau_v)$$

where ΔT_{air} represents the temperature changes in the zonal mean tropical air temperature, ΔT_{SST} is the change in sea surface temperature averages over the region 140°W to the coast of S. America for the latitude band 0-5°S and $\Delta\tau_v$ is the change in atmospheric transmission. This equation was based on the 1949-73 record. Middle latitude free air temperature changes are also related to low latitude changes (see Navato et al., 1981). In 1982 there were extremes in both ΔT_{SST} and $\Delta(\tau_v)$ well beyond previous experience on which the regression relationship was based: ΔT_{SST} reached 4.0°C and τ_v went from a normal value of 93.5% down to 80% (see Figs. 3 and 4). The interesting point is that the 1982 extremes were in opposite directions and it is hard to find the resultant change in tropospheric temperature in the observations because of this cancellation. These sea temperature changes are also related to the Southern Oscillation, an index of which appears in Fig. 3. Figure 5 shows the tropical tropospheric free air temperature changes up to October 1982; there is no systematic change after the volcano but it should be noted that the tropospheric cooling after Mt. Agung was delayed by over a year (Newell and Weare, 1976b; Newell, 1981). By contrast there is a substantial increase in stratospheric temperature (~5-6°C; see Labitzke et al., 1983 and Parker and Brownscombe, 1983) as occurred also for the Mt. Agung eruption in 1963 (Newell, 1970).

As noted above, the situation that produces the highest temperatures during the recent record is the El Niño mode. What is the background physics involved in this mode? From a 100-year set of Pacific sea surface temperature, wind and pressure data, a sequence of events in the El Niño phenomenon has been determined: first changes occur near Wake Island and consist of anti-cyclonic gyre modulation there and then near Easter Island before the sea surface temperature changes in the eastern equatorial Pacific and the change generally spreads westwards towards the central Pacific. The point to note here is that middle latitude changes occur first, although the mechanisms by which the process interacts with the tropics is not yet known (Newell et al., 1982).

In the 1982-83 El Niño phase with positive anomalies throughout the eastern and central tropical Pacific there were substantial

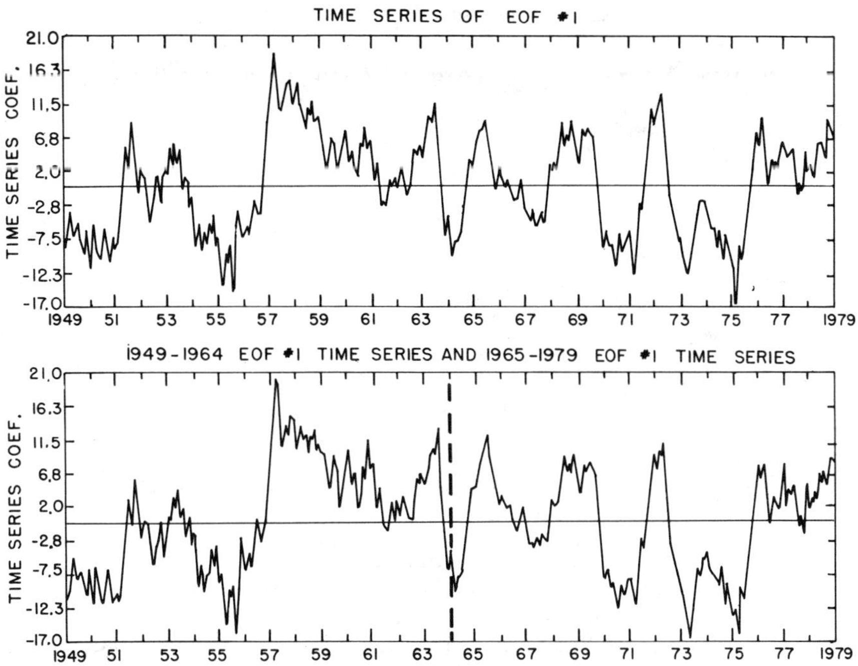

Figure 2b
Time series of the first eigenvector pattern
(after Hsiung and Newell, 1983)

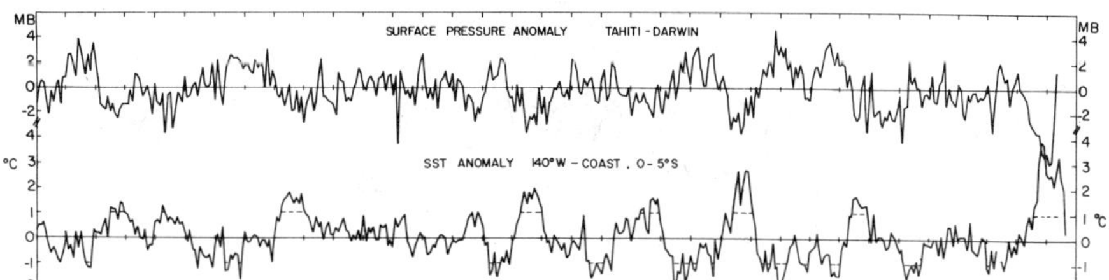

Figure 3
Time series of difference between nonseasonal
normalized surface pressure anomalies for Tahiti and Darwin,
and of zonal mean sea surface temperature, 0° - 5°S,
140°W to coast (from Rennie Selkirk, MIT)

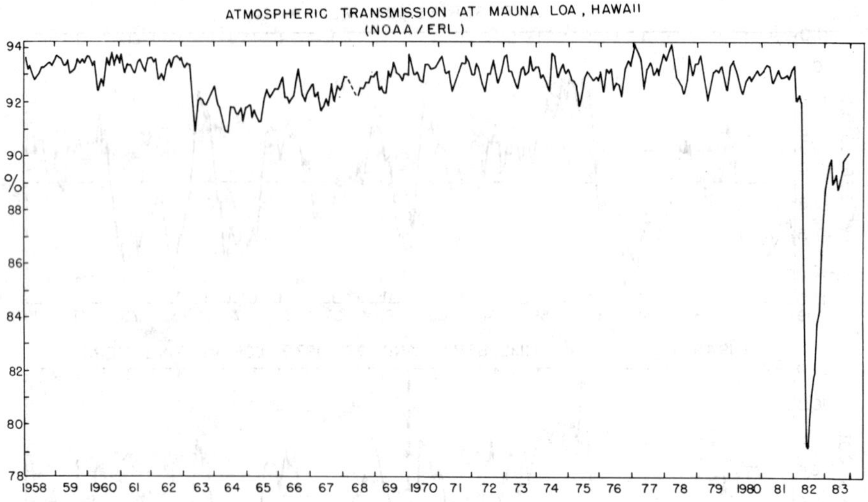

Figure 4
Monthly mean atmospheric transmission
in percent at Mauna Loa, Hawaii
(data supplied by J. DeLuisi and B.G. Mendonca, NOAA, ERL,
Geophysical Monitoring for Climate Change, Boulder, Colorado)

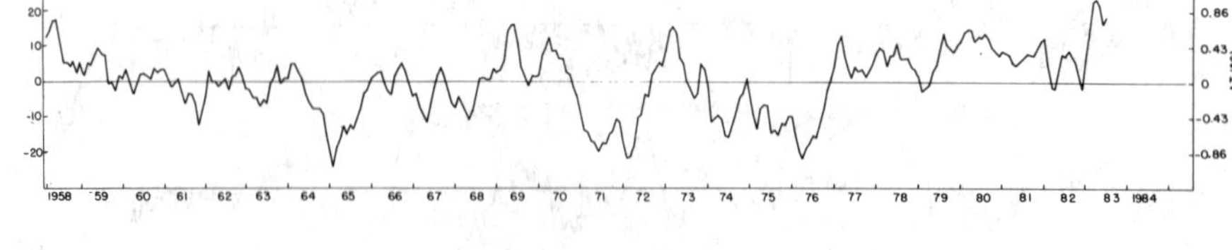

Figure 5
Nonseasonal tropical tropospheric temperature
from 300-700 mb geopotential differences

regions where temperature was near to 30°C and where anomalies of up to 5-6°C occurred. As might be expected, winds in these regions were much below normal strength. Instead of relatively strong easterlies there were weak westerlies in the central equatorial Pacific and maximum westerly anomalies propagated from 160°E in July 1982 to ~110°W in June 1983 (NOAA, 1983). Not only were normal wind speeds decreased but upwelling was diminished. We believe these changes were somehow related to middle latitude events. Overall this pattern yields an increase in evaporation. Accompanying this will be enhanced infrared cooling from water vapor and (somewhere) additional precipitation to offset the additional evaporation. The extra precipitation will liberate latent heat which will offset the infrared cooling.

The conservation of energy applied to an air column may be written:

$$Q_{RAD} + Q_{SH} + Q_{LH} = 0$$

where Q_{RAD} represents the heating by radiative processes dominated by infrared cooling, Q_{SH} represents heating due to sensible heat transfer from the lower boundary, and Q_{LH} is heating due to latent heat liberation. Simulations of Q_{RAD} for the ice age showed decreased cooling of about 10% leading to the deduction that rainfall must have decreased too by about 10% (Newell et al., 1975). General circulation model experiments gave quite similar results (Gates, 1976; Manabe and Hahn, 1977). Qualitatively we may expect this situation to reverse for the Holocene: weaker meridional temperature gradients (because of the higher energy input over the northern polar cap), weaker winds, warmer equatorial Pacific, more evaporation and heating by latent heat liberation, higher specific humidity and greater radiative cooling. Overall therefore rainfall would be expected to be larger.

The contrast with the last ice age may also be applied to the overall energy cycle of the atmosphere. As noted elsewhere for the ice age, the meridional temperature gradient increased and thus the conversion from zonal available potential energy to eddy available potential energy, which depends on the square of this gradient, also increased, and therefore the wind speed (Newell, 1974; Newell et al., 1981). For the Holocene a decrease of the meridional temperature gradient brings about a decrease in energy conversion rate and thus weaker winds.

If there are weaker middle latitude westerly winds then by conservation of angular momentum there will be weaker easterlies at low latitudes and therefore less upwelling and a warmer equatorial ocean, although not exceeding the evaporational limit of 30°C. We propose that the Milankovitch radiational changes initiate this sequence by providing more energy to the polar cap and more energy to the oceanic storage regions (~45°N) in summer in the Holocene. These two processes act together to produce a tendency for high latitude warming in the Holocene as is evident from Table 2. This approach is similar to that we expressed for the ice age, namely that the Milankovitch effects control middle latitude changes and that these in turn control low latitudes (Newell et al., 1981). From Table 2 is is clear that both processes give a cooling tendency in the 30,000-20,000 BP period.

Figure 6 shows sea surface temperature values during the 1982-83 El Niño and Fig. 7 some examples of anomalies. Actual temperatures do not exceed 30°C which sets a limit on the anomalies of about 6°C - and this is achieved in several regions. It is clear from Fig. 6 that the 30°C regions are not brought about by simple warm advection. We suggest that they occur where upwelling and/or cold water advection is reduced until the evaporational limit may be obtained. Maximum possible anomalies are shown in Fig. 8 for the August and February climatologies. In fact we may treat an El Niño as a tendency for eastern tropical Pacific temperature to return to the evaporational limit once the wind-stress-forced upwelling is diminished.

Note that 30°C is above the SST limit for hurricanes to form-is there evidence of more hurricanes than usual in the Holocene tropical regions? Hurricane frequency in the eastern Pacific was higher in 1982 than normal (Gunther et al., 1983).

The coupling between Milankovitch forcing and low latitude changes is quite non-linear, because the tropical temperatures are buffered by evaporation. Thus maximum air temperature in the tropics changes very little while the area covered by 30°C water expands as upwelling is reduced. A lid of ≈30°C is set on the maximum temperature. But in an ice age there is no analogous lower limit in equatorial regions and stronger winds can produce more upwelling and colder water and thus colder tropical air temperatures (an "anti-El Niño"). There is a limit in the subtropical gyre centers where there is no upwelling; here temperature actually increases due to Ekman convergence and suppression of boundary layer mixing. Solar radiation here prevents the temperatures from dropping and the limit is again constrained to be less than 30°C. The lower limit on high latitude temperatures is probably set by the dynamic constraint that meridional heat transport by the atmosphere depends on the square of the meridional temperature gradient. This gradient increases as the ice moves equatorward. At 18,000 BP the meridional flux required to balance radiative imbalance is not much different from the present but as F_O decreases, $F_A + F_Q$ must increase. These parameters were defined following equation (1). Figure 9 may be used to estimate the equilibrium values of the temperature gradient at this time (from Newell et al., 1974). It could be argued that

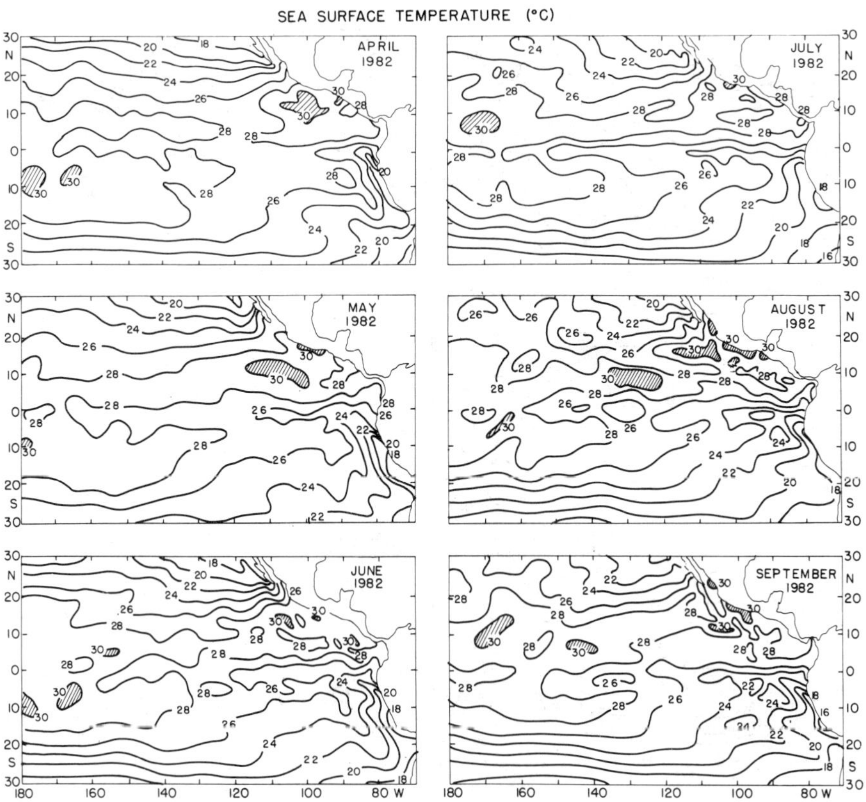

Figure 6
Time evolution of sea surface temperature from April, 1982 to September, 1982 (from Forrest Miller, Inter-American Tropical Tuna Commission, Scripps Institution of Oceanography)

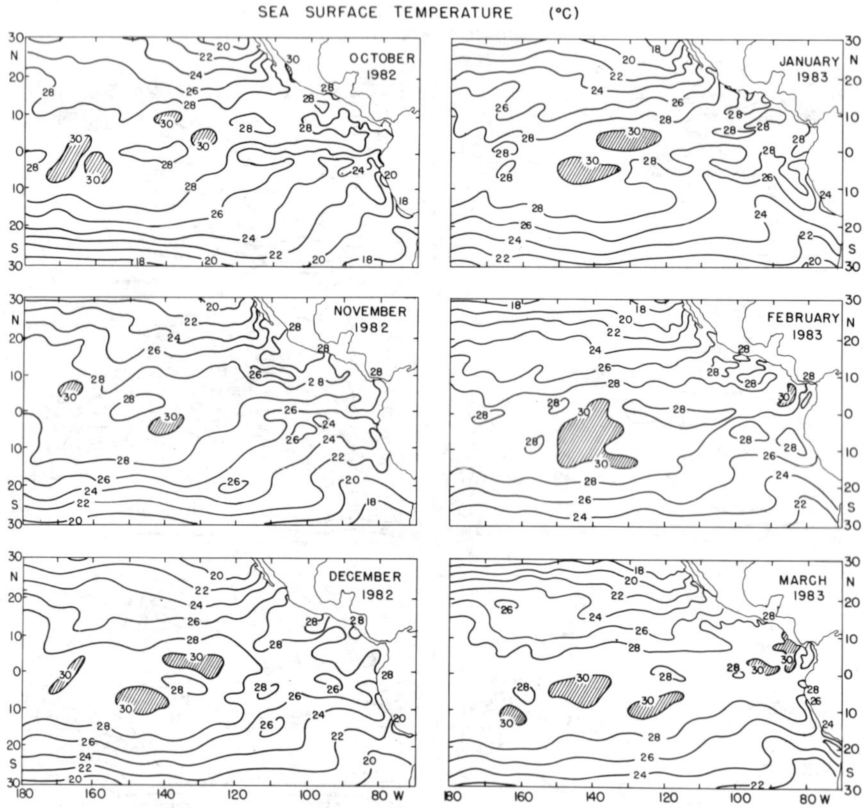

Figure 6 continued
Time evolution of sea surface temperature from October, 1982 to March, 1983 (from Forrest Miller, Inter-American Tropical Tuna Commission, Scripps Institution of Oceanography)

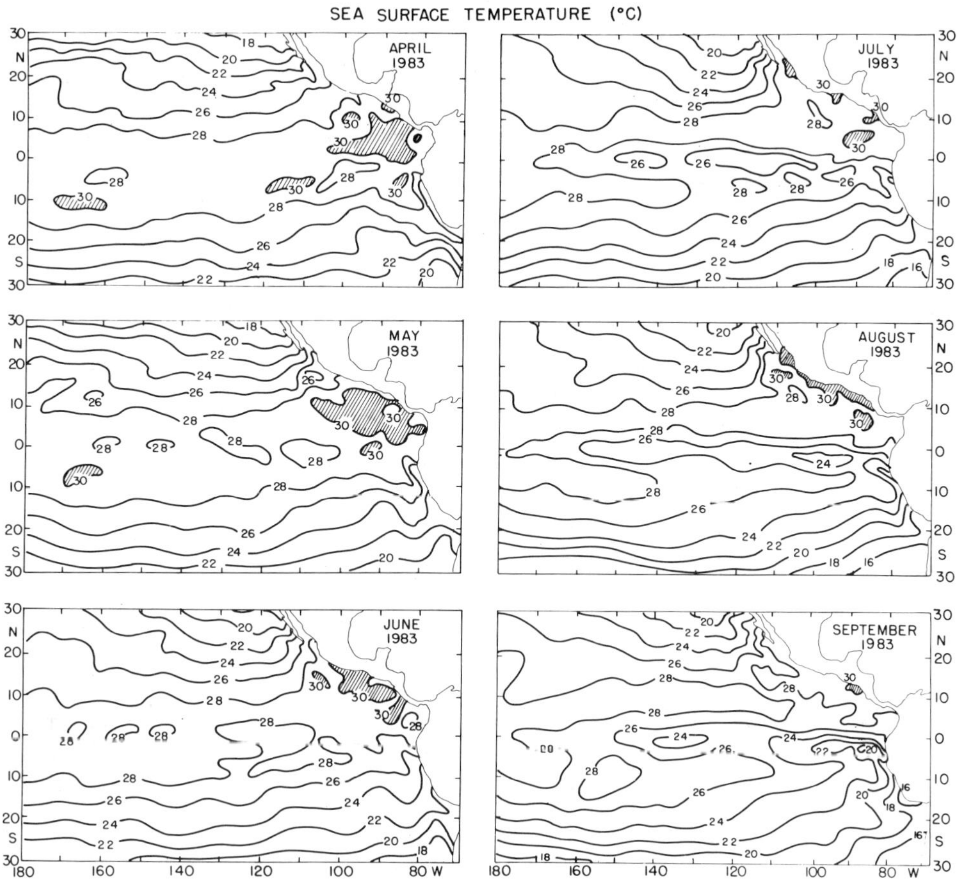

Figure 6 continued
Time evolution of sea surface temperature from April, 1983 to September, 1983 (from Forrest Miller, Inter-American Tropical Tuna Commission, Scripps Institution of Oceanography)

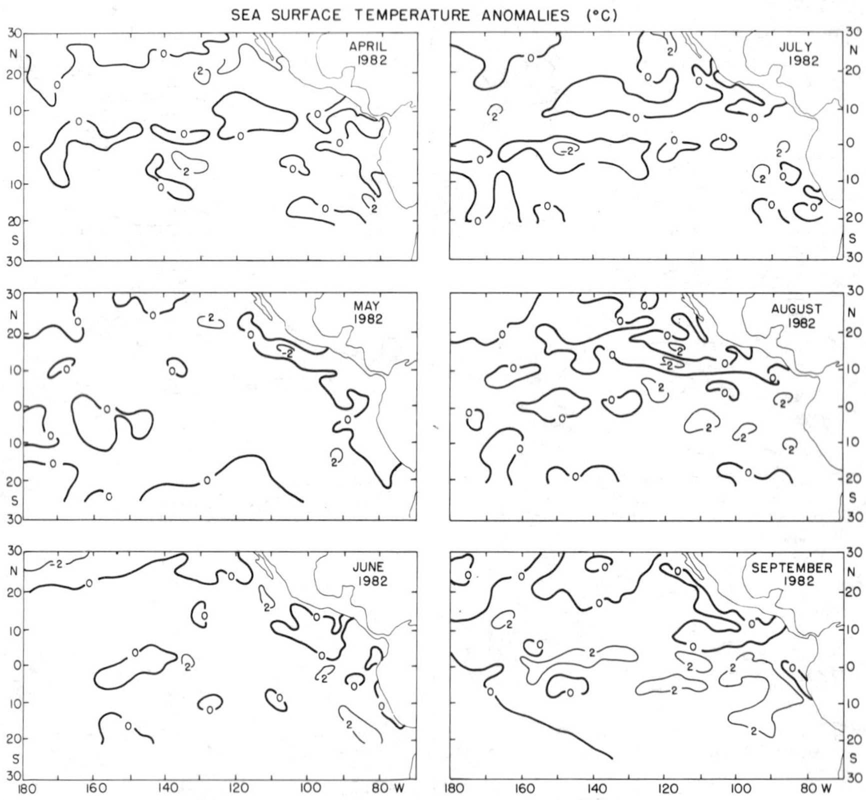

Figure 7
Time evolution of sea surface temperature anomalies from April, 1982 to September, 1982 (from Forrest Miller, Inter-American Tropical Tuna Commission, Scripps Institution of Oceanography)

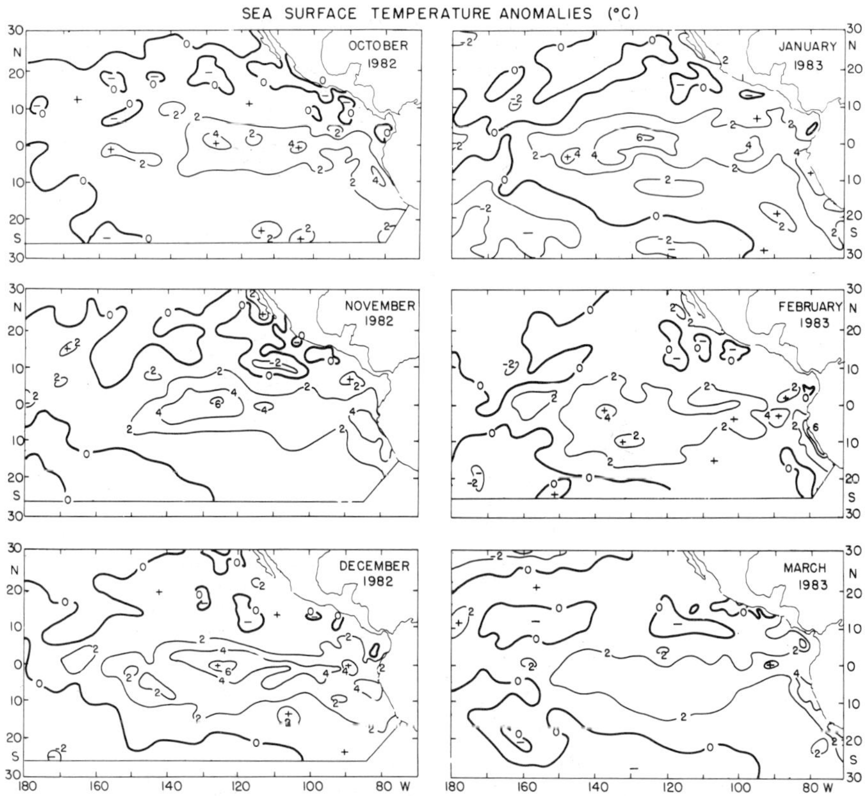

Figure 7 continued
Time evolution of sea surface temperature anomalies from October, 1982 to March, 1983 (from Forrest Miller, Inter-American Tropical Tuna Commission, Scripps Institution of Oceanography)

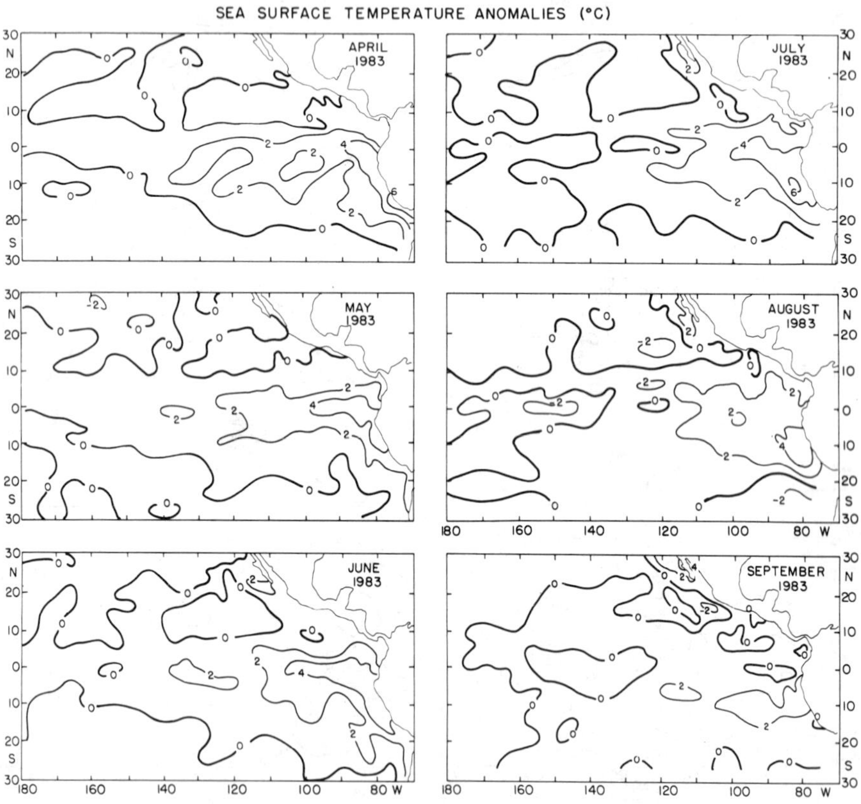

Figure 7 continued
Time evolution of sea surface temperature anomalies from April, 1983 to September, 1983 (from Forrest Miller, Inter-American Tropical Tuna Commission, Scripps Institution of Oceanography)

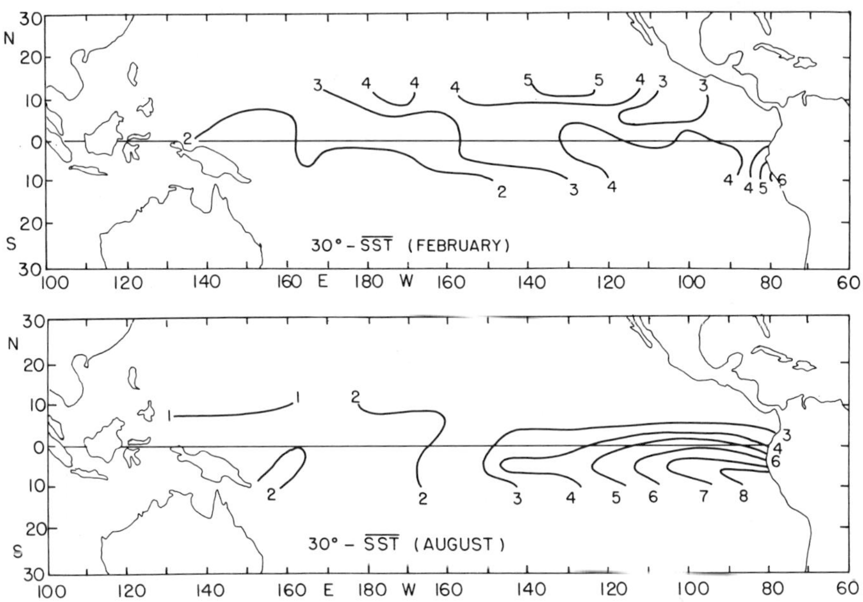

Figure 8

Maximum possible anomalies for August and February obtained by subtracting long-term mean temperature from 30°C

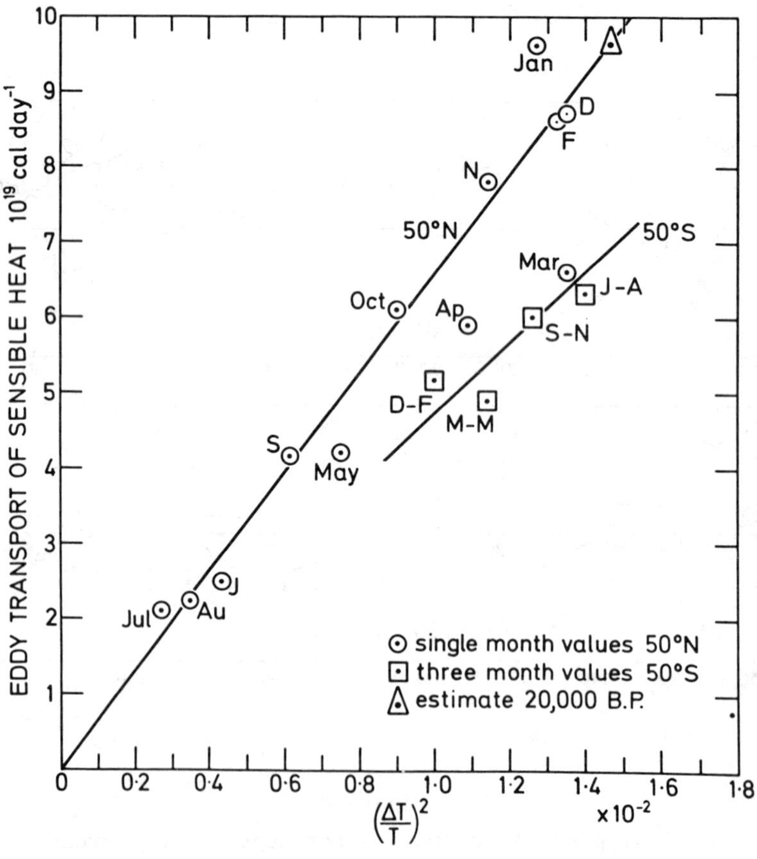

Figure 9
Eddy transport of sensible heat at 50° latitude plotted against temperature gradient at 700 mb (from Newell, 1974; Reprinted with permission)
Units in 10^{19} cal/day.
To convert to watts, multiply by 4.84×10^{-5}

F_0 increased in the Holocene associated with the warmer tropical ocean but the oceanic flux depends on both wind strength and temperature so that it is impossible to deduce its value in the past - it is even difficult to evaluate from present data.

The limit of 30°C in the low latitudes would also have prevailed in the distant past provided the solar output was the same and it is noteworthy that the concept has been used to interpret δ018 data in ice cores back to 50 My (Matthews and Poore, 1980).

Figure 10 shows the recent temperature changes in the equatorial eastern Pacific and at 35°N in the North Pacific. In this case the largest positive anomalies occurred first away from the coast in October 1982 and it was not until the spring of 1983 that large coastal anomalies occurred. Negative anomalies have persisted throughout at 35°N. Factors controlling temperature in this region have been discussed by Newell and Hsiung (1983); they conclude that enhanced cloudiness is the cause of this cool middle latitude region associated with El Niño.

4. Factors controlling atmospheric CO_2

Six years ago we presented what we termed a nutrient hypothesis for the control of atmospheric CO_2 (Newell, Navato and Hsiung, 1978). We noted that the rate of change of atmospheric CO_2, after seasonal variations and trends had been removed, following work by Bacastow (1976), was positively correlated with SST in the eastern equatorial Pacific (see Fig. 11). This region is normally a major source for CO_2 from the ocean as shown by the partial pressure maps of Keeling (1968); the North and South Pacific and the North and South Atlantic are the sink regions. For the Pacific we showed that the pattern of partial pressure difference between sea and air and the pattern of SST variability were similar (see Fig. 6 of Newell et al., 1978). We argued that a good deal of the similarity could be due to nutrient patterns. Increased upwelling in the eastern equatorial Pacific would bring up additional nutrients so that more carbon would be used in the surface layers and find its way via organisms to the sea floor. Hence less CO_2 would be exuded to the atmosphere - hence the fossil fuel increase would stand alone, or may be lower than average if take-up by the ocean continued at higher latitudes. On the other hand, decreased upwelling would result in fewer nutrients and a less extensive oceanic biosphere so that CO_2 could be exuded to the atmosphere rather than used in the surface waters. Upwelling regions are sources for CO_2 because water at depth contains more CO_2 than surface water. This source is essentially infinite as long as upwelling occurs. However, if upwelling is cut off completely (as may have happened in 1982-83) then the middle latitude sinks still operate (as indicated by the lower

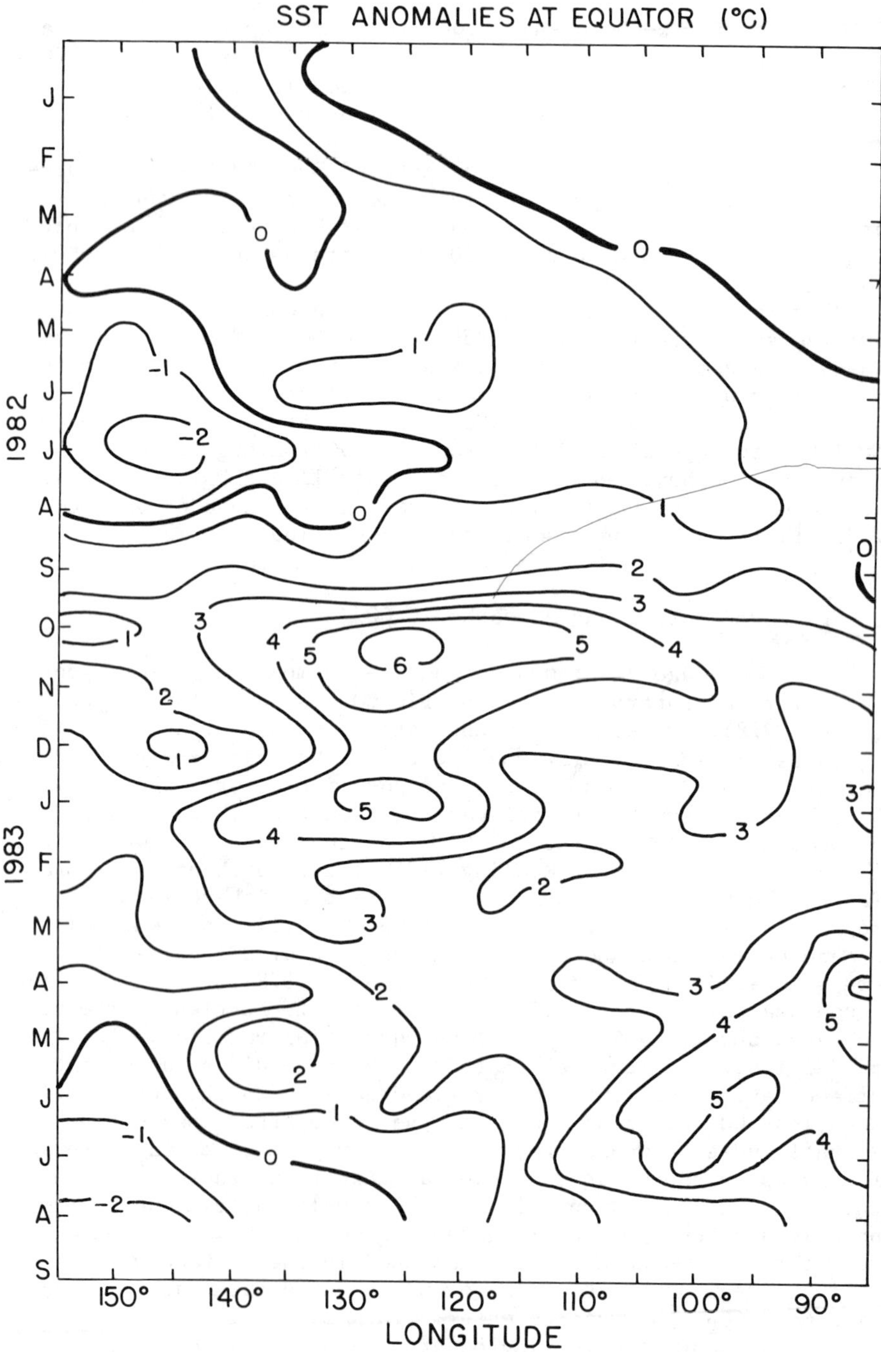

Figure 10
Recent temperature anomalies in the equatorial eastern Pacific

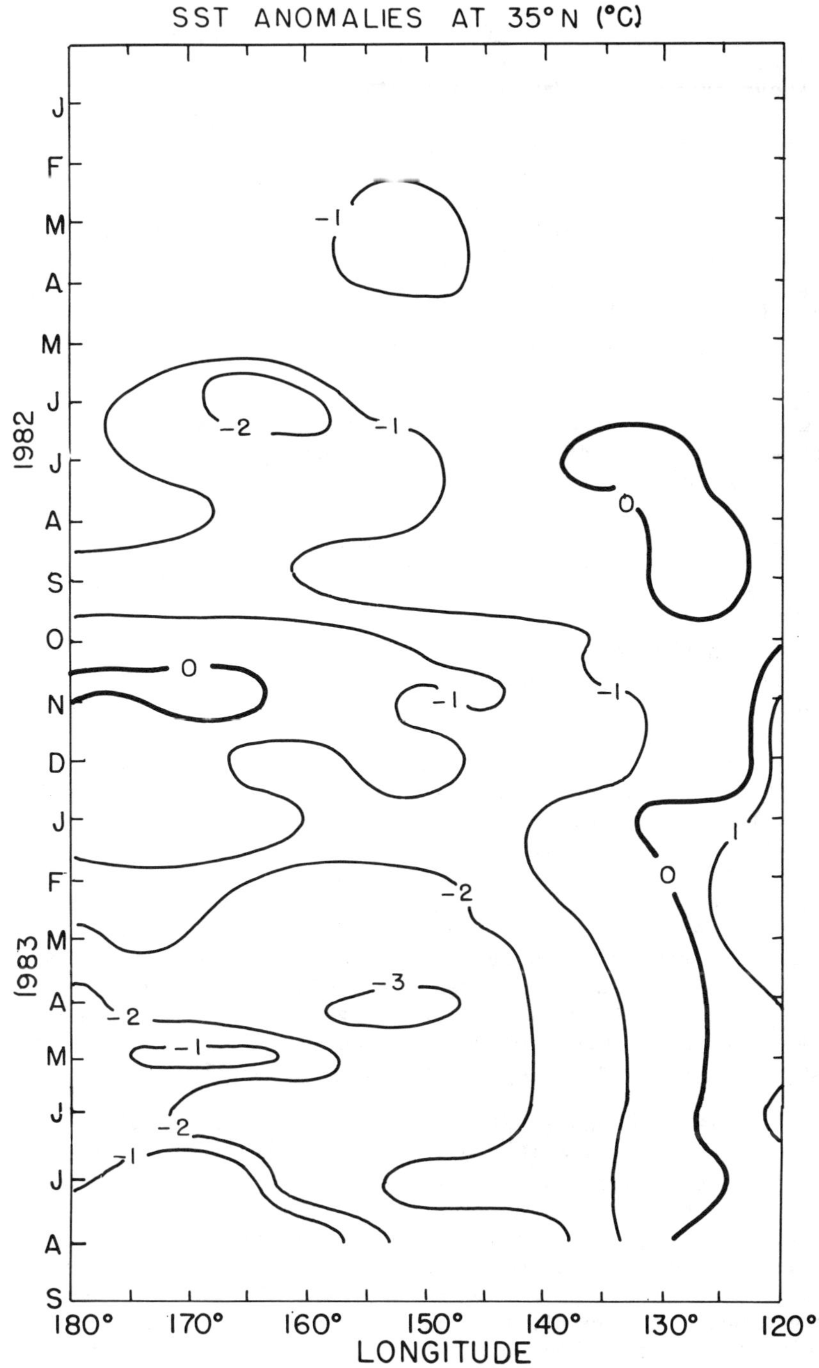

Figure 10 continued
Recent temperature anomalies at 35°N in the Eastern North Pacific

temperatures shown in Fig. 10) but there is no source of CO_2 so once surface waters are in equilibrium there is no further supply and atmospheric CO_2 can decrease.

We suggested that during the ice age, because of greater upwelling and more nutrients, CO_2 would be used up by the biosphere and that there was therefore less CO_2 in the air at this time (Newell et al., 1978). Since this suggestion, measurements have been made which show that there was less atmospheric CO_2 during the ice age (Delmas et al., 1980; Berner et al., 1980 and Neftel et al., 1982). We claim that these measurements support our hypothesis (Newell and Dopplick, 1981).

Since it was put forward, the nutrient hypothesis has received some support from Bacastow et al.(1980). Another nutrient hypothesis has been put forward by Broecker (1981, 1982) who argues that when continental shelves are uncovered in the ice age phosphate is mobilized and the ocean biosphere blooms and uses up CO_2 that would other wise have gone into the atmosphere. A third nutrient hypothesis has been made by McElroy (1983) based on consideration about nitrate.

If our hypothesis or Broecker's hypothesis is correct, CO_2 in the last ice age decreased because climate changed and not vice versa. But here again there is a non-linearity for the Holocene. Up to a point, CO_2 would increase as upwelling brought CO_2-rich water to the surface, and with few nutrients little of the carbon would be used and lost to the bottom. But a complete cut-off of upwelling would of course also cut off the CO_2 source and as waters at higher latitudes would continue to take up CO_2 until equilibrium was reached this would move towards values lower than at present. A little warmer climate, through an enhanced El Niño, could be expected to yield higher CO_2 by the nutrient hypothesis, but a really extensive El Niño and a warmer climate could yield lower CO_2 because of decreased upwelling. Yet we observe that in the Holocene CO_2 values were higher suggesting that middle latitude oceanic warming tended to reduce the middle sinks, a circumstance that did not occur in 1982-83 (Fig. 10).

CO_2 in the atmosphere presently amounts to about 710 Gt as carbon (1 Gt = 10^{12} kg = 10^9 tons). Fossil fuel injection occurs at about 5.2 Gt C per year or 2.4 ppmv/yr, with just under half of this amount remaining in the atmosphere. The steady increase due to fossil fuel injection is itself modulated by the oceanic source and sink variability. As seen from Fig. 11, the "steady" increase varies by ± 1 ppmv/yr or ± 2 Gt C per year. In 1972-73 when the eastern equatorial Pacific was warm there was an extra 2 Gt C released that was not taken up by the ocean and in 1973-74 the opposite situation occurred. In the latter case could the additional CO_2 have been taken up by the oceanic biosphere due to additional

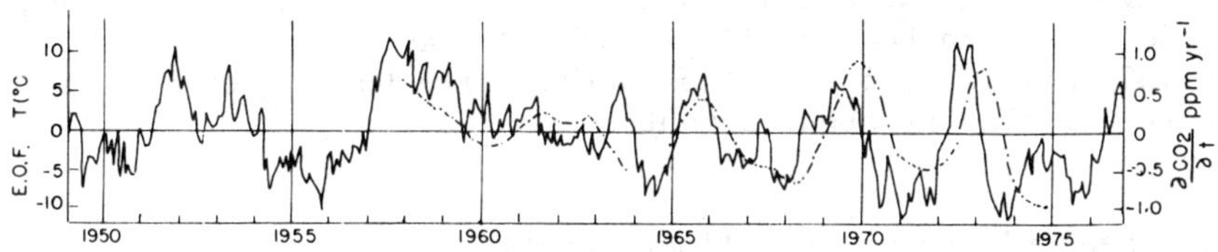

Figure 11
Pacific SST Empirical Orthogonal Function #1 time series (full line with scale on left) vs. $\partial CO_2/\partial t$ (dashed line with scale on right).
Ticks denote January of indicated year (from Newell, Hsiung and Navato, 1978)

nutrients being present in the upwelling? The oceanic biosphere contains 1.8 Gt of C and has a net primary productivity of 24.7 Gt of C per year or a turnover time of 1.8/24.7 or 27 days (Revelle and Munk, 1977). Hence it would be comparatively easy for the additional 2 Gt of C to be sequestered and removed to the depths in the course of a year. The continental biomass contains 828 Gt of C and has a net primary productivity of 52.9 Gt of C per year, or a turnover time of 15.6 years. It could not therefore be responsible for the rather rapid change in atmospheric increase. As additional support for our equatorial Pacific nutrient hypothesis, Newell and Weare (1977) report that the changes of CO_2 at Mauna Loa Hawaii lag behind changes in SST in the Equatorial Pacific by one to two months while the CO_2 change at the South Pole lag by about 5 months.

5. Sensitivity of the climate system

The sensitivity is defined as the temperature change associated with an additional energy received at the ground of 1 Wm^{-2}. Most numerical models yield values of about 1 K $(Wm^{-2})^{-1}$ when used to assess the influence of doubled CO_2 on temperature. Our analysis (Newell and Dopplick, 1979, 1981) and that by Idso (1980) yield about 0.1 K $(Wm^{-2})^{-1}$. Our value was obtained from the slope of the curve of net energy loss from the surface versus temperature; this slope is governed by the Clausius-Clapeyron relation. Based on the different sensitivities the actual temperature changes predicted for CO_2 doubling range up to 4°C on down to about 0.2°C (the work by Newell and Dopplick, 1979 and Idso, 1980); the range of values has been summarized recently by Schlesinger (1983). Gates et al. (1981) believe their value for surface air temperature change, 0.2°C, is too low because of "unrealistic suppression of water vapor and albedo temperature feedback." Mitchell (1983) also finds 0.2°C and as this is smaller than Manabe and Wetherald's (1975) results, he increases all sea temperatures in the model by 2°C and obtains a new value of 2.25°C. We believe that the 2°C artifical increase is actually physically impossible in the tropics because of the 30°C limit set by evaporation.

6. Conclusions

We make five main points:

1) Milankovitch radiation changes at high latitudes (controlling ice extent) and at middle latitudes (controlling ocean storage in summer), coupled together, can account for middle latitude temperature changes. These in turn control low latitude temperature changes through the influence of low latitude surface winds on upwelling. These joint influences on the temperature could have acted in phase both for the last ice age and the Holocene.

Thus a radiation surplus over the polar cap, and an increase in summer oceanic storage could reduce middle latitude temperature gradients in the Holocene. A concomitant decrease in westerly winds would be accompanied by a decrease in easterlies (conservation of angular momentum) and therefore of upwelling cold tropical water, leading to an increase in tropical temperatures.

2) The El Niño pattern is the main mode of recent nonseasonal variability of global SST. It results from a tendency for a decrease in upwelling in equatorial regions which leads solar radiation to heat the water locally. The heating can only progress until the sea reaches 30°C above which evaporation limits the temperature.

3) In the recent record equatorial sea surface temperature changes and atmospheric transmission changes produced by volcanic aerosol are together the main cause of tropical air temperature variability.

4) Atmospheric carbon dioxide content is controlled by nutrients in the upwelling region in the equatorial Pacific. In the Holocene with less upwelling and therefore fewer nutrients, less CO_2 is used up by organisms in the tropics. At the same time there is less take-up at middle latitudes as the sea temperatures there are also higher. This is different to the recent El Niño where we showed a cooling occurred at 35°N. Hence atmospheric CO_2 increases as shown by the ice cores. CO_2 is controlled by the climate and not *vice versa*.

5) Because ocean temperatures are buffered by evaporation to remain below about 30°C, small additional amounts of radiation in the tropics or doubled CO_2 cannot influence maximum tropical temperatures. In the Holocene there was probably an *increase in area* of the 30°C tropical ocean.

Acknowledgements

Support for this work was received from the Climate Dynamics Section of the National Science Foundation under Grant ATM-8211645 and the United States Department of of Energy under Contract DE-AC02-76EV12195. We are grateful to Dr. Tamara Ledley for use of her Milankovitch solar radiation calculations as given in Table 2.

References

Bacastow, R.B., 1976: Modulation of atmospheric carbon dioxide by the Southern Oscillation, *Nature*, *261*, 116-118.

Bacastow, R.B., J.A. Adams, C.D. Keeling, D.J. Moos, T.P. Whorf and C.S. Wong, 1980: Atmospheric carbon dioxide, the Southern

Oscillation, and the weak 1975 El Niño, Science, 210, 66-68.
Berner, W., H. Oeschger and B. Stauffer, 1980: Information on the CO_2 cycle from ice core studies, Radiocarbon, 22, 227-235.
Broecker, W.S., 1981: Glacial to interglacial changes in ocean and atmosphere chemistry, In Climatic Variations and Variability: Facts and theories. Ed. A. Berger. D. Reidel Pub. Co., Dordrecht, Holland, 111-121.
Broecker, W.S., 1982: Glacial to interglacial changes in ocean chemistry, Prog. Oceanog., 11, 151-197.
Delmas, R.J., J.M. Ascencio and M. Legrand, 1980: Polar ice evidence that atmospheric CO_2 20,000 BP was 50% of present, Nature, 284, 155-157.
Flint, R.F., 1971: Glacial and Quaternary Geology, John Wiley and Sons, Inc., 892 pp.
Gates, W.L., 1976: The numerical simulation of ice-age climate with a global general circulation model, J. Atmos. Sci., 33, 1844-1873.
Gates, W.L., K.H. Cook and M.E. Schlesinger, 1981: Preliminary analysis of experiments on the climatic effects of increased CO_2 with an atmospheric general circulation model and a climatological ocean, J. Geophys. Res., 86, 6385-6393.
Gunther, E.B., R.L. Cross and R.A. Wagoner, 1983: Eastern North Pacific tropical cyclones of 1982, Mon. Wea. Rev., 111, 1080-1101.
Hastenrath, S. and J.E. Kutzbach, 1983: Paleoclimatic estimates from water and energy budgets of East African lakes, Quat. Res., 19, 141-153.
Hsiung, J. and R.E. Newell, 1983: The principal non-seasonal modes of variation of global sea surface temperature, J. Phys. Ocean., 13, (in press).
Idso, S.B., 1980: The climatological significance of a doubling of earth's atmospheric carbon dioxide concentration, Science, 207, 1462-1463.
Manabe, S. and R.T. Weatherald, 1975: The effects of doubling the CO_2 concentration on the climate of a general circulation model, J. Atmos. Sci., 32, 3-15.
Manabe, S. and D.G. Hahn, 1977: Simulation of the tropical climate of an ice age, J. Geophys. Res., 82, 3889-3911.
Matthews, R.K. and R.Z. Poore, 1980: Tertiary δO^{18} record and glacio-eustatic sea-level fluctuations, Geology, 8, 501-504.
McElroy, M.B., 1983: Marine biological controls on atmospheric CO_2 and climate, Nature, 302, 328-329.
Mitchell, J.F.B., 1983: The seasonal response of a general circulation model to changes in CO_2 and sea temperatures, Quart. J.R. Met. Soc., 109, 113-152.
Neftel, A., H. Oeschger, J. Schwander, B. Stauffer and R. Zumbruhn, 1982: Ice core sample measurements give atmospheric CO_2 content during the past 40,000 years, Nature, 295, 220-223.

Newell, R.E., 1970: Stratospheric temperature change from the Mt. Agung volcanic eruption of 1963, J. Atmos. Sci., 27, 977-978.

Newell, R.E., 1974: Changes in the poleward energy flux by the atmosphere and ocean as a possible cause for ice ages, Quat. Res., 4, 117-127.

Newell, R.E., 1979: Climate and the ocean, American Scientist, 67, 405-416.

Newell, R.E., 1981: Further studies of the atmospheric temperature change produced by the Mt. Agung volcanic eruption in 1963, J. Volcanology and Geothermal Research, 11, 61-66.

Newell, R.E. and B.C. Weare, 1976a: Ocean temperatures and large scale atmospheric variations, Nature, 262, 40-41.

Newell, R.E. and B.C. Weare, 1976b: Factors governing tropospheric mean temperature, Science, 194, 1413-1414.

Newell, R.E. and T. Dopplick, 1979: Questions concerning the possible influence of anthropogenic CO_2 on atmospheric temperature, J. Appl. Meteor., 18, 822-825.

Newell, R.E. and T. Dopplick, 1981: Reply to Robert G. Watts' "Discussion of 'Questions concerning the possible influence of anthropogenic CO_2 on atmospheric temperature'", J. Appl. Meteor., 20, 114-117.

Newell, R.E., G.J. Boer, J.W. Kidson, and D.G. Vincent, 1974: The General Circulation of the Tropical Atmosphere and Interactions with Extra-tropical Latitudes, Vol. 2, M.I.T. Press, 371 pp.

Newell, R.E., G.F. Herman, S. Gould-Stewart, and M. Tanaka, 1975: Decreased global rainfall during the past Ice Age, Nature, 253, 33-34.

Newell, R.E., J. Hsiung, and A.R. Navato, 1978: Long-term global sea surface temperature fluctuations and their possible influence on atmospheric CO_2 concentrations, Pure and Applied Geophysics, 116, 351-371.

Newell, R.E., J.C. Chung and S. Gould-Stewart, 1981: A possible interpretation of paleoclimatic reconstructions for 18,000 BP for the region 60°N to 60°S, 60°W to 100°E, Paleoecology of Africa, 13, 1-19.

NOAA, 1983: Climate Diagnostics Bulletin, No. 83/5, NOAA, NMC, Climate Analysis Center, Washington, D.C., p.4.

Revelle, R. and W. Munk, 1977: The carbon dioxide cycle and the biosphere, In Energy and Climate, National Academy of Sciences, Washington, D.C., 140-158.

Schlesinger, M.E., 1983: A review of climate models and their simulation of CO_2 induced warming, Intern. J. Envir. Studies, 20, 103-114.

Street, F.A. and A.T. Grove, 1979: Global maps of lake-level fluctuations since 30,000 years BP, Quat. Res., 12, 83-118.

THE EFFECT OF ATMOSPHERIC AND OCEANIC ENERGY TRANSPORTS IN CLIMATIC BALANCE

Hannu Savijärvi

Department of Meteorology
University of Helsinki
Finland

ABSTRACT. The poleward transport of energy by the atmosphere and by the oceans plays a central role in the earth's climate. Calculations based on one month of global atmospheric observations (ECMWF analyses) are shown for the transports of sensible and latent heat and kinetic and geopotential energy. Their sum, the total energy transport by the atmosphere, is compared with satellite information of the radiational imbalance of the earth-atmosphere system. The accuracy of the calculations, the energy transport by the oceans and the eddy diffusivity indicated by the "observed" transports is discussed.

The effect of the heat transport is further studied in a simple energy balance climate model (Budyko-Sellers type EBM). With the present climate parameter values the simple model climate is found to be more sensitive to feedback effects in the net heating than to the energy transport. The eddy diffusivity needed for a steady state is found to be slightly higher than previously accepted, but the higher value is consistent with the "observed" data calculations.

1. ENERGY TRANSPORT CALCULATIONS BASED ON OBSERVATIONS

The climate is the (slowly changing) balance between energy input, output and energy transports in the earth-atmosphere system. If the evolution equations which govern the energy forms is the ocean-soil-

N.-A. Mörner and W. Karlén (eds.), Climatic Changes on a Yearly to Millennial Basis, 563–570.

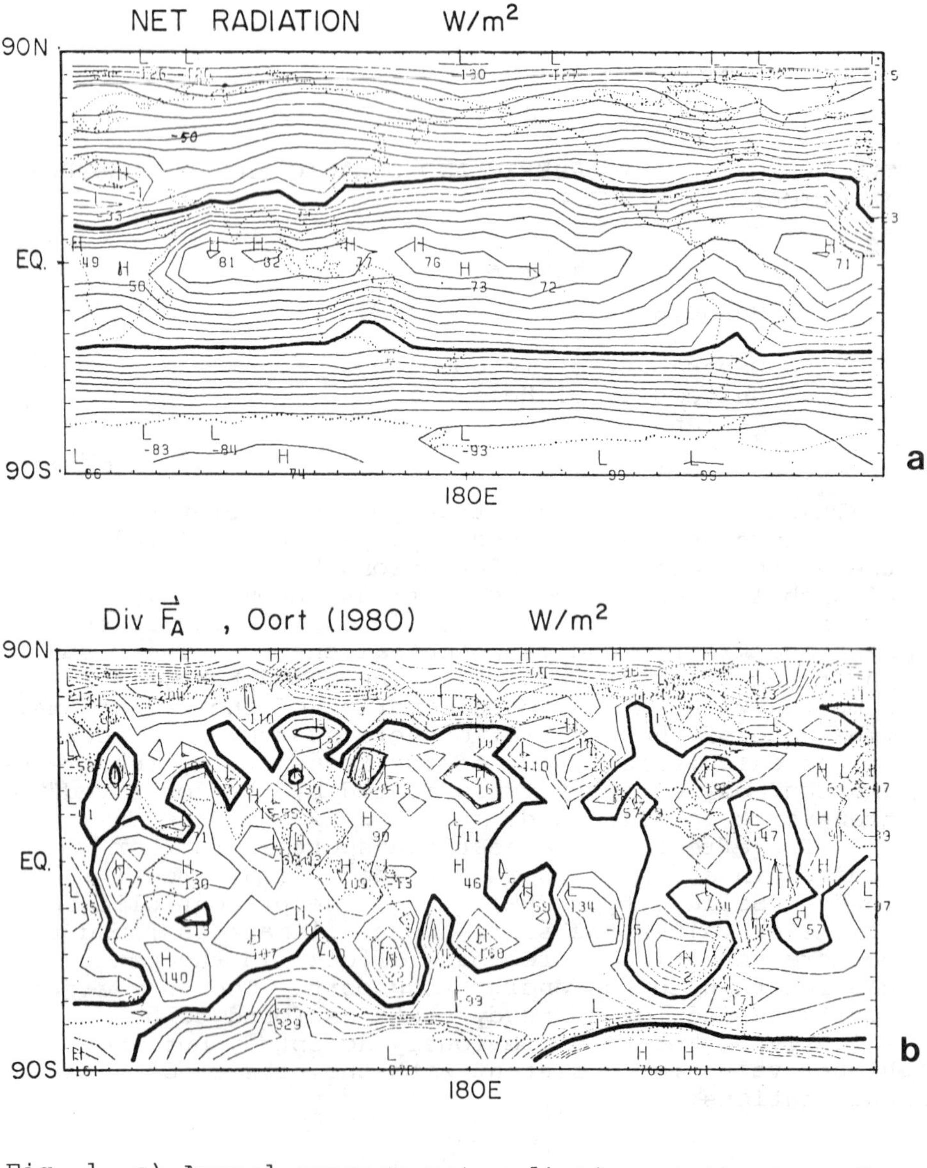

Fig. 1. a) Annual average net radiation at the top of the atmosphere from satellites. From CAGE-report (1982). Error 10 W/m^2.
b) The annual average divergence of atmospheric energy transport. From CAGE-report (1982).

cryosphere column and in the atmospheric column are integrated vertically, the sum is the equation of total energy (Oort and Vonder Haar, 1976):

$$S_a + S_o = -\nabla \cdot (\underline{F}_a + \underline{F}_o) + R_{ea} \tag{1}$$

This equation states that the energy storage S_a+S_o (the local rate of change of energy in the atmospheric and oceanic column) equals the local net energy flux convergences due to transports in the atmospheric and oceanic currents plus the net energy flux R_{ea} through the vertical boundary. R_{ea} is physically the net radiation at the top of the atmosphere and can be measured by satellites. Fig. 1a shows R_{ea} based on two years of recent satellite data. The standard deviation (error) is in this data about 10 W/m^2. The energy input/output imbalance in Fig. 1a, surplus in the tropics, deficit in the poles, calls for ventilation in the form of oceanic and atmospheric currents.

In a long time mean the storage terms are negligible and over land the oceanic transport disappears. The atmospheric energy is the sum of sensible (c_pT) and latent (Lq) heat, and geopotential (ϕ) and kinetic energy ($\frac{1}{2}\underline{V}^2$). Over land the atmospheric energy flux divergence should equal the net radiation:

$$\nabla \cdot \underline{F}_a = \nabla \cdot \int_0^{p_s} (c_pT + Lq + \phi + \tfrac{1}{2}\underline{V}^2)\,\underline{V}dp/g \quad (=R_{ea} \text{ over land}) \tag{2}$$

$\nabla \cdot \underline{F}_a$ can in principle be calculated from radiosonde data. Such a calculation is shown in Fig. 1b and the result clearly disagrees with the net radiation data from the same two year period with an error of ± 200 W/m^2 over land.

The local disagreement between the satellite and radiosonde data in Fig. 1 is probably due to:

1) difficulties in calculating wind divergence from sonde data
2) mass imbalance due to the wind divergence errors.

Fig. 2 shows the atmospheric energy flux divergence based on only one month (December 1980) of data from ECMWF objective analyses of wind, temperature and humidity. The mid-latitude wind divergence field is probably rather well represented in these analyses due to the normal mode initialization procedure applied. The mass balance has been corrected; this amounts up to ± 500 W/m^2 locally. The calculations are extensions of those in Savijärvi (1982, 1983). The corresponding net radiation field for winters can be taken from Stephens et al. (1981). The deviation over land between the two calculations is of the order of only

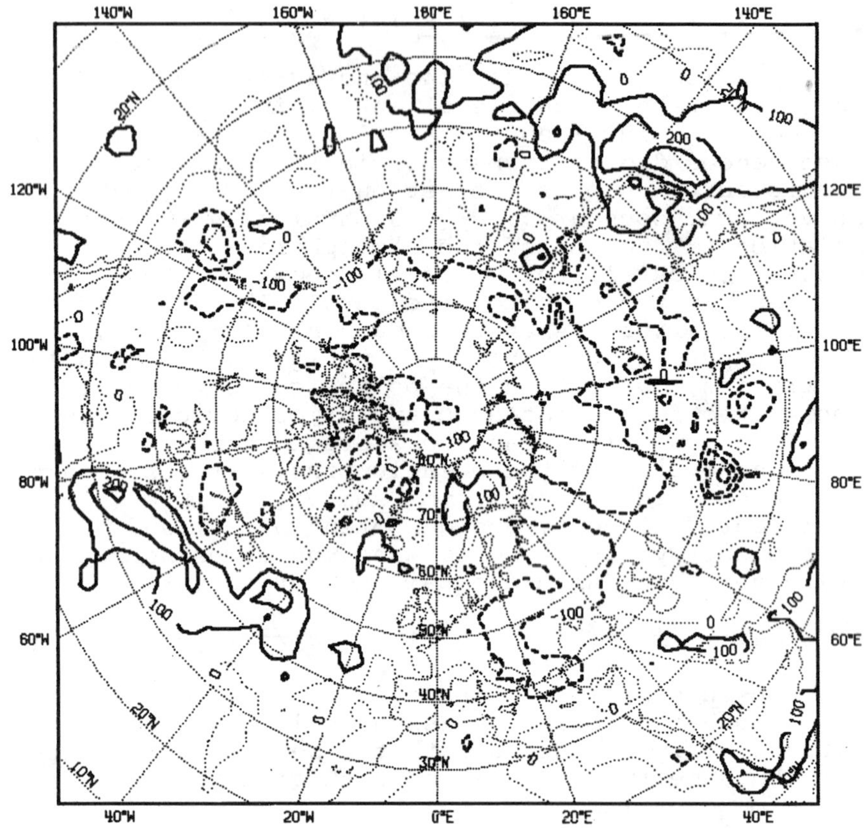

Fig. 2. Divergence of atmospheric energy transport. December 1980 ECMWF analyses (00 and 12 GMT data). Unit W/m^2.

$\pm$100 W/m^2 and probably less in a longer time series. It thus seems that initialization and mass balance correction are essential for local energy flux calculations.

According to Fig. 2 the oceanic heat flux convergence estimated as the somewhat inaccurate residual between $\nabla \cdot \underline{F}_a$ and R_{ea} is very large in the wintertime warm ocean currents (the Gulf Stream and Kuroshio) amounting up to +400 W/m^2. Note the strong oceanic heating indicated in Fig. 2 near Iceland and north of Scandinavia.

If the energy fluxes are assumed to be described by a diffusive transport mechanism, in which the fluid motions tend to smooth out energy (temperature) differences, then

$$\underline{F}_a+\underline{F}_o \sim -D\nabla T_s; \quad -\nabla\cdot(\underline{F}_a+\underline{F}_o) \sim D\nabla^2 T_s$$

where T_s is surface temperature and D a diffusivity coefficient. Former climatological data (e.g. Sellers, 1965) suggested for D a value of 0.7 W/m^2K (correlating e.g. R_{ea} with $\nabla^2 T_s$). The present data in Fig. 2 as well as the satellite data for R_{ea} suggests a slightly larger value of $D \sim 1.0\ W/m^2K$, making the atmosphere-ocean fluid a slightly more effective energy transport agent than previously thought. The importance of the suggested increase in D will now be studied with a simple climate model.

2. ENERGY BALANCE CLIMATE MODEL RESULTS

The sensitivity of the earth's climate to various mechanisms in the energy input-output-transport system has been studied extensively with climate models. The most simple of these are the energy balance models pioneered by Budyko and Sellers. Energy balance models (EBM) are zonally averaged simple differential equations for surface temperature (cf. eq. (1)):

$$C\frac{\partial T_s}{\partial t} = D\nabla^2 T_s + Q(1-\alpha) - \beta\sigma T_s^4 \quad . \quad (3)$$

Here C is heat capacity, Q solar constant (340 W/m^2), α the albedo and β the absorptivity of the earth-atmosphere system. The last two terms describe the solar heat input and longwave heat output radiation and include crudely, depending on the formulation of $\alpha = \alpha(T_s)$ and $\beta = \beta(T_s)$, the ice-albedo and temperature-absorptivity (greenhouse effect) feedbacks. The most important - and alarming - result in the early EBM experiments has been the possibility of two steady states, the "ice age" steady state being excited in the original Budyko-Sellers experiments by only a 2% decrease in the solar radiation. For a review of EBM's, see North et al (1981).

The energy balance model can be used to study the sensitivity of the climate to changes in the diffusivity D. As a first step, the net energy source

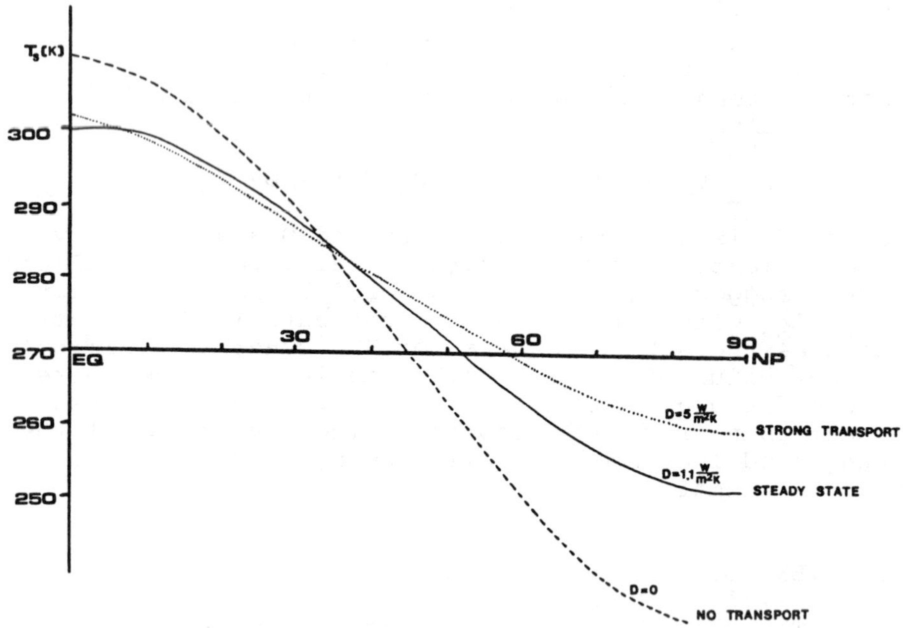

Fig. 3. The surface temperature as the function of latitude in an energy balance climate model, after 1000 years integration. Heat sources fixed to Fig. 1a values. Three diffusivities used as shown. The steady state curve ($D=1.1$ W/m^2) also shows the initial (present) temperature distribution.

$Q(1-\alpha) - \beta\sigma T_s^4$ was replaced by the zonal average of R_{ea} from Fig.1a and the model with the energy sources/sinks thus fixed to the present day values was integrated 1000 years with three diffusivity values D, starting from the present latitudinal distribution of T_s. The heat capacity of the ocean-atmosphere-earth was $C=10^9$ J/m^{2}K.

The result, in Fig. 3, shows that with no transports (D=0) the temperatures quickly rise in the tropics and sink deep to the ice age in the poles. With a strong diffusion ($D=5$ W/m^{2}K) the poles become warmer than presently but the tropical temperatures do not change much. The steady state of the model (i.e. no temperature change at all) was obtained with a value of $D=1.1$ W/m^{2}K, in agreement with calculations of D from the observed data presented in the previous section.

Some sensitivity calculations with the feedback effects included in an EBM have been reported (e.g. Oerlemans and van den Dool, 1978). Their results give the steady state mean surface temperature as the function of solar constant with three values of D and three values of outgoing radiation parametrization (Figs. 11 and 10 respectively in Oerlemans & van den Dool, 1978). The results indicate that the model is not very sensitive to small variations of D: with D=0.7 W/m^2K the drop of T_s (bifurcation) to the colder steady state takes place if Q decreases 9% while with D=1.0 W/m^2K the same climate change would take place with a drop of 8% in the solar radiation.

The results indicate further that the model is more sensitive to changes in the absorption-temperature feedback mechanism than to the diffusion.

3. CONCLUDING REMARKS

With new satellite data the radiation of the earth-atmosphere system is known with an accuracy of 10 W/m^2 as along time average. The meteorological observation-analysis schemes are now such that the atmospheric energy transport part is known locally with an accuracy of 100 W/m^2 on a monthly level. The error is probably smaller in a longer time series. A good initialization scheme and mass balance correction are, however, essential for the atmospheric transport calculations.

The total (atmospheric + oceanic) transport, taken as a diffusive mechanism, gives a diffusion coefficient which is 0.9-1.0 W/m^2K (higher than previously thought). This value was also needed to keep a Budyko-Sellers type energy balance climate model in a steady state with the radiative forcing fixed to the satellite observations. The models are more sensitive to the feedback effects in the radiation than to the diffusive transport mechanism. This may indicate that the fluid dynamics is a passive factor in climatic changes of a planet, compared to direct radiative effects, such as CO_2-warming through greenhouse effect.

REFERENCES

CAGE-experiment: A feasibility study. Report of the JSC/CCCO. WCP-22, 1982. WMO, Geneva, 95 pp.

North, G.R., Cahalan, R.F. and Coakley, J.A.,Jr., 1981: Energy balance climate models. Rev.Geophys. Space Phys. 19, 91-122.

Oerlemans, J. and H.M. van den Dool, 1978: Energy balance climate models: Stability experiments with a refined albedo and updated coefficients for infrared emission. J.Atmos.Sc. 35, 371-381.

Oort, A.H. and Vonder Haar, T.H., 1976: On the observed annual cycle in the ocean-atmosphere heat balance over the Northern Hemisphere. J.Phys.Oceanogr. 6, 781-800.

Savijärvi, H.I., 1982: The mass balance in diagnostic studies: an example of analysed and forecast data calculations. Tellus 34, 540-544.

Savijärvi, H.I., 1983: The atmospheric energy budgets over North America, the North Atlantic and Europe based on ECMWF analyses and forecasts. Tellus 35A, 39-50.

Sellers, W.D., 1965: Physical climatology. The University of Chicago Press. 272 pp.

Stephens, G.L., G.G. Campbell and T.H. Vonder Haar, 1981: Earth radiation budgets. J.Geophys.Res. 86, 9739-9760.

POSSIBLE CO_2-INDUCED WARMING EFFECTS ON THE CRYOSPHERE

Roger G. Barry

Cooperative Institute for Research in Environmental Sciences and Department of Geography, University of Colorado, Boulder

1. INTRODUCTION

Over the last few years, considerable attention has been given to the possible climatic effects of increasing atmospheric carbon dioxide concentrations. The present figure of 340 p.p.m. by volume is 15 to 20 percent above pre-industrial levels and a concentration of double the assumed 1860 value of 295 p.p.m. has been projected by the second half of the next century due to continued burning of fossil fuels (Laurmann and Rotty, 1983). Climate modelling simulations suggest that such a doubling will cause a global mean surface air temperature rise of about 3.0 ± 1.5K (National Research Council, 1982), with at least a two-to three-fold amplification in high latitudes. Hansen et al. (1981) estimate that the post-1890s CO_2 increase should already have caused a 0.5°C warming, but allowing for oceanic heat capacity and mixing effects this would be reduced to 0.2°C. Further modifications of climatic regimes will occur through changes in precipitation and circulation characteristics, although these are less well known (Schlesinger, 1983).

Preliminary assessments of the possible impacts of these projected climatic perturbations on global snow and ice cover have already been reported (Barry, 1978; Hollin and Barry, 1979). This study updates those reports and attempts to develop a composite picture from modelling results, analytical projections, and analogs based on other "warm" climatic intervals. Other analyses have focused primarily on the possible disappearance of the Arctic pack ice (Parkinson and Kellogg, 1979; Flohn, 1982) and on the possible disintegration of the West Antarctic ice sheet and its consequences for global sea level

N.-A. Mörner and W. Karlén (eds.), Climatic Changes on a Yearly to Millennial Basis, 571–604.

(Mercer, 1978; Thomas et al., 1979; Young, 1981; MacDonald, 1982; Bentley, 1983; Schneider and Chen, 1980). Other potential effects have received much less attention.

The components of the cryosphere are summarized in Table 1. Perennial ice covers 7 percent of the world oceans and 11 percent of the land surface, almost entirely in the polar regions. Seasonal snow and ice, however, occupies 15 percent of the earth's surface in January, at a time when Antarctic sea ice is near its minimum extent, and 9 percent in July when there is almost no snow cover in the Northern Hemisphere. The volume of water locked-up in the Antarctic and Greenland ice sheets represents 77 m of water potentially available to raise the world's mean sea level.

2. THE NATURE OF CO_2 EFFECTS ON THE ATMOSPHERE-OCEAN-CRYOSPHERE SYSTEM

An increase of atmospheric CO_2 content will mainly affect the cryosphere indirectly through meteorological factors. The primary effects involve modifications in the atmospheric radiation balance and air temperature. Radiative absorption by CO_2 is centered in the 15 μm band, although other CO_2 absorption bands contribute 30 percent to the computed surface temperature change (Augustsson and Ramanathan, 1977). However, the direct radiative component accounts for only a small part of the total surface warming due to CO_2 effects according to Ramanathan (1981); ninety percent of the predicted change is due to feedback effects. For example, the greenhouse effect of CO_2 is strongly reinforced by atmospheric water vapour. Greenhouse effects could be augmented further by increases in trace gases such as NO, CH_4, NH_3 and CCl_2F_2 (Wang et al., 1976). In low latitudes the tropospheric warming due to absorption by water vapor of the increased CO_2 infrared emission greatly outweighs direct radiative heating of the surface, whereas the reverse is true in high latitudes (Ramanathan et al., 1979; Crane, 1981). Nevertheless, the high-latitude amplification particularly in spring-early summer is primarily caused by the feedback between ice/snow cover and surface temperature. Schlesinger (1983) discusses the complexity of this question for the various GCM simulations of increased CO_2.

Choudhury and Kukla (1979) postulate that a CO_2-increase could reduce the shortwave near-IR absorption by several percent in the case of a snow surface, thereby delaying the onset of melt. Further investigation of this possibility is required although the net effect seems likely to be minor.

Table 1. Components of Global Ice and Snow Cover: Area, Volume and Sea Level Equivalent (after Hollin and Barry, 1979, with modifications).

	Area	Volume	Sea-level Equivalent
	10^6 km^2	10^6 km^3 water	m
Land ice:			
Antarctica[1, 2]	12.2	25	70
Greenland	1.8	2.7	7
Small ice caps and mountain glaciers	0.5	0.12	0.3
Ground ice (excluding Antarctica)[3]:			
Continuous	7.6	0.02	0.01
Discontinuous	17.3	0.01	0.01
Sea ice:			
Arctic: maximum	15		
Arctic: minimum	5		
Antarctic: maximum	20		
Antarctic: minimum	2.5		
Total land ice, sea ice, and snow:			
Jan: N. Hemisphere	58		
Jan: S. Hemisphere	18		
July: N. Hemisphere	14		
July: S. Hemisphere	25		
Global mean annual	59		

[1] Excludes peripheral, floating ice shelves (which do not affect sea-level).

[2] Roughly 10 percent of the Antarctic ice is in West Antarctica and 90 percent is in East Antarctica.

[3] The estimated ice volume in the permafrost areas of the USSR (6.96×10^6 km^2) is given as 19,000 km^3 by Vtyurin (1973). This has been used as the basis for calculating the water content of the continuous permafrost; ice volume in the discontinuous area has been weighted by a factor of 1/4.

The first modelling studies of the warming associated with a CO_2 increase evaluated changes in mean annual temperature values, but for purposes of assessing effects on snowfall, snowcover extent and duration, etc., it is essential to know the seasonal and geographical patterns and any changes in diurnal and annual ranges of temperature. Indications of these effects can best be obtained, at present, through the combined results of modelling studies, climatic scenarios, and historical or geological analogs.

Less-well defined are the changes in atmospheric water vapor content, cloudiness, and precipitation, resulting either directly from changes in atmospheric lapse rates, or indirectly from changes in atmospheric circulation and from feedbacks involving the surface hydrologic cycle. A simulation of CO_2 doubling and quadrupling effects with the GFDL general circulation model by Manabe and Wetherald (1980) incorporating an interactive lapse rate parameterization, an ocean swamp surface, a hydrologic cycle and cloud/radiation feedback shows a high-latitude increase in precipitation and runoff, as well as the retreat of the snow/ice limit. The latter gives rise to a 30-40 percent decrease in albedo at about 75°-85° N. Cloud feedback appears to have little effect on overall model sensitivity, but no seasonal variation was included in the model.

Later work by Manabe and Stouffer (1980) with a model version including seasonal solar radiation variations, realistic geography, and a simple ocean mixed-layer, for a quadrupling of CO_2, shows strong seasonal differences in warming effects. For their model, the Arctic Basin experiences the greatest effect in early winter, thereby reducing the annual temperature variation, with a secondary warming in April around 65° N. The sea ice cover is reduced in extent and thickness and snowmelt occurs earlier than in the control case. However, the experiment is still not fully realistic in several respects, particularly with regard to prescribed clouds and surface albedos. Analyses of cloud formation and radiation processes in the Goddard Laboratory for Atmospheric Science GCM suggest that greenhouse effects which warm the atmosphere produce less layered cloud (Herman et al., 1980). This type of feedback would result in opposing effects on the Arctic surface regime, giving cooling in winter and warming in summer.

Since geographical and seasonal climatic effects are still not determined in reliable detail through such experiments, these characteristics must also be inferred from scenario studies based on climatic data for warm seasons/years in high latitudes. Such analyses are described by Drozodov (cited by Lamb, 1971), Namias (1980), Wigley _et al._ (1980) and Jager and

Table 2. Estimated atmospheric CO_2 concentrations in the Tertiary (from Budyko and Yefimova, 1981) and excess of global annual mean temperatures above present.

Period	Paleogene	Eocene	Oligocene	Miocene	Pliocene
Age (10^6 yr)	66-58	58-37	37-25	25-9	9-2
CO_2 (ppm)	1200	1850	500	1100	550
ΔT (°C)	7.5	9.1	4.3	6.9	3.7

Kellogg (1983) for the northern hemisphere and Pittock (1980) for Australia. However, a comparison of the northern hemisphere results shows some regional differences according to the years and periods of record chosen (Sear and Lough, 1981). Moreover, these warm analogs are undoubtedly not attributable to CO_2 and the causes of each may have been different. The Hypsithermal interval in the mid-Holocene was probably related to increased solar radiation, although with a delayed response due to oceanic inertia and the slow melting of the residual ice sheet in North America. In contrast, the mid-20th century warming in the northern hemisphere appears to reflect a complex response to decreased volcanic aerosols, increased CO_2 effects, and changes in atmospheric circulation intensity and patterns (Hansen, et al., 1981; Gilliland, 1982; Vinnikov and Groisman, 1982; MacCracken, 1983). For this reason, any attempt to predict regional patterns of cryospheric response from such climatic distributions would be premature at this time.

Interestingly, analysis of an ice core recently taken at Dye 3 in southern Greenland, shows fluctuations in CO_2 concentration of 50-70 p.p.m. occurring within only a few hundred years during the Last Glaciation about 30,000-40,000 B.P. (Stauffer et al., in press). It is suggested that these may be a result of changes in ocean circulation rate affecting the carbonate chemistry. Stauffer et al. conclude that external causes produced changes in climate and CO_2 with the CO_2 contributing a positive feedback effect. The confirmation of such findings will provide major new insights into the CO_2 - climate linkages.

It has been proposed that the late Tertiary era may represent a true warm analog for a future CO_2-induced "super-interglacial." Flohn (1982) discusses the apparent global asymmetry of polar glaciation during this period, assuming an ice-free Arctic Ocean, but this is contested by Clark, 1982, 1983). Budyko and Yefimova (1981) estimate from inferred rates of carbonate deposition that atmospheric CO_2 levels were larger than now by a factor of between 2 and 10 during the Tertiary. Table 2 summarizes their calculations. However, the CO_2 estimates are speculative and the continental locations and ocean circulations were also increasingly different from now before about 3-5 million years ago. Maps of temperature departures from modern January and July values and annual precipitation anomalies for the northern hemisphere during the Pliocene "optimum" have been prepared (Grigoreva, 1982), but such reconstructions must be considered schematic in view of the uncertain temporal dating and limited spatial coverage of data.

Secondary effects on the cryosphere due to a global warming will involve changes in oceanic parameters. Such changes could be caused by local modifications of the surface energy balance and air temperature, or by changed rates of vertical mixing associated with modified salinity gradients. The latter could result from changes in continental runoff, as well as from feedback effects on the surface salinity if sea-ice cover were reduced. The treatment of ocean conditions in the GFDL experiments is much simplified since the ocean model ignores vertical mixing and horizontal advection. The reduced meridional atmospheric temperature gradient indicated by the GFDL results should weaken the atmospheric and ocean circulations, thereby tending towards decreased high-latitude warming (i.e., a negative feedback effect, cf. Kellogg, 1983).

3. CRYOSPHERIC RESPONSES

The response of snow and ice covers to climatic factors varies greatly in terms of time scale. Residence times of solid precipitation in the various reservoirs are approximately: <1 year for seasonal snow cover, 1-10 years for sea ice, and 10^3 - 10^5 years for ground ice and ice sheets. In each case certain phases of the seasonal regime are particularly critical for the occurrence of snow or ice and their response to climatic variations. Of primary importance are the times of seasonal temperature transition across the 0°C threshold (-1.8°C in the case of sea water). Other threshold effects influencing energy exchanges result from the large albedo differences between snow cover (about 0.80) and snow-free ground (0.10-0.25) (Kukla, 1981), or between ice (0.65) and water (0.05-0.10). However, these differences are greatly reduced by cloud cover in so far as the planetary (surface + atmosphere) albedo is concerned. Cloudiness itself is also significantly affected by the presence of snow and ice margins in terms of both cyclone systems and boundary layer stratus (Barry, this volume).

The time scales of natural variability of the individual cryospheric parameters are summarized in Table 3, together with an indication of the factors involved and the space scale of the variations. Clearly, many more environmental factors are involved in the slower-responding phenomena so that it will be less easy to identify possible CO_2-induced effects. Nevertheless, it is important to monitor many of these phenomena from the standpoint of the critical significance of any possible changes which may be detected in them. This is obviously the case with the extent of arctic sea ice and the stability of the Ross Ice Shelf. Each snow and ice component is now discussed in turn.

Table 3. Characteristics of Snow and Ice Variables and Factors Involved in Their Variability.

Parameter	Main Time Scale of Variation	Space Scale	Meteorological Factors	Other Factors
Snowfall/ Snow depth	Days	Point	Weather systems; air temp. <1°C	
	Annual	Point	Air temperature, freq. of cyclones	
Snow-cover extent	Days	10^2-10^4km^2	Weather systems, air temp. <0°C	
	Seasonal	10^2-10^6km^2	Air temp., freq. of cyclones	
	Decadal trends	10^4-10^6km^2	Air temp., storm tracks	
Lake ice-formation	Season	Point	Air temperature	Lake size
break-up			Solar radiation, air temperature; wind	Land runoff
Sea-ice extent	Days - Month	10^3-10^4km^2	Air temperature wind	Sea temp., salinity, currents
	Decadal trends	10^4-10^6km^2	Wind	Sea temp., salinity, currents, upwelling
Glacier terminus	Yr-10^2Yr	Point	Mass balance (winter accum., summer ablation)	Ice dynamics, bed profile
Ground ice	10^2 Yr	Point 10^2-10^4km^2	Air temperature, solar radiation	Snow cover, vegetation, geothermal heat

3.1 Snow cover

The duration of snow cover in high latitudes is determined primarily by summer temperatures since the depth is not highly variable from year-to-year. Information on the duration of snow cover is available for Canada (Potter, 1965), Europe (Kuchle-Scheidemental, 1956), the USSR (Kopanev, 1978) and for the northern hemisphere in monthly maps of snow cover probability (Dickson and Posey, 1967). Matson and Wiesnet (1981) have recently compiled satellite information on monthly snow limits, but such data are needed with a higher time and space resolution for monitoring future CO_2-induced climate perturbations.

The snow cover maps prepared by Dickson and Posey (1967) indicate that, in all months, extensive areas of North America and Eurasia have a probability of snow cover (exceeding 2.5 cm) ranging from 0 to 50 percent. This wide range is apparent in the large year-to-year variability in northern hemisphere snow cover, especially in Eurasia (Wiesnet and Matson, 1976). Interannual variability of the Eurasian snow boundary is largest in the transition seasons, and in spring the variability is pronounced in higher latitudes (Karapet'iants, 1978).

The effects of a warming on snowfall and snowcover will differ according to latitude. In low and middle latitudes, where the occurrence of snow rather than rain is frequently marginal, warming will decrease the frequency of snowfall and the duration of snowcover on the ground thereby shifting the wide zone of low frequency of snow cover depicted by Dickson and Posey well to the north. This zone has also been displayed in a somewhat different way by Kukla (1981, 1982). He refers to the Snow Transition Zone (zonally averaged round the earth) delimited by the highest (lowest) latitude without (with) snow cover on land. Figure 1 indicates that the warming effects of a CO_2 doubling will be most important for early spring snowmelt in middle and high latitudes. Observations show that the period with snow lying in Britain was a minimum in the 1920s and 1930s, corresponding to the period of maximum warmth (Manley, 1969). Records from four European stations since the 1890s indicate some regional variations in the timing of decades when maximum and minimum snow cover have occurred (Uttinger, 1963), but the causes of such variations remain to be investigated.

The snowline will also rise in mountain areas. Studies in the equatorial mountains of New Guinea show that the rise in snowline of about 100 m, over the last century, has caused a retreat of the glaciers on Mt. Jaya and disappearance of several small ice bodies on other mountains in Irian Jaya (Allison and Kruss, 1977). Calculations with a glacier-flow model indicate

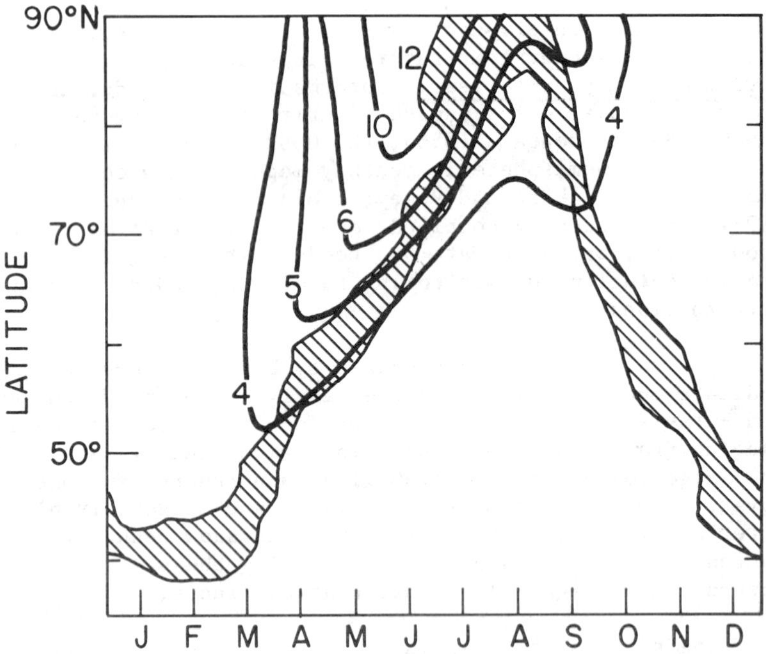

Figure 1. Increases in zonal mean surface-air temperature due to a CO_2 doubling (from the model of Ramanathan et al., 1979) in relation to the mean position of the present Snow Transition Zone (after Kukla, 1982).

that a 0.6C warming could account for the observed changes. For the Sonnblick glacier in the eastern Alps, Stelzer (1958) calculates a 200m rise in snowline between 1850 and 1969. Longer-term variations in the Alps, reported by Pfister (this volume) show a snow cover duration in the Swiss lowlands (400-500m) of only 45 days during the last two decades compared with 50-55 days between 1721-38 and 70 days for 1683-1700, the coldest interval of the Little Ice Age.

In high latitudes, where snowfall amounts are related to the frequency of cyclonic incursions and the moisture flux, there is a tendency for warmer winters to be more snowy. At Barrow, Alaska, for example, for the 30 winters 1946-76 there is a -0.57 correlation between freezing degree-days and snowfall during December-February. At Yakutsk, January snowfalls for 1941-60 are above the mean in warmer than average months (6/20) and below the mean in the colder than average months (7/20), although monthly totals range only from 2 to 14 mm (Gavrilova, 1962). An analysis of temperature and precipitation anomalies in the three winter months by Krames (1951) also shows generally positive correlations over middle and higher latitude land areas, especially in areas of continental climate. The correlations for January exceed +0.6 in eastern subarctic Canada and central-eastern Siberia, where almost all of the precipitation is snowfall. In Labrador-Ungava and Keewatin, snowy winter months tend to be associated with an average temperature departure of +1° to +3° C (Brinkmann and Barry, 1972), although this relationship is apparently non-linear and it does not hold on the time-scale of daily data (A.M. Carleton and R.G. Barry, unpub.). While increased snowfall will provide a deeper snowpack, the duration of snowcover in the Arctic is likely to be only marginally affected by this. At Barrow, Alaska, the 30-40 cm snow cover disappears in about 10 days from the start of melt (Weller and Holmgren, 1974). Hence, a CO_2-induced warming is likely to result in an overall increase in the length of the snow-free season, as implied by Figure 1, despite some regional increases in winter snowfall amounts.

3.2. Sea ice

Sea ice extent is affected by both meteorological and oceanographic conditions, but there exists a useful historical record of ice conditions which provides a significant advantage for long-term assessments (Barry, in press). In the southern hemisphere sea ice displays a wide seasonal variation in extent, as a result of the well-mixed structure of the waters on the shelf around Antarctica. In contrast, in the largely closed Arctic Ocean, a shallow layer of low salinity, maintained by continental runoff, rainfall, and ice melt in summer, forms a pronounced pycnocline at 25-30 m depth. Hence freezing occurs readily and there is rather limited summer breakup (Aagaard and Coachman, 1975; Aagaard et al., 1981). Estimates of the possible magnitudes of changes in sea-ice thickness and extent can be derived from empirical and modelling studies.

Recent studies indicate a decreasing trend in Antarctic sea ice extent for 1958 to 1978, although the changes are still within 1 standard deviation of the long term mean (Budd, 1980). Zwally et al. (1983) show that there is no long-term decrease in Antarctic sea ice extent over the 1973-81 satellite record, but regional interannual variability is found to be large. Kukla and Gavin (1981) suggest that the apparent decrease in summertime sea ice extent in the 1970s in the sector 35°W-115°E compared with the 1930s may be associated with a warming trend in Antarctic temperatures. These changes are at least consistent with the postulated CO_2-induced warming effect, although they are still within the range of natural variability of ice extent. Nevertheless, changes in atmospheric circulation or other climatic factors are more likely to be responsible for the changes in ice conditions. For Laurie Island (60°S), Budd (1975a) calculates that a ±1°C change in annual mean temperature corresponds to a ±70-day variation in the duration of sea ice (about a mean of 180 days). Assuming this relationship is representative of other Antarctic locations, Budd estimates that a 1°C annual mean temperature anomaly is associated with a 2.5° latitude variation in maximum ice extent at 67°S. It is possible, however, that non-linear feedbacks would act to modify the simple temperature effect.

The interannual variability of total ice extent in the Northern Hemisphere is relatively small (Zakharov, 1981), although it is possible that regional analysis of ice fluctuations will identify important longer term changes. Figure 2 shows the variation of Arctic ice area in late August since 1925. There is a broad minimum in the 1930-60s reflecting the general warmth of the Northern Hemisphere at that time. Zakharov and Strokina (1978) report a negative correlation (-0.59) of ice extent for July 1936-76 and annual mean temperatures in the Northern Hemisphere extratropics. This suggests a decrease of 0.8×10^6 km^2 of ice per 1°C annual temperature increase. However, the annual mean temperatures are strongly affected by winter values which have no effect on the summer ice. Re-analysis of the same ice data, for late August from the 1950s to the present, versus summer temperatures in the Arctic from Kelly and Jones (1981) shows a -0.81 correlation for 5-year running means (binomially weighted). Correlations are lower if the earlier years are included, which may reflect inadequacies of both the ice extent and temperature data. The regression relationship obtained for recent summers would imply an average decrease of 1.0×10^6 km^2 in the minimum ice area per 1°C rise in temperature. Another factor to consider in seeking ice-climate relationships is the variability shown by Zakharov (1981) in the timing of maximum and minimum ice extent in the trends for individual months. Further research into the causes of changes in sea-ice extent will be essential in order to distinguish the role of climatic and oceanographic factors.

The Arctic pack ice regime has been extensively studied. Observational evidence suggests that snow cover persisting on the ice through early summer, until the solar radiation has passed its maximum and incipient melt has resulted in a seasonal increase in stratus cloud cover, can considerably delay the summer melt. Chernigovskiy (1967) shows that the net radiation available to a bare ice surface in the Arctic is five times that of ice with an only partially-melted snow cover. Nevertheless, the 1-dimensional thermodynamic sea ice model of Maykut and Untersteiner (1971) demonstrates that the equilibrium thickness

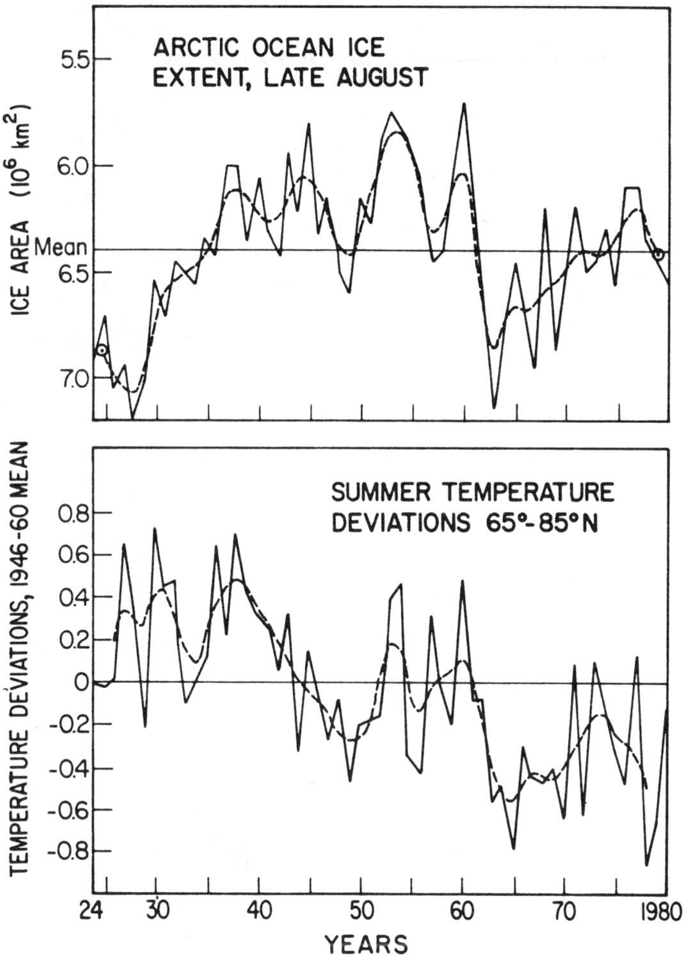

Figure 2. Arctic Ocean ice area in late August, 1924-1980 (after Zakharov, 1981) and summer temperature deviations from the 1946-60 mean for latitudes 65°-85°N (after Kelly and Jones, 1981). Dashed lines show 5-year binomially weighted running means.

of sea ice remains essentially unchanged for snow depths between 0 and 70 cm. Although summer ablation losses on bare ice are large, the ice simultaneously grows by bottom accretion and many undrained melt ponds refreeze in late summer (Maykut, 1982).

Simple estimates of changes in sea ice thickness due to warming can be based on empirical relationships with freezing degree-days (Bilello, 1961). Under present winter conditions at Barrow, Alaska, with a freezing degree-day total of 4833°C,

Billello's equation gives an ice thickness of 247 cm. Assuming an ice growth period of 270 days, a 5°C rise would lead to reduction of 15 percent, and a 10°C rise a reduction of 34 percent, in ice thickness. Analogous empirical relationships have been developed for summer retreat of the ice margin. There is a correlation of 0.81 between distance to the limit of 4/8 ice concentration north of Point Barrow on 15 September and the thawing degree-day total (see Barry, 1977). The mean distance for 1953-75 was 155 km. A rise in mean summer temperature of 5°C would increase this to 600 km, if the relationship is linear, and a 600 km retreat around the Arctic Basin would leave a core of pack ice only about 1500 km in diameter. Results of a similar calculation (Zubenok, 1963) for a 2°C summer warming are shown in figure 3.

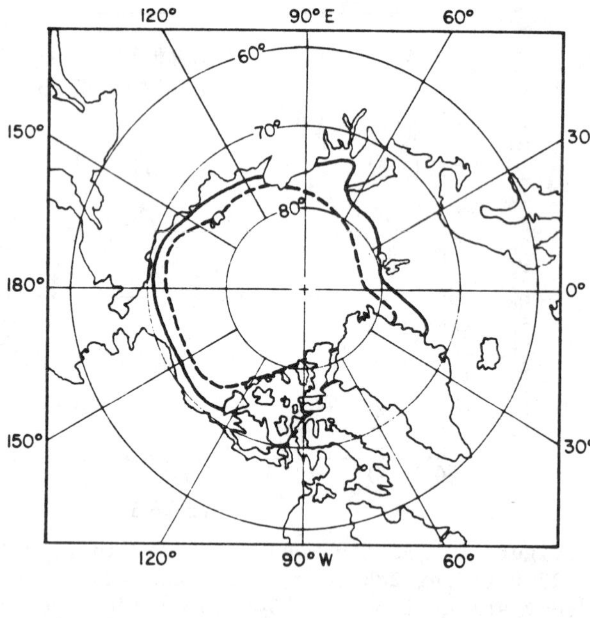

Figure 3. The present limit of summer ice (solid line) and the estimated position for a 2°C summer warming (dashed line) (after Zubenok, 1963).

Considerable uncertainty exists over the changes needed to remove the pack and maintain open water. Sedimentary records in the Arctic Basin have been interpreted as showing that a pack-ice cover has been present for the last 700,000 years (Herman and Hopkins, 1980; Worsley and Herman, 1980) or the last 5 million years (Clark 1983). In the Norwegian Sea, Belanger (1981) finds paleo-oceanographic evidence of three climate response patterns (glacial, intermediate, and interglacial).

Isotopic stages 2 and 3 (late and mid-Last Glacial, nominally about 10,000-70,000 years ago) appear to have had year-round ice-cover, stages 1 (Holocene), 5 a-d and part of 4 (early Glacial, nominally about 70,000-115,000 years ago) are "intermediate," whereas during stage 5e around 125,000 years ago there was significant convective overturning and bottom water formation in the Norwegian-Greenland Sea. These phases may also be represented by differences in Arctic Ocean ice conditions. Intervals of driftwood transport across the Arctic Basin from Siberia to Svalbard, coastal Alaska and Ellesmere Island during the Holocene point to rather open pack ice conditions between about 6000 and 4200 BP and particularly so within the last 500 years according to Stewart and England (1983), although Haggblom (1982) proposes the opposite relationship. He argues that driftwood floats for less than 18 months and must therefore be ice-rafted - implying close pack-ice conditions - in order to reach Svalbard. Nevertheless, drift rates in very open pack can be two to three times greater than the values of 600-800 km yr^{-1} assumed by Häggblom. Accordingly, the nature of Arctic ice conditions during both the Holocene and Pleistocene remains an open question.

The long-term stability of the Arctic pack-ice to climatic change has been considered from a modelling standpoint by Parkinson (1978; Parkinson and Kellogg, 1979). By coupling a simple thermodynamic sea-ice model with the NCAR general circulation model, a reasonable simulation of the annual ice cycle has been achieved. When a 5°C warming is incorporated, the pack-ice is reduced (after five model years) to a seasonal ice cover which reforms in the winter. However, in this model the ice limit is unrealistically forced by the monthly-mean air temperature.

Theoretical consideration of this topic will need to take many factors into account. For example, experiments with a dynamical model of the upper layer of the Arctic Ocean indicate that a 30-50 percent reduction in freshwater inflow could result in the disappearance of the pack ice if it were combined with less transport of Pacific water through the Bering Strait and increased export of ice via the East Greenland Current (Stigebrandt, 1981). However, CO_2 doubling experiments with the GFDL model (Manabe et al., 1981) show that zonal mean precipitation and runoff in high latitudes may increase.

Several model experiments have been performed to assess the possible atmospheric effects of an ice-free Arctic Ocean (Donn and Shaw, 1966; MacCracken, 1970; Fletcher et al., 1971; Newson, 1973; Warshaw and Rapp, 1973). The results all show major warming of the lower troposphere over the Arctic Basin and cooling over continental areas in middle latitudes, but these simulations do not take account of oceanic advection effects.

3.4 Freshwater Ice

Freeze-up depends on the heat storage in the water (i.e., water depth), the rate and temperature of any inflow, and energy exchange with the atmosphere. For small lakes, empirical relationships between freezing degree-days and the date of ice formation can be developed (Williams, 1971), but they are less satisfactory for the break-up date, especially where river inflow and wind are important factors, or where there is a high variability in year-to-year snow pack depth on the ice. Predictions of breakup by this method typically give standard deviations of 3-8 days for continental lakes. Williams (1965) shows that for a lake near Ottawa, the heat gain at the melting surface over a 10-year period was provided mainly by absorbed short-wave radiation. A complicating factor in areas of heavy snowfall is that the weight of snow depresses the ice causing overflow and the freezing of additional ice on the ice cover.

Historical data do exist on the freeze-up/break-up of lakes in North America (Allen and Cudbird, 1971; Bilello, 1980) although their spatial representativness is not well-known. A study of these records and their potential significance as a climatic index is currently being carried out by the author. Records of the freeze-up of Lake Suwa, Japan, (available since the mid-15th century) have been related to mean winter temperatures in Tokyo by Tanaka and Yoshino (1982). There is a range of about 60 days in freeze-up date, corresponding to a 4°C range in December-January temperature. For lakes located in middle latitudes, the effect of a 1°C temperature rise, uniformly distributed over the year, would have the effect of delaying the freezing date by approximately 10-15 days and advancing the breakup date by about 10-15 days.

3.5 Permafrost and Ground Ice

Permanently frozen ground, or permafrost, occurs continuously where mean annual air temperatures are below about -8°C and discontinuously in areas with temperature less than -1°C. Based on average latitudinal temperature gradients, a 2°C temperature oscillation might cause a 250-350 km displacement of the southern boundary of permafrost.

Quantitative assessments of the effects of a warming on temperatures in permanently-frozen ground are provided by Lachenbruch et al. (1966). With an equilibrium profile of $\partial T/\partial z$ for the present surface temperature of -5°C, 25 percent of the present 400 m thickness at Cape Thompson, Alaska, is unstable.

A warming of 2°C over 75-100 years is inferred. However, Lachenbruch notes that an imbalance of 3.35 MJ m^{-2} yr^{-1} (80 cal cm^{-2} yr^{-1}) over 60 years would melt only about 0.6 m of ice at the surface. Smith and Riseborough (in press), using a microclimtic model to determine permafrost sensitivity to climatic changes, report that degradation would be most rapid for a uniform warming in all months. A winter warming is less effective because of the insulating effect of snow cover.

A general melting and retreat of permafrost has been reported in the upper Mackenzie Valley (MacKay, 1975), but a warming trend would be most significant in the areas of discontinuous permafrost along the southern margins (Thie, 1974). A 3°C warming could lead to the disappearance of 25 percent or more of the permafrost in Canada (Judge et al., in press; Harris, in press). However, Table 1 indicates that even melting all the ground ice would have a negligible effect on world sea level.

3.6 Ice Sheets

The effects of a warming on large ice sheets are likely to be complex. As outlined above, warming may augment precipitation in polar latitudes giving more accumulation but, if the net effect were to cause a negative mass balance, world sea levels would begin to rise. Mercer (1978, 1981) has suggested that higher summer temperatures due to CO_2 increase could mimic conditions during the Last Interglacial (125,000 BP) when it postulated there was disintegration of most of the West Antarctic ice sheet, due to warmer conditions than now, and a 5-6 m rise in sea level. The temperature increase during this period was 1°C in the western North Atlantic (Sancetta et al., 1972), but up to 6°-7°C for mean annual temperature in Russia and central Siberia according to pollen evidence (Frenzel, 1973; p. 123). Simulations of the effect of a temperature perturbation on a thick ice sheet suggest that the time required to reach equilibrium is very long, up to 20,000 to 50,000 yr. (see Barry, 1978). However, Hughes (1975) has examined possible mechanisms for disintegration in West Antarctica where much of the base of the ice sheet is below sea level (Figure 4). Field evidence shows that two-thirds of the West Antarctic ice sheet collapsed between 17,000 and 7,000 yr BP, leaving the remnant Ross and Filchner-Ronne ice shelves according to Hughes (1983). He argues further that continuing disintegration, through the discharge and calving of surging ice streams, is already in progress and will be accelerated by the surface melting associated with a climatic warming. Disintegration could also be triggered by interaction of the rising sea level with the grounded ice, or by internal ice processes. Hollin (1972) postulated that such major ice sheet "surges" would lead to

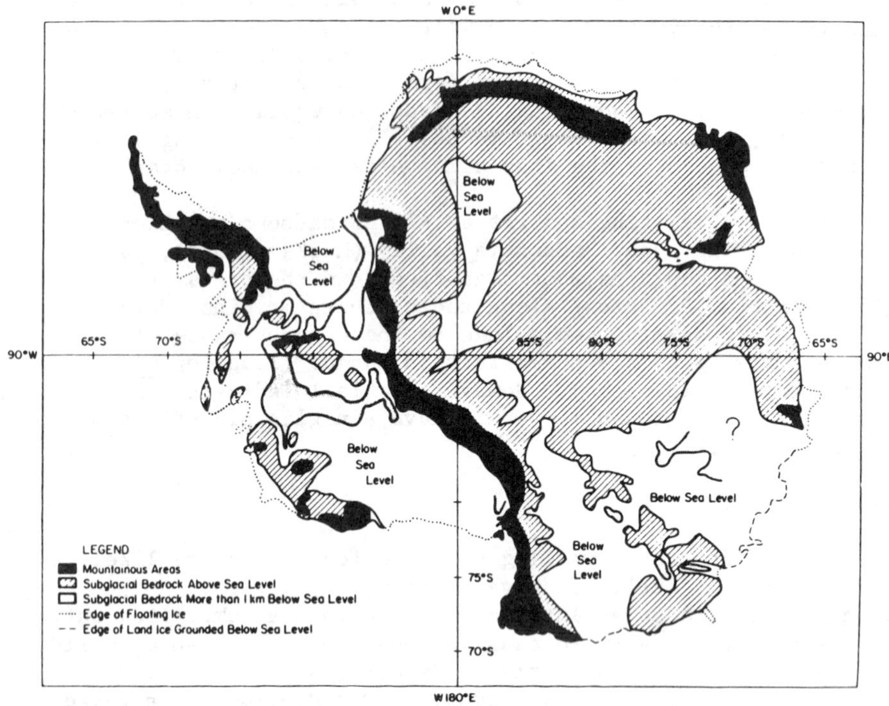

Figure 4. Generalized map of Antarctic subglacial topography showing the extensive areas below sea level (from Bentley, 1983).

10-30 m rises of sea level in less than 100 years and reports geological evidence for such fluctuations in the past (Hollin, 1980).

The question of the stability of the West Antarctic Ice Sheet and Ross Ice Shelf has been examined analytically and from field data. Paterson (1980) summarizes several studies suggesting that the ice sheet is in a near-steady state and there is no evidence of major changes in flow pattern during the last 30,000 years. The ice shelf also appears to be close to a steady state and Thomas et al. (1979, 1980) calculate that an ice shelf will buffer against possible ice sheet collapse, at least by lengthening the time scale to something of the order of 400 years. In contrast to Hughes (1983), Bentley (1983) concludes that the West Antarctic ice sheet appears to be growing in the Ross Sea and Pine Island Glacier drainage areas and Young (1981) notes that natural thickening of the Ross Ice Shelf, or slow uplift of the underlying bedrock, could cancel or reverse any CO_2-induced warming effects. Nevertheless, dynamical models indicate that ice sheets may be subject to surge behavior (Budd and McInnes, 1979) and it should not be assumed that they are inherently stable. Considerably more work on this question is required to resolve many of these questions and also to account for the claims of a recent sea level rise which Etkins and Epstein (1982) attribute to increased ice melt. On this point, it is worth noting that Gornitz et al. (1982) and Barnett (1983) find no significant recent increase in sea level rise from the average rate of 1 mm yr.$^{-1}$ over the last century.

Much less attention has been given to the Greenland Ice Sheet. Ambach (1980) proposes that a temperature rise of 1.5°C could give a substantial change in its mass balance although the time scale of response would be long. However, recent ablation estimates for Greenland are much lower that those assumed by Ambach (Radok et al., 1982). A major problem in estimating total net ablation is caused by the refreezing of meltwater and its reappearance as superimposed ice below the equilibrium line.

3.7 Alpine and Polar Glaciers

The relationships between glacier mass balance and meteorological parameters are complex (Krenke, 1974; Kuhn, 1981). Glaciers in mid-latitudes are controlled largely by summer ablation although small ice caps in the Canadian Arctic are mostly accumulation-controlled (Koerner, 1979). Summer ablation, which is determined by solar radiation, warm air advection, and condensation, and also accumulation will all be affected by direct and secondary effects of a CO_2-induced warming. If we utilise simple empirical correlations between temperature and glacier behavior (Hoinkes and Steinacker, 1975; Posamentier, 1977), quantitative estimates of the effect of any particular degree of warming can be made. Grosval'd and Kotliakov (1980) conclude that, on alpine glaciers in the Caucasus and Soviet Central Asia, the shorter accumulation period will be counter-balanced by increased winter precipitation, at least up to about AD 2020. However, they consider that small glaciers and ice caps on the Arctic islands may disappear due to increased summer melting. Using temperature predictions of Budyko et al. (1979), for a 12°-15°C rise in high-latitude temperatures by AD 2020, and an ablation formula of Krenke and Khodakov (1966):

$$\text{ablation (mm w.e.)} = (\overline{T} + 9.5)^3$$

where $\overline{T}$ = mean July temperature (°C),

ablation rates of 4-6 m yr^{-1} of water equivalent are estimated for Franz Josef Land and Novaya Zemlya. Such rates, if sustained, could lead to the disappearance of most small Arctic ice bodies (150-250 m thick) within about 50 years.

Observational evidence for the Alps shows that glaciers did respond strongly to the general warming between 1920 and 1950 (Hoinkes, 1968). Results of Gamper and Suter (1978) and Orombelli and Porter (1982) indicate that 1°C changes in mean temperature give rise to fluctuations of the order of 1 km in Alpine glacier termini locations. Similar evidence exists in other glacierized mountain ranges. Nevertheless, the retreat and advance of individual glaciers may be asynchronous to the same climatic forcing due to differences in glacier length, slope, and speed of motion (Paterson, 1982), although surge behavior in glaciers does not appear to be climatically induced. Glacier flow regimes have been analyzed using a time-dependent model by Smith and Budd (1981). Using changes in observed parameters for four European glaciers of differing characteristics and responses, they estimate that average summer temperatures have increased between 0.5° and 1.0°C since the Little Ice Age about 300 yr. ago. Such models can be used in a reverse sense to predict possible future changes in individual glaciers, but regional patterns of ice response are probably best assessed via simpler, more general regression relationships of the types described above.

3. CANDIDATE CRYOSPHERIC INDICATORS FOR DETECTING CO_2 EFFECTS

The concern over rising CO_2 levels and their potential effects on the global climate system has led the atmospheric science community to estimate the climatic signals that will characterize the occurrence of such effects. Related components of the climate system (hydrologic, oceanographic, cryspheric) that will be involved in these changes are also being investigated in order to identify a characteristic "climatic fingerprint" of CO_2-induced changes distinguishable from other external or internal processes (MacCracken, 1983). The final section of this study deals with the selection of cryospheric phenomena that might serve as indicators of CO_2 effects. Such a selection involves a number of compromises. The individual criteria that need to be considered in selecting candidate indicators are as follows:

(1) high sensitivity to a single meteorological variable that is affected by the increasing CO_2 concentration; a predictable threshold effect would be particularly useful;

(2) short time-constant in response to a CO_2-induced warming effect;

(3) the variable itself should have a high signal/noise ratio, at least in some geographical locations that can be pre-selected;

(4) the selected variable(s) should be readily observable in terms of techniques, frequency of observation, and accessibility of location (in the case of site measurements);

(5) ideally there should be historical information on the long-term (natural) variability of the selected indicator(s).

Four potential candidate indicators can be identified, although each has its limitations in terms of the desiderata listed above. They are: snow cover, lake ice freeze-up/break-up, ice cap mass-balance changes, and sea ice. The ordering reflects my assessment of their priority in this context.

Each of the four has an appropriately short time-constant, although the signal: noise ratio is likely to be relatively low as a result of the control of synoptic weather activity, especially on snowfall and ablation events. Snow cover duration and lake ice both respond to a limited number of meteorological variables, while sea ice and ice cap mass-balance are more complex parameters. The first three are also readily observable and are amenable to repetitive monitoring for satellites. One advantage of ice cap situations is the availability of snow/ice stratigraphic records on past seasonal climates (e.g. WDC-A for Glaciology, 1980). With each of these indicators it will be essential to pursue basic research aimed at a more complete understanding of natural causes of variability and their respective contributions to past and future changes in snow/ice extent and thickness.

Because a strong warming effect is anticipated in high latitudes, it would be worthwhile to investigate direct (meteorological) and indirect (cryospheric) indicators of CO_2-effects near sea level and on an elevated ice-free plateau, above about 600-700 m, in the north polar region. The high ground should be free of permanent ice to remove any local cooling effect. Variables to examine include temperature lapse rates, IR fluxes, and snow cover. However, while the climatic signal may be large in high latitudes, there is also a higher noise level in the data (Wigley and Jones, 1981).

Monitoring of the mass balance of thin polar ice caps will provide a further indicator of net, integrated changes in snowfall and summer temperature. On high-latitude mountain ice-caps there is little or no summer melt in most years. Snow pits or shallow ice cores can be used to determine the annual stratigraphic layers over decades to several hundreds/thousands of years (Giovinneto and Schwerdtfeger, 1966; Koerner, 1977). Hattersley-Smith and Serson (1973) report on the rapid response of small ice caps in eastern Ellesmere Island to a cooling in the 1960s, for example.

4. CONCLUSIONS

Based on our present knowledge, it appears that a CO_2-induced warming on the century time-scale will have only minor consequences for ice sheets and ground ice or permafrost. Nevertheless, possible changes in ice sheet accumulation rates should be monitored at regular intervals, at least every 5 years, for large-scale transects across key areas in Antarctica and Greenland using satellite laser-altimetry. A major effort in data acquisition and analysis will be required to meet this objective. Changes in permafrost or ground-ice conditions in relation to climatic parameters are rather poorly known and are considerably complicated by effects of surface cover and geothermal heat flux differences. No attempt has yet been made to coordinate quantitative data on an international basis, apart from general mapping of permafrost limits, although work on Canadian data is well advanced. Also, much information is proprietary and inaccessible for scientific research. Changes in sea ice concentration or extent due to a CO_2-induced warming may be anticipated on the 10-50 year time-scale, but any major destabilization of the Arctic pack ice will probably occur only in association with major changes in oceanic circulation. Major weaknesses exist in our understanding of Arctic ice dynamics of Arctic Basin water exchanges, particularly in the shelf areas, and of summer stratus cloud conditions. The processes in the marginal ice zones are also little known (Anderson et al., 1980; Wadhams et al., 1982).

Changes in cryospheric variables due to a warming are most likely to be first detected in seasonal snow-cover duration and the dates of lake ice break-up in middle latitudes as a result of the larger signal/noise ratio in middle lattitudes. Nevertheless, possible trends in Antarctic sea ice extent and in the mass-balance of small polar ice caps warrant careful monitoring. Much information is now routinely available from satellite data, but careful climatic analysis of the causes of any changes that are detected will be essential if erroneous interpretations are to be avoided. A better understanding of the natural variability of cryospheric components and their role in the global climate system must be developed in order to achieve this objective.

5. ACKNOWLEDGMENTS

This paper was completed during my tenure of a University of Colorado Faculty Fellowship and a Fellowship from the John Simon Guggenheim Memorial Foundation; this support is gratefully acknowledged. The typing was patiently undertaken by Margaret Strauch. Thanks are due to the Association of American Geographers for permission to reproduce figures 2 and 3, which first appeared in the December 1983 issue of the Annals AAG.

REFERENCES

Aagaard, K. and Coachman, L.K., 1975. Toward an ice-free Arctic Ocean. EOS 56, 484-486.

Aagaard, K., Coachman, L.K. and Carmack, E., 1981. On the halocline of the Arctic Ocean. Deep Sea Res. 28A, 529-545.

Allen, W.T.R. and Cudbird, B.S.V., 1971. Freeze-up and break-up dates of water bodies in Canada. Report CLI-1-71. Meteorological Branch, Dept. of Transport, Toronto, Canada.

Allison, I. and Kruss, P., 1977. Estimation of recent climate change in Irian Jaya by numerical modelling of its tropical glaciers. Arct. Alp. Res. 9, 49-60.

Ambach, W., 1980. Anstieg der CO_2-Konzentration in der Atmosphäre und Klimaänderung: Mögliche Auswirkungen auf dem Grönlandischen Eisschild. Wetter u. Leben 32, 135-142 (translated by P.P. Weidhas. Lawrence Livermore Laboratory, UCRL-Trans-11767, 1982).

Anderson, B.G., Weeks, W.F. and Newton, J.L. (eds.), 1980. The seasonal sea ice zone. Cold Reg. Sci. Technol. 2, 1-357.

Augustsson, T. and Ramanathan, V., 1977. A radiative-convective model study of the CO_2 climate problem. J. Atmos. Sci. 34, 448-451.

Barry, R.G., 1977. Study of climatic effects on fast ice extent and its seasonal decay along the Beaufort-Chukchi coasts. In: Environmental Assessment of the Alaskan Continental Shelf, V. 14. National Oceanic and Atmospheric Administration, Environmental Research Laboratories, Boulder, Colorado, 574-743.

Barry, R.G., 1978. Cryospheric responses to a global temperature increase. In: Carbon Dioxide, Climate and Society (ed. J. Williams). Oxford: Pergamon, 169-180.

Barry, R.G., 1983. The sea ice data base. In: The Geophysics of Sea Ice(eds. N. Untersteiner, et al.), Plenum, New York, (in press).

Barry, R.G., 1984. The nature and significance of cloud-cryosphere interactions in the marginal snow and ice zones. This volume.

Belanger, P.E., 1981. Paleo-oceanography of the Norwegian Sea during the past 130,000 years: Coccolithophorid and foraminiferal data. Boreas 11, 29-36.

Bentley, C.R., 1983. The West Antarctic ice sheet: Diagnosis and prognosis. In: Proceedings of the Carbon Dioxide Research Conference: Carbon Dioxide, Science and Consensus, CONF-820970, Institute for Energy Analysis, Oak Ridge Associated Universities, Washington, DC, pp. IV, 3-50.

Bilello, M.A., 1961. Formation, growth and decay of sea ice in the Canadian Arctic Archipelago, Arctic 14, 2-25.

Bilello, M.A., 1980. Maximum thickness and subsequent decay of lake, river, and fast sea ice in Canada and Alaska. USACRREL Report 80-6. U.S. Army Cold Reg. Res. Eng. Lab., Hanover, NH.

Brinkmann, W.A.R. and Barry, R.G., 1972. Palaeoclimatological aspects of the synoptic climatology of Keewatin, North Territories, Canada. Palaeogeogr. Palaeoclimatol. Palaeoecol. 11, 77-91.

Budd, W.F., 1975. Antarctic sea-ice variations from satellite sensing in relation to climate. J. Glaciol. 15, 417-427.

Budd, W.F., 1980. The importance of the Antarctic region for studies of the atmospheric carbon dioxide concentration. In: Carbon Dioxide and Climate: Antarctic Research (ed. G.I. Pearman), Australian Academy of Science, Canberra, Australia, 115-128.

Budd, W.F. and McInnes, B.J., 1979. Periodic surging of the Antarctic ice sheet--an assessment by modeling. Hydrol. Sci. Bull. 24, 95-104.

Budyko, M.I. and Yefimova, N.A., 1981. Impact of carbon dioxide on climate. Met. and Hydrol., 2, 1-10.

Budyko, M.I., Vinnikov, K. Ya., Drozdov, O.A., and Yefimova, N.A., 1979. Impending climatic change. Soviet Geographer 29, 395-411.

Chernigovskiy, N.T., 1967. Radiation Regime of the Central Arctic Basin. WMO Tech. Note 87, World Met. Organ., Geneva, Switzerland, 107-115.

Choudbury, B. and Kukla, G., 1979. Impact of CO_2 on cooling of snow and water surfaces. Nature 280, 668-671.

Clark, D.L., 1982a. Commentary. In: Carbon Dioxide Review 1982 (ed., W.C. Clark), Clarendon Press, Oxford, 181-185.

Clark, D.L., 1982b. Origin, nature and world climate effect of Arctic Ocean ice-cover. Nature 300, 321-325.

Crane, A.J., 1981. Comments on recent doubts about the CO_2 greenhouse effect. J. Appl. Met. 20, 1547-1549.

Dickson, R.R. and Posey, J., 1967. Maps of snow-cover probability for the northern hemisphere. Mon. Wea. Rev. 95, 347-353.

Donn, W.L. and Shaw, D.M., 1966. The heat budget of an ice-free and an ice-covered Arctic ocean. J. Geophys. Res. 71, 1087-1093.

Etkins, R. and Epstein, E.S., 1982. The rise of global mean sea level as an indication of climate change. Science 215, 287-289.

Fletcher, J.O., Mintz, Y., Arakawa, A., and Fox, T., 1971. Numerical simulation of the influence of Arctic sea ice on climate. WMO Tech. Note 129, World Met. Organ., Geneva, Switzerland, 181-218.

Flohn, H., 1982. Climate change and an ice-free Arctic Ocean. In: Carbon Dioxide Review 1982 (ed. W.C. Clark), Clarendon Press, Oxford, 145-179.

Frenzel, B., 1973. Climatic Fluctuations of the Ice Age. (Translated by A.E.M. Nairn.) Cleveland: The Press of the Case Western Reserve University.

Gamper, M. and Suter, J., 1978. Der Einfluss von Temperaturänderungen auf die Länge von Gletscherzungen. Geogr. Helvet., 33, 183-189.

Gavrilova, M.K., 1962. Klimat tsentral'noi Yakutii, Akad. Nauk SSSR, Sibirskoe Otdelenie Institut. Merzlotovedeniya, Yatusk, 63 pp.

Gilliland, R., 1982. Solar, volcanic and CO_2 forcing of recent climatic changes. Climatic Change 4, 111-131.

Giovinetto, M.B. and Schwerdtfeger, W., 1966. Analysis of a 200-year snow accumulation series from the South Pole. Archiv. Meteorol. Geophys. und Biokl. A15, 227-250.

Gornitz, V., Lebedeff, S. and Hansen, J., 1982. Global sea level trend in the past century. Science 215, 1611-1614.

Grigor'eva, T.A. (ed.), 1982. Vliianie uvelichenia kolichestva uglekislogo gaza v atmosfere na klimat. Gidrometeoizdat, Leningrad, 56 pp.

Grosval'd, M.G. and Kotlyakov, V.M., 1980. Impending climatic change and the fate of glaciers. Polar Geogr. Geol. 4, 65-78.

Häggblom, A., 1982. Driftwood in Svalbard as an indicator of sea ice conditions. Geograf. Annal. G4A, 81-94.

Hansen, J.E., Johnson, D., Lacis, A., Lebedeff, S., Lee, P., Rind, D. and Russel, G., 1981. Climatic impact of increasing atmospheric carbon dioxide. Science 213, 957-966.

Hanson, A.M., 1980. The snow cover of sea ice during the Arctic Ice Dynamics Joint Experiment, 1975 to 1976. Arct. Alp. Res. 12, 215-226.

Harris, S.A., in press. Distribution and zonation of permafrost along the eastern ranges of the Cordillera of North America. Biuletin Peryglacjalny, in press.

Hattersley-Smith, G.F. and Serson, H., 1973. Reconnaissance of a small ice cap near St. Patrick Bay, Robeson Channel, Northern Ellesmere Island. J. Glaciol. 12, 417-421.

Herman, G.F., Wu, M-L. C. and Johnson, W.T., 1980. The effect of clouds on the earth's solar and infrared radiation budgets. J. Atmos. Sci. 37, 1251-1261.

Herman, Y. and Hopkins, D.M., 1930. Arctic Ocean climate since Late Cenozoic time. Science 209, 557-562.

Hoinkes, H.C., 1968. Glacier variations and weather. J. Glaciol. 7, 3-19.

Hoinkes, H.C. and Steinacker, R., 1975. Zur Paramaterisierung der Beziehung Klima-Gletscher. Rivista Italiana di Geofisica e Science Affini 1 (Spec.), 97-104.

Hollin, J.T., 1972. Interglacial climates and Antarctic ice surges. Quat. Res. 2, 401-408.

Hollin, J.T., 1980. Climate and sea level in isotope stage 5: An East Antarctic ice surge at 95,000 BP? Nature 283, 629-633.

Hollin, J.T. and Barry, R.G., 1979. Empirical and theoretical evidence concerning the response of the earth's ice and snow cover to a global temperature increase. Env. Int. 2, 437-444.

Hughes, T., 1975. The West Antarctic ice sheet: Instability, disintegration and initiation of ice ages. Rev. Geophys.-Space Phys. 13, 502-526.

Hughes, T.J., 1983. The stability of the West Antarctic ice sheet: What has happened and what will happen. In: Proceedings of the Carbon Dioxide Research Conference: Carbon Dioxide, Science and Consensus, CONF-820970, Institute for Energy Analysis, Oak Ridge Associated Universities, Washington, DC, pp. IV, 51-73.

Jäger, J. and Kellogg, W.W., 1983. Anomalies in rainfall and temperature during warm Arctic seasons. Climatic Change 5, 39-60.

Jenssen, M.D., 1977. Personnal communication.

Judge, A. (chairman), 1983. Climate change and geothermal regime. Permafrost. Fourth International Conference. Abstracts and Programs, Univ. of Alaska, Fairbanks, 51-55.

Karapet'iants, E.H., 1978. Klimaticheskie osobennosti polozheniia granitsy rasprostraneniia snezhnogo pokrova territorii SSSR. Glav. Geofiz. Obs.,Trudy 404, 16-29.

Kellogg, W.W., 1983. Feedback mechanisms in the climate system affecting future levels of carbon dioxide. J. Geophys. Res. 88, 1263-1270.

Kelly, P.M. and Jones, P.D., 1981. Summer temperatures in the Arctic, 1881-1981, Climate Monitor 10, 66-67.

Koerner, R.M., 1977. Devon Island ice cap: Core stratigraphy and paleoclimate. Science 196, 15-18.

Kopanev, I.D., 1978. Snezhnii pokrov na territorii SSSR. Gidrometeoizdat, Leningrad, 181 pp.

Krames, K., 1951. Die Korrelation zwischen Temperatur- und Niederschlags-anomalien im Winter in der Nordhemisphäre. Geograf. Annal. 33, 210-229.

Krenke, A.N., 1974. Climatic existence conditions for glaciers and the shaping of glacial climates. In: Meteorology and Climatology, V. 2 (eds. J.P. Danilina and A.P. Kapitsa) (Transl. of ITOGI NAUKI, 1971), Boston, G.K. Hall and Co., 179-224.

Krenke, A.N. and Khodakov, V.G., 1966. Relation between the surface melting on glaciers and air temperature. Materialy Glatsiologicheskikh Issledovanii, Khronika, Obsuzhedeniia 12, Moscow, USSR, 153-164.

Kuchle-Scheidemantel, I., 1956. Die Dauer der Schneedecke in Europa. Petermanns Geogr. Mitt., 100, 185-192.

Kuhn, M., 1981. Climate and glaciers. In: Sea Level, Ice and Climatic Change (ed. I. Allison), Internatl. Assoc. of Hydrol. Sci., Pub. 131, IAHS, Wallingford, Oxfordshire, UK, 3-20.

Kukla, G., 1981. Climatic role of snow covers. In: Sea Level, Ice and Climatic Change (ed. I. Allison), Internatl. Assoc. of Hydrol. Sci., Pub. 131, IAHS, Wallingford, Oxfordshire, UK, 79-109.

Kukla, G., 1982. Carbon dioxide and polar climates. In: Proceedings of the Workshop on First Detection of Carbon Dioxide Effects, DOE/CONF-8106214, U.S. Department of Energy, Washington, DC, 237-288.

Kukla, G. and Gavin, J., 1981. Summer ice and carbon dioxide. Science 214, 497-503.

Lachenbruch, A.H., Greene, G.W., and Marshall, B.V., 1966. Permafrost and the geothermal regimes. In: Environment of the Cape Thompson Region, Alaska (eds. N.J. Wilimovksy and J.N. Wolfe), U.S. Atomic Energy Commission, Washington, DC, 149-164.

Lamb, H.H., 1971. Climate-engineering schemes to meet a climate emergency. Earth Sci. Rev. 7, 87-95.

Laurmann, J.A. and Rotty, R.M., 1983. Exponential growth and atmospheric carbon dioxide. J. Geophys. Res. 88(C2), 1295-1300.

MacCracken, M.C., 1970. Tests of ice age theories using a zonal atmospheric model. UCRL-72803. Lawrence Radiation Laboratory, Livermore, California.

MacCracken, M.C., 1983. Have we detected CO_2-induced climate change? Problems and prospects. In: Proceedings of the Carbon Dioxide Research Conference: and Consensus, CONF-820970, Institute for Energy Analysis, Oak Ridge Associated Universities, Washington, DC, pp.

MacDonald, G.J., 1982. Effect of warming on West Antarctic ice sheets. In: Long-term Impacts of Increasing Atmospheric Carbon Dioxide Levels (ed. G.J. MacDonald), Ballinger Publishing, Cambridge, Mass., 211-235.

Manabe, S. and Stouffer, R.J., 1980. Sensitivity of a global climate model to an increase of CO_2 concentration in the atmosphere. J. Geophys. Res. 85, 5529-5554.

Manabe, S. and Wetherald, R.T., 1980. On the distribution of climate change resulting from an increase in CO_2 content of the atmosphere. J. Atmos. Sci. 37, 99-118.

Manabe, S., Wetherald, R.T. and Stouffer, R.J., 1981. Summer dryness due to an increase of atmospheric CO_2-concentration. Climatic Change 3, 347-386.

Manley, G., 1969. Snowfall in Britain over the past 300 years. Weather 24, 428-437.

Matson, M. and Wiesnet, D.R., 1981. New data base for climate studies. Nature 289, 451-456.

Maykut, G.A. and Untersteiner, N., 1971. Some results from a time dependent thermodynamic model of sea ice. J. Geophys. Res. 76, 1550-1575.

Maykut, G.A., 1982. Large-scale heat exchange and ice production in the Central Arctic. J. Geophys. Res. 87, 7971-7984.

McKay, J.R., 1975. The stability of permafrost and recent climatic change in the Mackenzie Valley, N.W.T. Geol. Surv. Canada, Paper 75-1, 173-176.

Mercer, J.H., 1978. West Antarctic ice sheet and CO_2 greenhouse effect: A threat of disaster. Nature 271, 321-325.

Mercer, J.H., 1981. West Antarctic ice volume: The interplay of sea level and temperature, and a strandline test for absence of the ice sheet during the Last Interglacial. In: Sea Level, Ice and Climatic Change (ed. I. Allison), Internatl. Assoc. of Hydrol. Sci., Publ. 131, IAHS, Wallingford, Oxfordshire, UK, 323-330.

Namias, J., 1980. Some concomitant regional anomalies associated with hemispherically average temperature variations. J. Geophys. Res. 85, 1585-1590.

National Research Council, 1982. Carbon dioxide and climate: A second assessment, National Academy Press, Washington, DC, 72 pp.

Newson, R.L., 1973. Response of a general circulation model of the atmosphere to removal of the Arctic ice cap. Nature 241, 39-40.

Orombelli, G. and Porter, S.C., 1982. Late Holocene fluctuations of Brenva Glacier. Geogr. Fiz. Dinam. Quat. 5, 14-37.

Parkinson, C.L., 1978. A numerical simulation of the annual cycle of sea ice in the Arctic and Antarctic. NCAR Cooperative Thesis 46. National Center for Atmospheric Research, Boulder, Colorado.

Parkinson, C.L. and Kellogg, W.W., 1979. Arctic sea ice decay simulated for a CO_2-induced temperature rise. Climatic Change 2, 149-162.

Paterson, W.S.B., 1980. Ice sheets and ice shelves. In: Dynamics of Snow and Ice Masses (ed. S.C. Colbeck), Academic Press, New York, 1-78.

Paterson, W.S.B., 1982. The Physics of Glaciers, 2nd ed., Pergamon, Oxford.

Pfister, Ch., 1984. Climatic changes in central Europe over the last century as reflected in the duration of snow-cover in the lowlands and fluctuations in the temporary snow-lines in the Alps. This volume.

Pittock, A.B., 1980. Towards a warm earth scenario for Australia. In: Carbon Dioxide and Climate: Antarctic Research (ed. G.I. Pearman), Australian Academy of Science, Canberra, Australia, 197-209.

Posamentier, H.W., 1977. A new climatic model for glacier behavior of the Austrian Alps. J. Glaciol. 18, 57-65.

Potter, J.G., 1965. Snow cover. Climatological Studies 3. Meteorological Branch, Dept. of Transport, Toronto, Canada.

Radok, U., Barry, R.G., Jenssen, D., Keen, R.A., Kiladis, G.N. and McInnes, B., 1982. Climatic and physical characteristics of the Greenland Ice Sheet, Parts I, II and III. Cooperative Institute for Research in Environmental Sciences, University of Colorado, Boulder, 193 pp. + appendices, Maps and Cross Section Diagrams.

Ramanathan, V., Lian, M.S. and Cess, R.D., 1979. Increased atmospheric CO_2: Zonal and seasonal estimates of the effect on the radiation energy balance and surface temperature. J. Geophys. Res. 84, 4949-4958.

Ramanathan, V., 1981. The role of ocean-atmosphere interactions in the CO_2 climate problem. J. Atmos. Sci. 38, 918-930.

Sancetta, C., Imbrie, J., Kipp, N.G., McIntyre, A. and Ruddiman, W.F., 1972. Climate record in the North Atlantic deep-sea core V23-82: Comparison of the last and present interglacials based on quantitative time series. Quat. Res. 22, 363-367.

Schlesinger, M.E., 1983. Simulating CO_2-induced climatic change with mathematical models: Capabilities, limitations and prospects. In: Proceedings of the Carbon Dioxide Research Conference: Carbon Dioxide, Science and Consensus, CONF-820970, Institute for Energy Analysis, Oak Ridge Associated Universities, Washington, DC, pp. III, 3-140.

Schneider, S.H. and Chen, R.S., 1980. Carbon dioxide warming and coastal flooding: A problem review and exploratory climatic impact assessment. Annual Rev. of Energy, 107-140.

Sear, C.B. and Lough, J.M., 1981. Three scenarios for a warm, high-CO_2 world. Unpub. manuscript.

Smith, I.N. and Budd, W.F., 1981. The derivation of past climate changes from observed changes of glaciers. In: Sea Level, Ice and Climatic Change (ed. I. Allison), Internatl. Assoc. of Hydrol. Sci., Pub. 131, IAHS, Wallingford, Oxfordshire, UK, 31-52.

Smith, P.W. and Riseborough, D.W., 1983. Permafrost sensitivity to climatic change. Proceedings. Permafrost. Fourth International Conference. Academic Press, New York, in press.

Stauffer, B., Hofer, H., Oeschger, H., Schwander, J. and Siegenthaler, U., 1983. Atmospheric CO_2 concentration during the Last Glaciation. Paper presented at the Symposium on Ice and Climate Modelling, Evanston, Illinois.

Stelzer, F., 1958. Die Hebung der Schneegrenze im Sonnblickgebiet im Zeitraum 1850-1949. Wetter und Leben 10, 171-175.

Stewart, T.G. and England, J., 1983. Holocene sea-ice variations and paleoenvironmental change, northernmost Ellesmere Island, N.W.T., Canada, Arct. Alp. Res. 15, 1-17.

Stigebrandt, A., 1981. A model for thickness and salinity of the upper layer in the Arctic Basin and the relationship between ice thickness and some external parameters. J. Phys. Ocean. 11, 1407-1422.

Tanaka, M. and Yoshino, M.M., 1982. Re-examination of the climatic change in central Japan based on freezing dates of Lake Suwa. Weather 37, 252-259.

Thie, J., 1974. Distribution and thawing of permafrost in the southern part of the discontinuous permafrost zone in Manitoba. Arctic 27, 189-200.

Thomas, R.H.T., Sanderson, J.O., and Rose, K.E., 1979. Effect of climatic warming on the West Antarctic ice sheet. Nature 277, 355-358. (Also, discussion: Nature 287 (1980), 759.)

Untersteiner, N., 1981. Potential effects on Arctic sea ice of a CO_2-induced global warming. Unpubl. manuscript.

Uttinger, H., 1963. Die Dauer der Schneedecke in Zurich. Archiv. Met. Geophys. Biokl., B12, 404-421.

Vinnikov, K. Ya. and Groisman, P. Ya., 1982. The empirical study of climate sensitivity. Izvestiya, Atmos. Ocean. Phys., 18(11).

Vyturin, B.I., 1978. Patterns of distribution and quantitative estimate of the ground ice in the USSR. In: USSR Contribution. Permafrost. Second International Conference, 1973 (eds. F.J. Sanger and P.J. Hyde), National Academy of Sciences, Washington, DC, 159-164.

Wadhams, P., Martin, S., Johannesen, O.M., Hibler, W.D., III, Campbell, W.J. (eds.), 1982. MIZEX - A program for mesoscale air-ice-ocean interaction: Experiments in Arctic marginal ice zones. 1. Research strategy. U.S. Army CRREL, Spec. Rep. 81-19, Hanover, NH, 20 pp.

Wang, W.C., Yung, Y.L., Lacis, T.M., and Hansen, J.E., 1976. Greenhouse effects due to man-made perturbations of trace gases. Science 194, 685.

Warshaw, M and Rapp, R.A., 1973. An experiment on the sensitivity of a global circulation model. J. Appl. Met. 12, 43-49.

Weller, G. and Holmgren, B., 1974. The microclimates of the Arctic tundra. J. Appl. Met. 13, 854-862.

Wiesnet, D.R. and Matson, M., 1976. A possible forecasting technique for winter snow cover in the northern hemisphere and Eurasia. Mon. Wea. Rev. 104, 828-835.

Wigley, T.M.L., Jones, P.D. and Kelly, P.M., 1980. Scenario for a warm high-CO_2 world. Nature 283, 17-21.

Wigley, T.M.L. and Jones, P.D., 1981. Detecting CO_2-induced climate change. Nature 292, 205-208.

Williams, G.P., 1965. Correlating freeze-up and break-up with weather conditions. Can. Geotech. J. 2, 313-326.

Williams, G.P., 1971. Predicting the date of lake-ice break-up. Water Resources Res. 7, 323-333.

World Data Center-A for Glaciology, 1980. Ice Cores. Glaciol. Data, Report GD-8. World Data Center-A for Glaciology, Boulder, CO.

Worsley, T.R. and Herman, Y., 1980. Episodic ice-free Arctic Ocean in Pliocene and Pleistocene time: Calcareous nanofossil evidence. Science 210, 323-325.

Young, N.W., 1981. Responses of ice sheets to environmental changes. In: Sea Level, Ice and Climatic Change (ed. I. Allison), Internatl. Assoc. of Hydrol. Sci., Pub. 131, IAHS, Wallingford, Oxfordshire, UK, 331-360.

Zakharov, V.F. and Strokina, L.A., 1978. Recent variations in extent of Arctic Ocean ice cover. Met. and Hydrol. no. 7, 35-43.

Zakharov, V.F., 1981. L'dy Arkiki i Sovremennie Prirodnye Protsessy, Gidrometeoizdat, Leningrad, 136 pp.

Zubenok, L.I., 1963. Vliianie anomalii temperatury na ledianoi pokrov Arktiki. Met. i Gidrol., no. 6, 25-30.

Zwally, H.J., Parkinson, C.L. and Comiso, J.C., 1983. Variability of Antarctic sea ice and changes in carbon dioxide. Science 220, 1005-1012.

THE NATURE AND SIGNIFICANCE OF CLOUD-CRYOSPHERE INTERACTIONS IN THE MARGINAL SNOW AND ICE ZONES

Roger G. Barry

Cooperative Institute for Research in Environmental Sciences and Department of Geography, University of Colorado, Boulder

1. SUMMARY

Climate modelers have emphasized the potential significance of ice-albedo feedback processes for the enhancement of glacial inception conditions as well as short-term climatic anomalies. Such considerations commonly neglect the important role of cloud cover for planetary albedo and surface energy budgets. There are also few data sets with which to examine cloud conditions in relation to the state of the underlying surface, especially in high latitudes (Crane and Barry, 1983).

This work, which is fully reported elsewhere (Barry et al., 1983; Shine et al., 1983), utilizes DMSP satellite visible and infrared imagery, the U.S. Air Force 3-D nephanalysis product, and ground observations, to analyze cloudiness variations across the boundaries of snow-covered land areas, and arctic sea ice. Both synoptic-scale cloud vortices, which may be influenced by the baroclinic effects of the cryospheric boundaries, and boundary layer stratiform cloud are considered. Based on synoptic case-studies of DMSP imagery, via manual interpretation (Scharfen, 1982) and statistics of monthly cloudiness from the 3-D nephanalysis data, certain seasonal patterns seem to emerge. In July and September, at least, average total cloud amounts decrease poleward across the snow cover or sea ice boundary, in contrast to the summer mean maps prepared by Huschke (1969) and Vowinckel and Orvig (1970) which show an increase towards the central Arctic.

N.-A. Mörner and W. Karlén (eds.), Climatic Changes on a Yearly to Millennial Basis, 605–607.

The effect of more cloud south of the snow/ice margins will be to lessen the gradient in planetary albedo (the albedo is high over snow/ice and cloud, low over bare land or open ocean). There also appear to be concomitant changes in cloud optical depth. Such effects have been modeled using a one-dimensional radiative column model (Shine et al., 1983) and it is concluded that some of the early parameterizations used in Energy Balance Models (EBMs) over-estimated the sensitivity of climate to the ice-albedo feedback effect through their neglect of the role of cloudiness. However, more recent EBM experiments may have resulted in under-estimation owing to the formulation of parameterizations based on satellite-derived cloud data. Considerably more investigation of the properties of arctic surfaces and cloud cover, and their seasonal and longer term variations is necessary for improved modeling of climate processes.

2. ACKNOWLEDGMENTS

This study has been supported under NASA grant NAG-5-142 and National Science Foundation grant ATM-80-18898. The work has been performed in collaboration with A. Henderson-Sellers and K.P. Shine at the Geography Department, University of Liverpool, England.

References

Barry, R.G., Henderson-Sellers, A. and Shine, K., 1983. Climate sensitivity and the marginal cryosphere. In Climate Processes: Sensitivity to Solar Irradiance and CO_2, J. Hanson and T. Takahashi, eds., Fourth Biennial Ewing Symposium, American Geophysical Union, in press.

Crane, R.G. and Barry, R.G., 1983. The influence of clouds on climate with a focus on high latitude interactions. J. Climatol. 3, in press.

Huschke, R.E., 1969. Arctic cloud statistics from "air-calibrated" surface weather observations. Rand Corporation Memorandum, RM-6173, Santa Monica, California.

Scharfen, G.R., 1982. Snow cover, sea ice and cloudiness characteristics from satellite data. M.A. thesis, Department of Geography, University of Colorado, Boulder.

Shine, K.P., Henderson-Sellers, A. and Barry, R.G., 1983. Albedo-climate feedback: the importance of cloud and cryosphere variability. In: New Perspectives in Climate Modeling, A. Berger and C. Nicolis, eds., Elsevier, Amsterdam, in press.

Vowinckel, E. and Orvig, S., 1970. The climate of the north polar basin. In: Climates of the Polar Regions, S. Orvig, ed., Vol. 14 of World Survey of Climatology, H.E. Landsberg, ed., Elsevier, Amsterdam, pp. 129-252.

CLIMATIC BELTS IN THE CASE OF A UNIPOLAR GLACIATION

H. Flohn

University of Bonn

The recent discussion of the future evolution of our climate under increasing anthropogenic stress has been initiated and invigorated by the observed continuous increase of carbon dioxide (CO_2) and other infrared absorbing trace gases (nitrous oxide, methane, chlorofluoromethanes, tropospheric ozone) (W. Bach 1982; W.C. Clark (Ed.) 1982; G.F.M. McDonald (Ed.) 1982; H. Flohn 1981; Kellogg-Schware 1981). Advanced models of the climatic system - consisting of atmosphere, ocean, snow and ice, the biosphere and their multiple interactions - have been developed, with remarkable success in simulating some large-scale features of the general atmospheric circulation (Manabe 1983; Schlesinger 1983). The model simulation of regional (and local) details of the present climate, however, has not as yet been achieved with sufficient accuracy - this would be a necessary prerequisite for any responsible forecasting of future climatic evolution.

In this situation it is necessary to go back into the past in reconstructing the earlier evolution of our climate, especially the warm periods of the past. Some examples have already been given (W.C. Clark (Ed.) 1982; Kellogg-Schware 1981; A. Velichko 1982; Zubakov-Borsenkova 1983; Keller-Barron 1983; Sarnthein et al. 1982). While the available climate models simplify the climate system by neglecting (as yet) ocean circulation and interactions with the biosphere and parametrizing small-scale processes in the atmosphere, comparisons with paleoclimatic stages are hampered by the slow change of boundary conditions, such as landsea-mountain distribution (at the geological timescale of ka = 10^3 years and Ma = 10^6 years) or of the composition of the atmosphere. The latter effect is especially important in the cases of CO_2 and volcanogenic particles in the stratosphere, which recently have been thoroughly measured by physical laboratories in Bern, Grenoble

N.-A. Mörner and W. Karlén (eds.), Climatic Changes on a Yearly to Millennial Basis, 609–620.

and Copenhagen in ice-cores from Greenland and Antarctica (Dansgaard et al. 1982; Neftel et al. 1982; Lorius 1983). Unfortunately the (even more important) H_2O content of past climates can only be very roughly estimated (Flohn 1983a); it depends on the assumed quasi-constancy of the solar radiation (at a timescale up to ka) and on the relative area of the oceans as the main H_2O source.

One of the most fascinating paleoclimatic evolutions is the spatially different history of the large continental ice-caps of Antarctica and Greenland and of the Arctic and Antarctic sea-ice. Convincing evidence from both continental sites and ocean bottom drilling indicates that the evolution of the Antarctic ice-sheet was initiated as early as 38 Ma ago at the boundary between Eocene and Oligocene, accompanied by a sudden world-wide cooling of ocean bottom water. As early as during the middle Miocene (between 16 and 14 Ma ago: (Zubakov-Borzenkova 1983; Keller-Barron 1983) the large continent of East Antarctica was completely ice-covered; this ice-sheet extended to West-Antarctica about 8 Ma ago and reached its maximum volume - most probably now as a "cold" ice-cap, i.e. free of melting-regelation processes - at the end of Miocene (Messinian 6.5-5 Ma ago). The accompanying glacial-eustatic sea-level drop led to a closing of the Mediterranean and its desiccation, repeated many times until as much as 6 percent of the total salt content of world oceans was stored at its bottom.

The Mediterranean reopened to the world ocean at the beginning of the Pliocene leading to a climatic optimum during the Gilbert magnetic epoch (5.1-3.3 Ma ago) (Sarnthein et al. 1982; Zubako-Borzenkova 1983). During this period annual temperatures at the northern coast of Iceland had been estimated to 11-12°C (Strauch 1973), as now in the British Channel. Convincing evidence is available (Frenzel 1968; Zubakow-Borzenkova 1983; somewhat earlier Hopkins et al. 1971) of mixed boreal forest at northern Siberia, the islands at the Siberian shelf, at the Alaskan coast and in the Canadian archipelago. These boreal forests suggest summer`temperatures of at least 12-13°C along the coast. This and the non-existence of tundra is certainly imcompatible with a permanent sea-ice cover of the Arctic ocean, but does not exclude marginal seasonal sea-ice along the coasts during winter. The fact that during this period of few isolated erratica have been found at the bottom of the central Arctic Ocean (D.L. Clark 1980, 1982) cannot be used as evidence of a closed ice-cover - they may have been dropped from mountain glaciers or local ice-caps descending into the sea. Shackleton and Opdyke (1977) have found rather convincing evidence - from the isotopic composition of benthic foraminifera of the equatorial Pacific representing mainly global ice volume - of the initiation of northern hemisphere glaciations after 3.5-3.3 Ma ago. About 2.5 Ma ago, marked changes of the vegetation (Frenzel 1968; Zubakov-Borzenkova 1983) in wide areas including northwestern Europe and Iceland indicated the initiation of large-scale glaciations at the northern continents. At the bottom of the Arctic Ocean, wide-spread ice-rafted material has been found - if the chronology is correct - by D.L. Clark (1980, 1982) shortly before the end of the Gauss magnetic epoch (2.7 Ma).

From this evidence we have to conclude that during the period between about 14 and 3 Ma ago, i.e. during more than 10 Ma, a unipolar glaciation existed. We shall come back later to a weaker asymmetry of today (Flohn 1978). During the World Climate Conference, several authors (Gerassimov, Flohn 1979) pointed out the need to evaluate the climatic pattern during this period, as a model to possible future evolution of our climate (in the case of a prolonged CO_2-induced warming). Recent papers (Budyko 1982; Flohn 1982; Keller-Barron 1983; Zubakov-Borzenkova 1983) contributed to this task. In this paper it is intended to outline from our present knowledge of the mechanism of atmospheric circulation, the climatic zonation during such an asymmetric state, as a base for discussion and improvement.

This relatively recent climatic asymmetry should have been no surprise: during the later Paleozoic the giant Gondwana continent - now in fragments - carried wide-spread glacial evidence, correlated with the wandering of this supercontinent over the area of the southern pole. At the same time the northern continents carried a warm-temperate or subtropical vegetation which later produced our main coal deposits. It may be mentioned that during the Eocene, 55-38 Ma ago, a similar vegetation has existed fossilized as coal at Spitzbergen (Schweitzer 1980), while crocodiles were found at Ellesmere-Land (McKenna 1980).

In contrast to the time-span since the Late Paleozoic, the displacement of our present continents since early Pliocene is quite minor, nowhere more than around 200 kms in latitude. Nevertheless, the climatic boundary conditions have changed considerably: during late Miocene and early Pliocene the isthmus of Panama finally closed, caused by tectonic uplift and volcanic eruptions; since about 3.5 Ma years the mammals of South and North America migrated back and forth freely (Webb 1976), while the evolution of marine life in the Caribbean and in the Gulf of Panama went along separate lines (Keigwin 1978). The most important consequences of this gradual process - probably lasting since the Late Miocene - was the blocking of the great North Equatorial Ocean Current running before from West Africa towards Indonesia over a distance of more than 250° Longitude. These warm water masses were gradually deflected northward and intensified the Gulf Stream, thus transporting heat and atmospheric moisture towards the Greenland-Iceland area. It is generally agreed that this evolution initiated the early glaciation of Iceland (Berggren-Hollister 1977); most probably this is also valid for Greenland and Baffinland, where the multitude of later glaciations removed all earlier traces. The evolution of the cold Labrador current during this time (Gradstein-Srivastava 1980) finalized the development of the present situation in the arctic.

In addition to these changes of the oceanic circulation,

major changes of the land surface occurred. Many of our present high mountains were uplifted to their present altitude only during the Late Pliocene and the Pleistocene, i.e. during the last 3-3.5 Ma. This is especially true for the Himalayas (starting just before the Olduvai paleomagnetic event, 1.8 Ma ago. (Li et al. 1979; Opdyke et al. 1979)) and for the Alps, probably for most "young mountains", uplifted after the continuous northward thrust of the plates of Africa and India into Eurasia. In both mountains, extended "old" surfaces had been uplifted since this time to altitudes of 2 to 4-5 kms; the effects on the atmospheric circulation and thus on climate are striking. (It should be mentioned that the uplift of the mountains in western North America is probably older, at least middle or late Miocene.) These mountains act as elevated heat sources (for sensible and latent heat) and as mechanical obstacles; position and intensity of the quasi-stationary Rossby waves are mainly determined by them, only modified by the seasonal heating contrast of land and sea. Since these Rossby waves extend, in the upper troposphere (Flohn 1971, 1975), far into the subtropical and even tropical latitudes, they are responsible for many regional anomalies of climatic zonation.

Even more important is the evolution of the large monsoon system consisting of weak "equatorial" westerlies and of a powerful Tropical Easterly Jet (TEJ) centered at an altitude of 14 kms. During northern summer this system extends to between about 3°S and 20-28°N, and reaches from east of Philippines (about Long. 150°E) across India and Africa to Long. 25°W. Its crosscirculations - as caused by the elevated Tibetan heat source (Flohn 1968; Yeh and Gao 1979, 1981) now control the anomalous pattern of summer rainfall. Without entering into dynamical properties of the TEJ (Flohn 1964a, 1965), it should only be noted that in the exit region of this jet W of Long. 75°E, i.e. above northern Africa. subsidence and supression of convective rainfall necessarily prevails on its northern flank in the latitude belt 14-25°N, while uplift and rainfall is maintained south of its central core, in the belt 5-14°N. The slow evolution of this system since the latest Pliocene has been evidenced in the gradual desiccation of the Sahara (McCauley et al. 1982). In an opposite sense, in the entrance region of the TEJ east of Long. 90°E uplift and consequently rainfall should have increased with time north of (about) 15°N Lat., i.e. above NE India, E Tibet and large areas of SE Asia up to Lat. 35°N, including southern China. Unfortunately, no verification of the convincing model results of Hahn and Manabe (1975) can be given.

Intending to estimate the climatic zonation of a unipolar glaciated earth, we have to look first at the actually observed asymmetry. Table 1 shows the temperature, averaged for the layer 700-300 mb or about 3-9 kms, above both poles during the extreme months and in the annual average (Flohn 1978). Here the layer between 700 mb and sea-level had to be neglected because there is no way to extrapolate temperature below the surface at the South

Table 1: Temperature differences (layer 700-300 mb) between both poles (N=North Pole, S=South Pole) and Equator (E) in °C.

	E	N	S	E - N	E - S
January		-41.5	-38.3	32-9	29.7
July	8.6	-25.9	-52.7	17.3	44.1
Year		-35.9	-47.7	27.3	39.1

Pole station (2700 m, pressure ∿680 mb).

Estimates of the different terms of the heat budget have been given earlier (Flohn 1978). According to the classic theorem derived by V. Bjerknes more than 80 years ago, the intensity of the atmospheric circulation depends on the number of isobaric/isothermic solenoids, i.e. largely on the meridional (and vertical) temperature differences. Since the vertical tropospheric lapse rate varies only little around 6.5°C/km (except above the polar regions), the most essential parameter is the meriodional difference equator-pole. It has been shown earlier (Flohn 1981, 1982, 1983) that the seasonal migration of the (zonally averaged) latitude of the subtropical anticyclones ϕ_{STA} depends strongly on the values E-N and E-S for each month. The correlation between 24 simultaneous values of the equator-pole temperature difference Γ and ϕ_{STA} at both hemispheres is 0.85, with a time lag of 1-2 months even 0.91; a linear regression equation yields finally $\cot \phi_{STA} = 0.0242\ \Gamma + 0.726$ with Γ in °C and ϕ_{STA} in °Lat. This empirical relation (Fig. 1) can be used to describe the seasonal migration of ϕ_{STA} as a function of Γ; it gives also realistic results of the present asymmetry where the annual averages of ϕ_{STA} are situated near Lat. 36°N and 31°S.[1]) The annual averaged position of the intertropical convergence zone (ITCZ) is now 6°N; the distances towards ϕ_{STA} at both hemispheres are not quite symmetric, due to the greater intensity of the southern hemisphere circulation (see Flohn 1982, p. 158f).

An extrapolation of this relation towards the conditions of unipolar glaciation needs an extrapolation of the tropospheric lapse rate above an ice-free Arctic Ocean, which could be based on model results based on some unknown assumptions (e.g. H_2O and CO_2 content, cloudiness) which cannot be verified. A reasonable estimate could be based on the results of a comprehensive circulation model (Manabe-Wetherald 1980, 1981; Manabe 1983), which gives the expected temperature increase after doubling or quadrupling of the CO_2 content as a function of latitude. While this model does not include an open Arctic Ocean, it has been shown (Flohn 1981) that this situation may occur after a rise of the CO_2 content to 600-650 ppm, if one takes into account the additional greenhouse effect of other trace gases (Kellogg-Schware 1981; McDonald 1982, Bach 1982). Manabe-Wetherald (1980, Fig. 19) suggest a decrease of Γ with increasing global surface temperature

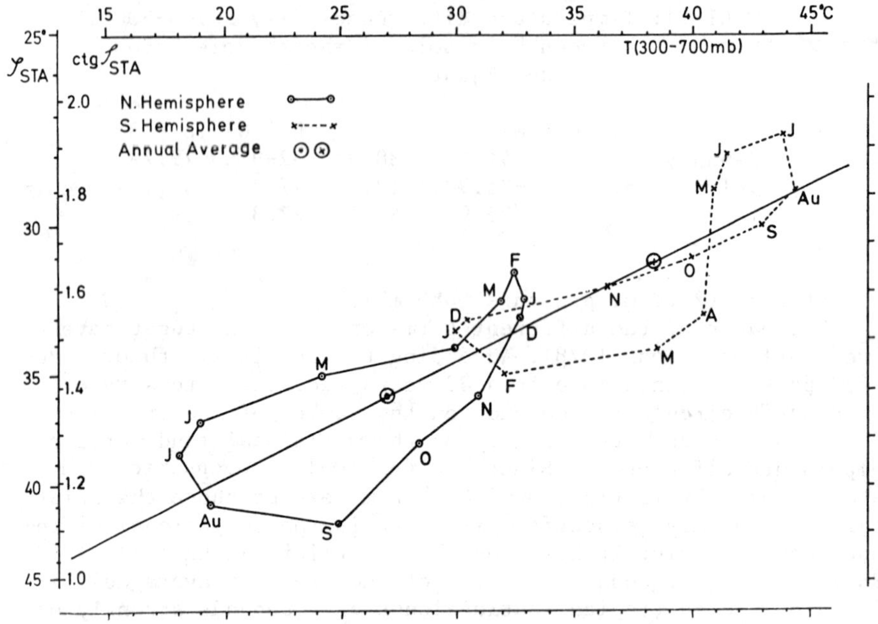

Figure 1. Seasonal Course of temperature difference Equator-Pole (300/700 mb layer) Γ and latitude ϕ_{STA} of subtropical anticyclones at both hemispheres (zonal average) (Flohn 1983b).

T_s - as caused by the well-known snow-ice-albedo feedback mechanism - as follows:

$$\Delta\Gamma = -K\Delta T_s$$

The amplification factor -K can be estimated to be at least 1.4; if one takes into account the expected disappearance of the surface inversion during most of the year (except summer), values of 1.5-1.7 seem to be more realistic. ΔT_s is estimated to be +4 or 5°C; arguments for such values see Flohn (1982).

Under these assumptions the annual averaged ϕ_{STA} should be displaced from 36°N to 39-40°N. Similar considerations for the seasonal values lead to the result, that the subtropical anticyclones can be expected to migrate 4-6° Lat. towards north during winter, but only 2-3° Lat. during summer. Since the heat budget variations of the Antarctic can be assumed to be much smaller than above the Arctic, the hemispheric asymmetry must increase; we assume therefore a displacment of the ITCZ towards the vicinity of Lat. 10°N.

This gives a background for an estimate of the shift of all climatic belts towards north. One of the most important consequenses should be a narrowing of the (interrupted) northern belt of subtropical winter rains, which is controlled (in the zonal average) by the position of the STA during summer and winter. Indeed the STA is now situated near 33°N in winter and 40°N in summer; this should shift to 38-39°N in winter, but only to about 42°N in summer.

This semi-empirical result is in principle coherent with the model results of the GFDL group (Manabe-Wetherald 1980, Fig. 14) in spite of the fact that this swamp ocean model is based on a idealized geography without an Arctic Ocean; the (geophysically more realistic) model with a mixed-layer (but static) ocean (Manabe-Stouffer 1980) simulates a seasonally ice-free Arctic. However, the latter model assumes a mixed layer which is designed only for seasonal storage of heat; (also in Parkinson-Kellog 1979); the heat added to the Arctic Ocean is not large enough to provide the difference between now and an ice-free status. Both model assumptions are insufficient to provide for a permanently ice-free Arctic Ocean. We can use the model results for a fourfold CO_2 content; since the combined greenhouse effect of CO_2 and of other infrared absorbing trace gase is now probably 70 (Hansen et al. 1981) or even 100 percent higher than that of CO_2 alone. Our result is also coherent with the paleoclimatic facts as outlined in an unpublished paper by Budyko and his co-workers (1981). Figure 2 combines for precipitation (resp for the hydrological balance precipitation - evaporation), these three completely different approaches: their coincidence within the belt around Lat. 40°N is striking. In view of these results, Revelle (1982) has stressed the possible role of such climatic change for water budget and water supply in the highly climate-sensitive areas of the subtropical winter rains (e.g. California/Nevada, the Mediterranean, Near and Middle East). Other available results (see Schlesinger 1983) should be disregarded, so far as they do not take into account the fundamental change of the heat and hydrological budget pattern as caused by an ice-free Arctic Ocean.

Using these three quite different approaches in comparison, estimates of the future distribution of T_s as a function of (northern) latitude have been given elsewhere (Flohn 1983b). Most probably the temperatures in the Tropics cannot be much higher - due to the upper limits of $\sim$30°C of the undisturbed sea surface temperature in the Tropics at constant solar irradiance (Newell et al. 1981) - , i.e. not more than 2°C. At Lat. 50°N and increase of the annual surface temperatures of 4-5°C can be expected - as evidenced e.g. by the Pliocene vegetation in Central Europe and the European part of the USSR. In the coastal areas of the Arctic Ocean around Lat. 70-80°N a temperature rise of 7-12°C has been estimated. However, such latitudinal annual averages are of little value because of the observed large longitudinal and seasonal variations of surface (and upper air) temperatures, which must also exist in the case of an ice-free Arctic Ocean.

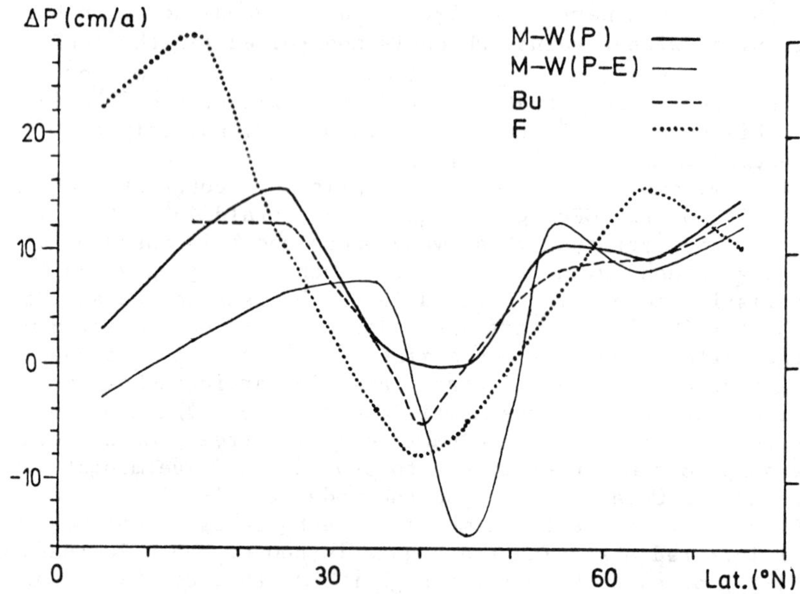

Figure 2. Expected changes of precipitation resp precipitation-evaporation (P-E) from model results (M-W), from paleoclimatic data (Bu, unpublished results) and semi-emperical methods (Flohn 1982, 1983b; Manabe-Wetherald 1980; cf. also Budyko 1982).

Figure 3 shows an early version of the expected pattern of climatic zones. It is based on the actually observed longitudinal contrast (left side); the turniplike shape of an idealized continent has already been suggested by W. Köppen (1931). A numerically more realistic version of it has been used by C. Troll (1947, 1948). Since it represents rather well the latitude-dependent distribution of land and sea, it may be advisable to use an appropriate version also in a 3-dimensional general circulation model, possibly adding a mountain chain at the western flank of the continent comparable to the Rocky Mountains-Andes in the Americas. In our context this is nothing more than a scenario which could be improved (or altered) in some details. At the eastern flank of each continent a marked trough in the upper westerlies penetrates the belt of subtropical anticyclones into the Tropics. Here the arid belt is interrupted: this should also be the case (albeit weakened) in Figure 3 (Late Tertiary) at the northern continents. At the southern hemisphere these quasi-permanent troughs are now well developed in the southern Atlantic and (to a lesser degree) in the southwestern Indian Ocean. A particularly strong and permanent trough (South Pacific Convergence Zone) is regularly observed in the mid-Pacific without any recognizable

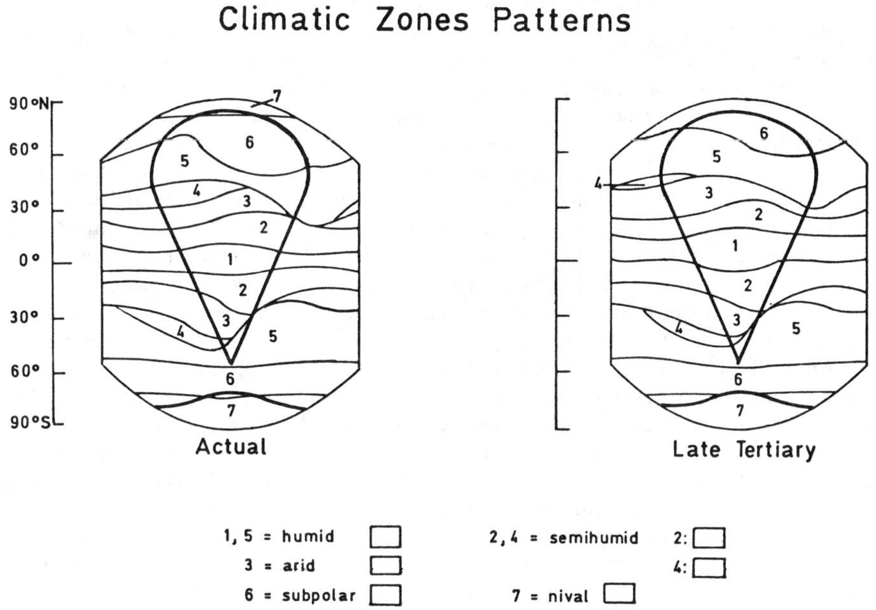

Figure 3. Latitude of climatic belts at an idealized earth with one continent. Zone 1 equatorial, 5 extratropical humid climates, 2 semihumid (summer rains), 3 arid, 4 semihumid (winter rains), 6 subpolar climates, 7 snow and ice.

relation to Australia (Flohn 1975). In high southern latitudes the expected changes are rather small, even if the greenhouse effect of increased CO_2 (and H_2O) will cause a weak atmospheric warming and also retreat of the seasonal ice-cover at the subantarctic oceans. At any rate, model experiments with a realistic ocean-atmosphere interaction - which is responsible for evaporation, heat storage and transport - are needed to understand the important regional climatic pattern.

CONCLUSIONS AND SUMMARY

It is fairly well established that during late Miocene and early Pliocene, between about 14 and 3 Ma ago, a stronger asymmetric climate pattern existed on a unipolar glaciated earth, with a remarkable stability superimposed to relative minor fluctuations. Numerical estimates of the climatic zonation during this rather recent period, based on the available quantitative techniques, are unquestionably of high interest, in spite of the fact that some fundamental boundary conditions of climate (land-sea pattern, mountain height) have changed during and since this time.

A possible future return to a simular situation - after a significant change in the composition of the atmosphere but under present geological boundary conditions - has been expected by several authors (Gerassimov, Flohn 1979; Budyko 1982; Flohn 1982; Revelle 1982) perhaps in the second half of next century or in the 22nd century. This depends critically on the future evolution of the CO_2-induced greenhouse effect; it is one of the greatest challenges to climatologists. It necessitates an effective cooperation at least among the geophysical disciplines including terrestrial and marine geology. The persistence of the (East) Antarctic ice-shield is ascertained for many ka to come; a possible disintegration of the marinebased parts, mainly but not exclusively in the West Antarctic, has been discussed elsewhere (Denton-Hughes 1981; Bentley 1983). What are the conditions under which the evolution of an ice-free Arctic Ocean could occur? It has been suggested (Budyko 1982; Bach 1982; Flohn 1981, 1983b) that an increase of atmospheric CO_2 content to 600-650 ppm, together with an 50-100 percent contribution of infrared absorbing gases (Hansen 1981 and others) would be sufficient to reach an equilibirum state coherent with an ice-free Arctic. The latter contribution suggests that in model computations a 3-4 fold increase of CO_2 is more indicative than a mere doubling. A negative (damping) feedback of such an evolution should be the expected increase of cold season snowfall along the coast of the then warm Arctic Ocean. This would include Greenland, where the rising melting rate could be reduced because of this effect. A rise of ocean temperatures by a few degrees ^oC due to advection of warmer North Atlantic water is expected by all comprehensive climate models; this and the accompanying rise of air temperature should elongate the summer melting period, with the result of an exponentially accelerated melting of the thin (2-3 m) broken Arctic sea-ice cover. The decrease of the surface albedo from about 40-60 percent (during melting season) to 10 percent and less (open water) causes, during the polar summer, efficient absorption of solar radiation and storage of heat in the upper 30-50 m layer of the ocean. Thus after a short time (estimated to be a few decades or even less) this new equilibrium temperature can be expected to remain stable for many centuries, before the unsaturated deep ocean had had sufficient time to swallow the surplus of atmospheric CO_2

REFERENCES

Bach, W., 1982: Gefahr für unser Klima. C.F. Müller, Karlsruhe.
Bentley, C.R., 1983 in: Proc. Carbon Dioxide Research Conference, Sept. 1982, U.S. Dept. of Energy 021 (1983), pp. IV. 1-88.
Berggren, W.A., Hollister, C.D., 1977: Tectonophysics 38, pp. 11-48.

Budyko, M.J., 1982: The Eart's Climate: Past and Future. Intern. Geophys. Ser. 29, New York, 307 pp.
Clark, D.L., 1980: Spec. Paper Geol. Soc. 181; Nature 300(1982), 321-325.
Clark, W.C. (Ed.), 1982: Carbon Dioxide Review 1982. Clarenden Press Oxford, XIX + 469 pp.
Dansgaard, W. et al., 1982: Science, 218, pp. 1273-1278.
Denton, G.H., Hughes, T.J. (Ed.) 1981: The Last Great Ice Sheets. J. Wiley, New York, 2 Vol.
Flohn, H., 1964a: Bonner Meteor. Abhandl. 4; Z.f. Meteor 17(1965), 316-320, and enlarged in H. Mensching (Ed.): Physische Geographie der Trockengebiete, Darmstadt 1982, pp. 55-69.
Flohn, H., 1964b: Geol. Rundschau 54, pp. 504-515.
Flohn, H., 1968: Colorado State University, Atmospheric Science Paper 130, 120 pp.
Flohn, H., 1971: Bonner Meteor. Abhandl. 15, 55 pp. and 21(1975), 82 pp.
Flohn, H., 1978 in: E.M. van Zinderen Bakker (Ed.): Antarctic Glacial History and World Palaeoenvironment. Balkema, Rotterdam, pp. 3-13.
Flohn, H., 1981: Physik. Blätter 37, pp. 184-190.
Flohn, H., 1982 in: W.C. Clark (Ed.): Carbon Dioxide Review 1982, pp. 143-179.
Flohn, H., 1983 in: A. Street-Perrott (Ed.): Variations in the Global Water Budget. D. Reidel, Dordrecht, pp. 403-418.
Flohn, H., 1983b in: W. Bach et al. (Ed.): Carbon Dioxide Current Views and Developments in Energy/Climate Research. D. Reidel, Dordrecht, pp. 299-314.
Frenzel, B., 1968 in: Erdwissenschaftliche Forschung 1, Wiesbaden; Science 161(1968), pp. 637-649.
Gerassimov, J.P., Flohn, H., 1979 in: Proceedings of the World Climate Conference, Geneva. 1979. WMO-No. 537, pp. 88-111, 243-266.
Gradstein, F.M., Srivastava, S.P.. 1980: Palaeogeography, Palaeoclimatology, Palaeoecology 30, pp. 261-295.
Hahn, D.G., Manabe, S., 1975: Journ. Atmos. Sci. 32, pp. 1514-1541.
Hansen, J. et al., 1981: Science 213, pp. 957-966.
Hopkins, D.M. et al., 1971: Palaeogeography, Palaeoclimatology, Palaeoecology 9, pp. 211-231.
Keigwin jr., L.D., 1978: Geology 6, pp. 630-634.
Keller, G., Barron, J.A., 1983: Geol. Soc. Amer. Bull. 94, pp 590-613.
Kellogg, W.W., Schware, R., 1981: Climate Change and Society. Westview Press, Boulder Col., XII + 178 pp.
Köppen, W., 1931: Grundriss der Klimakunde. Berlin, Leipzig (1st edition 1923), see p. 137 f.
Li, J.J. et al., 1979: Scientia Sinica 22, pp. 1314-1328.
Lorius, G., 1983 in: W. Bach et al. (Ed.): Carbon Dioxide: Current Views and Developments in Energy/Climate Research. D. Rei-

del, Dordrecht.
Manabe, S., 1983: Advances in Geophysics 25, pp. 39-82.
Manabe, S., Stouffer, R.J., 1980: Journ. Geophys. Res. 85, pp. 5529-5554.
Manabe, S., Wetherald, R.T., 1980: Journ. Atmos, Sci. 37. pp. 99-118.
McCauley, J.F. et al., 1982: Science 218, pp. 1004-1020.
McDonald, G.J.M. (Ed.), 1982: The Long-Term Impact of Increasing Atmospheric Dioxide Levels. Ballinger, Cambridge (Mass.).
McKenna, M.C., 1980: Palaeogeography, Palaeoclimatology, Palaeoecology 30, 349-362.
Neftel, A. et al., 1982: Nature 295, pp. 220-223.
Newell, R.E., Gould-Stewart, E., 1981: Journ. Atmos. Sci. 38, pp. 2789-2796.
Opdyke, N.D. et al., 1979: Palaeogeography, Palaeoclimatology, Palaeoecology 27, 1-34.
Parkinson, C.L., Kellogg, W.W., 1979: Climatic Change 2, pp. 149-162.
Revelle, R., 1982: Scientific American: German Edition, October 1982, pp. 16-25.
Sarnthein, W. et al. 1982 in: U.V. Rad et al. (Ed.): Geology of Northwest African Continental Margin. Springer Verlag, Berlin, pp. 545-604.
Schlesinger, M.E., 1983: Journ. Environm. Studies, pp. 103-114.
Schweitzer, H.J., 1980: Palaeogeography, Palaeoclimatology, Palaeoecology 30, pp. 297-311.
Shackleton, N.J., Opdyke, N.D., 1977: Nature 270, pp. 216-219.
Strauch, F., 1972: Z. Dt. Geol. Ges. 123, pp. 163-177.
Troll, C., 1947: Ber. Geobot. Forsch. Inst. Rübel 1947, pp. 46-63; Göttinger Geogr. Abhandl. 1948, pp. 11-27.
Velichko, A., 1982: Izvest. Acad. Nauk USSR, Ser. Geogr. 1982, No. 1, pp. 15-27.
Webb, S.D., 1976: Palaeobiology 2, pp. 220-234.
Yeh, T.C., Gao, Y.X., 1979: Meteorology of Qinghai-Xizang Plateau (in Chinese), Beijing, 279 pp.; Bull. Amer. Meteor. Soc. 62(1981), pp. 14-19, 31-35.
Zubakov, W.A., Borzenkova, I.I., 1983: Paleoklimaty Poznego Kainozoya. Leningrad, Gidrometeoizdat, 216 pp.

[1]) Extrapolating this relationship to the very small Γ during the early Tertiary and Mesozoic, the results for $\Gamma = 5^{o}$ (0°) would be 50°N (54°N); a rather similar result has already been given from theoretical considerations (Flohn 1964b).

ENERGY-FLOW BUDGETS IN AQUATIC ECOSYSTEMS AND THE CONFLICT BETWEEN BIOLOGY AND GEOPHYSICS ABOUT EARTH-AXIS TILT.

G. Høpner Petersen

Zoological Museum, University of Copenhagen

ABSTRACT.

Quantitative studies on energy flows in arctic, temperate and tropical ecosystems on the marine shelf and in lakes indicate that in the arctic the benthic energy flow dominates, and in the tropics the pelagic flow dominates. This is confirmed by the different structures of the ecosystems. The variation between midnight sun and polar darkness, the timelag between the steps in the energy flow, the synchronization of the organism's life histories, and the low temperature all favour the benthic energy flow in the subarctic. Polar darkness seems to have appeared 20 - 30 million years ago. This is in conflict with the geophysical and astronomical theories, which assume an almost constant earth axis tilt through the ages. The data may be taken to suggest that the earth axis' tilt is changed from the present angle of about 23° to approximately perpendicular to the ecliptic by unknown forces two times during a cosmic year.

INTRODUCTION.

The original aim of these studies was to check the estimates of benthic production against those of other types of production (Curtis 1977, Petersen 1964, 1978, Petersen & Curtis 1980). It was later extended to studies of other comparable ecosystems (see e.g. Harvey & Geen 1974, Jonasson 1972, Petersen 1982, Serruya 1978, Steele 1974). A number of papers and personal communications were also obtained and budgets on energy flow were set up according to the simplified model of a total flow in an ecosystem (Fig. 1). The ecosystems considered were marine shelf areas or lakes with

N.-A. Mörner and W. Karlén (eds.), Climatic Changes on a Yearly to Millennial Basis, 621–633.

the benthic layer at or just below the photosynthetic layer, thus no benthic plants, and only minor exchange with other ecosystems. The marine environment was classified as: 1) Eupolar seas (high arctic, shelf ice areas), where sea ice is present both summer and winter. The production measurements are too few to allow energy budgets for these areas. 2) Subpolar seas (subarctic, subantarctic), where sea ice is present during winter and spring but not in summer. Besides large areas in the North Atlantic, the North Pacific and around Antarctics, these seas also include the northern Baltic and the northern Caspian Sea. 3) Tropical seas, where the temperature is 25-30°C year round and a permanent thermocline is present below the photosynthetic layer. 4) Intermediate seas, areas with conditions intermediate between those of subpolar and tropical areas.

RESULTS.

Figs. 2, 3, and 4 give the energy budgets for 12 selected ecosystems. In the arctic-subarctic ecosystems (Char Lake, Disko Bugt) the benthic flow dominates, while the pelagic flow dominate in the subtropical-tropical ecosystems (Phangnga Bay, Israeli Shelf, Lake Kinnereth). The temperate ecosystems (North Sea, Scotian Shelf, Lake Esrom) are intermediate. These different energy flows are reflected in the composition of species and life forms present. An independent confirmation of this trend is given by the occurrence of top predators such as marine mammals and marine birds. Their production is not included in the energy flow budgets because it is not known, and it probably plays a very small part. In the subarctic seas are walrus, bearded seal, common eider, and king eider, all feeding on benthos, and in subarctic lakes are "ducklike" birds. These species or species with equivalent life forms are rare or absent in tropical waters. The subarctic marine mammals and marine birds which feed on pelagic organisms are mainly migratory. To explain the difference between subarctic and tropical ecosystems, the temperature difference alone is not sufficient. The very different periods in the cycles of light and darkness seem important (see Figs. 5 and 6). In the subarctic the period for a complete cycle is 365 days, which is long enough to allow the water mass to be cooled down enough to give a vertical mixing and to freeze to sea ice. There is a considerable time lag from solar energy input to heating and primary production. In the equatorial belt the period is 24 hours, which is short enough to allow the water to keep a constant temperature (25-30°C) and thus to have a permanent termocline, and there is almost no time lag from solar energy input to primary production.

A first step in the energy calculation is that photosynthesis is almost independent of temperature, while metabolism is temperature dependent. The algae's own metabolism is probably a serious competitor for the berbivores, especially in warm waters

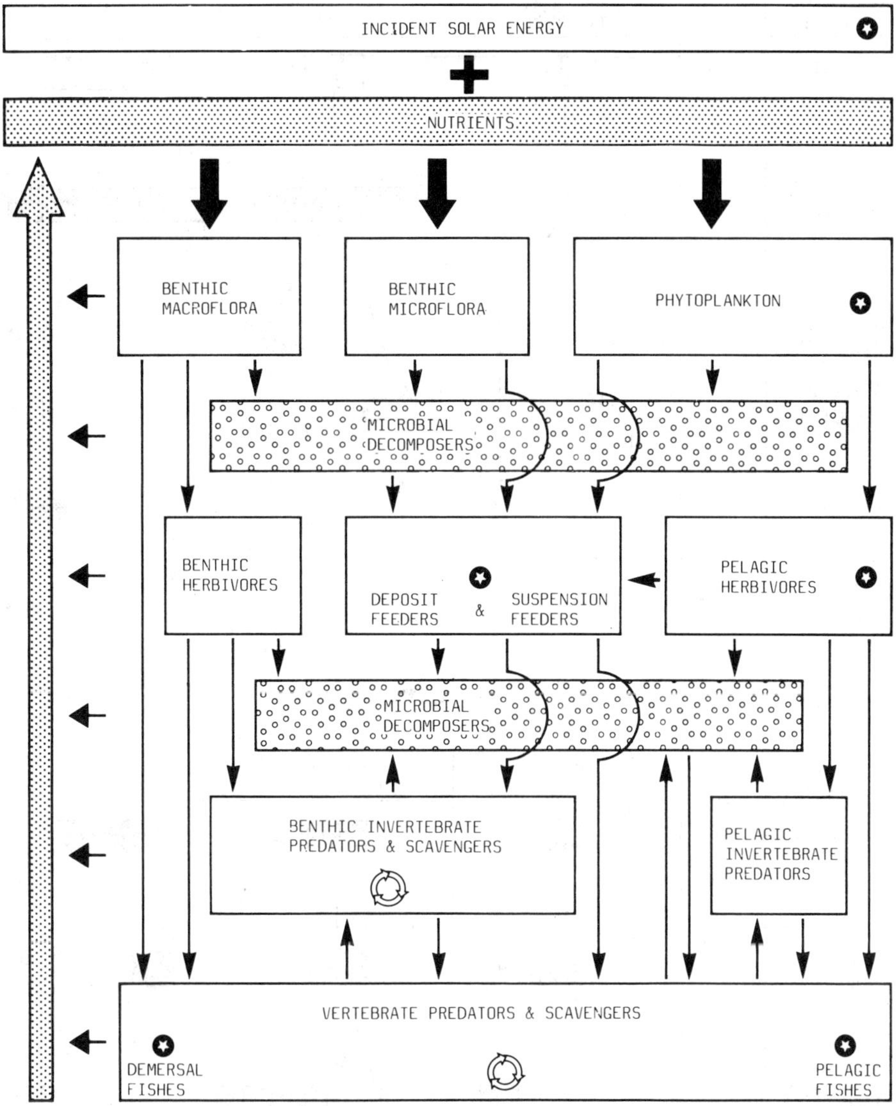

Figure 1. A flow model of an aquatic ecosystem. Three thick vertical arrows: input of organic energy. Six thin horizontal arrows: release of nutrients. Remaining arrows: main transfer pathways. Recycling symbol: recycling within the box through predation and/or cannibalism. The steps selected for the energy flow models in this paper are marked with a star in a black circle.

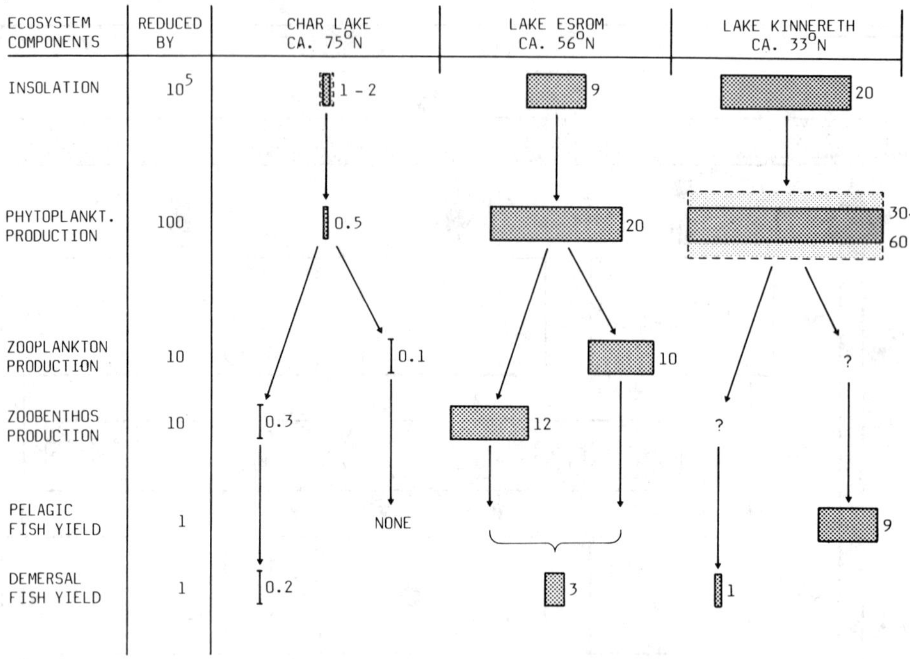

ECOSYSTEM COMPONENTS
REDUCED BY
CHAR LAKE CA. 75°N
LAKE ESROM CA. 56°N
LAKE KINNERETH CA. 33°N
INSOLATION
10^5
1 - 2
9
20
PHYTOPLANKT. PRODUCTION
100
0.5
20
30-60
ZOOPLANKTON PRODUCTION
10
0.1
10
?
ZOOBENTHOS PRODUCTION
10
0.3
12
?
PELAGIC FISH YIELD
1
NONE
9
DEMERSAL FISH YIELD
1
0.2
3
1

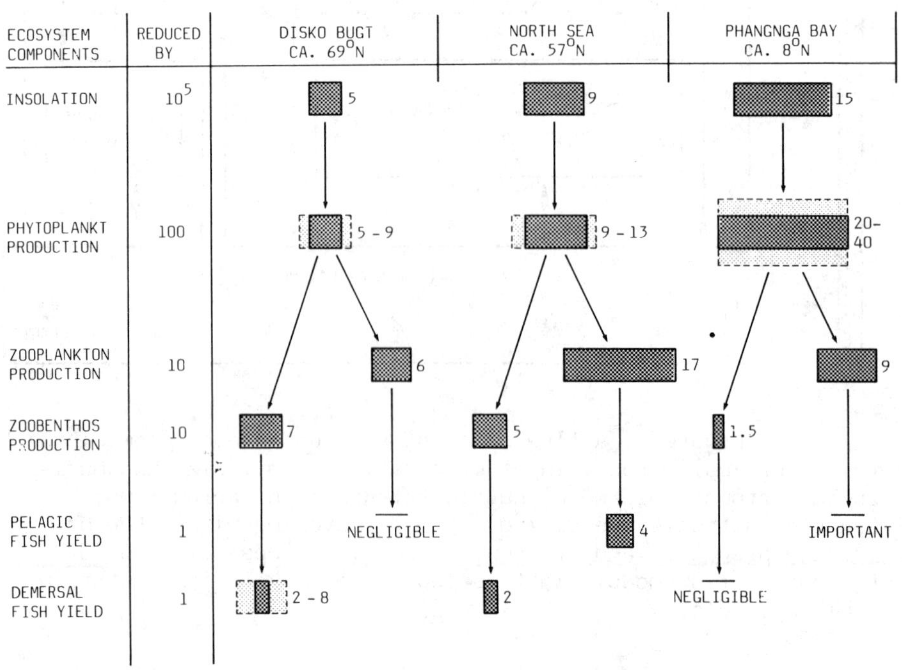

ECOSYSTEM COMPONENTS
REDUCED BY
DISKO BUGT CA. 69°N
NORTH SEA CA. 57°N
PHANGNGA BAY CA. 8°N
INSOLATION
10^5
5
9
15
PHYTOPLANKT PRODUCTION
100
5 - 9
9 - 13
20-40
ZOOPLANKTON PRODUCTION
10
6
17
9
ZOOBENTHOS PRODUCTION
10
7
5
1.5
PELAGIC FISH YIELD
1
NEGLIGIBLE
4
IMPORTANT
DEMERSAL FISH YIELD
1
2 - 8
2
NEGLIGIBLE

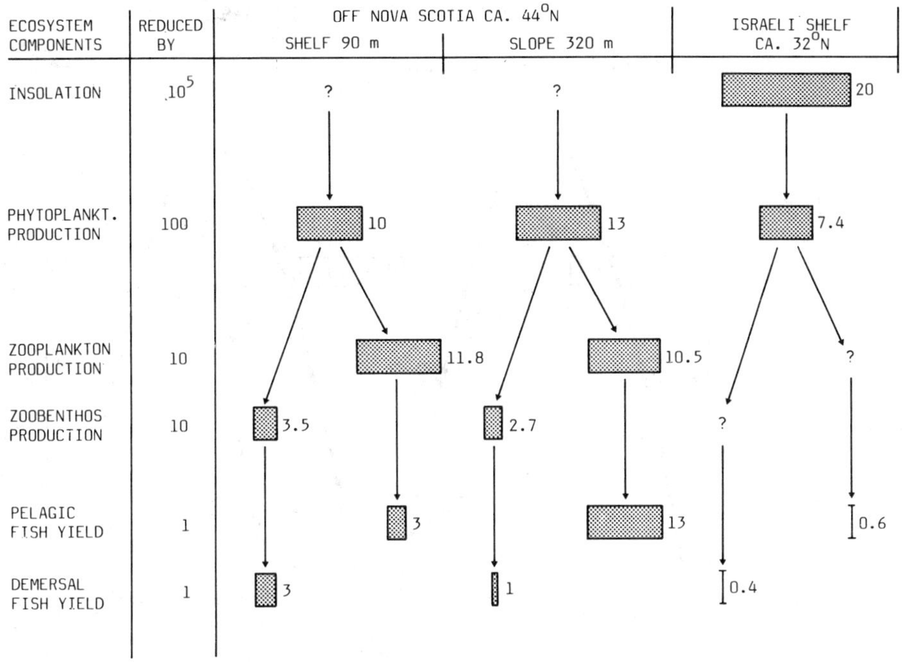

Figure 4. Energy flow models for a shelf locality and for a slope locality, and for the Israeli shelf in the eastern Mediterranean. All values expressed as kcal/m^2/yr.

Figure 2 (opposite page, above). Energy flow models for lakes (20-40 meters deep) from arctic, temperate and subtropical regions. All values expressed as kcal/m^2/yr.

Figure 3 (opposite page, below). Energy flow models for shelf ecosystems (about 50 meters deep) from subarctic, temperate, and tropical regions. All values expressed as kcal/m^2/yr.

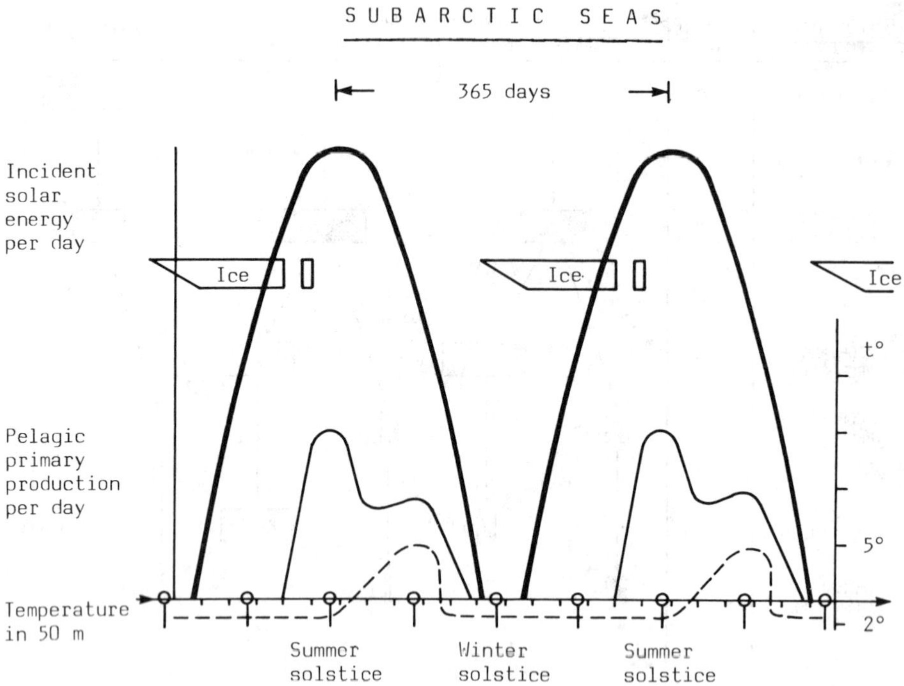

Figure 5. The subarctic sea. The variation in incident solar energy (thick line) during a twoyear period, showing the subsequent variation in pelagic primary production (thin line), sea-ice cover and temperature (broken line) at about 50 meters depth.

(see Fig. 7). A complete balance sheet for naturally occurring populations and life forms with all the connections to each population's or life form's neighbours and environment is not yet possible. Figs. 8 and 9 give a few of the typical features of subarctic and tropical ecosystems. In the subarctic, when the thick layer of sea-ice disappears, the pelagic primary production bursts into the spring diatom bloom, even though the temperature in the water mass is still low. The spawning, feeding, migration etc. of the animals are synchronized, directly or indirectly, to this bloom. The low temperature causes a low metabolism and a low feeding activity for the poikilotherms. This may to be compensated for by utilizing passive food collecting, for example, by means of hydrodynamic forces created around the suspension-feeding animals (Petersen, 1978, Vogel 1978). In tropical waters it is necessary to be close to the food source, e.g. to have symbiosis with algae (zooxanthellae) or to utilize the high metabolism caused by

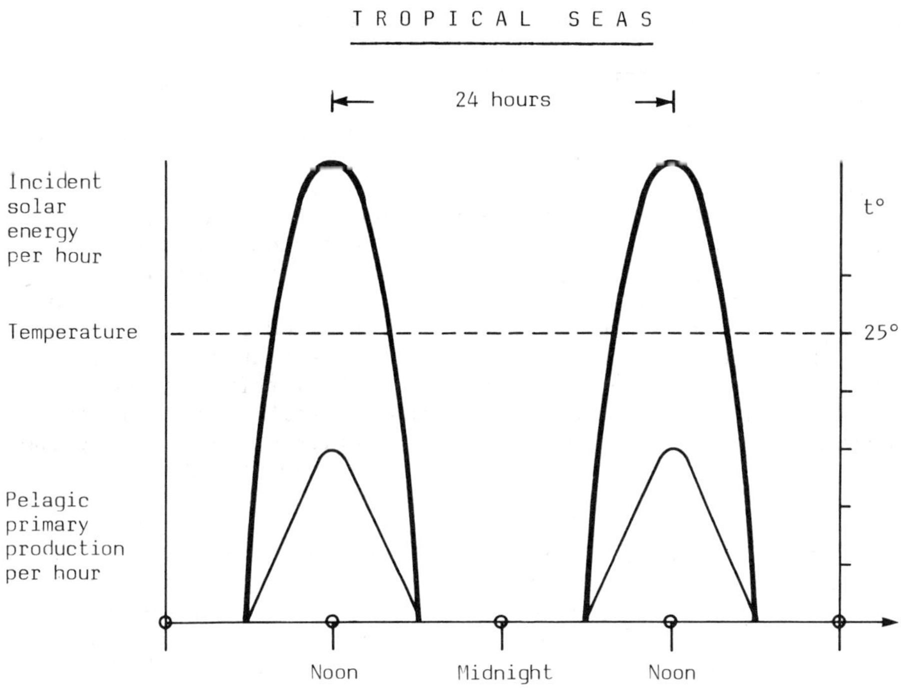

Figure 6. The tropical sea. The variation in incident solar energy (thick line) during a twoday period, showing the almost constant high temperature (broken line) and negligible time lag for photosynthesis (thin line).

the high temperature and be fast and active. Cryptobiological life forms like tube dwellers, cave dwellers, deep burrowers, and borers are very common in the tropical seas, but rare or absent in subarctic ones. These energetic positions seem profitable in tropical waters, but impossible in the arctic. The presence of such life forms in fossil material may be indicators of the paleoclimate.

The energetically profitable positions are qualitatively different in the different environments; in the tropical waters the number of profitable positions seems larger and the biomass necessary for the survival of the species seems smaller. It is here postulated that there are a limited number of energetically profitable positions in each ecosystem, and that each position is exploited by a number of taxonomic groups, whereby analogous life forms and life strategies are developed. It is also postulated that roughly all possible positions are occupied. (The current concepts of niche, habitat, diversity, stability, evolution are not appropriate in this connection.)

The age of the different ecosystems.

Members of the Buccinidae (Gastropoda) and Amphipoda, important invertebrate predators and scavengers on the subarctic benthos, developed during the Tertiary (Golikov & Tzvetkova 1972). Information from palaeontologists reveals that other species utilizing the subarctic benthos, e.g. walrus, eiders, codfish, and flatfish were also developed in the Tertiary, while the tropical fauna usually is described as more "conservative". If it is true that the subarctic ecosystem's design is determined by a longer photoperiod than that in the tropical and that the predators utilizing the subarctic benthos developed during the Tertiary, then the arctic light regime, i.e., the earth-axis tilt, also must have developed in this period. Attempts to verify or falsify this conclusion lead to a discussion of climatic changes and their causes. The word ice age is used both for the cold periods in the Quaternary, and for the 6 or 7 ice ages that have occurred since Precambrium. The observed climatic changes have been linked both to terrestrial and to astronomical events, for instance to magnetic reversals, mountain-building periods, etc. The following references give a survey of the different opinions: Beaty 1978, Bucha 1980, Dawson 1980, Donn 1982, Frakes 1979, Frydendahl 1978, Hays et al. 1976, John 1979, Mörner 1978, Pearson 1978, and Wolfe 1978.

The present paper's conclusions concerning the climatic influence on the differences between recent polar and equatorial ecosystems place it, together with many other biological papers, among the "tilters": those who believe in large changes of earth's axis' tilt.

This conclusion is in conflict with the modern geophysical and astronomical theories ("non-tilters"): those who believe in unchanging earth's axis' tilt. In order to preserve the geophysical-astronomical model the palaeomagnetic measurements and the

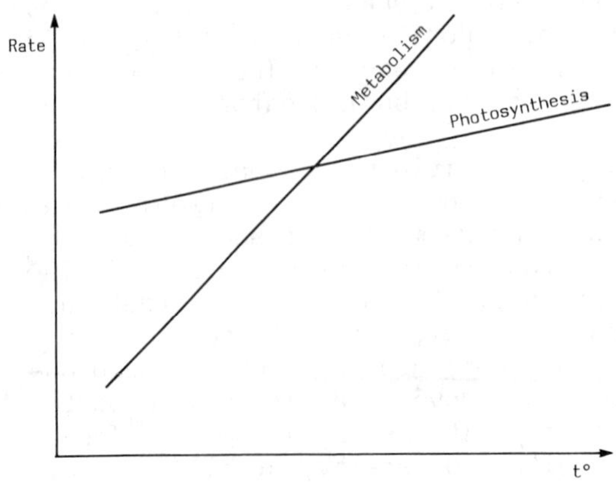

Figure 7.

The influence of temperature on the rates of photosynthesis and metabolism. The units and the crossing point are arbitrary, but the lines give the difference between the two processes.

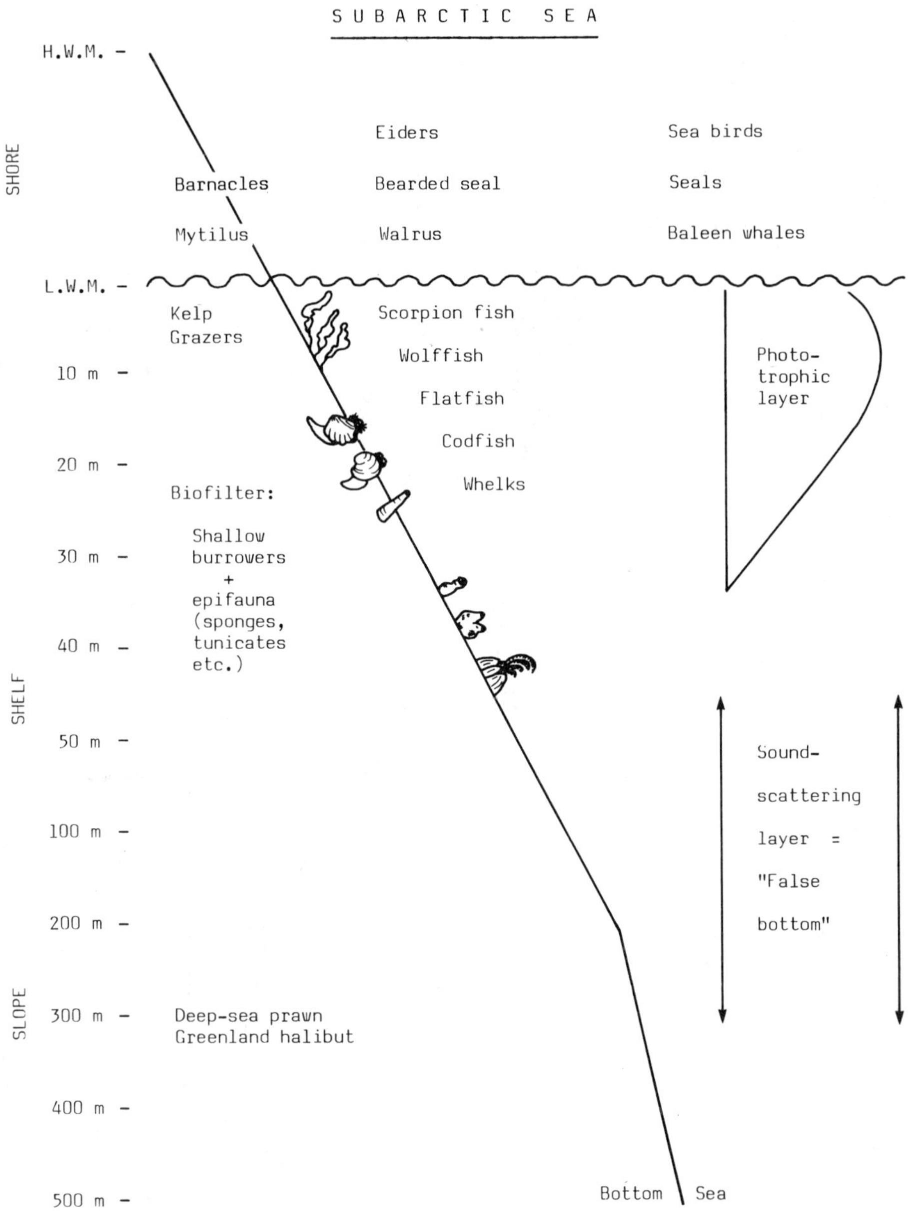

Figure 8. The design of a subarctic marine ecosystem, showing some of the more important species and life forms.

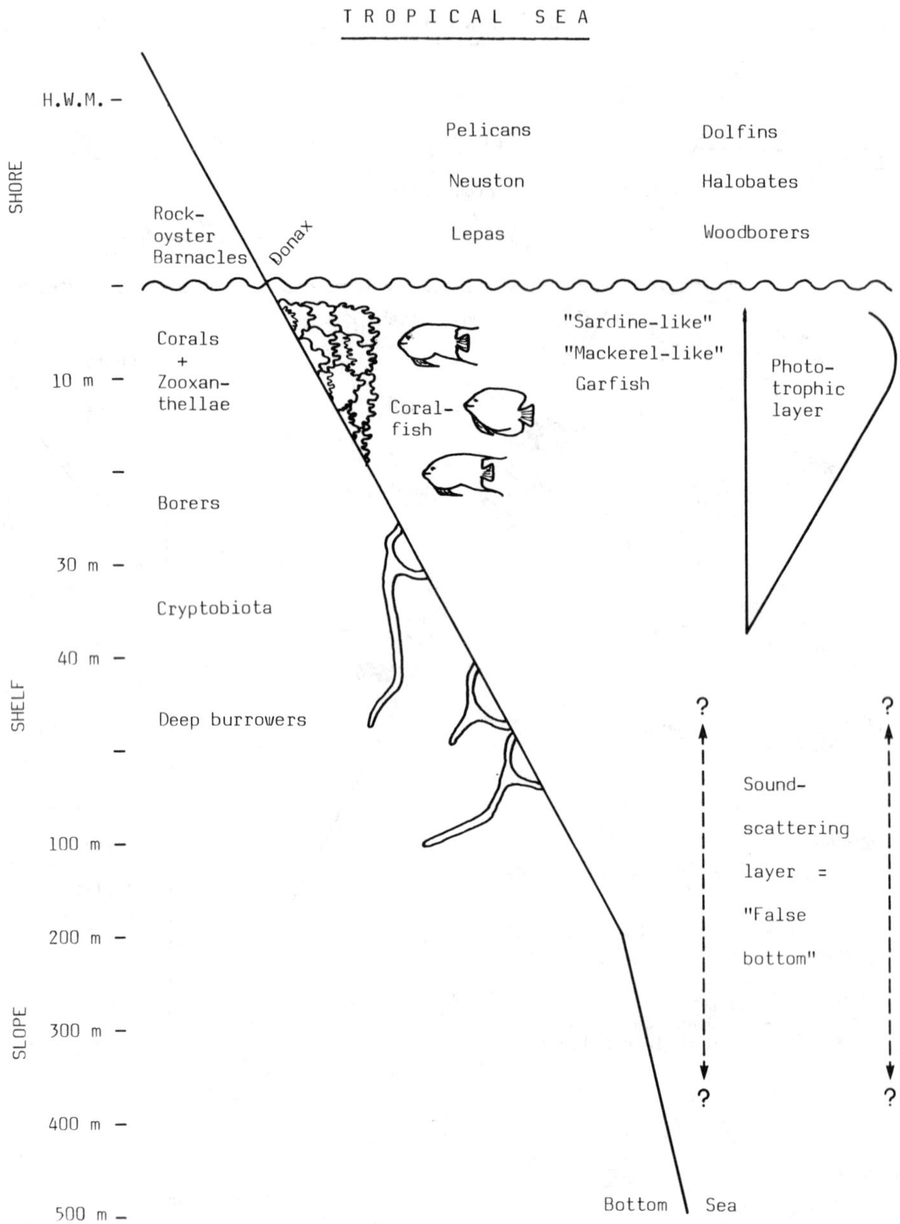

Figure 9. The design of a tropical ecosystem, showing some of the more important species and life forms.

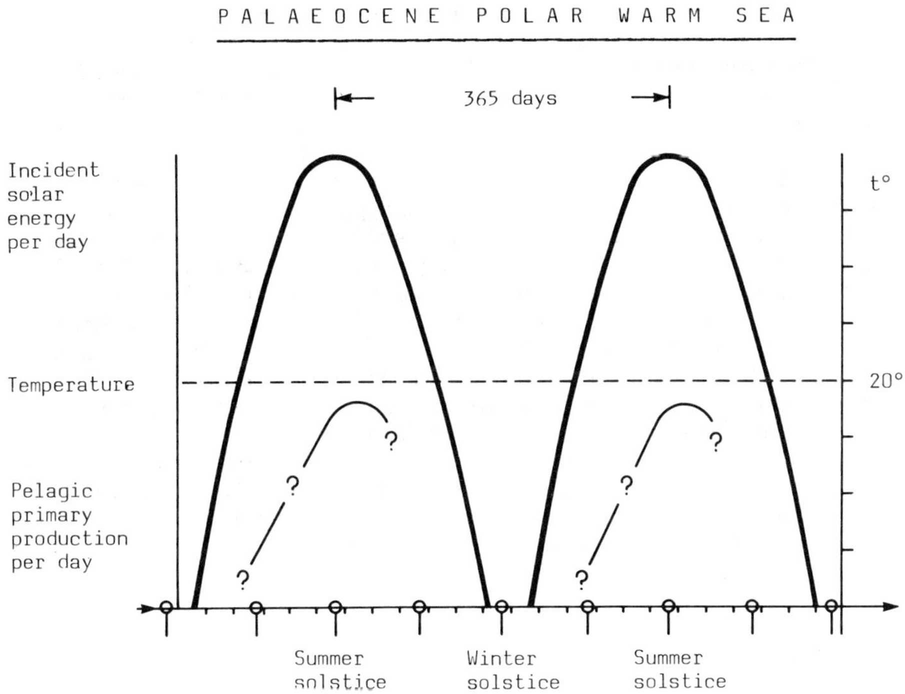

Figure 10. Hypothetical situation for a Palaeocene polar warm sea with arctic light conditions (thick line). The temperature (broken line) is high (20°C) due to heat flow from the core or to heat transport from the equatorial belt. Pelagic primary production (thin line) could occur during summer and be grazed by migratory species. Sessile plants might depend on heterotrophy during winter. Fossil ecosystems adapted to such conditions are not known.

plate-tectonic theory have been questioned. Consequently in order to fit their data to the geophysical model, some palaeontologists have expressed a very large confidence in the adaptability of organisms, which in my opinion will lead to impossible energetic positions for the species.

Let us imagine a polar ocean with a high temperature, say 20°C, and a photoperiod with long dark periods (Fig. 10). This situation would probably lead to deserts both on land and in the sea due to rapid mineralisation. Deep-sea animals near the thermal vents cope with similar situations today, and some plants and animals utilize dissolved organic matter, but the design of the ecosystems and life forms would be very different from what is known from recent and fossil records. The temperature increase in a polar sea could come either from an increase in solar radiation or its absorption, or an increase in the heat flow from the earth's core. However, with a tilted earth axis there would still be a greater heat exchange between the equatorial belt and the polar regions, by means of air and ocean currents.

Although I cannot discuss the geophysical and astronomical models professionally, I cannot accept them when they are contradicted by a great many biological and palaeontological observations. I can just make the following suggestion. The time scale involved in the evolution of modern arctic ecosystems (John 1979) indicate that the earth's axis is tilted two times during the cosmic year (ca. 220-240 million years) by some as yet unknown mechanisms or forces. The hypothesis might be studied by "bio-astronomy", a discipline that describe how life processes depend on astronomical parameters. Winter rings, tidal rings, day and night rings are well known. The models presented here show that also the tropical and the arctic light regime influence the design of life forms,of ecosystems,of diversity etc. Such large scale features can also bee traced in the fossil records and may give informations on some astronomical parameters during geological periods.

REFERENCES.

Beaty, Chester B., 1978. The causes of glaciation. American Scientist, 66, 452-459.

Bucha, Vaclav, 1980. Model of the world plate tectonics since the early Palaeozoic. Journal of Geomagnetism and Geoelectricity 32, Supplement III, 47-60.

Curtis, Mark A., 1977. Life cycles and population dynamics of marine benthic polychaetes from the Disko Bay area of West Greenland. Ophelia, 16 (1), 9-58.

Dawson, M.R. (Ed.), 1980. Cenozoic history in and around the northern Atlantic and arctic oceans. Palaeogeography, Palaeoclimatology, Palaeoecology, 30 (3-4), 217-363.

Donn, William L., 1982. The enigma of high-latitude paleoclimate. Palaeogeography, Palaeoclimatology, Palaeoecology, 40, 192-221.

Estes, R. & Hutchinson, J. Howard, 1980. Eocene lower vertebrates from Ellesmere Island, Canadian arctic archipelago. Palaeogeography, Palaeoclimatology, Palaeoecology, 30 (3-4), 325-347.

Frakes, L.A., 1979. Climates throughout geologic time. Elsevier, Amsterdam. 310 pp.

Frydendahl, Knud (Ed.), 1978. Proceedings of the Nordic Symposium on Climatic changes and related problems, Copenhagen, 24-28 April 1978. Danish Meteorological Institute, Climatological Papers No. 4, 260 pp.

Golikov, A.N. & Tzvetkova, N.L., 1972. The ecological principle of evolutionary reconstruction as illustrated by marine animals. Marine Biology, 14, 1-9.

Harvey, Harold H. & Geen, Glen H. (Eds.), 1974. Limnology in Canada. Journal of the Fisheries Research Board of Canada, 31 (5), 1020 pp.

Hays, J.D., Imbrie, J. & Shackleton, N.J., 1976. Variations in the Earth's orbit: Pacemaker of the ice ages. Science, 194, 1121-1123.

Jonasson, Petur M., 1972. Ecology and production of the profundal benthos in relation to phytoplankton in Lake Esrom. Oikos, Supplement 14, 148 pp.

John, Brian S. (Ed.), 1979. The winters of the world. David & Charles, Newton Abbot, London. 256 pp.

McKenna, Malcolm C., 1980. Eocene paleolatitude, climate and the mammals of Ellesmere Island. Palaeogeography, Palaeoclimatology, Palaeoecology, 30 (3-4), 349-362.

Mörner, Nils-Axel, 1978. Recorded paleoclimatic changes: cyclicity, correlations, and relation to changes in the orbital geometry and to paleo-magnetic and paleogeoid changes. Danish Meteorological Institute, Climatological Papers, 4, 45-52.

Pearson, Ronald, 1978. Climate and evolution. Academic Press, London. 274 pp.

Petersen, G. Høpner, 1964. The hydrography, primary production, bathymetry and "Tagsaq" of Disko Bugt, West Greenland. Meddelelser om Grønland, 159 (10), 1-45.

- 1978. Life cycles and population dynamics of marine benthic bivalves from the Disko Bugt area of West Greenland. Ophelia, 17 (1), 95-120.

- 1982. Energy flow in comparable aquatic ecosystems from different climatic zones. International Council for the Exploration of the Sea. Symposium on Biological Productivity in the Temperate Zone of the North Atlantic. (Kiel, 2-5 March 1982). In press.

Petersen, G. Høpner & Curtis, Mark A., 1980. Differences in energy flow through major components of subarctic, temperate and tropical marine shelf ecosystems. Dana, 1, 53-64.

Serruya, C. (Ed.), 1978. Lake Kinnereth. Monographiae Biologicae, Vol. 32, 501 pp. W. Junk.

Steele, H., 1974. The structure of marine ecosystems. Blackwell Scientific Publications, Oxford, 128 pp.

Vogel, Steven, 1978. Organisms that capture currents. Scientific American, Aug. 1978, 108-117.

Wolfe, J.A., 1978. A paleobotanical interpretation of Tertiary climates in the northern hemisphere. American Scinetist, 66, 694-703.

- 1980. Tertiary climates and floristic relationships at high latitudes in the northern hemisphere. Palaeogeography, Palaeoclimatology, Palaeoecology, 30 (3-4), 313-323.

Epilogue

CLIMATIC CHANGES ON A YEARLY TO MILLENNIAL BASIS. CONCLUDING REMARKS

Nils-Axel Mörner

Stockholm University, Sweden

The Stockholm symposium brought forward a well of new observational data and new ideas and theories of their interpretation, all of which was discussed - applauded and/or criticized - at the symposium. The authors in this book have been given time to consider the material presented and discussed before they submitted their papers. Consequently, this book is a true proceedings of the symposium. This means that each paper must stay for itself and form its own conclusions (which also applies to this paper).

In this paper, I will only take up the point which the symposium was focused on, viz. the geographic extension (validity over the globe) of climatic changes in the time range inbetween the Milankovitch variables and the sunspot cycles (i.e. the typical Late Glacial and Postglacial fluctuations). I will also discuss the implications, with respect to possible causation mechanisms, of the results obtained.

1. GLOBALITY OR REGIONALITY

In the 60-ies and 70-ies several of us attempted global correlations. The idea of firm global correlations (in temperature, glaciation, sea level, etc.) became stabilized. Only a few critical voices were heard (e.g. Mercer, 1969). With the idea of geoid changes with time (Mörner, 1976, 1983), the concept of global correlations was seriously challanged, at least, when it concerns sea level. Further analyses including paleoclimatology (e.g. Mörner et al., 1982) lead me to propose that the 80-ies will be the time of non-correlations and/or negative correlations. This symposium and book have provided new - and, I would say, conclusive - evidence on the question of a globality or a regionality of

N.-A. Mörner and W. Karlén (eds.), Climatic Changes on a Yearly to Millennial Basis, 637–651.

climatic changes shorter than the Milankovitch variables (cf. Figs 1-2 of the introduction; Mörner, this volume).

a) The Late Weichselian

The Late Weichselian ice marginal fluctuations within the Nordic countries form a classical basis for the reconstruction of past climatic changes (see the introduction; Mörner, this volume). In this book, several papers have shown that even the Scandinavian records are in need of revision and improvements (Berglund & Mörner; Berglund et al.; Björck & Digerfeldt; Petersen). The Younger Dryas stands out as a most prominent climatic event; the beginning at 11,000 C14-years BP is sharp but the end becomes increasingly more problematic, and still this is to mark the Pleistocene/Holocene boundary. The Alleröd is a very distinct warm period with its maximum temperature at the end of the period. The Older Dryas is a problem both in age and characteristics. The Bölling interstadial has a problematic onset; the land records are poorly dated within the Nordic countries and seem in general to place the onset considerably younger than the marine records do, and the marine records show a clear and sudden invasion of a more temperate, boreo-arctic, fauna at around 12,700 BP. The proposed onset of a general "Bölling chronozone" (Mangerud et al., 1974) is places at the approximate and even age of 13,000 BP without direct geological substantiation (at least, within the Nordic countries). However, the 13,000-12,700 BP period marks a drastic change in several parameters, viz. in the mode of uplift, in the oceanic circulation and probably in the mode of deglaciation, too. Somewhere between 14,000 and 13,000 BP (maybe, at around 13,500-13,200 BP), there was a significant interstadial (the Vintapper or Lockarp interstadial). During this inverval, the ice seems to have left the North Sea region proper and withdrawn to the coastal side of the Norwegian Channel. It was followed by a cold phase during which the formation of the Lista moraines in Norway, the Low Baltic readvance and the formation of the Luga moraine in the East Baltic region took place.

The older part of the Late Weichselian is virtually undated within the Nordic region. It is therefore important that areas outside the glaciated region can provide additional information. The well-dated paleoclimatic record from Grande Pile in NE France (Mörner, 1979, 1981) record an interstadial/stadial change at 15,300 BP and a 3-parted main glaciation maximum; a cold peak 22,500-20,600 BP, an interstadial 20,600-19,300 BP and a second cold peak 19,300-18,600 BP.

The younger records from southern and central Europe do exhibit a drastic Younger Dryas period (Ammann et al.; de Beaulieu et al.; Couteaux, all in this volume). The Older Dryas is, however, absent in these areas (the pre-Alleröd chronostratigraphy of Couteaux, this volume, is hypothetical and gives no firm registration of an Older Dryas), which indicates that the Older Dryas

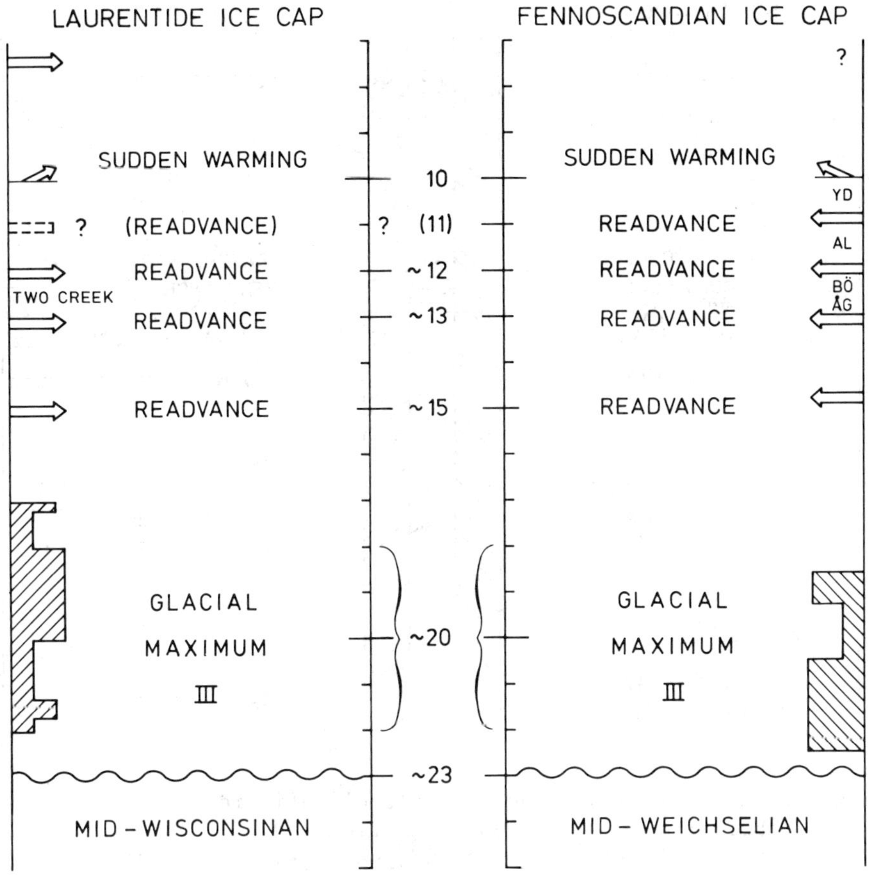

Fig. 1. Comparisons between main features in the ice marginal changes of the Fennoscandian ice cap in southern Scandinavia and of the Laurentide ice cap in central and eastern North America. The general tendency is good trans-Atlantic correlations. Vertical time scale in 10^3 years BP.

is a local Scandinavian event. This agrees with the findings of Coop (1977) in England that there was a long unbroken interstadial from about 13,000-11,000 BP (see Fig. 1 of Berglund et al., this volume). The amelioration correlated with the onset of the Bölling chronozone is here put at around 13,300 BP.

In the North Atlantic (Ruddinam & McIntyre, 1981; Duplessy et al., 1981; Kellogg, this volume), there is a marked warming at around 13,000-13,300 BP, a very distinct Younger Dryas cooling at about 11,000-10,000 BP, a second warming event at around 10,000 BP and a final amelioration in the range around 9000-8000 BP. The most important thing is the drastic Younger Dryas cooling which indi-

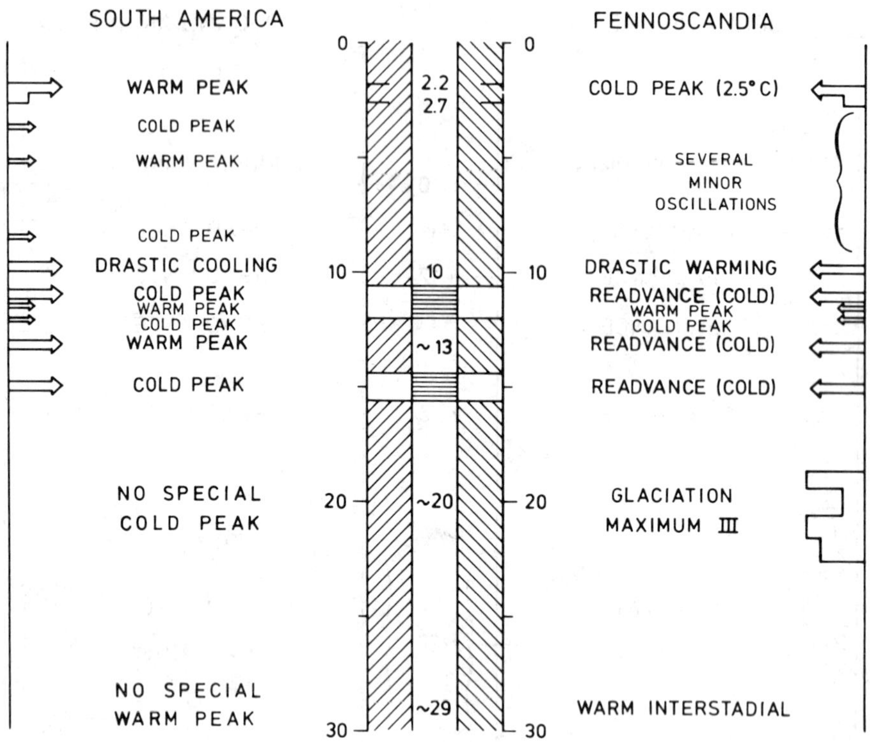

Fig. 2. Comparisons between the paleoclimatic records of the Bolivian region in South America and of Fennoscandia. Vertical time scale in 10^3 years BP. The two areas seem to be completely opposite each other (diagonal hatching) except for short periods around 15,000 and 11,000 BP. This gives evidence of a non-globality of these short-term climatic changes with compensational changes on a regional or hemispherical basis.

cates that it represents a major paleoceanographic event; the polar front was displaced to a position similar to that prior to 13,000 BP (Ruddiman & McIntyre, 1981).

The trans-Atlantic correlations/comparisons have been debated for a long time. As could be expected, there are examples both of similarities and dissimilarities; i.e. no perfect correlation but some correlations. In the St. Lawrence Valley, the initial phase of the Younger Dryas stadial may be represented by the oscilliation at the St. Narcisse moraine. In the Great Lakes region, there are readvances at around 12,000 BP, 13,000 BP, 15,000 BP and 18,000-22,000 BP which seems to match the Fennoscandian records very well (Fig. 1). Even in Alaska (Hamilton, 1982, Fig. 13),

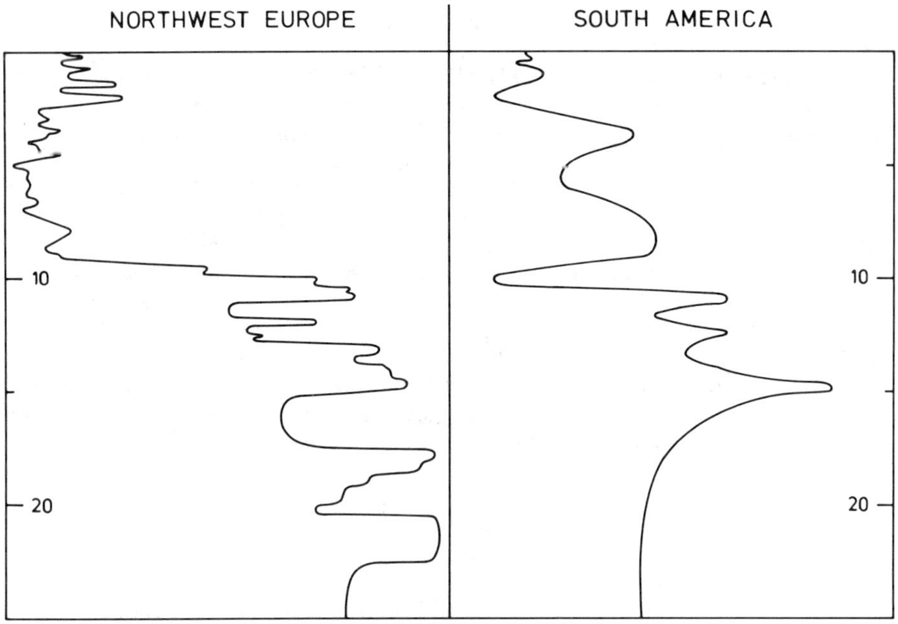

Fig. 3. Curves of ice marginal fluctuations and paleoclimate in Southern Scandinavia and in the Bolivian region. Warm to the left and cold to the right. Time in 10^3 years BP. The dissimilarity between the two corves is clear.

there are records that exhibit great similarities with the European records (Mörner, 1981).

A comparison of the main trend in the ice marginal fluctuations of the Fennoscandian and Laurentide ice caps is given in Fig. 1. The general conclusion is that the main trend exhibit good trans-Atlantic correlations, which may suggest that there are also intra-hemispherical correlations.

The low latitude record of Heine (this volume) form the mountains in Mexico is very important. Superimposed on a main 22,000-23,000-year, Milankovitch-like, temperature cycle, there are a few minor temperature oscillations and some significant changes in precipitation. Major glacial advances (induced by precipitation) are recorded at around 12,000 BP, 8,500-10,000 BP, 2500 BP and in historical time. It is of great importance that the Younger Dryas cooling is totally absent and this period rather is represented by "interstadial" climatic conditions. Similarly, the glacial advances starting at 10,000 BP are exactly in opposite to the Fennoscandian record. This indicates that the Fennoscandian records do not represent global changes in climate as often claimed. On the other hand, the glacier advance at 2500 BP perfectly matches the climatic deterioration at the Suboreal/Subatlantic boundary in Europe.

Inter-hemispheral correlations have often been advocated. The difference in opinion and interpretation between Mercer (e.g. 1976) and Heusser (1961) is well-known. The temperature curve of Heusser & Streeter (1980) from southern Chile does not exhibit any clear similarities with the northwest European records. The analysis of Servant (this volume), though preliminary, seems conclusively to indicate that the South American paleoclimatic changes were, in general, totally different to those recorded in Europe and North America. This is illustrated in Figs. 2 and 3. The only periods that are not directly opposite to the Fennoscandian records are cold peaks at around 15,000 BO and 11,000 BP (with warming shortly after, however). The cooling at 10,000 BP and warming at 2500 BP give clear indications of a non-globality of climatic changes and rather compensational climatic changes between the two hemispheres.

b) The Holocene

The Holocene climatic changes within the Nordic countries are characterized by the rapid, high-amplitude warming at 10,000 BP, the climatic optimum around 8000-2500 BP, the distinct climatic deterioration at 2500 BP and the sequence of many low-amplitude oscillations in climate and sea level (Mörner, this volume, Fig. 6).

The drastic 10,000 BP warming (cf. above) is seen in the whole of Europe, in the North Atlantic and, in general, in North America, too. It is directly counter-balanced by significant cooling events in, for example, Mexico and Bolivia, and can therefore not represent a global (similarly induced) event. This is of course, very important also in the light of the proposed position of the Pleistocene/Holocene boundary. It seems to represent a large regional event (Europe, North Atlantic, North America) or a northern hemisphere event.

The main Holocene trend in Europe of numerous low-amplitude oscillations is a regional feature as discussed by Mörner (this volume). It is interesting that Taira (e.g. 1981) have found a similar trend in Japan, which may indicate simultaneous rhythmic "vibrations" in the Gulf Stream and Kuro Siwo drift (Mörner, this volume). In other areas, the picture is different suggesting that each region had its own characteristic fluctuations. In Brazil, for example, there were 3 major "cycles" during the last 7000 years (cf. the Bolivian record in Fig. 3).

The 2500 BP deterioration in climate apply to the northwest European region and, maybe, parts of the North Atlantic, too (cf. Mörner, 1971, 1973). In Greenland (Fredskild, this volume) and Canada (Payette, this volume; Nicolls, 1970), the main deterioration and end of the Hypsithermal or Holocene Optimum, seem to have occurred earlier viz. at about 3500-3000 BP. In South America (Figs. 2-3; Servant, this volume), the trend is directly opposite; i.e. a compensational warming.

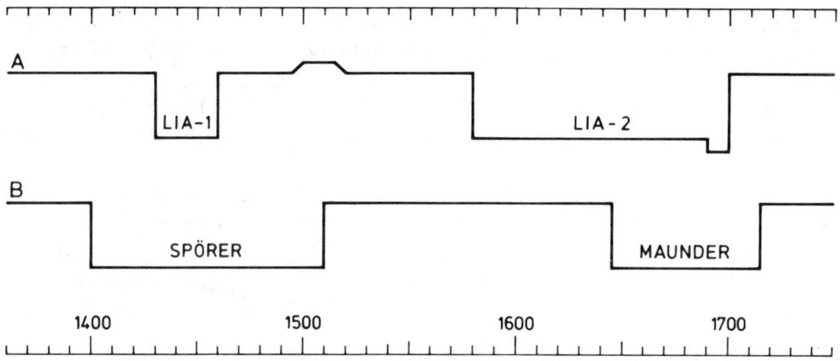

Fig. 4. Generalized temperature curve from England (A) according to Lamb (this volume, Fig. 1) with the two Little Ice Ages, LIA-1 and LIA-2, and the position of the two sunspot minima (B), the Maunder Minimum and the Spörer Minimum, according to Eddy (1976). The correlations and lead-lag relations are not clear and do not permit conclusions about a causal origin (as claimed by Eddy, 1976).

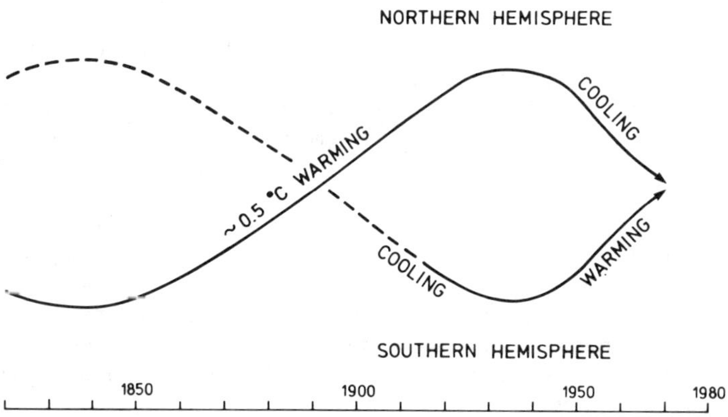

Fig. 5. Generalized trends of temperature changes in the northern and southern hemispheres over the past 160 years. Opposed, compensational changes are indicated

Consequently, the 2500 BP cooling is a regional event with compensational warming, at least, in South America.

c) The Little Ice Ages

"Little Ice Ages" have been assigned to very many different events and periods in late Holocene time. I think it is necessary to restrict these terms to those periods which have become the most well-known examples of historical cooling epochs, viz. those periods defined by Lamb (e.g. 1982; this volume, Fig. 1) and occurring about 1430-1460 AD and 1580-1700 AD (Fig. 4). If "Little Ice Age" should be assigned to any other period, it should be named "a" Little Ice Age, not "the" Little Ice Age.

The two Little Ice Ages of Lamb are typical regional to local climatic events that affected the sea around Iceland, England, Scandinavia and parts of the rest of Europe. Even within this area there are large local differences. The origin seems to be a change in the North Atlantic circulation leading to a drop in sea surface temperature (Mörner, this volume, Fig. 12; Lamb, 1979).

Eddy (1976) claimed that these Little Ice Ages coincided with periods of sunspot minima; viz. the Maunder Minimum and the Spörer Minimum, and that this indicated an origin in changes in the solar activity. As illustrated in Fig. 4, however, there is no straight forward correlation and no logical lead-lag relationship between the sunspot record and the English temperature record. The local validity of these events, of course, precludes a general solar vaorigin (cf. Table 1).

d) The 1840-1940 period

The main trend during this period is a warming in the order of 0.5°C on the northern hemisphere and a similar cooling on the southern hemisphere (e.g. Mitchell, 1961). This is illustrated in Fig. 5 and suggests that the two hemispheres in general compensated each other's temperature changes. This implies that the changes had hemispherical validity.

e) The 1940-1980 period

The main trend during this period is a cooling in the order of 0.2°C and a compensational warming in the southern hemisphere (Fig. 5). This seems to be another evidence of compensational hemispherical changes.

f) Conclusions about the geographic extension or space range

The present analysis of the geographic extension of especially important climatic changes on a yearly (or rather decadal) to millennial basis have led to the following observations:
(1) No climatic event on a strictly global basis; not even the distinct 20,000 BP glaciation maximum on the northern hemisphere is, for sure, present on the southern hemisphere as a separate and distinct cold maximum (at least, not in South America).
(2) Examples of climatic changes that seem to have a hemispherical validity with opposite, compensational, changes in the other hemisphere (e.g. the 1840-1940 period, the 1940-1980 period, the

10,000 BP warming, maybe the 11,000 BP cooling).
(3) Examples of climatic changes that seem to have had a validity over a geographic region of larger of smaller extension (e.g. the 13,000 BP warming, the 2500 BP cooling); even here there seems to have been compensational changes of opposed sign in other regions.
(4) Examples of climatic changes that seem to have had a validity over very large regions or a hemisphere (e.g. the 11,000 BP cooling); i.e. something inbetween points 2 and 3.
(5) A few examples of climatic changes that were typical for a limited region or were directly local (e.g. the Older Dryas, the two Little Ice Ages).

At the Stockholm symposium (and in this book), the main question was the character of significant climatic changes of time ranges inbetween the Milankovitch variables and the sunspot cycles.

Our conclusion that these climatic changes were not of global extension but rather of regional or hemispherical extension is therefore based on a very careful analysis. The conclusion is very important for the understanding of the physical background of these climatic changes. It is the first time that this non-globality has been demonstrated. Previously, many colleagues took it more or less for granted that these changes had a global validity. Global contra regional validity calls for quite different causation mechanisms (cf. Table 1).

2. TIME RANGE AND AMPLITUDES

All the climatic changes (Late Weichselian, Holocene, historical, instrumental) discussed in this book represent fairly rapid changes and shifts in climate with time ranges in the order of several decades to about 100 years, but at least less than 200 years. This is illustrated in Fig. 6. Here are compared (1) the 0.5°C event 1840-1940, (2) the about 1.0°C event of the Little Ice Ages, (3) the about 2.5°C event at 2500 BP, (4) the, at least, 5°C (probably more) event at 11,000 BP, (5) the local Older Dryas event, and (6) the 10°C or more event of substage 5d (and 5b).

The Fig. 6 curves are rather similar and seem to represent some sort of "family" of curves. What differs, are the amplitudes. One might, therefore, expect that all these changes had similar origins, only the strength (and hence the amplitude) differing. If the amplitude differences follow a strict rhythmicity or rather are stocastic, we do at present not know.

If the various abrupt short-term changes in climate in Fig. 6 really have the same or similar origin, it would imply that we in predicting future changes cannot foresee whether an expected change or shift will be a type 1, 2, 3, 4 or 6 event.

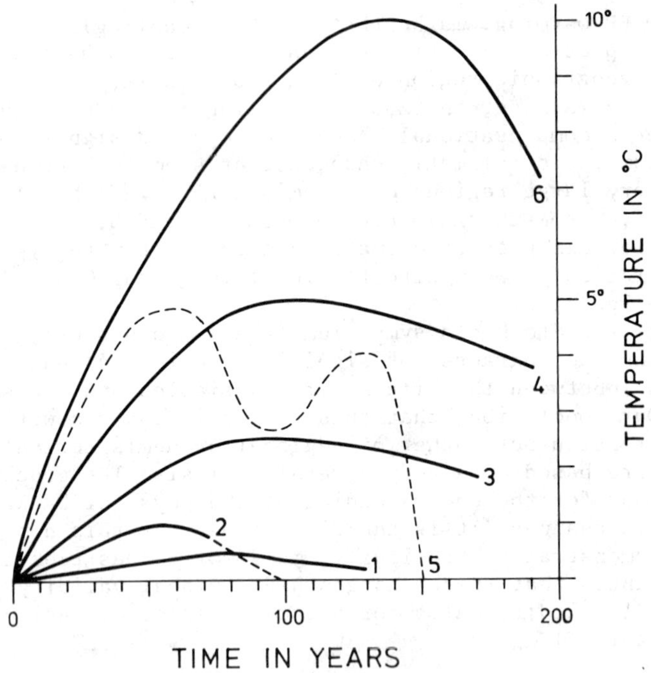

Fig. 6. Comparisons of some of the main short-term changes analyzed, arranging themselves in increasing amplitudes but with a time range remaining within some 50-150 years. (1) the 1840-1940 record, (2) the Little Ice Age coolings, (3) the 2500 BP cooling, (4) the Younger Dryas cooling at 11,000 BP, (5) the local Older Dryas cooling (added for comparison), and (6) the isotopic substage 5d (and 5b) cooling (as indicated by the Grande Pile record). The similarity in time range is important and may indicate a similar causation mechanism. The differences in amplitudes may reflect rhythmical or stocastical changes of the basic driving forces. The non-globality of these changes (discussed in the text) indicates that they represent rapid changes in the distribution of heat over the globe (and not global rises and falls in temperature as generally assumed). These changes and time ranges can hardly represent anything else than changes in the oceanographic circulation (Mörner, this volume, Figs. 8-13).

3. IMPLICATIONS AND CONCLUSIONS

Combining the conclusions arrived at from the analysis of the geographic extension (part 1) and from the analysis of the time range and amplitudes (part 2), we arrive at the novel and fundamental

TABLE 1. Major groups of variables affecting and controlling the Earth's heat budget and their approximate time and space application. M = millions, T = thousands, H = hundreds, Y = years; G = global, H = hemispherical, R = regional, L = local; ⊕ = very important, + = important, – = some effects, ? = possible effects. The suggested time and space distribution offer a possibility of distinguishing between possible causation mechanisms.

VARIABLES	TIME				SPACE			
	M	T	H	Y	G	H	R	L
HEAT PRODUCTION								
in the Sun	⊕	–	?	+	⊕	?		
within the Earth	⊕	+	?	?	⊕	+	–	–
SOLAR RADIATION								
Sun-Earth constellation	+	⊕	?	–	+	⊕		
inter-planetary shielding	⊕				⊕	+		
atmospheric shielding	+	+	+	+	⊕	+	?	
albedo	+	+	+		–	+	⊕	+
REDISTRIBUTION (of heat)								
atmospheric processes			+	⊕		–	⊕	–
oceanographic processes		+	⊕	⊕		+	⊕	–
terrestrial processes	⊕	+			–	–	⊕	+
STORING, TRANSFORMATION								
in the atmosphere			+	⊕	–	+	⊕	–
in the oceans		–	⊕	+		+	⊕	
on land		⊕	+		–	+	⊕	+

conclusion that none of the rapid climatic changes and shifts of the types illustrated in Fig. 6 during the last 20,000 years have had a global extension, but only a hemispherical or regional extension of a type where the change is counter-balanced by compensational changes of opposed sign in other regions. This applies even to the changes at the Pleistocene/Holocene boundary and the Younger Dryas cooling, and probably also to the substage 5d "aborted full-glacial" cooling.

This means that the climatic changes analysed cannot originate from global rises and falls in temperature, but must represent *redistributions of heat over the globe*; regionally or hemispherically, and sometimes even locally. This is our first main conclusion.

The duration of these redistributions of heat over the globe is in all cases analysed (e.g. Fig. 6) given at *about 50-150 years*. This is our second main conclusion.

Redistribution of heat, including storing and transformation of heat, which lasts for 50-150 years seems only to be possible via *ocean circulation changes*. This is our third main conclusion.

Circulation changes can be driven by several mechanisms (Mörner, this volume). This duration precludes a wind-driven mechanism (as proposed by Lamb, 1979) and indicates a combined *rotational-gravitational-oceanographic origin* as proposed above (Mörner, this volume, Figs. 8-13). This is our fourth main conclusion.

This opens the possibility of discriminating between different possible causation mechanisms. In Table 1, the major groups of variables that affect and control the Earth's heat budget are compared with the expected global space range and time range of their effects on weather and climate.

The Table 1 relations mean that many variables can be directly dismissed as a possible origin of the short-term climatic changes and shifts here discussed (see Fig. 1 in the introduction; Mörner, this volume). We can exclude changes in heat production in the Sun and within the Earth (except for local lithospheric changes as discussed above; Mörner, this volume), long-term changes in the Sun-Earth constellation (excluding possible short-term changes as illustrated in Fig. 13 of Mörner, this volume), changes in inter-planetary shielding, changes in atmospheric shielding (including dust from major volcanic eruption as proposed by Flohn, 1979) and redistribution of storing and transformation of heat due to terrestrial processes. We are left with changes in albedo, redistribution of heat in the atmosphere and the oceans, and storing and transformation of heat in the atmosphere and the oceans. The albedo can hardly change that rapidly and with such geographical distribution that it can explain the recorded changes. The atmospheric processes can hardly cause changes that last so long and have such geographical distribution. We, therefore, seem to be left only with *redistribution, storing and transformation of heat due to oceanographic processes*.

The origin and driving mechanisms of these oceanographic changes seem to me to be *differential rotational changes* as proposed above (Mörner, this volume, Fig. 10). These differential rotational changes are closely linked to *gravitational changes* (Mörner, this volume, Fig. 8). Both these mechanisms affect weather and climate along different causational lines. This is illustrated in Fig. 7.

The ultimate origin of these rotational-gravitational-oceanographic changes on the Earth seems to me to be *a planetary beat* according to Fig. 13 in Mörner (this volume).

If the causation mechanism may still be debated and subjectively evaluated (as in this paper), the main conslusions of (1) *a non-globality*, (2) *a 50-150 year time range*, and (3) an about *0.5-10.0 °C amplitude range* should remain a firm basis and probably the main product of the Stockholm Symposium.

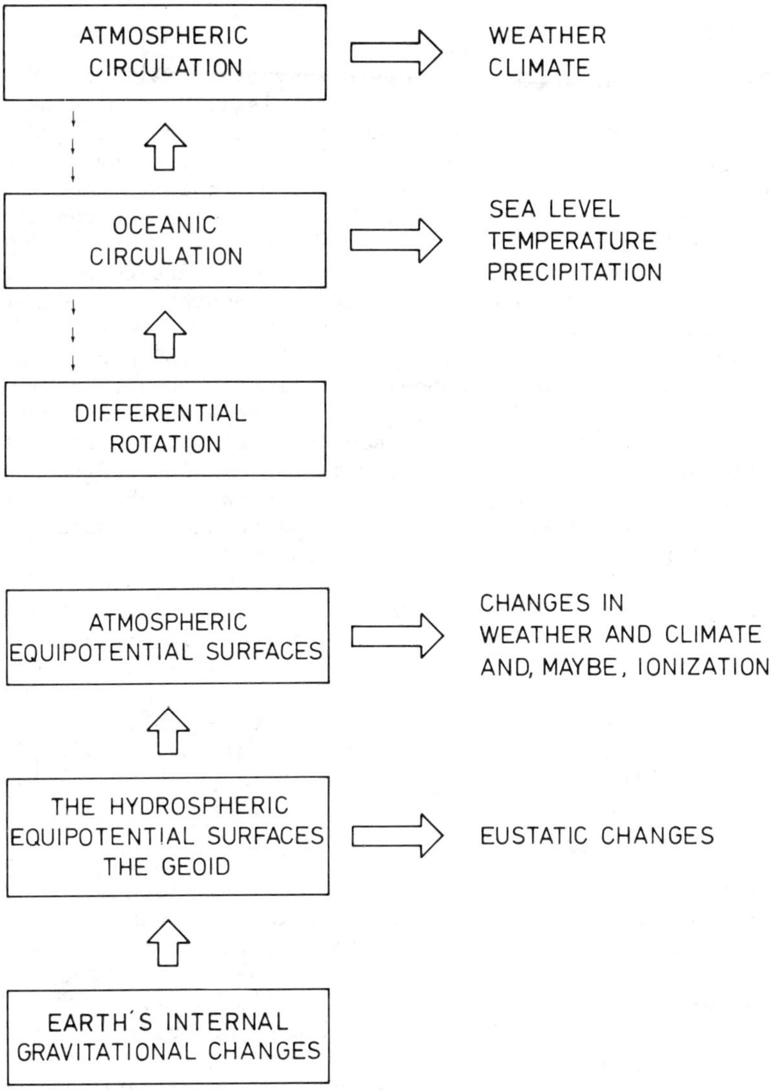

Fig. 7. Multiple effects of differential rotational changes behind the changes in oceanic circulation (above), and of related gravitational changes (below) as discussed by Mörner (this volume, Figs. 8 and 10). The possible ultimate driving mechanism from a planetary beat (and its multiple effects) is illustrated in Fig. 13 of Mörner (this volume).

REFERENCES

Amman, B., Chaix, L., Eicher, U., Elias, S.A., Gaillard, M.-J., Hofmann, W., Siegenthaler, U., Tobolski, K. & Wilkinson, B., this volume. Flora, fauna and stable isotopes in Late-Wurm deposits at Lobsigensee (Swiss Plateau).

de Beaulieu, J.-L., Clerc, J. & Reille, M., this volume. Late Weichselian fluctuations in the French Alps and Massif Central from pollenanalyses.

Berglund, B.E. & Mörner, N.-A., this volume. Late Weichselian deglaciation and chronostratigraphy of Southern Scandinavia: Problems and present "state of the art".

Berglund, B.E., Lemdahl, G., Liedberg-Jönsson, B. & Persson, T., this volume. Biotic response to climatic changes during the time span 13,000-10,000 BP - A case study from SW Sweden.

Björck, S. & Digerfeldt, G., this volume. Climatic changes at Pleistocene/Holocene boundary in the Middle Swedish endmoraine zone, mainly inferred from stratigraphic indications.

Coop, G.R., 1977. Fossil coleopteran assemblages as sensitive indicators of climatic changes during the Devensian (last) cold stage. Phil. Trans. R. Soc. Lond. B 280, 313-340.

Couteaux, M., this volume. Climatic and glacial fluctuations during the Late-glacial and the Holocene in Disans.

Duplessy, J.C., Delibras, G., Turon, J.L., Pujol, C. & Duprat, J., 1981. Deglacial warming of the northeastern Atlantic Ocean: correlation with the paleoclimatic evolution of the European continent. Palaeogeogr. Palaeoclim. Palaeoecol. 35, 121-144.

Eddy, J.A., 1976. The Maunder Minimum. Science 192, 1189-1202.

Flohn, H., 1979. On time scales and causes of abrupt paleoclimatic events. Quatern. Res. 12, 135-149.

Fredskild, G., this volume. Holocene palaeo-winds and climatic changes in West Greenland as indicated by long-distance transported and local pollen in lake sediments.

Hamilton, T.D.l 1982. A late Pleistocene glacial chronology for the southern Brooks Range: Stratigraphic record and regional significance. Geol. Soc. Am. Bull. 93, 700-716.

Heine, K., this volume. The classical Late Weichselian climatic fluctuations in Mexico.

Heusser, C.J., 1961. Some comparisons between climatic changes in northwestern North America and Patagonia. Ann. N.Y., Acad. Sci. 95, 642-657.

Heusser, C.J. & Streeter, S.S., 1980. A temperature and precipitation record of the past 16,000 years in southern Chile. Science 210, 1345-1347.

Kellogg, T.B., this volume. Late-Glacial - Holocene high-frequency climatic changes in deep-sea cores from the Denmark Strait.

Lamb, H.H., 1979. Climatic variation and changes in the wind and ocean circulation: The Little Ice Age in the Northeast Atlantic. Quatern. Res. 11, 1-20.

Lamb, H.H., 1982. The development of climate and its relevance to human affaires. Proc. Roy Inst. 54, 65-95.
Lamb, H.H., this volume. Climate and history in northern Europe and elsewhere.
Mangerud, J., Andersen, S.T., Berglund, B.E. & Donner, J.J., 1974. Quaternary stratigraphy of Norden, a proposal for terminology and classification. Boreas 3, 109-128.
Mercer, J.H., 1969. The Alleröd oscillation: A European climatic anomay? Arctic Alpine Res. 1, 227-234.
Mercer, J.H., 1976. Glacial history of southernmost South America. Quatern. Res. 6, 125-166.
Mitchell Jr., J.M., 1961. Recent secular changes of global temperature. Ann. N.Y. Acad. Sci. 95, 235-250.
Mörner, N.-A., 1971. Eustatic and climatic oscillations. Arctic Alpine Res. 3, 167-171.
Mörner, N.-A., 1973. Climatic changes during the last 35,000 years as indicated by land, sea and air data. Boreas 2, 33-53.
Mörner, N.-A., 1976. Eustasy and geoid changes. J. Geol. 84, 123-151.
Mörner, N.-A., 1979. The Grande Pile paleomagnetic/paleoclimatic record and the European glacial history of the last 130,000 years. Intern. Proj. Paleolimnol. Late Cenozoic Clim. 2, 19-24.
Mörner, N.-A., 1981. Weichselian chronostratigraphy and correlations. Boreas 10, 463-470.
Mörner, N.-A., 1983. Sea Levels. In: Large-scale geomorphology (R.A.M. Gardner & H. Scoging, Eds.), Oxford Univ. Press, 73-91.
Mörner, N.-A., this volume. Climatic changes on a yearly to millennial basis. An introduction.
Mörner, N.-A., this volume. Planetary, solar, atmospheric and endogene processes as origin of climatic changes on the Earth.
Mörner, N.-A., Minka, D. & Caroline, R., 1982. Trans-Atlantic and trans-hemispheric correlations during the Late Quaternary. Abstracts, INQUA XI Congr. Moscow 1982, vol. 11, 201.
Nichols, H., 1970. Late Quaternary pollen diagrams from Canadian Arctic barren grounds at Pelly Lake, northern Keewatin. Arctic Alpine Res. 2, 43-61.
Payette, S., this volume. Peat inception and climatic changes in northern Quebec.
Petersen, K.S., this volume. Late Weichselian sea-levels and fauna communities in northern Vendsyssel, Jutland, Denmark.
Ruddiman, W.F. & McIntyre, A., 1981: The North Atlantic Ocean during the last deglaciation. Palaeogr. Palaeoclim. Palaeoecol. 35, 145-214.
Servant, M., this volume. Climatic variations in the low continental latitudes during the last 30,000 years,
Taira, K., 1981. Holocene tectonism in eastern Asia and geoidal changes. Palaeogeogr. Palaeoclim. Palaeoecol. 36, 75-85.

INDEX